# Lecture Notes in Civil Engineering

Volume 175

**Lecture Notes in Civil Engineering** (LNCE) publishes the latest developments in Civil Engineering - quickly, informally and in top quality. Though original research reported in proceedings and post-proceedings represents the core of LNCE, edited volumes of exceptionally high quality and interest may also be considered for publication. Volumes published in LNCE embrace all aspects and subfields of, as well as new challenges in, Civil Engineering. Topics in the series include:

- Construction and Structural Mechanics
- Building Materials
- Concrete, Steel and Timber Structures
- Geotechnical Engineering
- Earthquake Engineering
- Coastal Engineering
- Ocean and Offshore Engineering; Ships and Floating Structures
- Hydraulics, Hydrology and Water Resources Engineering
- Environmental Engineering and Sustainability
- Structural Health and Monitoring
- Surveying and Geographical Information Systems
- Indoor Environments
- Transportation and Traffic
- Risk Analysis
- Safety and Security

To submit a proposal or request further information, please contact the appropriate Springer Editor:

- Pierpaolo Riva at pierpaolo.riva@springer.com (Europe and Americas);
- Swati Meherishi at swati.meherishi@springer.com (Asia - except China, and Australia, New Zealand);
- Wayne Hu at wayne.hu@springer.com (China).

**All books in the series now indexed by Scopus and EI Compendex database!**

More information about this series at http://www.springer.com/series/15087

Sreevalsa Kolathayar · Siau Chen Chian
Editors

# Recent Advances in Earthquake Engineering

## Select Proceedings of VCDRR 2021

Springer

*Editors*
Sreevalsa Kolathayar
Department of Civil Engineering
National Institute of Technology Karnataka
Surahtkal, India

Siau Chen Chian
Department of Civil and Environmental Engineering
National University of Singapore
Singapore, Singapore

ISSN 2366-2557 ISSN 2366-2565 (electronic)
Lecture Notes in Civil Engineering
ISBN 978-981-16-4616-4 ISBN 978-981-16-4617-1 (eBook)
https://doi.org/10.1007/978-981-16-4617-1



This Springer imprint is published by the registered company Springer Nature Singapore Pte Ltd.
The registered company address is: 152 Beach Road, #21-01/04 Gateway East, Singapore 189721, Singapore

# Contents

**Recent Advances in Earthquake Engineering—An Introduction** ....... 1
Siau Chen Chian and Sreevalsa Kolathayar

**Seismic Ground Motion Studies**

**Wavelet Analysis of Far-Field Ground Motions from the $M_w$ 7.6 2005 Kashmir Earthquake** ....... 11
Mohammed Ayub Ifan, Shalin Mathew, and Jayaprakash Vemuri

**Correlation Between Seismic Intensity Measures and Response of Skewed Bridges** ....... 25
Van-Tien Phan and Duy-Duan Nguyen

**Structural Vulnerability Assessment**

**Evaluation on Seismic Performance of "Dhajji-Dewari" Timber Wall Panel** ....... 39
M. Sushma, Vijayalakshmi Akella, B. K. Raghu Prasad, and G. Deepu

**Variation in Seismic Behaviour of R.C Shaft Supported Elevated Water Towers with Change in Proportion of the Shaft Staging** ....... 53
Pronoy Roy Chowdhury

**Seismic Assessment of RC Framed Staging of Elevated Water Tanks** ... 69
Trishit Chandra and Saraswati Setia

**Effect of Pier Height on Behavior of Integrated Bridge Under Far and Near-Field Earthquake** ....... 77
Kedar Ramesh Kumbhojkar, M. K. Shrimali, S. D. Bharti, and Vishal Kamble

**Seismic Performance Evaluation of Integral Bridges Considering Soil-Structure Interaction** ....... 91
Duy-Duan Nguyen and Huu-Cuong Nguyen

**Non Linear Seismic Analysis of RC Framed Structure by Extended N2 Method** ........ 103
T. Sakthi and V. Vasugi

**Assessment of Seismic Performance and Vulnerability of a 20-Storied Reinforced Concrete Moment-Resisting Frame Designed by Direct Displacement-Based Design and Force-Based Design** ........ 115
Sumit Ganvit, Sejal Purvang Dalal, Ronak Motiani, and Purvang Dalal

**Comparative Study on Seismic Behaviour of G+41 Storey Frame with X, K and V Bracings** ........ 127
Rosemary K. Thomas, Swetha Elizabeth Philip, and V. Vasugi

**Influence of Mass Irregularity on the Response of RC Frame with Stiffness Irregularity by Non-linear Time History Analysis** ....... 135
Athar Tauheed, Mehtab Alam, and T. K. Datta

**Comparative Study on Static and Dynamic Analysis of RC Buildings of Different Heights in Different Seismic Zones** ........ 161
Rajdeep Saharia, Debashis Bhuyan, Saunak Chowdhury, and Karabi Bharadwaj

**Effect of Dog-Legged Staircase on the Seismic Response of Hill Buildings** ........ 175
Ahmed Bilal, Zaid Mohammad, and Abdul Baqi

**Effect of Supporting Structure's Torsion on Floor Acceleration Demands in Buildings on Slopes** ........ 185
Ankur Jain and Mitesh Surana

**Seismic Safety of Buildings on Indian Hill Slopes—A Review** ........ 197
Prateek Roshan and Shilpa Pal

**Seismic Mainshock—Aftershock (MS-AS) Vulnerability Assessment of Reinforced Concrete Bridge Exposed to Flood Induced Scour** ........ 211
K. K. Jithiya, Muhamed Safeer Pandikkadavath, Sujith Mangalathu, and Praveen Nagarajan

**Effect of Fire Hazard on Seismic Capacity of RC Frame Building** ...... 223
Praveen Oggu, Swati Raman, and K. Gopikrishna

**Geotechnical Vulnerability Assessment**

**Effect of Soil-Structure Interaction on the Seismic Response of Elevated Water Tank** ........ 237
Pranitha Jogi and B. R. Jayalekshmi

**A Study on Settlement Variation in Piled Raft Foundation Under Seismic Loads** ........ 249
S. Vinoda Krishna and B. R. Jayalekshmi

**Seismic Behavior of Tunnel Structure Under Varying Surrounding Soil, Water Table, and Overburden Condition** ........ 259
Chiranjib Sarkar and Sibapriya Mukherjee

**Seismic Response Study on Kaswati Dam** ........ 271
Barnali Ghosh and S. K. Prasad

**Influence of Reservoir Level on the Dynamic Behaviour of Concrete Gravity Dam** ........ 281
P. N. Biju and Glory Joseph

**Assessment of Seismic Displacement of Quay Walls** ........ 291
K. Pushpa, S. K. Prasad, and P. Nanjundaswamy

**Retrofitting and Rehabilitation**

**Seismic Response Elimination of Structure by Using Passive Devices: An Overview** ........ 303
Rushikesh Jadhav, Digambar Patil, and Prachi Sohoni

**Response Reduction of Secondary Piping Systems in Base-Isolated Buildings** ........ 319
Vishal Kamble, Shiv Dayal Bharti, Mahendra Kumar Shrimali, and Kedar Kumbhojkar

**Seismic Analysis of Isolated and Non-isolated Railway Bridges with Slab Track—A Comparative Study** ........ 333
A. Shivaraj, G. Sudarshan, G. Sridevi, and B. Umesh

**Seismic Performance Analysis of Regular and Irregular RCC Framed Building with Dampers** ........ 345
Laxmi M. Ramdas and M. Helen Santhi

**Seismic Control of Structures with Mass Irregularities Through Optimally Placed Active Tendons Using Multi-objective Genetic Algorithm** ........ 357
F. Rather and M. Alam

**Seismic Performance of UHPFRC-Strengthened RC Beam–Column Joints Using Damage Plasticity Model—A Numerical Study** ........ 371
K. Sai Kubair and J. S. Kalyana Rama

**Assessment and Retrofitting Augmentation Methods of Seismic Performance on Existing RC Buildings Across India** ........ 385
Ankeeta Karmakar

**Review of Building Codes**

**Scientific Perspectives to Earthquake Resistant Design of RC Buildings—A Global Approach** ... 399
B. M. Raisinghani, T. H. Bhoraniya, and E. Noroozinejad Farsangi

**Parametric Study on the Variation of Time Period of RC MRF Buildings** ... 415
Sasanka Sarma, Tridib Sundar Das, Ankita Bora, and Karabi Bharadwaj

**Parametric Study of Performance-Based Seismic Design of Plan Irregular RC Frames—Indian Scenario** ... 427
Rajat Abhay Sirsikar, Ganesh D. Awchat, and J. S. Kalyana Rama

**Role of Openings in Seismic Resistance of Infill Walls: A Critical Review of Indian Codal Provision** ... 439
Karismita Pathak and Atanu Kumar Dutta

**Simplified Macro Modeling Approach for Estimation of Nonlinear Response of Infilled RC Frames** ... 451
Panna Lal Kurmi and Putul Haldar

**Field Surveys, Geospatial Tools and Others**

**Seismic Vulnerability of Buildings in Palghar Area: Observations and Remedial Measures** ... 467
Rohan M. Shinde and Ravi Sinha

**Evaluating the Seismic Performance of Domestic and Historical Masonry Structures in Himachal Pradesh Region of India** ... 477
Ashwani Kumar Sharma, Ashutosh Kumar, and Vasilis Sarhosis

**Mapping of Earthquake-Induced Land Deformation on Urban Area Using Interferometric Synthetic Aperture Radar Data of Sentinel-1** ... 491
Fathoni Usman, Agusril Syamsir, and Jihan Melasari

**Seismic Data Mapping of South India from 1820 to 2020 Using Geographic Information System** ... 503
Mrudula Madhukumar and M. Helen Santhi

**Earthquake Loss Estimation Using State-of-the-Art Tools** ... 517
Leanda J. Payyappilly and Surendra Nadh Somala

**Post-Earthquake Safe Shelter for Shimla City** ... 527
Hemant Kumar Vinayak, Abhinandan, and Shashidhara

# About the Editors

**Dr. Sreevalsa Kolathayar** pursued M.Tech. from IIT Kanpur, Ph.D. from Indian Institute of Science (IISc) and served as International Research Staff at UPC BarcelonaTech, Spain. He is presently Faculty in the Department of Civil Engineering, National Institute of Technology Karnataka, Surathkal, India. Dr. Sreevalsa has authored five books and over 80 research papers. He is Associate Editor of two international journals. His research interests include Geotechnical Earthquake Engineering, Seismic Hazard and Risk, Landslides, Disaster Risk Reduction, Reinforced Earth and Water Geotechnics. He is currently the Secretary Indian Chapter of International Association for Coastal Reservoir Research (IACRR) and Executive Committee Member of Indian Society of Earthquake Technology. In 2017, The New Indian Express honoured Dr. Sreevalsa with South India's Most Inspiring Young Teachers Award. He is the recipient of ISET DK Paul Research Award from Indian Society of Earthquake Technology in 2018. He is in the roster of two technical committees of ASCE Geo-Institute. He received "IEI Young Engineers Award" by The Institution of Engineers (India) in recognition of his contributions in the field of Civil Engineering. Recently, Dr. Sreevalsa was featured by the American Society of Civil Engineers in their Geostrata Magazine.

**Dr. Siau Chen Chian** is Associate Professor at the Department of Civil and Environmental Engineering, National University of Singapore (NUS). He obtained his Ph.D. and B.Eng. (First Class with Gold Medal) from Cambridge University and Nanyang Technological University, respectively. Dr. Chian's contribution in earthquake engineering lies in the field of catastrophe modelling, landslides and damage vulnerability of underground structures. He was funded by the UK Engineering and Physical Sciences Research Council (EPSRC) as the geotechnical specialist to carry out reconnaissance missions at the Mw7.6 Padang Earthquake in 2009, Mw9.0 Great East Japan Earthquake in 2011 and Mw7.8 Muisne Earthquake in 2016. Dr. Chian was named Top 10 Innovators Under 35 in Asia by the MIT Technology Review for his novel contribution in catastrophe modelling in 2016. Dr. Chian is a

Nominated Member of the Technical Committee on Earthquake Geotechnical Engineering TC203 under the International Society for Soil Mechanics and Geotechnical Engineering (ISSMGE). He is also an Integrated Research on Disaster Risk (IRDR) Young Scientist, supported by the International Science Council (ISC) and the United Nations Office for Disaster Risk Reduction (UNDRR).

# Recent Advances in Earthquake Engineering—An Introduction

**Siau Chen Chian and Sreevalsa Kolathayar**

## 1 Introduction and Background

Earthquakes remain one of the most destructive natural phenomena to date, leading to catastrophic damage to infrastructure and human casualty. Although the study of earthquake dates back to prehistoric ages, this phenomenon has not been thoroughly understood and addressed as evident in recent earthquakes, where infrastructure damage and loss of lives still prevail. Earthquake engineering is an interdisciplinary branch of engineering that involves making infrastructures more resistant to earthquakes, thereby limiting seismic risks to socio-economically acceptable levels.

Although earthquake engineering is traditionally the study of behavior of infrastructures' subject to seismic loadings, modern definition has expanded its scope to encompass wider field of engineering so long as they serve the same objective of limiting seismic risks. Specific to civil engineering, earthquake engineering consists of seismic ground motion studies, structural and geotechnical vulnerability assessment, retrofitting and rehabilitation, review of building codes, as well as field surveys and introduction of remote sensing in the form of geospatial tools and geographical information systems to facilitate identification of areas susceptible to earthquake damage. This book volume serves to compile the recent advances of earthquake engineering in these sections of civil engineering, in order to facilitate the contributions towards civil engineering for a disaster resilient society.

S. C. Chian
National University of Singapore, Singapore, Singapore

S. Kolathayar (✉)
National Institute of Technology Karnataka, Surathkal, India
e-mail: sreevalsa@nitk.edu.in



S. Kolathayar and S. C. Chian (eds.), *Recent Advances in Earthquake Engineering*,
Lecture Notes in Civil Engineering 175,
https://doi.org/10.1007/978-981-16-4617-1_1

## 2 Seismic Ground Motion Studies

A notable paper in this volume on ground motion analysis is from Mohammed et al., titled *Wavelet Analysis of Far-Field Ground Motions from the Mw 7.6 2005 Kashmir Earthquake*. In this paper, the adoption of wavelet analysis representing both time and frequency content has been useful in better understand the changing time–frequency characteristics of the ground motion. It was concluded that the observed high amplitude waves in the frequency domain correlate well with the distribution of damage to reinforced concrete and unreinforced masonry structures in the region.

Last but not least, Phan and Nguyen presented a study on the *Correlation between Seismic Intensity Measures and Response of Skewed Bridges*. Relationship between 23 earthquake intensity measures (IMs) and seismic responses of skewed bridges are investigated to identify the most appropriate IMs of damage of the bridge based on Pearson's correlation coefficients. A series of non-linear time history analyses were carried out to observe the bridge seismic response in terms of drift ratio and shear forces of its piers. It was found that for low-frequency ground motions, seismic responses such as Arias intensity, characteristic intensity, spectral acceleration, and spectral velocity at fundamental period can be best correlated to the IMs. For high-frequency motions, specific energy density, Arias intensity, root-mean-square of velocity, and characteristic intensity correlate best with IM. In addition, PGV/PGA ratio, mean period, and predominant period were showed to be weakly correlated with seismic responses of bridges for both low- and high-frequency seismic ground motions.

## 3 Structural Vulnerability Assessment

This book volume has a wealth of papers on the topic of structural vulnerability assessment. They range from analysing performances of specific structural members such wall panels and shafts to global structural analysis of low and high rise reinforced concrete buildings as well as elevated water tanks and bridges.

In Sushma et al.'s paper titled, *Evaluation on Seismic Performance of "Dhajji-Dewari" Timber Wall Panel*, native Acacia Nilotica wood species was used for the construction of the timber wall panel. Tension and compressive strength tests on the timber as well as load carrying capacity of joint connections were carried out. Thereafter, numerical analysis further demonstrates the panel's large deformability and good resilience under lateral loads. Chowdhury stressed the importance of ensuring functionality of water towers for continual water supply following an earthquake and the typical failures often at the shaft staging. The paper titled *Variation in Seismic Behaviour of R.C. Shaft Supported Elevated Water Towers with Change in Proportion of Shaft Staging* shed light on the influence of shaft diameter and height under simulated strong earthquakes such as the El Centro, San Fernando and Bhuji events. Another paper on water tank, titled *Seismic Assessment of RC Framed Staging of*

*Elevated Water Tanks* by Chandra and Setia discusses the modelling of fluid–structure interaction following the IS provision while avoiding coupled resonance effect. In the case of bridges, Kumbhojkar et al. discussed the *Effects of Pier Height on Behaviour of Integrated Bridge under Far and Near-field Earthquake*, where the deck moment and deflection are compared with conventional bridges. It was noted from the study that with the increase in the pier height, deck moments were decreased for far-field earthquakes and increased for near-field earthquakes whereas pier moments and deck displacement were observed to reduce and increase, respectively. In *Seismic Performance Evaluation of Integral Abutment Bridges considering Soil-Structure Interaction* by Nguyen and Nguyen, nonlinear static (pushover) analysis in both transverse and longitudinal directions of a railroad bridge in Korea was performed and a set of damage states of bridge piers were proposed.

This volume also includes the state-of-art conventional structural analysis such as *Non Linear Seismic Analysis of RC Framed Structure by Extended N2 Method* by Sakthi and Vasugi, *Assessment of Seismic Performance and Vulnerability of a 20 Storied Reinforced Concrete Moment Resisting Frame Designed by Direct Displacement Based Design and Force Based Design* by Ganvit et al., *Comparative Study on Seismic Behaviour of G+41 Storey Frame with X, K and V Bracings* by Thomas et al., *Influence of Mass Irregularity on the Response of RC Frame with Stiffness Irregularity by Nonlinear Time History Analysis* by Tauheed et al., and *Comparative Study on Static and Dynamic Analysis of RC Buildings of Different Heights in Difference Seismic Zones* by Saharia et al. In Sakthi and Vasugi's paper, the strength of the extended N2 Method is the ability to evaluate the seismic performance of asymmetric setback building when compared to the equivalent static analysis method of a regular building. In practice, force based design is traditionally used in India; however, it suffers from the limitation of addressing inelastic behaviour of structures under seismic loads. Hence, displacement-based designs have been continuously developed such damage is often related to displacement than force. Comparisons between the two approaches indicated potentially higher efficiency of the direct displacement based design (DDBD) for study frames as per Indian Standard codes (Ganvit et al.). X, K and V bracings used with concrete slabs were studied with STAAD Pro by response spectrum and equivalent static methods, which showed V bracing providing more efficient seismic resistance to a building (Thomas et al.). An in-depth study on the inelastic response of a 11-storey frame with irregularities introduced at the first, sixth and top stories was performed demonstrated the importance of stiffness on the response of the building which produced opposing effects to the frame and should be taken into account in design (Tauheed et al.). Lastly, equivalent static method (ESM) and response spectrum method (RSM) were applied to RC buildings of different heights and different seismic zone which showed that RSM gives higher values of responses for low rise whereas ESM generate higher values for high rise buildings.

Papers on buildings found on slopes such as *Effect of Dog-Legged Staircase on Seismic Response of Hill Buildings* by Bilal et al., *Effect of Supporting Structure's Torsion on Floor Acceleration Demands in Buildings on Slopes* by Jain and Surana, and *Seismic Safety of Buildings on Indian Hill Slopes—A Review* by Roshan and Pal, demonstrated the need to pay attention to such buildings where structural behaviour would differ as compared to those found on level terrain. Dog-legged staircase elements has the potential to impact lateral stiffness and work as efficient energy dissipators under earthquake loads when placed at the appropriate position of the building footprint. A correlation analysis demonstrated that the torsional amplification factors were well correlated to the ratio between the maximum to minimum edge displacements, as compared to other indices such as ratio of maximum to average edge displacements, floor rotation and angular acceleration of the floor. The review paper highlights the challenges and undesirable construction practices that can put buildings on hill slopes at high seismic risk.

Other exciting contributions looking at secondary effects prior to seismic loading include *Seismic Mainshock—Aftershock (MS-AS) Vulnerability Assessment of Reinforced Concrete Bridge Exposed to Flood Induced Scour* by Jithiya et al. and *Effect of Fire Hazard on Seismic Capacity of RC Frame Building* by Oggu et al. In the first item, the regional annual peak flood discharge data with different return period was used to quantity the scour depth activity. Additionally, the soil-structure interaction was considered using suitable spring models. A study bridge was incorporated in a modelling platform to simulate the scour effect as well as successively nonlinear dynamic (time history) analyses. The outcome were seismic fragility curves developed corresponding to different damage states of building under both mainshock and mainshock-aftershock events of these buildings affected by scour. The second item discussed the reduction in seismic structural capacity of a structure due to fire hazard and appropriate repair/retrofitting strategy to restore its functionality.

## 4 Geotechnical Vulnerability Assessment

This volume also contains several geotechnical vulnerability assessments of infrastructure which can be read in conjunction with the structural papers. For example, *Effect of Soil-Structure Interaction on the Seismic Response of Elevated Water Tank* by Jogi and Jayalekshmi supplements the two structural papers on elevated water tank design. One notable finding in this geotechnical paper is the greater maximum base shear and base moment of a tank resting on hard soil for a completely filled condition, while similar tanks resting on soft soils are expected to suffer from greater displacements. *A Study on Settlement Variation in Piled Raft Foundation under Seismic Loads* by Krishna and Jayalekshmi investigated a ten storey moment-resisting frame resting on piled raft foundation with different pile length by numerical simulation. It was found that the settlement at the edges of the raft is minimal for the model having the longest peripheral pile, while minimum settlement at the center observed for the model that has longer central pile compared to peripheral piles. This substantiates the logic that longer piles have higher bearing capacity which lowers settlement at those locations. This volume also contains a study on *Seismic Behaviour of Tunnel Structure under Varying Surrounding Soil, Water Table and Overburden Condition* by Sarkar and Mukherjee. A parametric study of the box tunnel at different buried depth and soil condition were performed. Results showed that distortion and bending moment reduced by up to 22% when compactness of subsoil surrounding the structure changes from loose to dense under same seismic conditions. Deformation also reduced by 7% when tunnel buried depth increased from 8 to 12 m. However, deformation almost doubled when seismic zone changes from III to IV for the same soil condition.

There are three papers on coastal protection in this book, namely, *Seismic Response Study on Kaswati Dam* by Ghosh and Prasad, *Influence of Reservoir Level on the Dynamic Behaviour of Concrete Gravity Dam* by Biju and Joseph, and *Assessment of Seismic Displacement of Quay Walls* by Prasad and Nanjundaswamy. The first paper investigated the performance of the Kaswati Dam during the Bhuj earthquake using PLAXIS 2D, which reasonably match the actual behaviour in the field. The second paper adopted the Koyna earthquake. Reservoir water storage height was varied to study its influence on the response of a typical dam cross section. The final paper presents the seismic performance of quay wall using GEOSTUDIO with emphasis on the permanent horizontal displacement of the quay wall. Effects of varying properties of backfill and foundation soils were investigated which identified soil liquefaction as the major culprit leading to large displacement of quay wall towards sea in the event of an earthquake.

## 5 Retrofitting and Rehabiliation

Retrofitting and rehabilitation of existing infrastructure is an important topic in earthquake engineering to mitigate significant collapse which in turn heighten human casualties. *Seismic Response Elimination of Structure by Using Passive Devices: An Overview* provides an introduction to passive devices such as viscous damper, X-Plate damper, visco-elastic damper, tuned mass damper. An overview of range of parameters in the application of lifelines such as piping systems, bridges and health facilities were discussed along with their economic and seismic reduction comparisons. Following up with base isolation, *Response Reduction of Secondary Piping Systems in Base Isolated Buildings* by Kamble et al. demonstrated in their study that such buildings give significant protection to the secondary pipe system which experienced peak stress reduction at critical sections by about 60–70%. The benefits of isolation system can be further emphasized which saw seismic response of base-isolated bridge considerably altered due to dissimilarity in the isolator properties following parametric studies conducted using different system configurations, isolation systems and frequency ratios as discussed by Sridevi et al. in their paper titled *Seismic Analysis of Isolated and Non Isolated Railway Bridges with Slab Track—A Comparative Study*. Similar to earlier structural vulnerability assessment of irregular buildings, a study on *Seismic Performance Analysis of Regular and Irregular RCC Framed Building with Dampers* by Ramdas and Santhi extended the analysis to buildings with dampers using the response spectrum method in ETABS. Results indicated displacement reduction of about 38% and 36% in irregular and regular building respectively with damper. The use of active tendons in the paper titled *Seismic Control of Structures with Mass Irregularities through Optimally Placed Active Tendons* using Multi-objective Genetic Algorithm by Rather and Alam successfully minimized base shear, storey drift and floor displacement of a structure. The adoption of ultra-high-performance fiber-reinforced concrete (UHPFRC) in strengthening the beam column joint against seismic loads was discussed in Kubair and Rama's paper titled *Seismic Performance of UHPFRC Strengthened RC Beam Column Joints using Damage Plasticity Model—A Numerical Study* which indicated that the proposed Concrete Damage Plasticity (CDP) model accurately predicted the behaviour of the RC beam column joint. It was observed that the UHPFRC material was more ductile as compared to normal concrete and a thicker strip of the material increased the resistance of the beam column joint against seismic loads. Finally, Karmakar's paper titled *Assessment and Retrofitting Augmentation Methods of Seismic Performance on Existing RC Buildings across India* provided a more broad-based investigation of failure mechanisms of various reinforced concrete buildings of four major earthquakes across the country and how retrofitting techniques applied to some buildings performed in these earthquakes. Concrete jacketing was found to be consistently in increasing the performance of buildings and one of the most suitable methods employing local construction materials.

## 6 Review of Building Codes

Building codes form the guidance to seismic design of buildings. It is therefore essential to review if the anticipated and actual performance of the design are comparable. This was studied by Raisinghani et al. in their paper titled *Scientific Perspectives to Earthquake Resistance Design of RC Buildings—A Global Approach*, where seismic codes ACI08, IS1893, EC8 and TEC07 were investigated. This book volume also includes studies for more specific applications such as reinforced concrete moment resisting frame (MRF) buildings (*Parametric Study on the Variation of Time Period of RC MRF Buildings* by Sarma et al.), reinforced concrete frames (*Parametric Study of Performance-Based Seismic Design of Plan Irregular RC Frames—Indian Scenario* by Sirsikar et al.), infill walls (*Role of Openings in Sesimic Resistance of Infill Walls: A Critical Review of Indian Codal Provision* by Pathak and Dutta; and *Simplified Macro Modelling Approach for Estimation of Nonlinear Response of Infilled RC Frames* by Kurmi and Haldar) which would be useful for practicing engineers. For MRF buildings, it was concluded that the empirical equations given in IS1893 gave safe results in most number of cases. However, if additional parameters apart from building height are provided, a more precise time period of the building can be obtained. In the case of RC frames, it is recommended that a more detailed study on all vertical irregularities be identified to offer complete understanding of critical dimensions for each performance objective. For infill walls, base shear capacity and maximum displacement as per FEMA-273 and ATC-40 were found to be almost half of those calculated as per Indian IS1893 code. Hence, IS codal provisions would result in a doubly stiffer structure. Numerically developed capacity curves with experimental envelopes showed good agreement of peak strength between the proposed model and experimental data. A proposed simplified macro-modelling approach was also suggested to be an improvement over the widely used ASCE 41–06 infill model in estimating the initial stiffness, peak strength and overall load-deformation behaviour according to the study.

## 7 Field Surveys, Geospatial Tools and Others

Translating to field performance, Shinde and Sinha (*Seismic Vulnerability of Buildings in Palghar Area: Observations and Remedial Measures*) and Sharma et al. (*Evaluating the Seismic Performance of Domestic and Historical Masonry Structures in Himachal Pradesh Region of India*) evaluated seismic vulnerability of buildings at Palghar in Maharashtra and the Himachal Pradesh respectively. These papers detail the seismic hazard of the region, rapid visual survey of the buildings, and recommendations for strengthening measures.

In the topic of remote sensing, the ability to determine the velocity of land deformation prior and after an earthquake was demonstrated at Padang City in Indonesia using the Synthetic Aperture Radar (SAR) images by Usman et al. in their paper titled *Mapping of Earthquake Induced Land Deformation on Urban Area using Interferometric Synthetic Aperture Radar Data of Sentinel-1*. Geographic Information System (GIS) also facilitates presentation of spatially distributed data such as seismic information, historical data, geographical conditions and structural types to enable easier decision making. An open source software called QGIS was developed providing such information of South India from 1820 to 2020 as discussed in Madhukumar and Santhi's paper titled *Seismic Data Mapping of South India from 1820 to 2020 using Geographical Information System.*

Other valuable papers available in this book volume includes a paper on *Earthquake Loss Estimation using State-of-the-Art Tools* by Payyappilly and Somala, who developed an extensible workflow application for normalized damage and loss models using Performance Based Engineering (PBE), as well as *Post Earthquake Safe Shelter for Shimla City* by Vinayak et al., who proposed integration of emergency support functions and alternate care site that consists of first aid facility, temporary shelter, makeshift hospital and a post-disaster management centre that caters to a multitude of disaster scenarios.

## 8 Summary

This chapter has provided a brief highlight of the papers in this book volume on Recent Advances in Earthquake Engineering. A wide range of topics are discussed, such as seismic ground motion studies, structural and geotechnical vulnerability assessment, retrofitting and rehabilitation, review of building codes, as well as field surveys and introduction of remote sensing in the form of geospatial tools and geographical information systems to facilitate identification of areas susceptible to earthquake damage. It is expected that readers would be able to take away some useful pointers in one way or another.

Earthquakes would persist to do damage to our infrastructure and take lives from us if we fail to continue to push for advancement in earthquake engineering. This book volume serves to offer this opportunity to disseminate the contributions from fellow academics, practitioners and government authorities so as to bring ourselves closer towards a more disaster resilient society.

# Seismic Ground Motion Studies

# Wavelet Analysis of Far-Field Ground Motions from the $M_w$ 7.6 2005 Kashmir Earthquake

**Mohammed Ayub Ifan, Shalin Mathew, and Jayaprakash Vemuri**

## 1 Introduction

The continuing collision of the Indian and Eurasian tectonic plates has caused high seismic activity across the entire Himalayan belt and lead to enormous damage to the built environment [1–4]. Kashmir lies in this region of collision and is known to be seismically active. The region has been previously struck by several earthquakes such as 1842 $M_w$ 7.5 Kunnar earthquake, 1878 $M_w$ 6.7 Abbottabad earthquake, 1885 $M_w$ 6.3 Srinagar earthquake, and 1905 $M_w$ 7.8 Kangra earthquake. In the year 2005, Kashmir was struck by a major earthquake having a moment magnitude of $M_w$ 7.6 with a focal depth of 30 km [5]. The $M_w$ 7.6 earthquake struck Kashmir on 8th October 2005 with its epicentre located at 34.45°N, 73.65°E, about 50 km from Abbottabad and a focal depth of 26 km. The earthquake was the result of reactivation of the northwest striking Balakot-Bagh fault, and the rupture on the surface extended from Bagh to Muzaffarabad till Balakot [6, 7]. Two aftershocks of magnitude 5.9 and 5.8 were also felt in Kashmir within 30 min of the earthquake. The fatalities were over 86,000, and the number of injured was over 1,00,000. The damage to lifelines and buildings was immense with over 6,00,000 buildings classified as "damaged or destroyed". Isoseismal maps for the earthquake have been drawn based on the reconnaissance studies [8] The earthquake triggered thousands of landslides, comprising of rockfalls and debris falls, over a large region spread over an area over 7,500 km$^2$ [9]. The earthquake damage exposed a lot of structural defaults and construction malpractices. The high fatalities and heavy destruction caused this earthquake to be the deadliest to ever strike the Indian subcontinent [10, 11].

M. A. Ifan (✉) · S. Mathew · J. Vemuri
Department of Civil Engineering, Ecole Centrale College of Engineering, Mahindra University, Hyderabad 500043, Telangana, India
e-mail: ayubifan170119@mechyd.ac.in



S. Kolathayar and S. C. Chian (eds.), *Recent Advances in Earthquake Engineering*, Lecture Notes in Civil Engineering 175,
https://doi.org/10.1007/978-981-16-4617-1_2

Himachal Pradesh is a state in north India, bordered by Jammu and Kashmir in the north, Tibet autonomous region of China and Uttarakhand in the east, Haryana in the south, and Punjab in the west. The geographic location of Himachal Pradesh is in the southern edge of the Himalayan mountain region. The state of Himachal Pradesh is located in the high-risk seismic zone IV and V of the Indian seismic zoning map. The state lies along the Himalayan mountain range with Himalayan Frontal Thrust (HFT), the Main Boundary Thrust (MBT), and the Krolm, the Giri, Jutogh, Nahan thrusts also lie in this region. This paper analyses the time–frequency effect of near and far-field ground motion data from the 2005 Kashmir earthquake recorded at eight stations (Abbottabad, Bhanjaru, Sundla, Dalhousie, Jawali, Dharmashala, Kangra, and Mandi). Figure 1 shows the locations of these eight stations.

Strong motion records were recorded at a few stations in Pakistan, namely, Abbottabad, Murree, and Nilore. The town of Abbottabad is a distance of 50 km from the epicentre and recorded a peak ground accelerations (PGA) of 0.231 g. The town of Murree lies at a distance of 34 km from the epicentre and recorded a PGA of 0.078 g. The town of Nilore is situated 54 km from the epicentre and recorded a PGA of 0.026 g. The vertical PGAs at these three towns were recorded as 0.087 g, 0.069 g, and 0.03 g, respectively [12]. Two ground motions were recorded at dam sites, Tarbela and Mangla. At Tarbela Dam, which is at an epicentral distance of 78 km, a horizontal PGA of 0.16 g was recorded at the crest and 0.1 g at the base. The Mangla dam is located at 90 km epicentral distance and recorded a PGA of 0.1 g

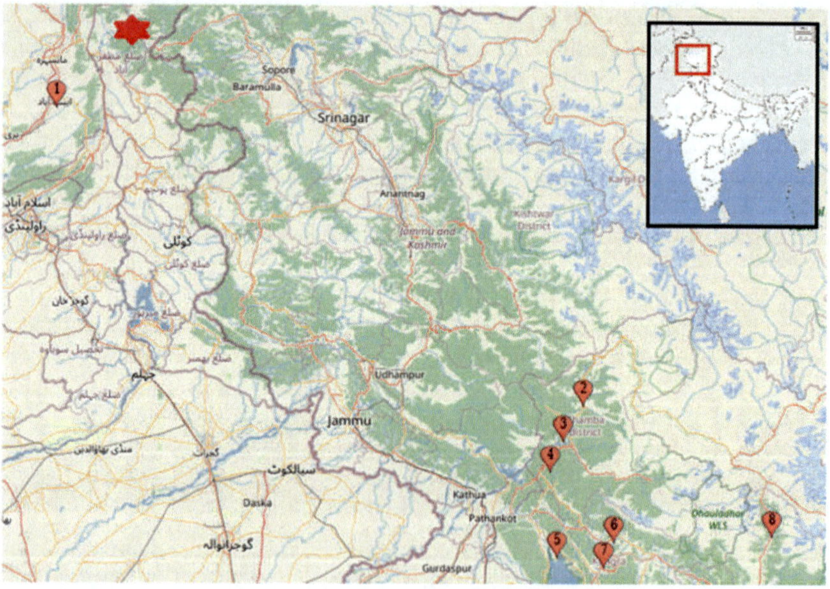


**Fig. 1** The location of the Earthquake epicentre along with the locations where data was recorded. (1) Abbottabad, (2) Bhanjraru, (3) Sundla, (4) Dalhousie, (5) Jawali, (6) Dharmashala, (7) Kangra (8) Mandi *Earthquake

at the downstream toe of the Dam [10]. However, only the Abbottabad record was made available to the authors of the present paper. This strong ground motion record is analysed in this paper. On the Indian side, far-field ground motions are available from seven-station in Himachal Pradesh. Table 1 shows the details of these stations along with key characteristics of the ground motions. Figure 2 shows the response spectra derived from the ground motions recorded at the seven stations. It is observed that the spectra show peaks in the low period ranges indicating the possible damage to low rise structures. The spectra of ground motions from the Abbottabad town (relatively near field) are far greater than the spectra of ground motions from other far-field towns. Further, the response spectra of ground motions from the Abbottabad town exhibit a wide acceleration sensitive region indicating possible damage to a large range to structures with varying natural periods.

## 2 Observed Damage to Buildings

It has been estimated that 84% of the total building stock in Pakistan Administrated Kashmir were damaged [13]. It has also been estimated that 98% of the highly damaged area lies within a topographically amplified seismic response area [14]. The damage to non-engineered construction was the maximum. Housing in the region is primarily built using Unreinforced Masonry (URM) with heavy Reinforced Concrete Slabs (RC) as roofs. Stone masonry with mud mortar is a common building typology. Such structures are inherently weak under lateral forces resulting from a major seismic event. Often there is a lack of confinement leading to extensive cracking in the masonry walls. Although engineered construction such as reinforced concrete frames performed better, it was observed that there was poor steel detailing leading to a lack of moment resisting action. Overall, it can be remarked that the poor construction practices in the region along with non-adherence to code standards lead to the deficient response of all types of structures.

## 3 Key Characteristics of Recorded Ground Motions

All three components of the strong ground motions, i.e., radial, transverse and vertical were recorded at the eight stations. Table 1 lists the key characteristics of all these ground motions, namely, Peak Ground Acceleration (PGA), Peak Ground Velocity (PGV), Sustained Maximum Acceleration (SMA), Sustained Maximum Velocity (SMV), Arias Intensity (AI), Mean Period ($T_m$) and Predominant Period ($T_p$).

**Table 1** Characteristics of recorded ground motions

| Sr. No. | Station | Direction | Distance (km) | PGA (g) | PGV (cm/s) | SMA (cm/s$^2$) | SMV (cm/s) | AI (m/s) | $T_p$ (s) | $T_m$ (s) |
|---|---|---|---|---|---|---|---|---|---|---|
| 1 | Abbottabad (near-field) | Radial | 50 | 0.26 | 45.05 | 0.20 | 32.60 | 2.44 | 0.78 | 0.99 |
| | | Transverse | | 0.18 | 31.02 | 0.16 | 22.79 | 1.66 | 0.98 | 0.86 |
| | | Vertical | | 0.08 | 9.89 | 0.08 | 8.46 | 0.36 | 0.64 | 0.50 |
| 2 | Bhanjraru | Radial | 293 | 0.03 | 2.96 | 0.02 | 2.83 | 0.01 | 0.32 | 0.76 |
| | | Transverse | | 0.03 | 2.62 | 0.02 | 2.08 | 0.01 | 0.36 | 0.67 |
| | | Vertical | | 0.01 | 2.05 | 0.01 | 1.61 | 0.003 | 0.30 | 0.88 |
| 3 | Sundla | Radial | 297 | 0.02 | 2.87 | 0.02 | 2.14 | 0.01 | 0.18 | 0.56 |
| | | Transverse | | 0.03 | 2.63 | 0.03 | 2.36 | 0.02 | 0.20 | 0.51 |
| | | Vertical | | 0.02 | 2.88 | 0.01 | 2.74 | 0.01 | 0.4 | 0.89 |
| 4 | Dalhousie | Radial | 302 | 0.02 | 5.33 | 0.01 | 4.19 | 0.03 | 1.08 | 1.59 |
| | | Transverse | | 0.02 | 4.86 | 0.02 | 4.84 | 0.03 | 1.70 | 1.72 |
| | | Vertical | | 0.01 | 4.42 | 0.01 | 3.71 | 0.01 | 0.26 | 1.95 |
| 5 | Jawali | Radial | 337 | 0.02 | 4.16 | 0.02 | 2.38 | 0.01 | 0.28 | 0.76 |
| | | Transverse | | 0.03 | 3.05 | 0.02 | 2.92 | 0.01 | 0.26 | 0.71 |
| | | Vertical | | 0.01 | 2.08 | 0.01 | 1.89 | 0.002 | 0.36 | 0.92 |
| 6 | Dharmashala | Radial | 351 | 0.02 | 5.82 | 0.02 | 4.57 | 0.05 | 0.64 | 1.11 |
| | | Transverse | | 0.02 | 6.79 | 0.02 | 4.95 | 0.05 | 0.48 | 1.07 |
| | | Vertical | | 0.01 | 3.09 | 0.01 | 2.30 | 0.01 | 2.46 | 1.24 |
| 7 | Kangra | Radial | 357 | 0.02 | 3.02 | 0.02 | 2.65 | 0.01 | 0.30 | 0.70 |
| | | Transverse | | 0.03 | 2.76 | 0.02 | 2.20 | 0.02 | 0.34 | 0.59 |
| | | Vertical | | 0.01 | 2.88 | 0.01 | 1.93 | 0.01 | 0.30 | 0.91 |

(continued)

**Table 1** (continued)

| Sr. No. | Station | Direction | Distance (km) | PGA (g) | PGV (cm/s) | SMA (cm/s$^2$) | SMV (cm/s) | AI (m/s) | $T_p$ (s) | $T_m$ (s) |
|---|---|---|---|---|---|---|---|---|---|---|
| 8 | Mandi | Radial | 410 | 0.04 | 3.33 | 0.03 | 2.78 | 0.07 | 0.54 | 0.60 |
| | | Transverse | | 0.02 | 2.35 | 0.01 | 2.11 | 0.02 | 0.48 | 0.67 |
| | | Vertical | | 0.01 | 1.70 | 0.01 | 1.39 | 0.01 | 0.48 | 0.60 |

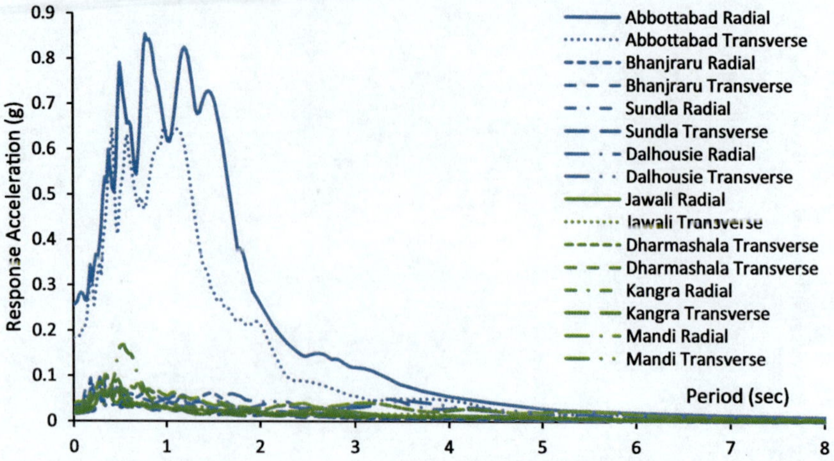


**Fig. 2** Response spectra for eight stations (radial and transverse) directions

# 4 Frequency Estimation Using Fast Fourier Transforms

An earthquake wave consists of several frequencies. This frequency distribution is understood from observations of the frequency content obtained using Fast Fourier Transforms (FFTs). FFTs are used to transform seismic time history into the frequency domain. FFTs highlight the frequency ranges at which seismic energy is concentrated. Figures 3, 4, 5, 6, 7, 8, 9 and 10a, b show the FFT's derived for the recorded ground motions. It is observed that most ground motions had low-frequency content (0–10 Hz) with a few stations, i.e., Sundla, Jawali, and Kangra exhibiting slightly higher frequency content (0–20 Hz). The Fourier amplitude of ground motions from the Abbottabad town is 10–20 higher than the ground motions from the other far-field stations.

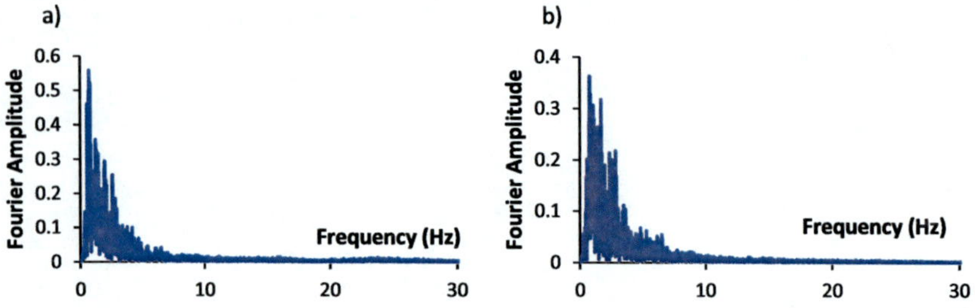


**Fig. 3** FFT's of Abbottabad. **a** Radial. **b** Transverse

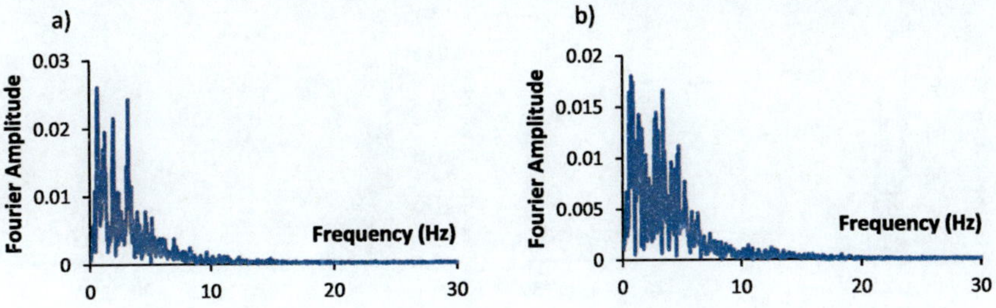


**Fig. 4** FFT's of Bhanjraru. **a** Radial. **b** Transverse

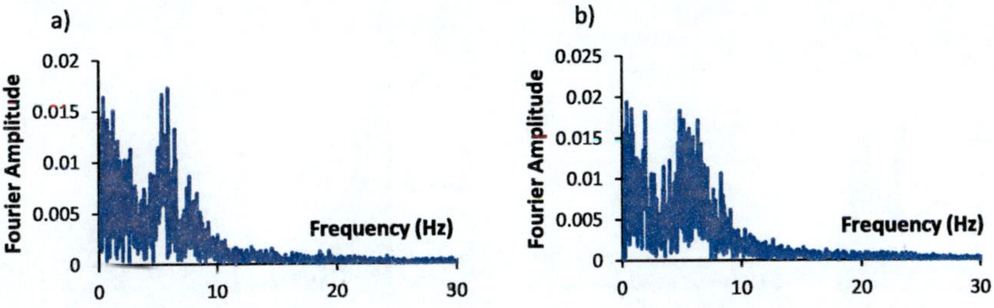


**Fig. 5** FFT's of Sundla. **a** Radial. **b** Transverse

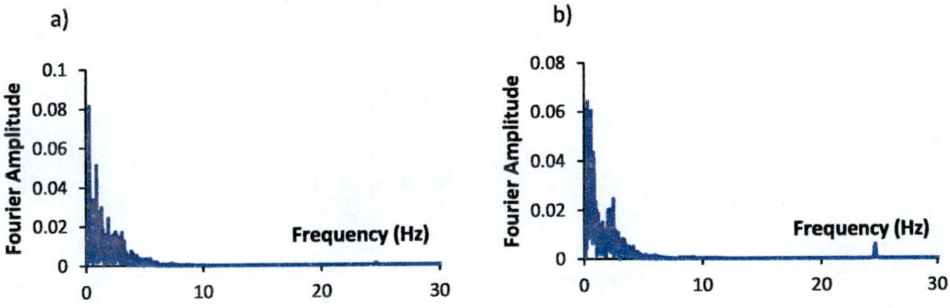


**Fig. 6** FFT's of Dalhousie. **a** Radial. **b** Transverse

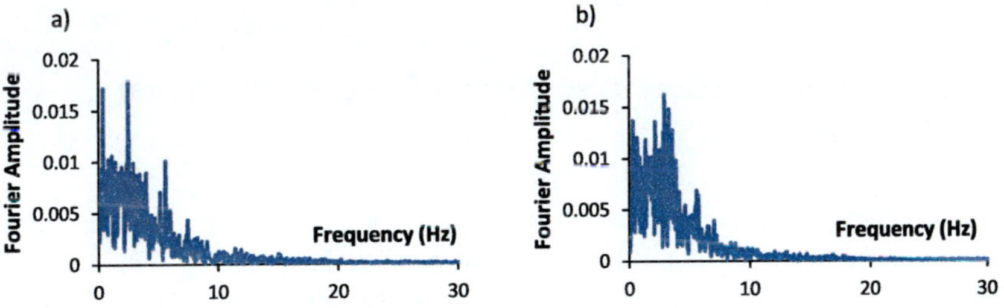


**Fig. 7** FFT's of Jawali. **a** Radial. **b** Transverse

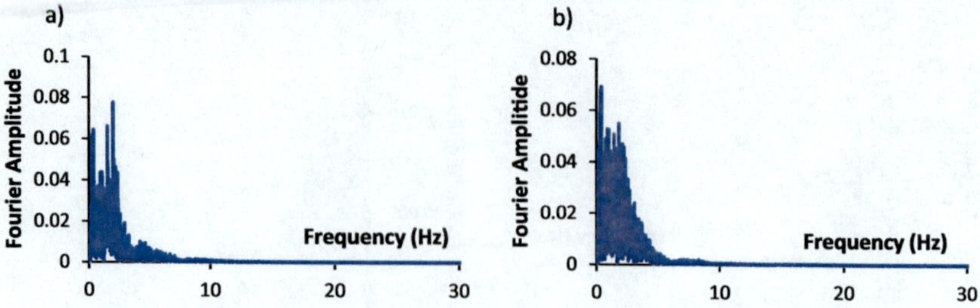


**Fig. 8** FFT's of Dharmashala. **a** Radial. **b** Transverse

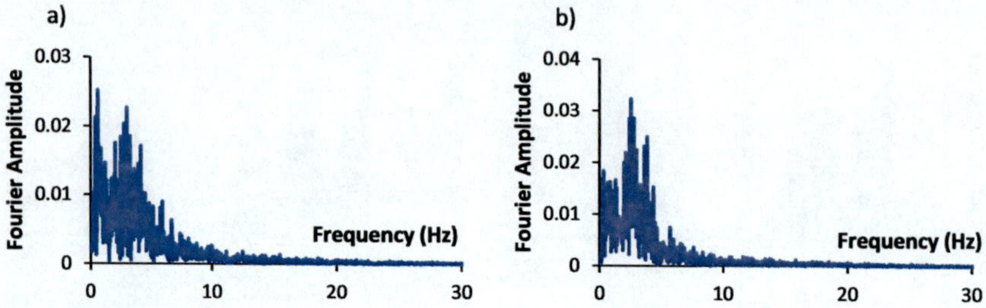


**Fig. 9** FFT's of Kangra. **a** Radial. **b** Transverse

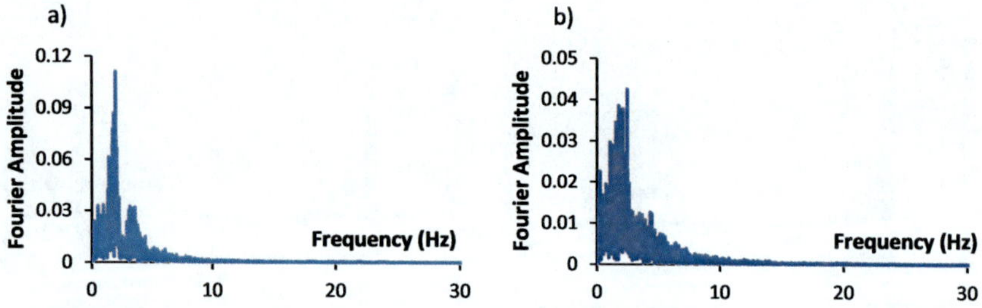


**Fig. 10** FFT's of Mandi. **a** Radial. **b** Transverse

## 5 Continuous Wavelet Transforms of Recorded Ground Motions

As discussed previously, while the time-acceleration representation hides the information on the frequency content of the ground motion, the frequency-Fourier amplitude representation hides the time information of the ground motion. Since earthquake ground motions are non-stationary, i.e., their frequency content varies with time, both these representations convey partial information. Figures 3, 4, 5, 6, 7, 8, 9 and 10 indicate that the frequency content in the ground motions is primarily in the range of 0–10 Hz. However, from these figures, the true damage potential of the seismic wave is unclear since it is neither apparent if a single damaging frequency was sustained in time nor is it apparent if a damaging lower frequency arrived after a damaging higher frequency. In both such cases, there is a possibility of high damage to structures. A structure when subjected to a seismic wave with a resonant frequency over a sustained duration of time is prone to large displacements. Similarly, during an earthquake, it is well known that the stiffness of the structure degrades with time, thereby lowering its period, i.e., increasing its frequency. Hence, the structure may be damaged twice during an earthquake if there are waves of lower frequency followed by waves of higher frequency, especially if these frequencies are close to the original and changed natural period of the structure.

The wavelet transforms, e.g., the continuous wavelet transform (CWT), provide information on both the time and frequency content of non-stationary signals. In this section, the CWT is utilized to understand the changing time–frequency characteristics of the earthquake ground motion. Figures 11, 12, 13, 14, 15, 16, 17 and 18a, b show the continuous wavelet transforms for the eight stations.

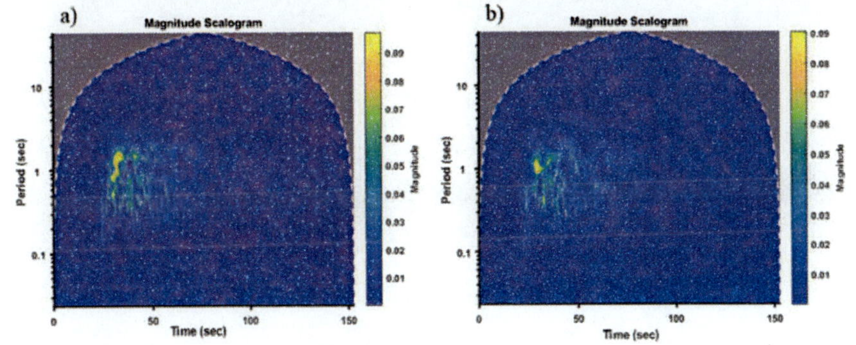


**Fig. 11** CWT's of Abbottabad. **a** Radial. **b** Transverse

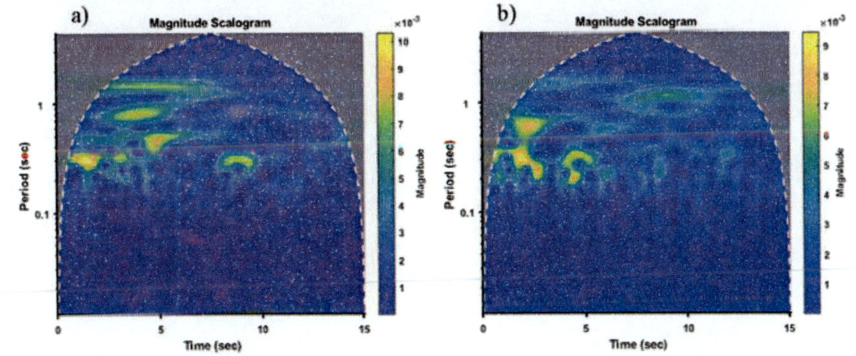


**Fig. 12** CWT's of Bhanjraru. **a** Radial. **b** Transverse

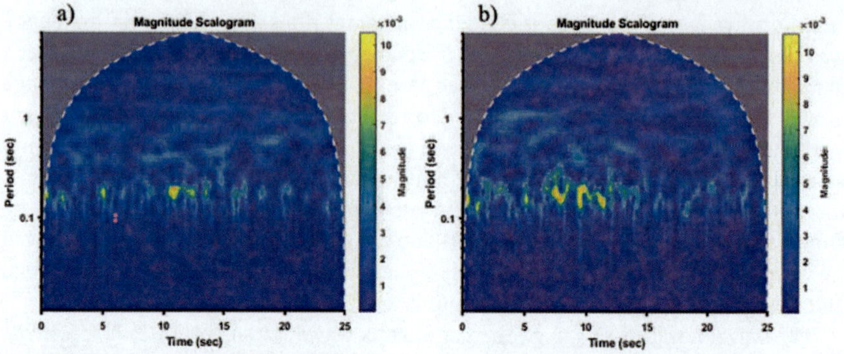


**Fig. 13** CWT's of Sundla. **a** Radial. **b** Transverse

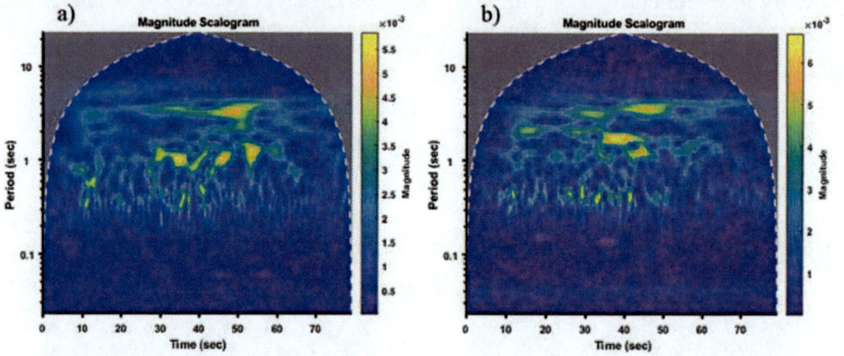


**Fig. 14** CWT's of Dalhousie. **a** Radial. **b** Transverse

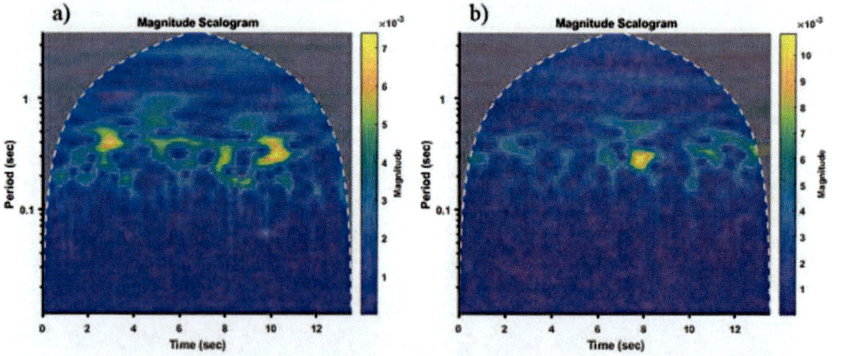


**Fig. 15** CWT's of Jawali. **a** Radial. **b** Transverse

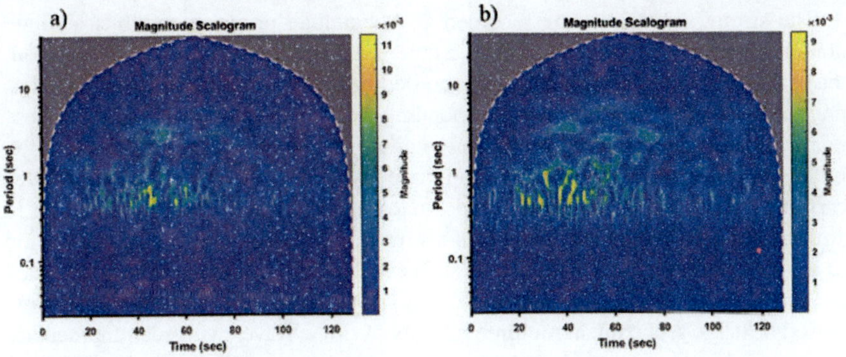


**Fig. 16** CWT's of Dharmashala. **a** Radial. **b** Transverse

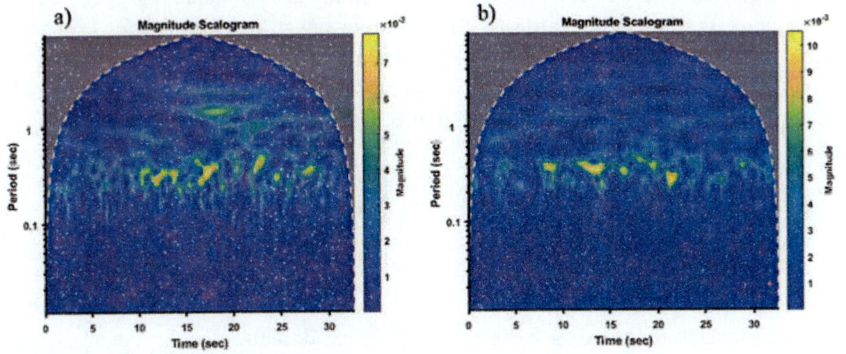


**Fig. 17** CWT's of Kangra. **a** Radial. **b** Transverse

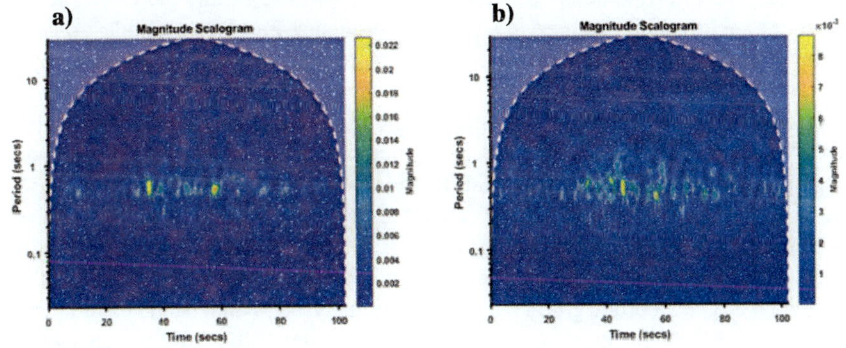


**Fig. 18** CWT's of Mandi. **a** Radial. **b** Transverse

The strong ground motion recorded at Abbottabad primarily exhibits a single dominant frequency corresponding to a time period of 1 s. Apart from Abbottabad, the remaining ground motions were recorded in the far-field range. Bhanjraru has multiple frequencies primarily corresponding to a time period of 0.3–0.4 s. These dominant frequencies were observed at multiple instants of time. At Bhanjraru, the initial seismic waves display a high amplitude-low period and are followed by waves carrying a high amplitude-high period: such a sequence is especially damaging to structures since the natural period of the structure is also altered from a low period to a higher period due to change in its stiffness. The strong ground motion recorded at Sundla primarily exhibits a single dominant frequency corresponding to a time period of 0.2 s, recurring at multiple instants of time. Waves with such frequencies (corresponding to a time period of 0.2 s) may cause structural damage to low rise (1–2 storeyed) structures, especially since these damaging frequencies are sustained over a long duration of time. The strong ground motions at Dalhousie exhibit multiple frequencies, some of which are sustained in time. Jawali has frequencies primarily corresponding to a time period of 0.3–0.4 s. Dharmashala and Mandi primarily exhibit a single high amplitude frequency for short bursts of time. Kangra has frequencies primarily corresponding to a time period of 0.3–0.4 s and recurring at multiple time instants.

## 6 Summary and Conclusions

Ground motions recorded in earthquakes are time-domain signals. For nonlinear dynamic analyses of structures, analysts use accelerograms, which are the time-acceleration representations of the ground motions. The time-acceleration description hides the information on the frequency content of the ground motions. The fast Fourier transform (FFT) is used to obtain the Frequency-Fourier amplitude representation of the accelerograms. However, in this representation, the time information is lost, and it is not possible to discern if the peak amplitudes or energies of the ground motion are reached at single or multiple time instances. Since earthquake ground motions are non-stationary, i.e., their frequency content varies with time, both these representations convey incomplete information. Wavelet analysis provides information on both the time and frequency content of non-stationary signals. In this paper, the continuous wavelet transforms (CWT) are utilized to understand the changing time–frequency characteristics of the ground motions from the $M_w$ 7.6 2005 Kashmir earthquake. The recorded ground motions were analysed to understand the influence of their critical characteristics on the observed damage to the building stock in the region. The observations from the CWT plots of ground motions recorded at various stations present a much-improved representation of the damage potential of strong ground motions. While some ground motions exhibit high amplitude waves at the same frequency content over a significant duration of time, other ground motions exhibit high amplitude waves in two to three frequency ranges. It is concluded that these observations correlate well with the observed non-uniform distribution of damage to reinforced concrete and unreinforced masonry structures in the region.

## References

1. Vemuri JP, Kolluru S (2017) Seismic analysis of unreinforced masonry walls. IDRiM J 6(2):102–115
2. Vemuri J, Ehteshamuddin S, Kolluru S (2018) Numerical simulation of soft brick unreinforced masonry walls subjected to lateral loads. Cogent Eng 5(1):1551503
3. Vemuri J, Ehteshamuddin S, Kolluru SV (2018) Evaluation of seismic displacement demand for unreinforced masonry shear walls. Cogent Eng 5(1):1480189
4. Garg R, Vemuri JP, Subramaniam KV (2019) Correlating peak ground A/V ratio with ground motion frequency content. In: Recent advances in structural engineering, vol 2. Springer, Singapore, pp 69–80
5. USGS (2005) https://earthquake.usgs.gov/earthquakes/eventpage/usp000e12e/executive
6. Kazmi ZA, Sodangi M (2019) The 2005 Kashmir Earthquake–devastation of infrastructures. Proc Inst Civil Eng-Struct Build 172(7):490–501
7. Hussain A, Yeats RS (2009) Geological setting of the 8 October 2005 Kashmir earthquake. J Seismolog 13(3):315–325
8. Mahajan AK, Kumar N, Arora BR (2006) Quick look isoseismal map of 8 October 2005 Kashmir earthquake. Curr Sci 356–361
9. Owen LA, Kamp U, Khattak GA, Harp EL, Keefer DK, Bauer MA (2008) Landslides triggered by the 8 October 2005 Kashmir earthquake. Geomorphology 94(1–2):1–9
10. Bothara JK, Hiçyılmaz KM (2008) General observations of building behaviour during the 8th October 2005 Pakistan earthquake. Bull N Z Soc Earthq Eng 41(4):209–233
11. Bilham R, Wallace K (2005) Future Mw > 8 earthquakes in the Himalaya: implications from the 26 Dec 2004 Mw = 9.0 earthquake on India's eastern plate margin. Geol Surv India, Supplement Pub 85:1–14
12. Mid America Earthquake Center (2005) A quick look report (#05–04) of the 2005 Kashmir Earthquake, University of Illinois at Urbana-Champaign
13. Mumtaz H, Mughal SH, Stephenson M, Bothara JK (2008) The challenges of reconstruction after the October 2005 Kashmir earthquake. Bull N Z Soc Earthq Eng 41(2):68–82
14. Khan S, van der Meijde M, van der Werff H, Shafique M (2020) The impact of topography on seismic amplification during the 2005 Kashmir earthquake. Nat Hazards Earth Syst Sci 20(2)

# Correlation Between Seismic Intensity Measures and Response of Skewed Bridges

**Van-Tien Phan and Duy-Duan Nguyen**

## 1 Introduction

In seismic design codes and structural analysis procedures, they have commonly used the peak ground acceleration (PGA) or spectral acceleration ($S_a$) as the seismic IMs. However, many studies demonstrated that PGA or $S_a$ are not the best selections for the seismic design and damage analyses of both on-ground and underground structures [1–9].

The interrelations between seismic IMs and seismic responses of reinforce concrete (RC) buildings were numerously studied [1, 2, 5, 7, 10–15]. Moreover, the correlation between seismic motion parameters and damage of other civil engineering structures such as tunnels [3, 16], storage tanks [17], nuclear power plants [18], and chimneys [19] were studied thoroughly. Numerous studies were performed to identify the good IMs for seismic damages of bridges. Padgett et al. [20] evaluated optimal IMs for deriving seismic fragility analysis models of multi-span steel girder bridges. Considering artificial and recorded earthquake motions, they concluded that PGA and $S_a(T_1)$ are efficient intensity measures for the artificial ground motions, while cumulative absolute velocity (CAV) is the most reliable for using recorded motions. Zhang et al. [9] identified the correlation between seismic IMs of far-field motions and the structural response of a cable-stayed bridge with a single pylon in China. The best correlated IMs exhibited were velocity spectral intensity (VSI), $S_a(T_1)$, and Housner intensity (HI). Jahangiri et al. [21] pointed out that root-mean-square of acceleration ($A_{rms}$) is the optimal intensity measure for seismic performance assessment of concrete arch bridges. In the study of Zelaschi et al. [22], by investigating a set of Italian RC bridges with 30 scaled motions they pointed out that Fajfar intensity ($I_v$), peak ground velocity (PGV), and $S_a$ are the optimal IMs. Avşar

V.-T. Phan · D.-D. Nguyen (✉)
Department of Civil Engineering, Vinh University, Vinh, Vietnam
e-mail: duyduankxd@vinhuni.edu.vn



S. Kolathayar and S. C. Chian (eds.), *Recent Advances in Earthquake Engineering*, Lecture Notes in Civil Engineering 175,
https://doi.org/10.1007/978-981-16-4617-1_3

et al. [23] highlighted that accelerated-related IMs are not strongly correlated with the response of seismic-isolated bridges. Additionally, the period-dependent IMs (e.g., VSI, HI, modified acceleration spectrum intensity, and $S_a$) are strongly correlated with the deformation base-isolated bridges subjected to normal earthquakes. Meanwhile, PGV and modified velocity spectrum intensity are good correlation for pulse-like ground motions.

Previously, the correlation between seismic IMs and damage of buildings and conventional bridges were well-studied, but it is still insufficient for skewed bridges. Specifically, the influence of earthquake frequency contents on the correlation analyses was not considered yet. This study aims to sufficiently recognize the relation between 23 ground motion IMs and seismic performances of skewed bridges accounting for the low- and high-frequency contents of earthquake. Accordingly, the strong and weak correlation IMs with structural performances of bridges also identified based on correlation analyses.

## 2 Seismic Intensity Measures and Ground Motions

To obtain the seismic IMs, a direct evaluation from earthquake accelerograms and a calculation by the software can be implemented [24]. This study accounts for 23 common ground motion IMs for correlation analyses. These parameters are calculated for every motion record using SeismoSignal software [25]. The considered IMs and its definitions are presented in Table 1.

Some undefined notations in Table 1 can be expressed as $t_{tot}$ which is the total duration of earthquake; $a(t)$, $v(t)$, and $d(t)$ are the time-history acceleration, velocity, and displacement of the record; $g$ is the gravity acceleration; $\xi$ is the damping ratio; $S_a$ is the spectral acceleration; $PS_v$ is the pseudo-spectral velocity; $T$ is the period; $T_1$ is the fundamental period; $C_i$ is the Fourier factor; and $f_i$ is the discrete frequency.

We selected 290 acceleration records from historic earthquakes. The data is available in PEER center [36] and KMA [37]. The magnitude of the earthquakes is ranged from 3.0 to 7.6. Low- and high-frequency content motions are divided with 212 records falling to the low-frequency set and 78 records belonging to the high-frequency set. It is important to note that the ground motion frequency content is normally recognized from the response spectra. The earthquake motions with large spectral accelerations fallen in the frequency range approximately larger than 10 Hz is considered as a high-frequency (HF) ground motion, otherwise discerned as low-frequency (LF) motions [38, 39]. Figure 1 shows the response spectra of two groups of ground motion records.

**Table 1** Selected earthquake IMs

| No. | Seismic IMs | Definition | Unit | Refs. |
|---|---|---|---|---|
| 1 | Peak ground acceleration | PGA = max \|a(t)\| | g | – |
| 2 | Peak ground velocity | PGV = max \|v(t)\| | m/s | – |
| 3 | Peak ground displacement | PGD = max \|d(t)\| | m | – |
| 4 | Ratio of PGV/PGA | PGV/PGA | s | [24] |
| 5 | Root-mean-square of acceleration | $A_{rms} = \sqrt{\frac{1}{t_{tot}} \int_0^{t_{tot}} a(t)^2 dt}$ | g | [26] |
| 6 | Root-mean-square of velocity | $V_{rms} = \sqrt{\frac{1}{t_{tot}} \int_0^{t_{tot}} v(t)^2 dt}$ | m/s | [24] |
| 7 | Root-mean-square of displacement | $D_{rms} = \sqrt{\frac{1}{t_{tot}} \int_0^{t_{tot}} d(t)^2 dt}$ | m | [24] |
| 8 | Arias intensity | $I_a = \frac{\pi}{2g} \int_0^{t_{tot}} a(t)^2 dt$ | m/s | [27] |
| 9 | Characteristic intensity | $I_c = (A_{rms})^{2/3} \sqrt{t_{tot}}$ | $m^{1.5}/s^{2.5}$ | [28] |
| 10 | Specific energy density | $\mathrm{SED} = \int_0^{t_{tot}} v(t)^2 dt$ | $m^2/s$ | – |
| 11 | Cumulative absolute velocity | $\mathrm{CAV} = \int_0^{t_{tot}} \|a(t)\| dt$ | m/s | [29] |
| 12 | Acceleration spectrum intensity | $\mathrm{ASI} = \int_{0.1}^{0.5} S_a(\xi = 0.05, T) dT$ | g*s | [30] |
| 13 | Velocity spectrum intensity | $\mathrm{VSI} = \int_{0.1}^{2.5} S_v(\xi = 0.05, T) dT$ | m | [31] |
| 14 | Housner spectrum intensity | $\mathrm{HI} = \int_{0.1}^{2.5} PS_v(\xi = 0.05, T) dT$ | m | [30] |
| 15 | Sustained maximum acceleration | SMA = the 3rd of PGA | g | [32] |
| 16 | Sustained maximum velocity | SMV = the 3rd of PGV | m/s | [32] |
| 17 | Effective peak acceleration | $\mathrm{EPA} = \frac{mean\left(S_a^{0.1-0.5}(\xi=0.05)\right)}{2.5}$ | g | [29] |
| 18 | Spectral acceleration at $T_1$ | $S_a(T_1)$ | g | [33] |
| 19 | Spectral velocity at $T_1$ | $S_v(T_1)$ | m/s | – |
| 20 | Spectral displacement at $T_1$ | $S_d(T_1)$ | m | – |
| 21 | A95 parameter | $A_{95} = 0.764\, I_a^{0.438}$ | g | [34] |
| 22 | Predominant period | $T_p$ | s | [24] |
| 23 | Mean period | $T_m = \frac{C_i^2/f_i)}{C_i^2}$ | s | [35] |

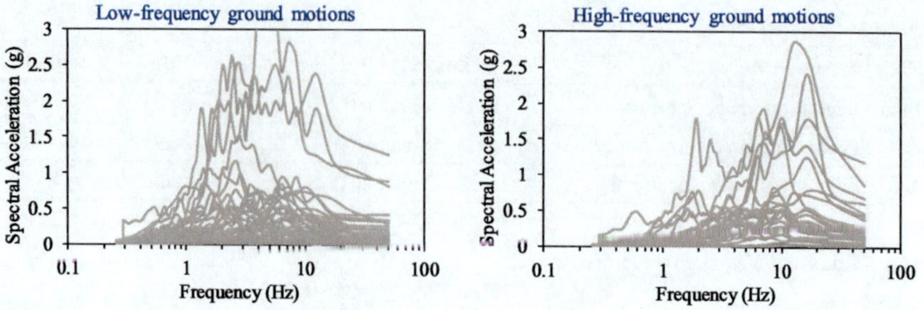


**Fig. 1** Response spectra of selected ground motion records

# 3 Bridge Modelling

The studied bridge is made of reinforced concrete (RC) with 10 triple-column bents. The circular cross-section of all piers has a diameter of 0.8 m. The bridge has 11 spans, and each span is 14.5 m length. The column height of bent P1 and P2 is 4.0 m, while that for bent P4–P6 is 6.0 m, and for other bents is 5.5 m. The bridge is skewed with an angle of 60°. The configuration and dimensions of the bridge are also shown in Fig. 2. For the foundation of bridge bents, 12 bored piles (D = 0.4 m) arranged in a double-row are connected with the pile-cap which its dimensions in the height, width, and length are 0.9 m, 2.5 m, and 10 m, respectively. The length of piles is

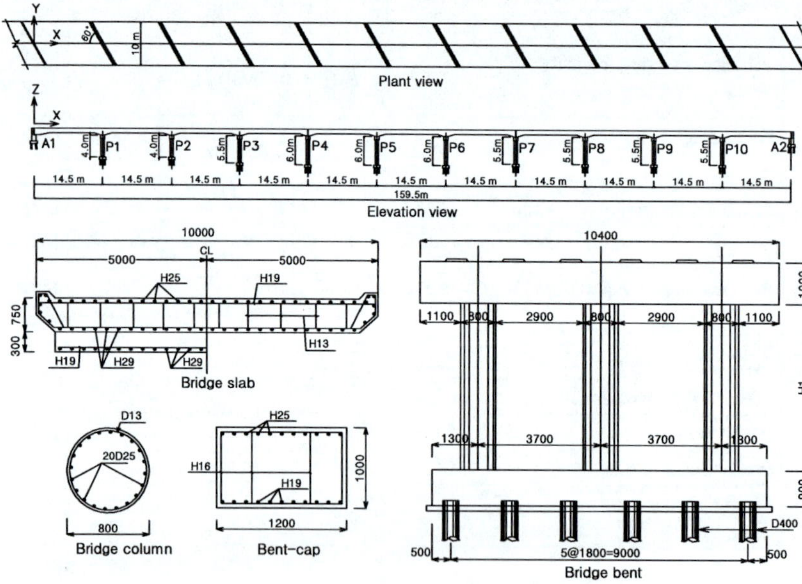


**Fig. 2** Configurations of the RC skewed bridges

reduced from 22 m (at the bent P1) to 10 m (at the bent P10). The piles are mostly embedded into sand, gravel-sand, amber gravel, weathered rock and soft rock layers.

The open-source platform, OpenSees [40], is employed to build the finite element model of the bridge. To model concrete material, the *concrete02* model is used [41], whereas the *steel02* model is applied for modelling reinforcement in OpenSees [42]. It is noted that those models are able to consider the nonlinearity of materials [43–47].

The girder of bridges is assumed to be elastic behaviour during seismic excitations. Thus, the *ElasticMembranePlateSection* element is assigned to the bridge slab. The column bridge is modelled using *nonlinearBeamColumn* elements. Figure 3 shows the fiber-section modelling scheme of the bridge column. Also, the moment–curvature relation of column cross-sections are presented in Fig. 3. Additionally, considering the soil-structure interaction in the bridge, the piles are modelled using elastic beam elements, in which a series of soil springs are attached to element nodes. To model soil springs, the *zerolength* model represented by *p-y* curve [48] is utilized, as illustrated in Fig. 4. Figure 5 shows the 3D finite element modelling of the bridges in OpenSees. The eigenvalue analysis result is described in Fig. 6.

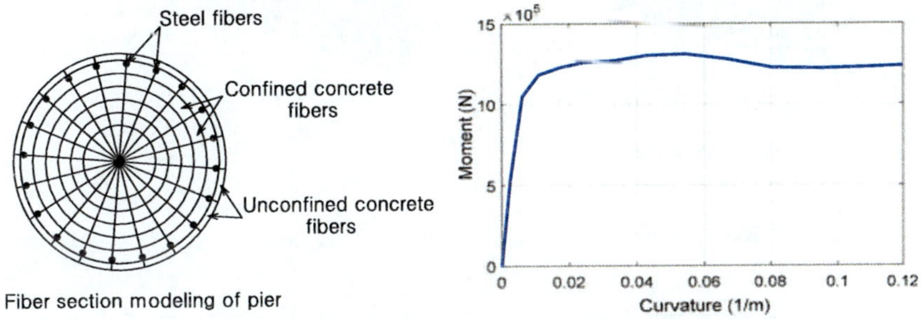


**Fig. 3** Fiber section model and moment–curvature relationship of the bridge piers

**Fig. 4** Illustration of the *p-y* curve for soil-pile interaction modeling

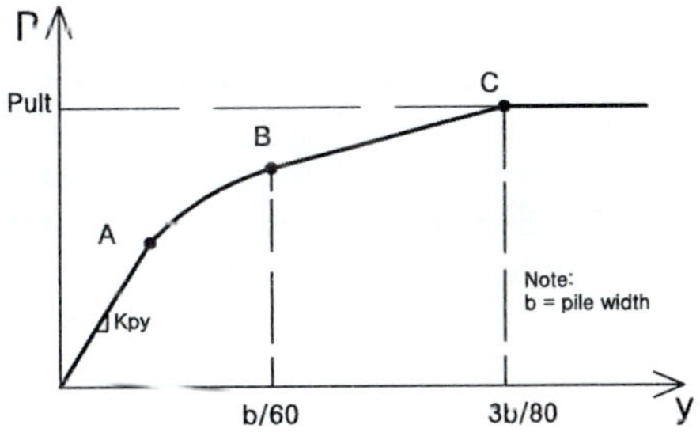

**Fig. 5** 3D finite element model of the bridge in OpenSees

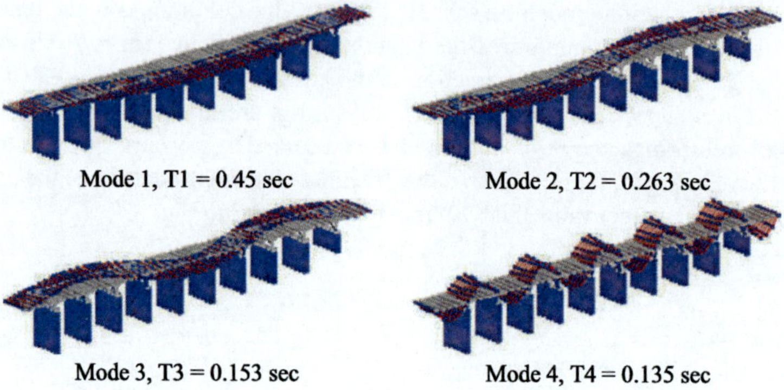


**Fig. 6** Eigenvalue analysis results

# 4 Seismic Response and Correlation Analysis

A series of nonlinear time-history analyses were performed. We imposed all ground motions in two groups on the bridge model in horizontal directions and captured the structural responses associated with each ground motion. Since the pier is one of the most crucial components of the bridge structures, seismic performances of the bridge are measured in terms of the drift ratio and shear forces of bridge piers. The use of displacement (or drift ratio) and internal forces is common and facilitated in the seismic design and fragility analyses [43–47, 49]. Figure 7 illustrates responses of the bridge subjected to the 1989 Loma Prieta and the 2016 Gyeongju earthquake, in which the displacement is measured at the top of the middle pier, while the shear force is monitored at the bottom of that pier. The responses of all bridge piers are obtained for every ground motion records.

The relationship between seismic responses of the structure and earthquake IMs is needed to identify the strong and the weak correlation indicators. For this study, the Pearson's coefficient is used to reflect the correlation between seismic responses of the bridge and earthquake IMs. The linear correlation coefficient given in Ang and Tang [50] is defined as

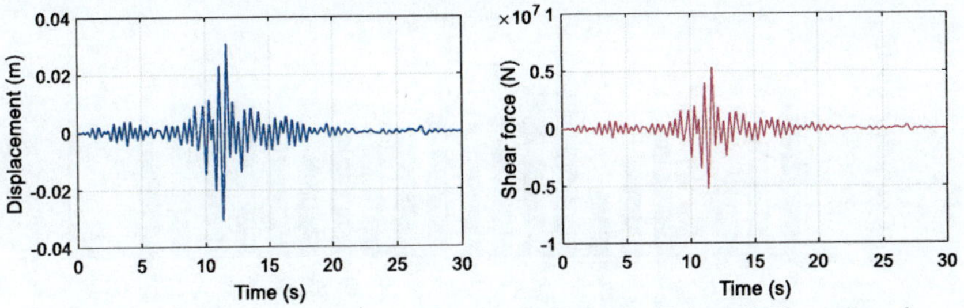


(a) Seismic responses of bridge under the 1989 Loma Prieta (LF) earthquake

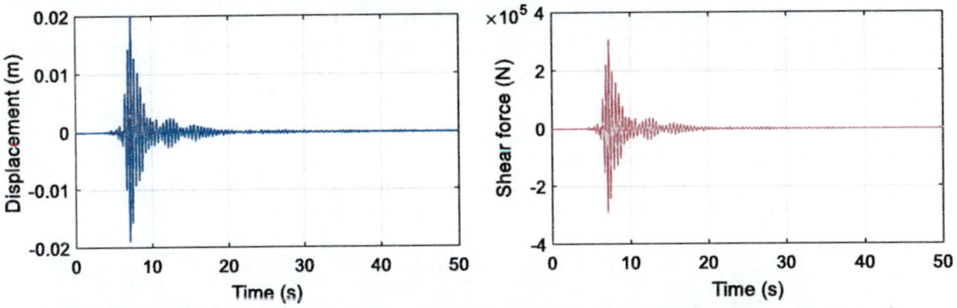


(b) Seismic responses of bridge under the 2016 Gyeongju (HF) earthquake

**Fig. 7** Example of seismic responses of the bridge under LF and HF earthquakes

$$\rho = \frac{1}{n-1}\frac{\sum(x_i - \overline{x})(y_i - \overline{y})}{\sigma_x \sigma_y} \tag{1}$$

where $\overline{x}$, $\overline{y}$, are the sample means of variables $x_i$ and $y_i$, and $x_i$ represents the seismic responses of bridges, while $y_i$ represents the intensity measures. $\sigma_x$,$\sigma_y$ are the sample standard deviations of $x$ and $y$, determined by

$$\sigma_x = \sqrt{\frac{1}{n-1}\left[(\Sigma x_i)^2 - n(\overline{x})^2\right]};\ \sigma_y = \sqrt{\frac{1}{n-1}\left[(\Sigma y_i)^2 - n(\overline{y})^2\right]} \tag{2}$$

We calculated the correlation coefficients for seismic responses of the bridge and IMs associated with two groups of ground motions. It should be noted that the results are showed here for a representative bridge pier because within a specific bridge the same trend is observed for all piers. Figures 8 and 9 show the calculated correlation coefficients corresponding to each earthquake IM for both LF and HF motions. In the case of low-frequency ground motions, it can be demonstrated that Arias intensity ($I_a$) has the strongest correlation with seismic damage, followed by characteristic intensity ($I_c$), $S_a(T_1)$, and spectral velocity at the fundamental period ($S_v(T_1)$). Whereas, PGV/PGA ratio, mean period ($T_m$), and predominant period ($T_p$) are weak correlated parameters with seismic damage of bridges. For the high-frequency ground motions, the strongest correlated measure with damage is specific energy density

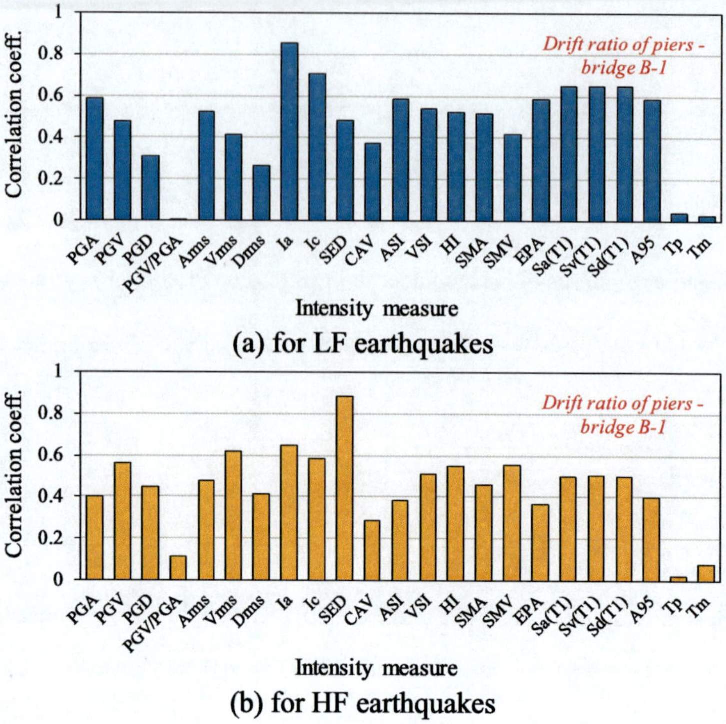


**Fig. 8** Correlation between the dift ratio of piers and seismic IMs

(SED), followed by $I_a$, root-mean-square of velocity ($V_{rms}$) and $I_c$. Similar to the low frequency motion group, three low-correlated IMs are PGV/PGA ratio, $T_m$, and $T_p$. The results also indicate that PGA and PGV have medium correlation and it may not always the best parameters for seismic design and seismic vulnerability assessment of skewed bridges.

This study identified the strongly correlated earthquake IMs for seismic responses of skewed RC bridges. Designers or analysts can use $S_a(T_1)$, $S_v(T_1)$, $I_a$ or $I_c$ for the design or performance evaluation process as well as fragility analyses of such bridges. Also, the findings of this paper imply that we should not use PGV/PGA, $T_m$ or $T_p$ for seismic design and analyses of the bridges. These findings are partially consistent with those of Padgett et al. [20] and Jahangiri et al. [21] since the bridge type is not similar.

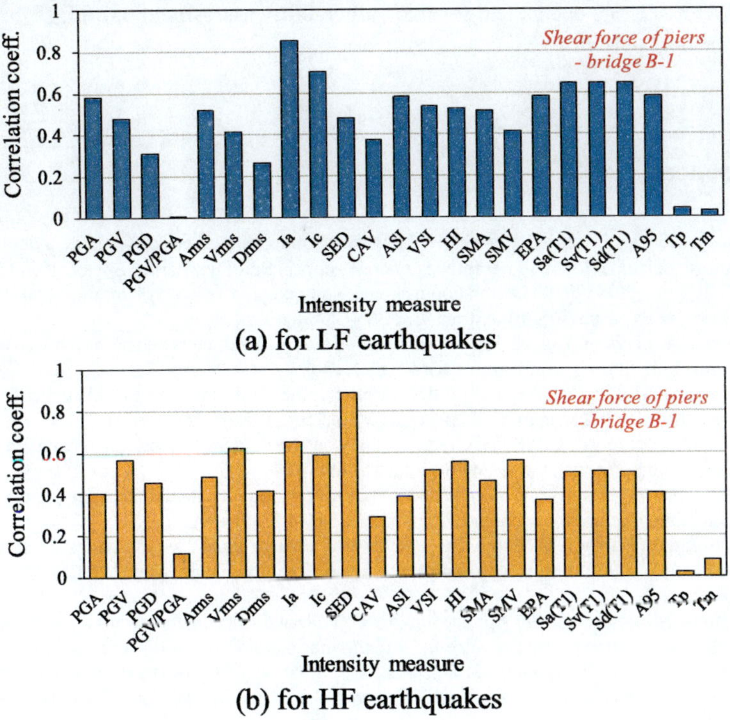


**Fig. 9** Correlation between the shear force of piers and seismic IMs

## 5 Conclusions

A series of nonlinear time-history analyses are performed for the skewed bridge. Two groups of earthquake ground motions classified into low- and high-frequency contents are used for analyses. For each ground motion, 23 seismic intensity measures are considered. A series of correlation coefficients between seismic responses of bridges and earthquake intensity measures are calculated. The following conclusions are drawn based on the analysis results.

- For the low frequency ground motions, the best intensity measure for correlating with seismic responses of bridges $I_a$, followed by $I_c$, $S_a(T_1)$, and $S_v(T_1)$.
- For the high frequency ground motions, the well-correlated IMs with seismic performances are SED, $I_a$, $V_{rms}$, and $I_c$.
- PGV/PGA, $T_m$, and $T_p$ show to be weak correlated measures if bridges subjected to low frequency motions. This trend is also observed for the case of high frequency motions.
- PGA and PGV have a medium correlation with responses of bridge structures during both low- and high-frequency earthquake motions. A selection of PGA or

PGV for seismic design and seismic vulnerability assessment of bridges may not be the best option.

## References


1. Cao V, Ronagh H (2014) Correlation between seismic parameters of far-fault motions and damage indices of low-rise reinforced concrete frames. Soil Dyn Earthq Eng 66:102–112
2. Cao V, Ronagh H (2014) Correlation between parameters of pulse-type motions and damage of low-rise RC frames. Earthq Struct 7(3):365–384
3. Chen Z, Wei J (2013) Correlation between ground motion parameters and lining damage indices for mountain tunnels. Nat Hazards 65(3):1683–1702
4. Corigliano M, Lai CG, Barla G (2007) Seismic vulnerability of rock tunnels using fragility curves. In: 11th ISRM congress. International Society for Rock Mechanics
5. Elenas A, Meskouris K (2001) Correlation study between seismic acceleration parameters and damage indices of structures. Eng Struct 23(6):698–704
6. Ghayoomi M, Dashti S (2015) Effect of ground motion characteristics on seismic soil-foundation-structure interaction. Earthq Spectra 31(3):1789–1812
7. Massumi A, Gholami F (2016) The influence of seismic intensity parameters on structural damage of RC buildings using principal components analysis. Appl Math Model 40(3):2161–2176
8. Yaghmaei-Sabegh S (2012) Application of wavelet transforms on characterization of inelastic displacement ratio spectra for pulse-like ground motions. J Earthq Eng 16(4):561–578
9. Zhang Y, Ding Y, Pang Y (2015) Selection of optimal intensity measures in seismic damage analysis of cable-stayed bridges subjected to far-fault ground motions. J Earthq Tsunami 9(01):1550003
10. Elenas A (2000) Correlation between seismic acceleration parameters and overall structural damage indices of buildings. Soil Dyn Earthq Eng 20(1–4):93–100
11. Nanos N, Elenas A, Ponterosso P (2008) Correlation of different strong motion duration parameters and damage indicators of reinforced concrete structures. In: The 14th word conference on earthquake engineering, Beijing, China
12. Kostinakis K, Athanatopoulou A, Morfidis K (2015) Correlation between ground motion intensity measures and seismic damage of 3D R/C buildings. Eng Struct 82:151–167
13. Pejovic J, Jankovic S (2015) Selection of ground motion intensity measure for reinforced concrete structure. Procedia Eng 117:588–595
14. Pejovic J, Serdar N, Pejovic R (2017) Optimal intensity measures for probabilistic seismic demand models of RC high-rise buildings. Earthq Struct 13(3):221–230
15. Lu X, Ye L, Lu X, Li M, Ma X (2013) An improved ground motion intensity measure for super high-rise buildings. Science China Technol Sci 56(6):1525–1533
16. Nguyen DD, Park D, Shamsher S, Nguyen VQ, Lee TH (2019) Seismic vulnerability assessment of rectangular cut-and-cover subway tunnels. Tunn Undergr Space Technol 86:247–261
17. Phan HN, Paolacci F (2016) Efficient intensity measures for probabilistic seismic response analysis of anchored above-ground liquid steel storage tanks. In: Pressure vessels and piping conference. American Society of Mechanical Engineers
18. Nguyen DD, Thusa B, Han TS, Lee TH (2020) Identifying significant earthquake intensity measures for evaluating seismic damage and fragility of nuclear power plant structures. Nucl Eng Technol 52(1):192–205
19. Qiu Y, Zhou C, Siha A (2020) Correlation between earthquake intensity parameters and damage indices of high-rise RC chimneys. Soil Dyn Earthq Eng 137:106282
20. Padgett JE, Nielson BG, DesRoches R (2008) Selection of optimal intensity measures in probabilistic seismic demand models of highway bridge portfolios. Earthq Eng Struct Dynam 37(5):711–725

21. Jahangiri V, Yazdani M, Marefat MS (2018) Intensity measures for the seismic response assessment of plain concrete arch bridges. Bull Earthq Eng 16(9):4225–4248
22. Zelaschi C, Monteiro R, Pinho R (2019) Critical assessment of intensity measures for seismic response of Italian RC bridge portfolios. J Earthq Eng 23(6):980–1000
23. Avşar Ö, Özdemir G (2011) Response of seismic-isolated bridges in relation to intensity measures of ordinary and pulselike ground motions. J Bridg Eng 18(3):250–260
24. Kramer SL (1996) Geotechnical earthquake engineering. Prentice Hall, Inc., Upper Saddle River, New Jersey, USA
25. SeismoSignal (2020) A computer program for signal processing of strong-motion data. http://www.seismosoft.com. Accessed June
26. Dobry R, Idriss IM, Ng E (1978) Duration characteristics of horizontal components of strong-motion earthquake records. Bull Seismol Soc Am 68(5):1487–1520
27. Arias A (1970) A measure of earthquake intensity. Massachusetts Inst of Tech, Cambridge. Univ of Chile, Santiago de Chile
28. Park Y, Ang AH, Wen YK (1985) Seismic damage analysis of reinforced concrete buildings. J Struct Eng 111(4):740–757
29. Benjamin JR (1988) A criterion for determining exceedance of the operating basis earthquake. Report EPRI NP-5930, Electrical Power Research Institute, Palo Alto, California
30. Housner GW (1952) Spectrum intensities of strong-motion earthquakes. In: Symposium on earthquake and blast effects on structures, Los Angeles, California, USA, pp 20–36
31. Thun JL (1988) Earthquake ground motions for design and analysis of dams. In: Earthquake engineering and soil dynamics II-recent advances in ground-motion evaluation
32. Nuttli OW (1979) The relation of sustained maximum ground acceleration and velocity to earthquake intensity and magnitude. Report 16, Misc. Paper S-73-1, US Army Waterways Experimental Station, Vicksburg, Mississippi
33. Shome N, Cornell CA, Bazzurro P, Carballo JE (1998) Earthquakes, records, and nonlinear responses. Earthq Spectra 14(3):469–500
34. Sarma SK, Yang KS (1987) An evaluation of strong motion records and a new parameter A95. Earthq Eng Struct Dyna 15(1):119–132
35. Rathje EM, Norman AA, Bray J (1998) Simplified frequency content estimates of earthquake ground motions. J Geotech Geoenviron Eng 124(2)
36. PEER (2019) Pacific Earthquake Engineering Research Center Database, http://peer.berkeley.edu/peer_ground_motion_database
37. KMA (2019) Korean Meteorological Administration, Korea
38. EPRI (2007) Program on technology innovation: the effects of high frequency ground motion on structures, components, and equipment in nuclear power plants. Report 1015108, Electrical Power Research Institute, Palo Alto, California, USA
39. EPRI (2017) Advanced nuclear technology: high-frequency seismic loading evaluation for standard nuclear power plants. Report 3002009429, Electrical Power Research Institute, Palo Alto, California, USA
40. Mazzoni S, McKenna F, Fenves GL (2005) OpenSees command language manual. Pacific Earthquake Engineering Research (PEER) Center
41. Kent DC, Park R (1971) Flexural members with confined concrete. J Struct Division
42. Menengotto M (1973) Method of analysis for cyclically loaded reinforced concrete plane frames including changes in geometry and nonelastic behavior of elements under combined normal force and bending. In: IABSE symposium on resistance and ultimate deformability of structures acted on by well-defined repeated loads, Lisbon
43. Lee TH, Nguyen DD (2018) Seismic vulnerability assessment of a continuous steel box girder bridge considering influence of LRB properties. Sādhanā 43(1):14
44. Nguyen DD, Lee TH (2018) Seismic fragility curves of bridge piers accounting for ground motions in Korea. IOP Conf Ser Earth Environ Sci 143(1):012029
45. Tran NL, Nguyen TH, Phan VT, Nguyen DD (2020) Seismic fragility analysis of reinforced concrete piers of steel box girder bridges: a parametric study. Mater Today Proc

46. Lee TH, Nguyen VH, Phan VT, Nguyen DD (2018) Seismic margin assessment of a reinforced concrete skewed bridge in a nuclear power plant. MATEC Web of Conferences 251:02019
47. Choi BH, Moreno LB, Lim CS, Nguyen DD Lee TH (2019) Seismic performance evaluation of a fully integral concrete bridge with end-restraining abutments. Adv Civil Eng 2019
48. Reese LC, Cox WR, Koop F D (1974) Field testing and analysis of laterally loaded piles in sand. In: The VI annual offshore technology conference, Houston, Texas, pp 473–485
49. Park HS, Nguyen DD, Lee TH (2016) Seismic fragilities of bridges and transmission towers considering recorded ground motions in South Korea. J Earthquake Eng Soc Korea 20(7_spc):435–441
50. Ang AH, Tang WH (2007) Probability concepts in engineering: emphasis on applications in civil & environmental engineering, vol 1. Wiley, New York

# Structural Vulnerability Assessment

# Evaluation on Seismic Performance of "Dhajji-Dewari" Timber Wall Panel

M. Sushma, Vijayalakshmi Akella, B. K. Raghu Prasad, and G. Deepu

## 1 Introduction

Vernacular architecture is a major category of architecture which is basically dependent on local needs, local traditions, and locally available construction material [1]. In recent years, technology has disrupted millennia-old building traditions. Several research works and thoughts are developed for the conventional building techniques in urban localities. Whereas, the suburban and rural areas are not taken care in this aspect, which stands as major part of any developing country [2]. UNESCO has emerged in identifying the combination of vernacular techniques with modern materials and technology.

Earthquakes are characterized as the major destructive natural hazard which leads to loss of life, economy, and property. Half-timbered vernacular architectural style popularly known as "Dhajji-Dewari" style of architecture in India combines the finest features of masonry and timber, which offers great resistance against seismic actions [3]. Taking this aspect into consideration, it is of great interest to enhance knowledge on the seismic behavior of such construction typology.

Dhajji-Dewari is the traditional timber-braced frame with random rubble masonry infill construction seen mainly in Kashmir, India. It is noted that vertical posts, horizontal and diagonal bracings are closely packed. The inherent property of timber to be flexible without breaking during an earthquake leads to the exceptional performance of such systems [4].

The research presented in this paper includes experimental evaluation of mechanical properties of timber and timber joints along with numerical investigations conducted on Dhajji-wall in order to compute the seismic capacity of Dhajji-Dewari

M. Sushma (✉) · V. Akella · G. Deepu
K S School of Engineering & Management, Affiliated to VTU Belgaum, Bengaluru, India

B. K. R. Prasad
Indian Institute of Science, Bengaluru, Bengaluru, India



S. Kolathayar and S. C. Chian (eds.), *Recent Advances in Earthquake Engineering*, Lecture Notes in Civil Engineering 175,
https://doi.org/10.1007/978-981-16-4617-1_4

timber-nailed wall panel. The masonry infill is neglected as no significant effect on the peak strength was seen except that of viscous damping. It is seen that wall with masonry infill has higher viscous damping value when compared to wall without infill. Infill just adds on to the initial stiffness of the wall [4]. Hence in the present work, analysis of timber frame is performed neglecting the effect of infill. For the experimental purpose, locally available Acacia wood species is used which has high durability and water-resistant qualities. The experimental work includes the testing of mechanical property of Acacia wood, tension and bending tests on possible tennon-mortise joints in the wall panel. The results obtained from the experiment is applied in the numerical model to perform nonlinear static pushover analysis.

## 2 Details of Dhajji Wall Panel

Dhajji wall panel considered for analysis has three sets of members with cross-sectional dimensions 4″ × 4″, 2″ × 4″, 4″ × 1″ as shown in Fig. 1. The 4″ × 4″ members are used as main posts and base plate. The 2″ × 4″ members are used as secondary post. The 4″ × 1″ is used as cross bracers. The length and height of the test specimen is taken as 6′5″ and 8′8″, respectively.

For the purpose of connections, nails with diamond pointed ends were used. Main vertical post is connected to top and bottom plate through Type-1 tennon and mortise connections. Secondary vertical post is connected to top and bottom plate through Type-2 tennon and mortise connections as shown in Fig. 2. These tennon and mortise joints were connected by nails designed as per IS: 2366-1983, Indian Standard Code of Practice for Nail-Jointed Timber Construction [5].

The lateral load response of the panel is majorly taken by the connections with respect to vertical posts and bottom plate [4]. Hence to find the capacity of such joints, tension and bending tests are conducted for the Type-1 and Type-2 connections. Member sizes of tennon mortise connection joints is kept same as the wall panel

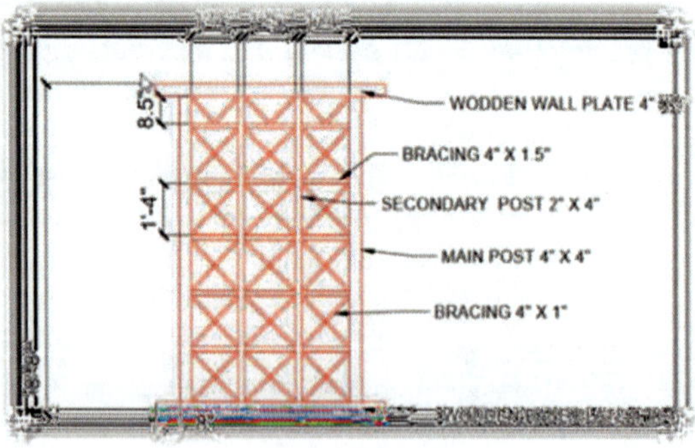


**Fig. 1** Cross-sectional dimensions of Dhajji-Dewari wall panel

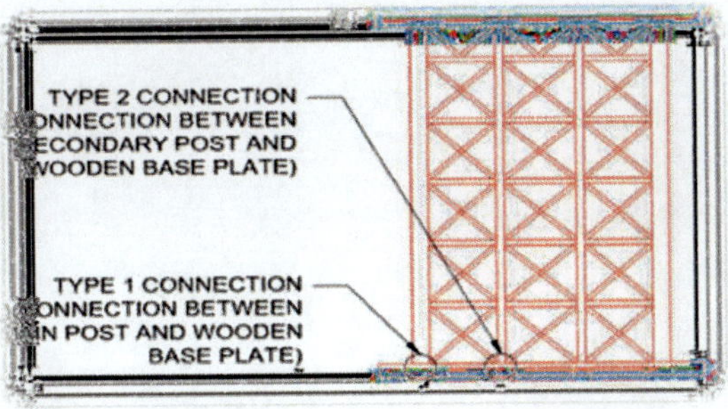


**Fig. 2** Type-1 and type-2 connection of timber wall panel

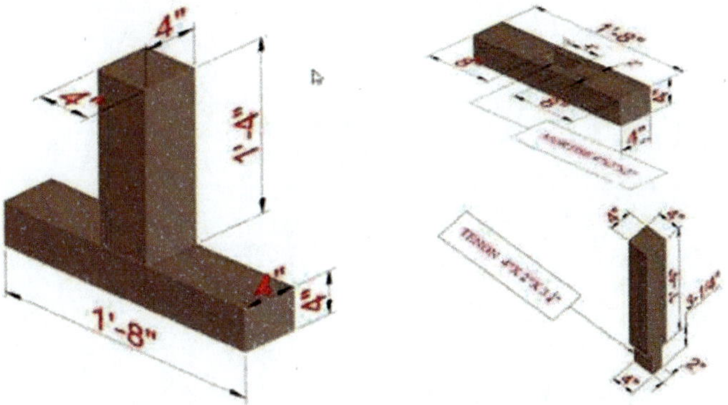


**Fig. 3** Joint details of type-1

taken for which height of member is taken as 1′8″. Tenon sizes for Type 1 and Type 2 joint were 4″ × 2″ × 3.25″ and 2″ × 1″× 3.25″. Figures 3 and 4 show joint details of Type-1 and Type-2 connections.

## 3 Details of Nail-Jointed Timber Construction

Nail with Diamond Pointed edge is used for the joints. As per the codal provision, the length and diameter of the nail used is 100 mm and 4 mm, respectively. Practically nails are arranged in such a way that line of force in the member passes through the centroid of the group of nails transmitting load to it. Nails are driven from opposite faces (adjacent nails are driven alternatively from either face of joint) to strengthen the nail-joint as shown in Figs. 5 and 6. The number of nails calculated as per IS: 2366-1983 for each type of joint is tabulated in Table 1.

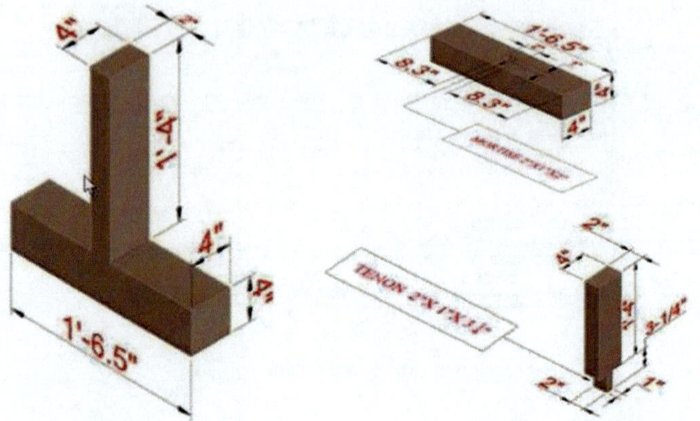


**Fig. 4** Joint details of type-2

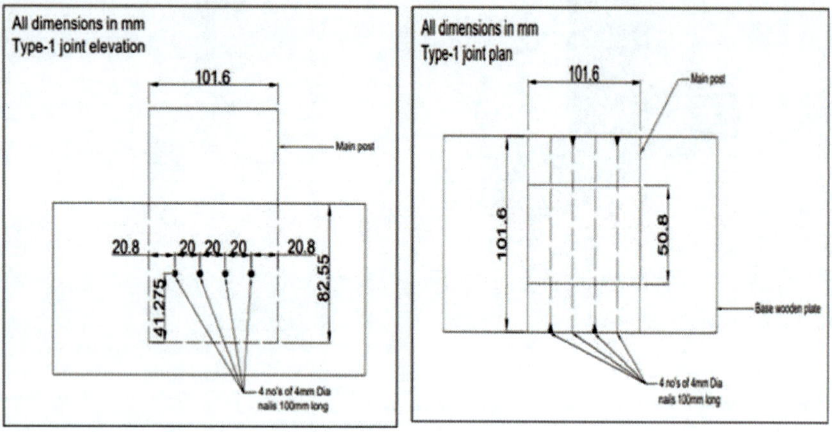


**Fig. 5** Elevation and plan view of type-1 joint

# 4 Test Setup and Procedure to Determine Mechanical Properties of Acacia Wood

## *4.1 Compressive Strength Parallel to Grain*

The timber "Acacia" selected for the present study is tested for compressive strength parallel to grain as shown in Fig. 7 using 20 × 20 mm cross-section and 80 mm in length with a rate of loading 0.8 mm/min as per IS: 1708 (Part 1 to 18)—2005, Methods of Testing of Small Clear Specimens of Timber [6].

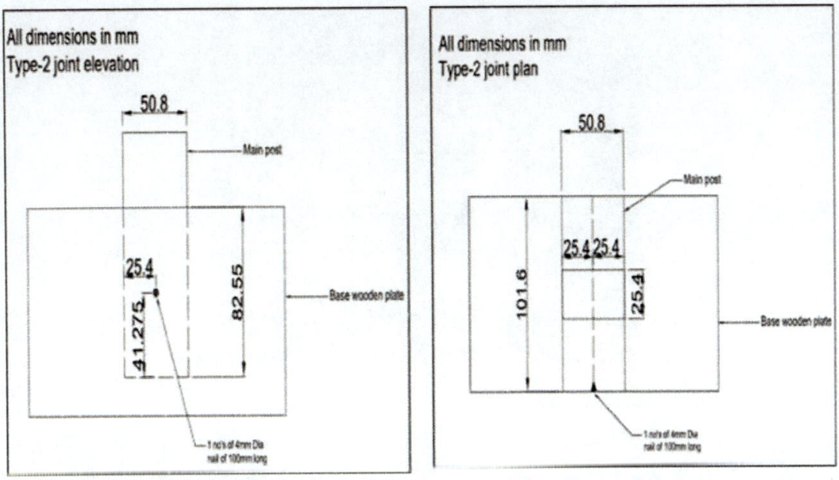


**Fig. 6** Elevation and plan view of type-2 joint

**Table 1** Number of nails calculated as per IS: 2366-1983

| Type of joint | Number of nails |
|---|---|
| Type 1 | 4 |
| Type 2 | 1 |

**Fig. 7** Test setup of compression test parallel to grain

## *4.2 Tensile Capacity Perpendicular to Grains*

The acacia wood specimen of size 50 mm × 50 mm × 56 mm is tested for tension capacity perpendicular to grains as shown in Fig. 8 with a rate of loading 6 mm/min as per IS:1708 (Part 1 to 18)-2005, Methods of Testing of Small Clear Specimens of Timber [6].

**Fig. 8** Test setup of tensile capacity test perpendicular to grain

# 5 Test Setup and Procedure to Determine Mechanical Properties of Timber Joints

## *5.1 Bending Test on T-joint*

Type-1 and Type-2 timber joints is subjected to bending test. The T-joints is loaded into the loading frame as shown in Fig. 9. Digital Protractor is placed at the connection to find the rotation. This rotation due to force applied is required to quantify the moment carrying capacity of the joint.

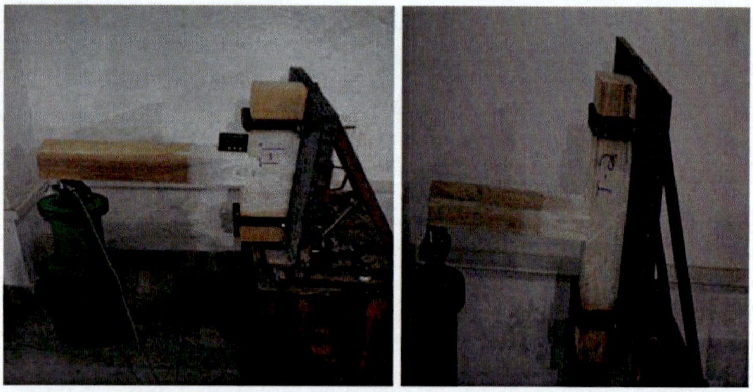

**Fig. 9** T-joints subjected to bending in loading frame

## 5.2 *Tension Test on T-Joint*

Type 1 and Type 2 timber joints is subjected to tension test as shown in Fig. 10. The T-joints is loaded in Universal Testing Machine. Extensometer is used to measure the displacement at joints. The force with corresponding deformation gives the tensile capacity of joints.

# 6 Numerical Modelling of Dhajji Wall

The finite element software ANSYS is used for the numerical modeling of the wall chosen. The result obtained from tension and bending elastoplastic curves are assigned to Type-1 and Type-2 connections in the numerical model with a function of "hinge property". At the cross bracings and interior horizontal members, moments are released to replicate the free rotation behavior at connections [7]. Image of the numerical model of wall in ANSYS is shown in Fig. 11. Link element is used to impart the nonlinear property to the linear elements. This non-linear property is induced by giving tension and bending parameters of the experiment to the hinge.

The horizontal main beams and vertical timber post is analyzed as elastic bending elements whereas horizontal bracing elements as a truss element to duplicate the pull-out behavior of joints as demonstrated in the experiment, with a limit on their tensile and compressive strengths. Downward force of 500 N is applied on each vertical post as per the calculation of roof load.

Two nodes having same coordinates for each of the connection between horizontal and vertical timber beams is the procedure used to create plastic hinges. Further link elements is used to connect these two nodes by applying the results of tension and bending test of joints. The node at bottom of the structure which is connected to bottom of horizontal beam is restrained against rotation and transition to model it as

**Fig. 10** T-joints subjected to tension test in universal testing machine

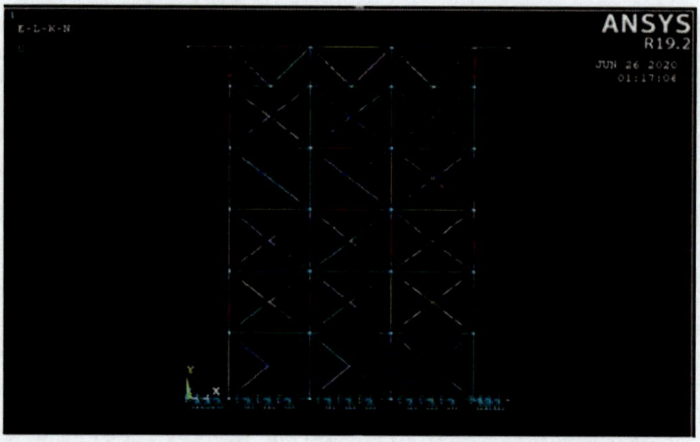


**Fig. 11** Numerical model of wall panel in ANSYS

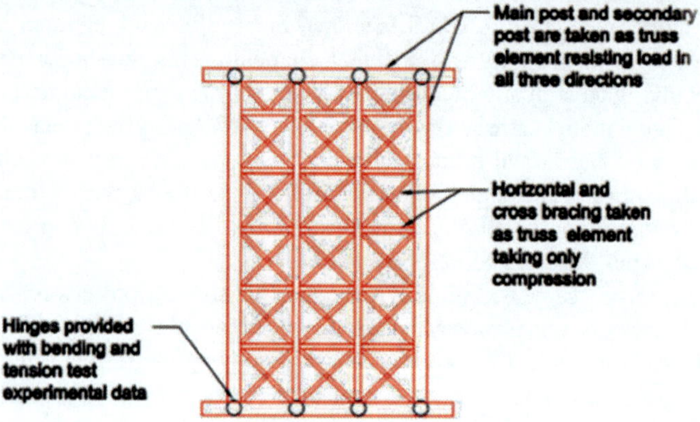


**Fig. 12** Details of the element types selected for numerical model

fixed support. Figure 12 represents the pictorial representation of wall panel model with their elemental details.

# 7 Experimental Test Results

## *7.1 Compressive and Tensile Strength Test on Acacia Wood Specimen*

From test conducted on acacia wood specimen based on IS: 1708 (Part 8)-1986, it is observed that compressive strength parallel to grains is 42.5 MPa. It is seen that the

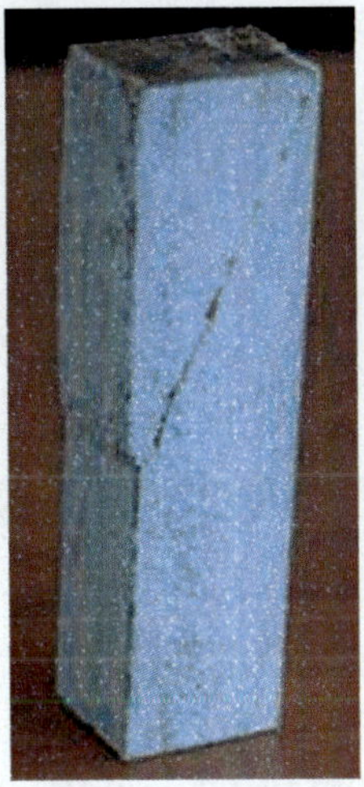

**Fig. 13** Failure pattern of Acacia wood test

**Fig. 14** Failure pattern of Acacia subjected to compressive strength wood subjected to tensile test

timber specimen subjected to compression parallel to grain failed due to shearing action as shown in Fig. 13. Modulus of elasticity of Acacia is found to be 12,000 MPa. The tensile strength perpendicular to grains is obtained as 5 kN. Failure pattern for timber specimen subjected to tensile strength test perpendicular grain is shown in Fig. 14.

## 7.2 *Tension Test on Type-1 and Type-2 Joint*

From the force–displacement graph plotted as shown in Fig. 15, it is observed that the ultimate load taken by Type-1 joint is 17.60 kN and Type-2 joint is 3.90 kN, for which corresponding displacements were 23 mm and 26.3 mm, respectively. The

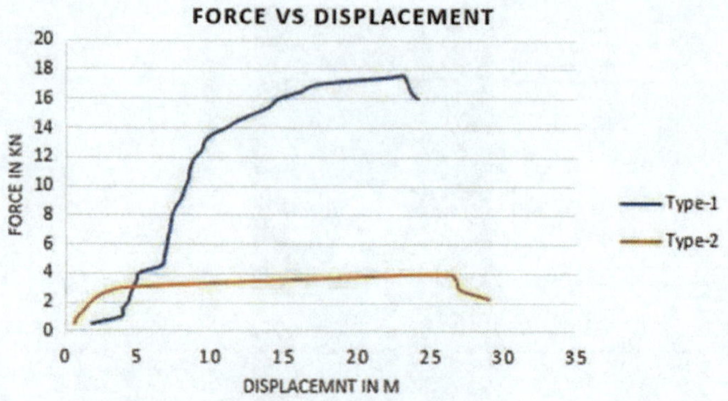


**Fig. 15** Force versus displacement plot of type-1 and type-2 joint

ultimate stress obtained with respect to ultimate load for Type-1 joint is 3.4 MPa and Type-2 joint is 3.02 MPa.

## *7.3 Bending Test on Type-1 and Type-2 Joint*

From the moment-rotation graph plotted as shown in Fig. 16, it is observed that the ultimate load taken by Type-1 joint is 0.62 kN and Type-2 joint is 0.49 kN, for which corresponding rotation is 0.0245 rad and 0.1015 rad respectively.

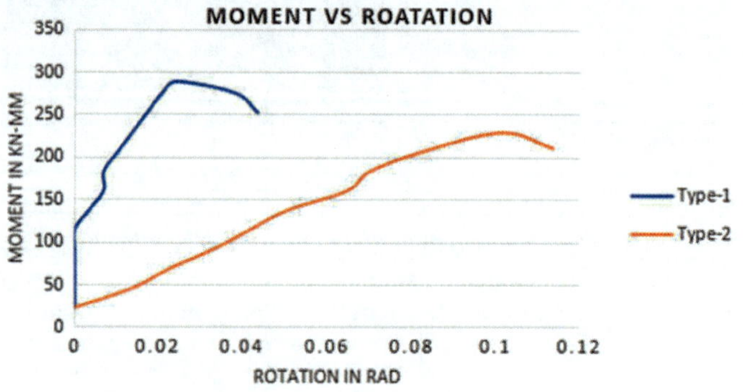


**Fig. 16** Moment versus rotation of type-1 and type-2 joint

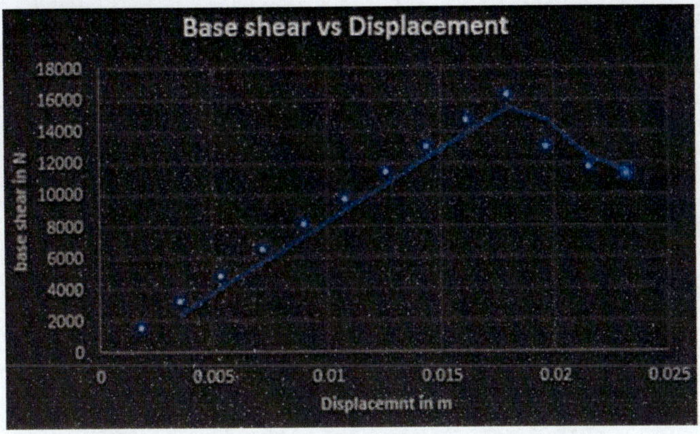


**Fig. 17** Base shear versus displacement graph for Dhajji-Dewari wall panel

## 8 Numerical Analysis of Wall Panel

Dhajji-Dewari wall panel considered for the study is subjected to pushover analysis. Properties of Acacia wood and joints obtained from experiment are assigned. The tension and bending results obtained from testing Type-1 and Type-2 is provided to the respective hinge.

Base shear versus displacement curve of the wall is plotted. The lateral capacity of the wall is found to be 20 kN. Maximum displacement is found to be 17.9 mm for the lateral load of 20 kN. Load, displacement, and base shear of the panel is shown in Fig. 17. The maximum displacement of the wall panel is shown in Fig. 18.

At 22 kN lateral load, the maximum compressive stress in horizontal bracing is 6.98 MPa and cross bracings is 7.46 MPa, which does not exceed the experimental results. Even though corner joints failed, bracings intact even at lateral load of 20 kN. The stress in horizontal and cross bracings is shown in Fig. 19.

When the lateral load of 22 kN is applied, the stress in type-1 joint is 4.3 MPa in the numerical model which exceeds the maximum stress of type-1 joint 3.4 MPa obtained from test results. Hence the failure of panel initiates at type-1 joint when the lateral load is more than 20 kN. The stress concentration in the frame elements is shown in Fig. 20 proves that the panel fails with the failure of Type-1 joint.

## 9 Conclusions

When the nails were provided with sufficient edge and pitch distance as per IS: 2366-1983, Code of Practice for Nail-Jointed Timber Construction, it was seen that nails failed during tensile capacity test and not the timber. The lateral capacity of wall was found to be 20 kN for the designed joints. Numerical analysis observations

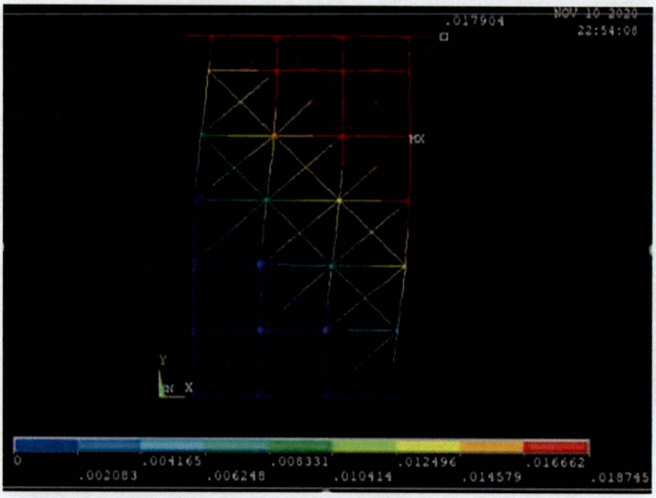


**Fig. 18** Maximum displacement of Dhajji-Dewari wall panel

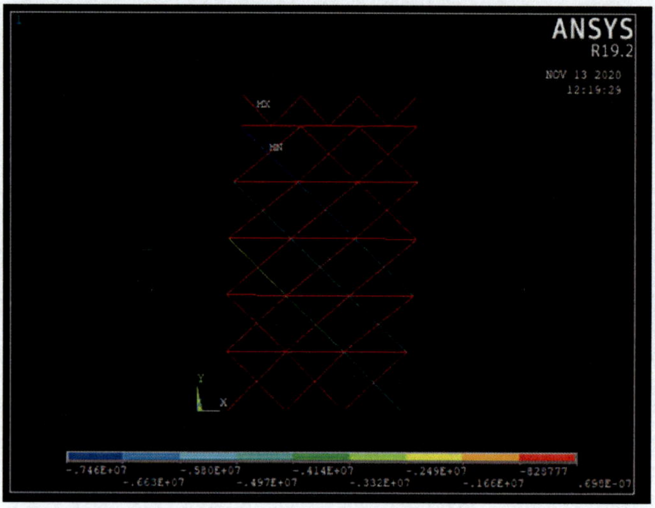


**Fig. 19** Stress in horizontal and cross bracings at 22 kN lateral load on wall panel

have demonstrated large deformability of Dhajji wall panel under lateral load. The displacement of the wall panel is found to be 17.9 mm which is within the maximum allowable deflection of L/120 as per International Code Council. Bracings being a member under compression gave good resistance even at the ultimate load. Maximum stress concentration with a value of 4.3 MPa was seen in the joint that connected main vertical post to top and bottom of the panel. The failure at this joint led to the

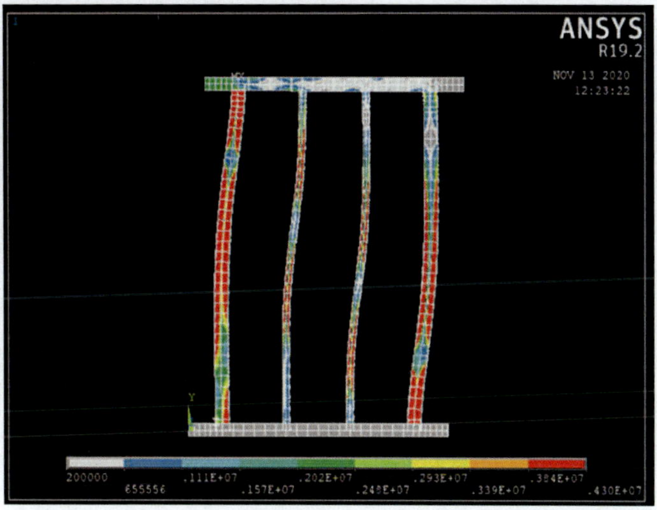


**Fig. 20** Stress concentration in Dhajji-Dewari wall panel

failure of the timber wall panel. Dhajji-Dewari wall panel showed inelastic deformation without reaching its breaking point. This behavior is responsible for good performance of the Dhajji-Dewari structures in case of an earthquake. Such style of vernacular architecture with traditional construction techniques and materials should be embraced and rediscovered. With little care and research these structures can be engineered to perform better.

## References

1. Fernandes J, Mateus R, Braganca L (2014) The potential of vernacular materials to the sustainable building design. In: Vernacular heritage and earthen architecture: contributions for sustainable development. ISBN 978-1-138-00083-4
2. Ali A (2015) Sustainability in vernacular architecture: Laurie Baker and Hassan Fathy's approach. Anthropol Bull 5(2):43–46
3. Vasconcelos G, Poletti E, Salavessa E, Jesus AMP, Lourence PB, Pilaon P (2013) In-plane shear behavior of traditional timber walls. Eng Struct 56:1028–1048
4. Ali Q, Schacher T, Ashraf M, Alam B, Naeem A, Ahmad N, Umar M (2012) In-plane behavior of the Dhajji-Dewari structural system (wooden braced frame with Masonry infill). Earthq Spectra 28:835–858
5. IS: 2366-1983, Indian Standard Code of practice for nail-jointed timber construction, Bureau of Indian Standards, New Delhi
6. IS: 1708 (Part 1 to 18)—2005, Methods of testing of small clear specimens of timber, Indian Standards Institution, New Delhi
7. Sheheryar, Ahmad N, Ashraf M, Ali Q (2014) Numerical modelling of timber braced frame Masonry structures (Dhajji Dewari). Numer Methods Civil Eng 1

# Variation in Seismic Behaviour of R.C Shaft Supported Elevated Water Towers with Change in Proportion of the Shaft Staging

**Pronoy Roy Chowdhury**

## 1 Introduction

Structurally elevated tanks are constructed either on frame or shaft type of staging system and such structures may be classified as top heavy inverted pendulum structures, and they have a tendency to overturn under the influence of lateral forces including wind and earthquakes. But it is found that the elevated water tank is generally more sensitive to seismic forces than wind forces. R.C shaft supported water tanks are more vulnerable towards damage or complete collapse due to seismic forces which has been reported in literature elsewhere [1]. Elevated water tanks are integral part of piped water supply schemes, and they are very significant structures from disaster management point of view. After a severe earthquake, generally grid power is out due to uprooting of electric poles and there are also events of break out of fire due to short circuit or rupture of gas pipelines. Under such inclement situation if the elevated water tank is standing erect without failure then water supply can be ensured for emergency firefighting and drinking water supply to the relief camps by water tanker until the grid power is restored. Structurally frame supported water tanks are more earthquake resistant due to the superior structural ductility in comparison to cylindrical shaft supported tanks. However it is found that shaft supported water tanks are popularly constructed in comparison to frame supported tanks because shaft supported water tanks are constructed with relatively lesser time using slip formwork and is also economic with respect to material consumption than frame supported tanks. Shaft supported tanks are also more aesthetic in comparison to frame supported tanks. Design and construction of R.C elevated water tank is covered in the Indian Standard code of practise IS 11682-1985, whereas seismic design of water tanks are covered in details in BIS code IS 1893 (part 2).

P. R. Chowdhury (✉)
Public Health Engineering Directorate Govt. of W.B, Kolkata 700001, India



S. Kolathayar and S. C. Chian (eds.), *Recent Advances in Earthquake Engineering*, Lecture Notes in Civil Engineering 175,
https://doi.org/10.1007/978-981-16-4617-1_5

The safety against overturning of elevated water tanks may be ensured through the provision of adequate factor of safety of the structure against overturning moment induced about the base of the tank support under lateral forces. If the diameter of the hollow cylindrical shaft support is small, the structure being slender is likely to have more lateral deflection at the tank level. Again if the diameter of the shaft is large it shall provide sufficient restoring couple against the lateral load induced overturning moment. More the diameter of the shaft supported tank, more is the material consumption and area of shuttering required for construction of the same. Thus it is required to strike a balance to the diameter of the shaft support which if adopted shall not make the structure unnecessarily costly and also prevent excessive lateral deflection. An important dynamic parameter of a structural system is the time period of the structural system. The fundamental period is dependent on the mass and the lateral stiffness of the structural system. Any variation in the structural proportions influences both the mass and the structural stiffness of the system. As the natural period of the system changes, the seismic behaviour of the structural system gets affected accordingly. This paper deals with the effect of change of structural proportion on the seismic behaviour of shaft supported elevated water tanks. Seismic force induced Bending Moment at the base of shaft staging shall be varied with the ratio of diameter of the tank to the shaft and the height of the shaft staging system. The outcome of this study is to understand the structural behaviour of the shaft supported elevated tank due to variation in the structural proportion of the shaft.

## 2 Study of Failure

The shaft support of elevated water tank is a thin cylindrical shell. From damage surveys of earthquakes [2–4], it has been revealed that thin shell shaft supports has generally failed by development of circumferential flexural tension crack approximately within a zone of about 1/3rd height from the base of the staging system. The shaft support deflects like a cantilever beam, and the maximum B.M due to lateral load is found to develop near the shaft base. The possibility for crack development near the base is further aggravated due to door opening in the shaft, which structurally weakens the shaft near the base location. Generally the height of the jumping formwork used to construct the shaft wall is 1.2 m height, for a 20 m height shaft staging about 18 nos. of subsequent casting operations are required. Thus 18 numbers of construction joints are formed at each lift of the slip form. These joints are the weakest points of the shaft staging. If adequate precautions are not taken while casting the shaft staging, under repetitive seismic jolts these joints opens up circumferentially under the influence of flexural tension leading to development of cracks in the shaft. The shaft is subjected to the combined effect of axial load from retained water and dead load of the tank, the seismic force induces B.M in the shaft cross-section. Under vertical load the thin shell shaft has a buckling tendency. While under lateral seismic load the shaft is subjected to severe bending stress near the base region. The flexural tension crack developed near the base zone of the shaft drastically reduces the load carrying capacity of the shaft and may cause collapse of the shaft staging. The strength of the shaft cross-section to accommodate bending stress is dependent on the shell thickness and the diameter of shaft staging. The tendency of the shaft to buckle under vertical load shall be studied with respect to the slenderness ratio ($l_{eff}/r_{min}$) of the staging. It is well known that there is paucity of literature and books on aseismic design of shaft supported water tanks. The BIS code IS 11682 is age old and requires immediate revision. As state of the art guideline is limited, consultants mostly design elevated water tank structures from their personal judgement and age old design provisions. Thus it is essential to re-examine seismic design aspects of R.C shaft supported water tanks in light of current research and development. If optimum diameter of the shaft of the elevated water tank is provided, it shall ensure adequate lateral stiffness to the shaft preventing excessive side sway and development of large bending stress in the shell concrete. The existing literature in this field does not enlighten this uncharted domain, although the matter is of great practical significance.

## 3 Structural Model

Different structural models have been studied by researchers for dynamic analysis of elevated water tanks against seismic forces. The erstwhile Indian seismic code IS 1893-1984 [5] vide cl no. 5.2 has recommended in favour of adoption of Single Degree of Freedom (SDOF) model. The SDOF model assumes that during seismic sway the water retained in the tank moves as a rigid body with the tank and causes the development of impulsive pressure and convective pressure, but since the convective pressure is meagre, it is neglected. The impulsive pressure is resisted by the support system of the elevated tank, it generates bending stress in the shaft cross-section, causing combined action of compressive and bending stress. G.W Housner in 1963 [6] propounded the two mass model for seismic analysis of water tanks this theory assumes that the portion of the liquid which moves with the tank container causes development of impulsive pressure and a portion of the liquid near the free board of the tank sloshes and causes the development of convective pressure. These two different modes of vibration are modelled as a two degree of freedom model. The current version of IS 1893 part 2 [7] has absorbed the two mass model as it is more rational and accurate in comparison to SDOF model, where convective component of vibration is totally neglected. Researchers have also modelled shaft supported elevated water tank using Multi Degree of Freedom Model (MDOF) [8] (lumped mass model) and Finite Element Model (FEM) [9]. While SDOF and Two DOF models are approximate models which hails from pre computer age, MDOF model and Finite Element Analysis of elevated water tank are done using structural analysis software, which are now easily accessible to the researcher. The MDOF model gives a better idea of the various higher modes of vibration of the tank. The Finite Element Model may be considered as a relatively most accurate structural model, which gives a more comprehensive idea of deflected shape in the structure with different load combinations. In this paper Finite Element Analysis of shaft supported elevated water tank shall be performed in SAP 2000 software [10]. This software has been particularly used, as the dynamic analysis module of the software is very comprehensive. However as there is no effective tool to model the sloshing behaviour of water, the philosophy of the two mass model has been incorporated in the analytical model. In absence of any finite element which may effectively model the sloshing behaviour and impulsive vibration of the water under seismic forces, the impulsive mass is approximately assumed to be attached with the tank wall as a rigid link and the convective stiffness $K_c$ has been adopted as per formula given in IS 1893 (Part-2) 2014, such modelling philosophy has been used by researchers elsewhere [11].

## 4 Problem Studied

An Intze type R.C water tank of 2250 CuM capacity supported on thin shell cylindrical shaft, having structural details as tabulated in Tables 1 and 2, of this paper (it is mentioned that the dimensions are adopted from practical water tank design, but the wall thickness of shaft staging has been kept as 150 mm to simulate the worst effect of forces on the tank structure.) The elevated tank is constructed with M30 grade concrete and Fe415 grade of steel. Keeping the capacity of the tank same the diameter of the shaft staging has been varied. The height of the staging is also varied as 20 m, 30 m and 40 m, respectively, the slenderness ratio of the shaft cross-section is accordingly modified. The tank structure is subjected to modal time history analysis in the elastic range of forces in SAP 2000 software using three different time histories of past earthquakes including (i) 1940 El-Centro Earthquake time history N-S component, (ii) San Fernando Earthquake time history of 1971 and (iii) Bhuj

**Table 1** Dimensions of elevated tank

| Sl. no. | Tank portion | Dimensions |
|---|---|---|
| 1 | Top dome | Diameter = 17.0 m, Rise = 1.5 m, Thickness = 100 mm |
| 2 | Top ring beam | Section 450 mm × 450 mm |
| 3 | Shell wall | 450 mm average thickness |
| 4 | Bottom ring | Section 750 mm × 1200 mm |
| 5 | Conical dome | 650 mm thickness of slab |
| 6 | Bottom dome | 300 mm thickness slab |
| 7 | Circular ring girder | Section 500 mm × 2000 mm |
| 8 | Thickness of shaft wall staging | 150 mm thick |
| 9 | Staging height | Varied by 20 m, 30 m and 40 m respectively |
| 10 | Diameter of shaft (d) | Varied by 14.9 m, 12.0 m, 10.0 m, 9.0 m, 7.5 m and 6.0 m respectively |

**Table 2** Details of water tank models

| Model no. | Shaft height (m) | Tank dia (m) | Dia of staging (m) | d/D | Thk. of shaft (mm) | $l_{eff}/R_{min}$ | Mass (KN) | Moment of inertia ($m^4$) | Stiffness (KN/m) |
|---|---|---|---|---|---|---|---|---|---|
| 1 | 20 | 17 | 14.9 | 0.876471 | 150 | 3.10 | 13,576 | 9.6E+13 | 899,795,786 |
| | 30 | 17 | 14.9 | 0.876471 | 150 | 4.66 | 15,329 | 9.6E+13 | 266,606,159 |
| | 40 | 17 | 14.9 | 0.876471 | 150 | 6.21 | 17,081 | 9.6E+13 | 112,474,473 |
| 2 | 20 | 17 | 12.0 | 0.705882 | 150 | 4.80 | 15,800 | 5E+13 | 468,318,242 |
| | 30 | 17 | 12.0 | 0.705882 | 150 | 7.21 | 17,208 | 5E+13 | 138,760,961 |
| | 40 | 17 | 12.0 | 0.705882 | 150 | 9.61 | 18,620 | 5E+13 | 58,539,780 |
| 3 | 20 | 17 | 10.0 | 0.588235 | 150 | 6.94 | 17,352 | 2.88E+13 | 270,000,817 |
| | 30 | 17 | 10.0 | 0.588235 | 150 | 10.42 | 19,462 | 2.88E+13 | 80,000,242 |
| | 40 | 17 | 10.0 | 0.588235 | 150 | 13.89 | 20,675 | 2.88E+13 | 33,750,102 |
| 4 | 20 | 17 | 9.0 | 0.529412 | 150 | 8.59 | 16,007 | 2.09E+13 | 196,337,870 |
| | 30 | 17 | 9.0 | 0.529412 | 150 | 12.89 | 17,065 | 2.09E+13 | 58,174,184 |
| | 40 | 17 | 9.0 | 0.529412 | 150 | 17.19 | 18,250 | 2.09E+13 | 24,542,234 |
| 5 | 20 | 17 | 7.5 | 0.441176 | 150 | 12.44 | 14,312 | 1.21E+13 | 113,053,085 |
| | 30 | 17 | 7.5 | 0.441176 | 150 | 18.66 | 15,200 | 1.21E+13 | 33,497,210 |
| | 40 | 17 | 7.5 | 0.441176 | 150 | 24.88 | 16,100 | 1.21E+13 | 14,131,636 |
| 6 | 20 | 17 | 6.0 | 0.352941 | 150 | 19.58 | 15,455 | 6.13E+12 | 57,449,120 |
| | 30 | 17 | 6.0 | 0.352941 | 150 | 29.37 | 16,200 | 6.13E+12 | 17,021,961 |
| | 40 | 17 | 6.0 | 0.352941 | 150 | 39.17 | 16,900 | 6.13E+12 | 7,181,140 |

Earthquake time history of 2001. The ground acceleration time history plots are shown in Fig. 1a, b and c respectively. Details of ground acceleration time histories of earthquakes such as Magnitude, PGA value, duration of strong ground motion, depth of focus, intensity and place of occurrence are tabulated in Table 3 only for reference. Time History Analysis (THA) is the most accurate method of dynamic analysis, by adopting different earthquake time histories it is possible to check the seismic response with different PGA values and durations of earthquakes. For the

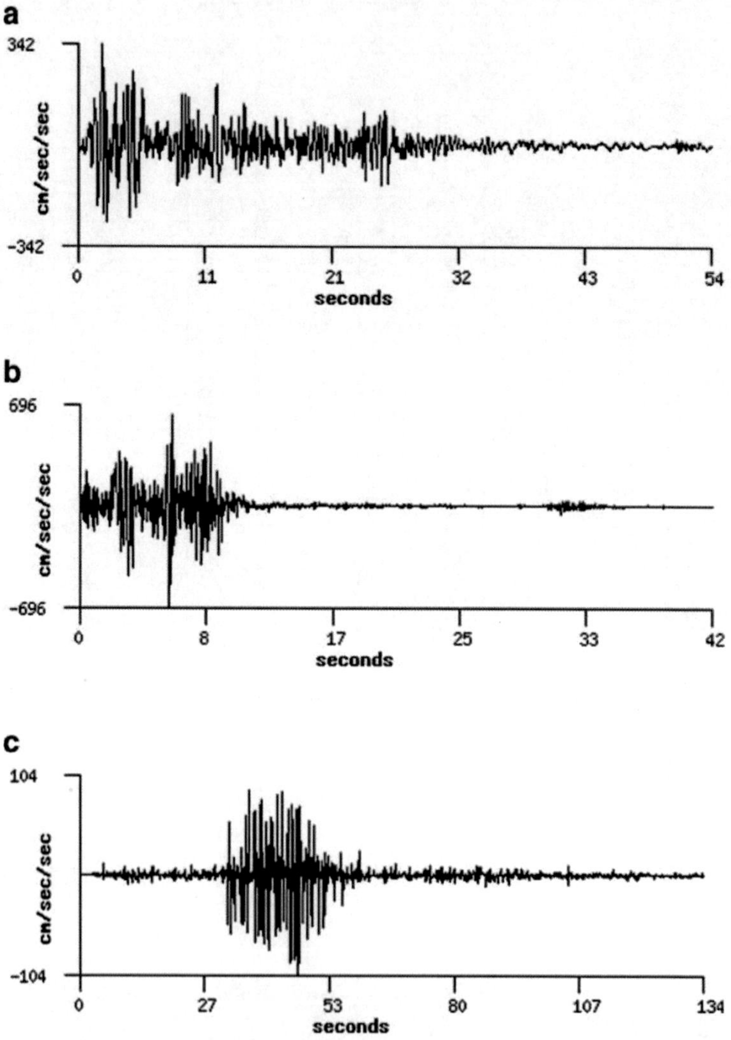


**Fig. 1** Ground acceleration related data of time histories of earthquake. **a** El-Centro Earthquake 1940, N–S component. **b** San Fernando earthquake 1971. **c** Bhuj earthquake 2001

**Table 3** Relevant information of earthquakes time history analysis

| Sl. no. | Name of the earthquake | PGA of earthquake (g) | Strong motion duration (s) | Focal depth (KM) | Magnitude (Mw) | Intensity (MMI scale) | Location of earthquake |
|---|---|---|---|---|---|---|---|
| 1. | El-Centro 1941 | 0.32 | 29 | 16 | 6.9 | X | California |
| 2. | San Fernando, 1971 | 1.25 | 12 | 8.4 | 6.5 | X | California |
| 3. | Bhuj, 2001 | 0.11 | 22 | 16 | 7.9 | X | India |

purpose of modelling the tank wall and shaft 4 noded quadrilateral shell element has been used. This type of shell element combines both the membrane action and plate bending behaviour which is observed in the shaft staging of the elevated tanks under the impact of vertical and lateral load. Modelling of the axisymmetric elevated water tank structure has been done with radial replication of the mesh of shell element. The Finite Element mesh size has been kept 1 m × 1 m, such mesh size is kept in many real life structural design problems. If the mesh size is made smaller it takes appreciable time for the desktop computer system to analysis the structure. The retained water in the tank container has been approximately modelled adopting the two mass model philosophy. As there is no available finite element to simulate impulsive and convective vibration of the retained water, so the retained water has been classified into two modes of vibration under seismic shaking as per guideline of the current BIS code IS 1893 (part-2) the impulsive mass and the convective mass respectively. The impulsive mass is assessed to be attached to the tank body with a linear rigid link while the convective mass is connected with convective springs whose stiffness value was obtained from Fig. 2a of IS 1893 (part 2). The impulsive mass is located at the height $h_i$ from the base of the tank, and the convective mass is located at the height $h_c$ from the tank base, the value of impulsive and convective height respectively may be obtained from the graph vide Fig. 2b of IS 1893 (part 2) depending on $h/D_t$ ratio, where h = tank height and $D_t$ = diameter of tank. The relative placement of the impulsive and the convective masses has been shown in Fig. 2 of this paper. The parameters of two mass model as given in IS 1893 (part 2) 2014 are valid for circular tanks with flat bottom. As the tank is of INTZE type the dimensions of impulsive height ($h_i$) and convective height ($h_c$) given in the paper are equivalent dimensions for cylindrical tanks. The elevated tank is assumed to be supported on fixed support. The shaft diameter has been varied from 14.9 m, 12.0 m, 10.0 m, 9.0 m, 7.5 m and 6.0 m respectively to change the structural proportion. Six different shaft diameters are given six model nos. whereas 14.9 m diameter shaft is model no. 1, 12.0 m diameter shaft is model no. 2, similarly 10.0 m diameter shaft model no. 3, 9.0 m diameter shaft is model no. 4, 7.5 m diameter shaft model no. 5 and 6.0 m diameter model no. 6 respectively. Structural data of different diameter shafts are given in Table 2. Shaft staging height has been changed through 20 m, 30 m, 40 m respectively. Shaft wall

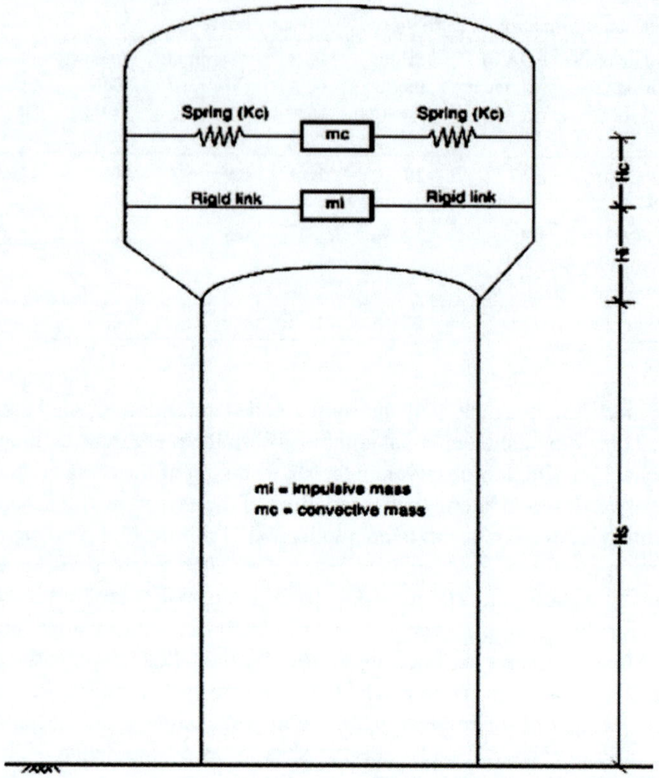


**Fig. 2** Schematic sketch showing the impulsive and convective masses within the tank as modelled in SAP 2000 (the dimensions $h_i$ and $h_c$ are equivalent dimensions of two mass model for flat bottom water tank)

thickness has been kept at 150 mm, which is the minimum thickness as per provisions of cl no. 8.2.1 of IS 11682 [12]. The shaft thickness has been kept at minimum, i.e., 150 mm, in a pursuit to study the worst effect of seismic force on the structure. The slenderness ratio ($l_{eff}/r_{min}$) of the shaft staging has been calculated for all the cases, the effective length is assumed to be 1.2 l, i.e. (the shaft staging is considered as a hollow compression member which is effectively held in position and restrained against rotation at one end, while restrained against rotation but not held in position at the other end). Such an end restraint has been conceived for Shaft support elevated water tank under lateral seismic force, as the deflected profile under lateral load for a large R.C elevated tank on shaft staging shall exhibit much stiffness against rotation. The analysis has been performed both for tank empty and tank full conditions (as per provisions of cl. no. 4.7.4 of IS 1893 (part-2)). The ratio of diameter of tank shaft staging (d) to that of tank container (D) i.e. (d/D) has been plotted along X-axis and the variation in seismic force induced B.M in the tank shaft has been plotted against Y-axis. Separate graphs (vide Fig. 3a–f of this paper) are plotted for tank empty and

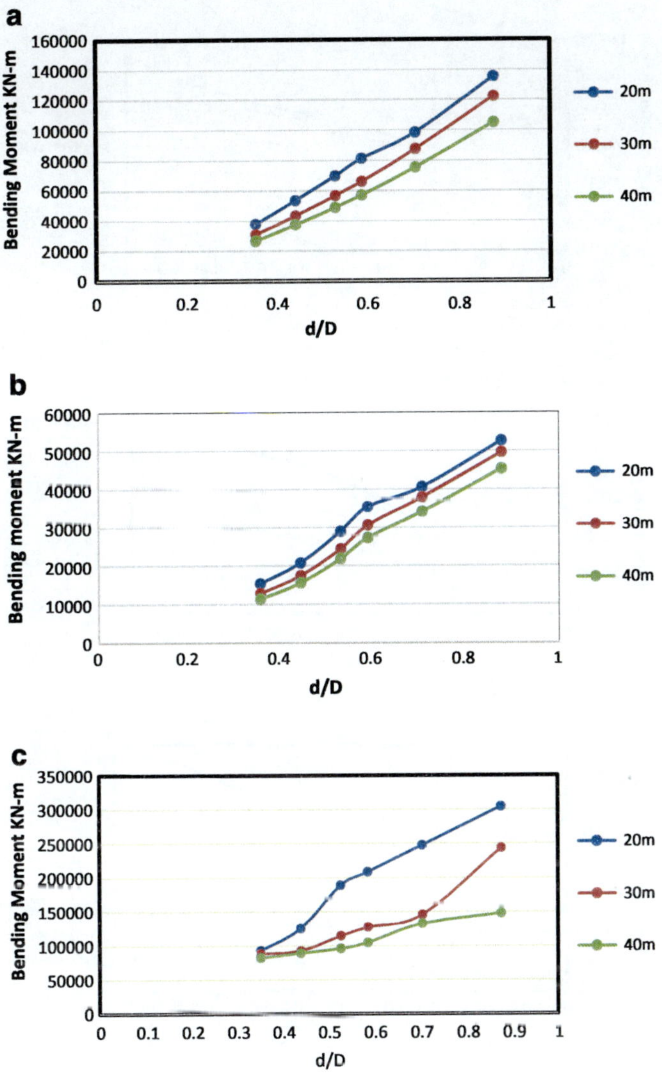


**Fig. 3** **a** Variation of bending moment at shaft base with change in d/D ratio in tank full condition for El-Centro earthquake time history. **b** Variation of bending moment at shaft base with change in d/D ratio in tank empty condition for El-Centro earthquake time history. **c** Variation of bending moment at shaft base with change in d/D ratio in tank full condition for San Fernando earthquake time history. **d** Variation of bending moment at shaft base with change in d/D ratio in tank empty condition for San Fernando earthquake time history. **e** Variation of bending moment at shaft base with change in d/D ratio in tank full condition for Bhuj earthquake time history. **f** Variation of bending moment at shaft base with change in d/D ratio in tank empty condition for Bhuj earthquake time history

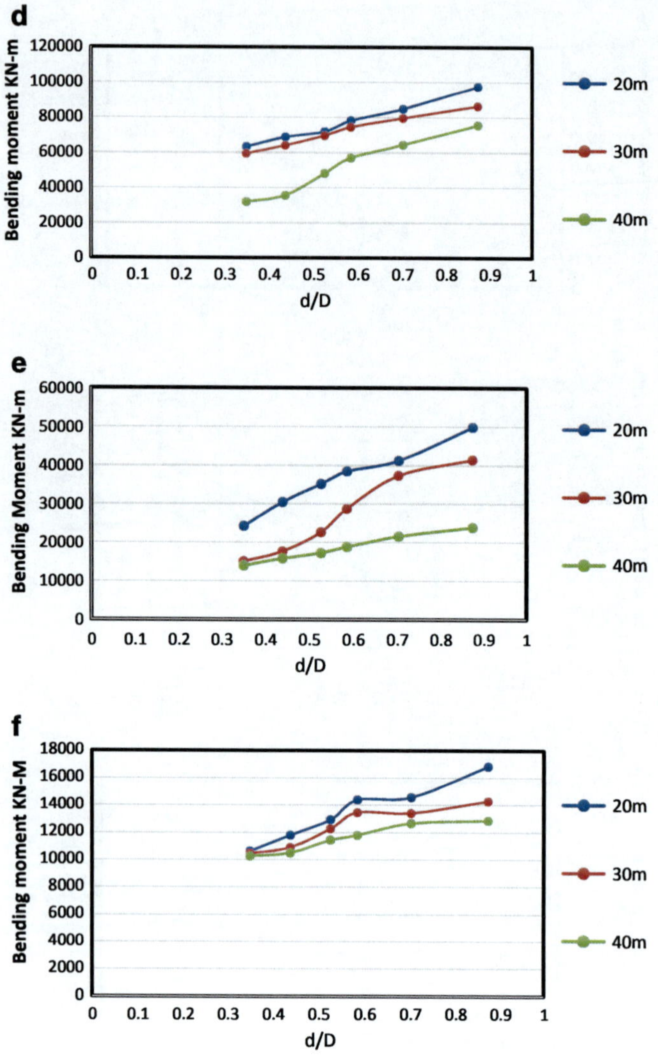


**Fig. 3** (continued)

tank full condition for all the three time histories. Trend-line plots are obtained for 20 m, 30 m, and 40 m high shaft staging respectively. Mode shapes for first three modes of vibration for tank full condition (as the worst case) of the shaft supported tank for shaft diameters 14.9 m, 9.0 m and 6.0 m respectively are shown side by side (vide Fig. 4 of this paper) with the respective time period of vibration to compare between the mode shapes with change in shaft diameter.

## 5 Results and Discussion

A study of the graph showing the variation of B.M at the base of shaft (along y-axis) versus d/D ratio (along x-axis) both for tank full and tank empty condition under the effect of all the three seismic time histories, the following general observations are indicated:

(i) The value of B.M at the base of the shaft increases continuously with increase in d/D ratio.
(ii) The maximum B.M at the base of the shaft is observed for tank full condition under San Fernando earthquake time history excitation in comparison with other earthquake time histories.
(iii) The B.M at the base of the shaft is higher for tank full condition than tank empty condition.
(iv) As the diameter of the shaft is reduced progressively the lateral drift of the tank increases under the same seismic excitation.

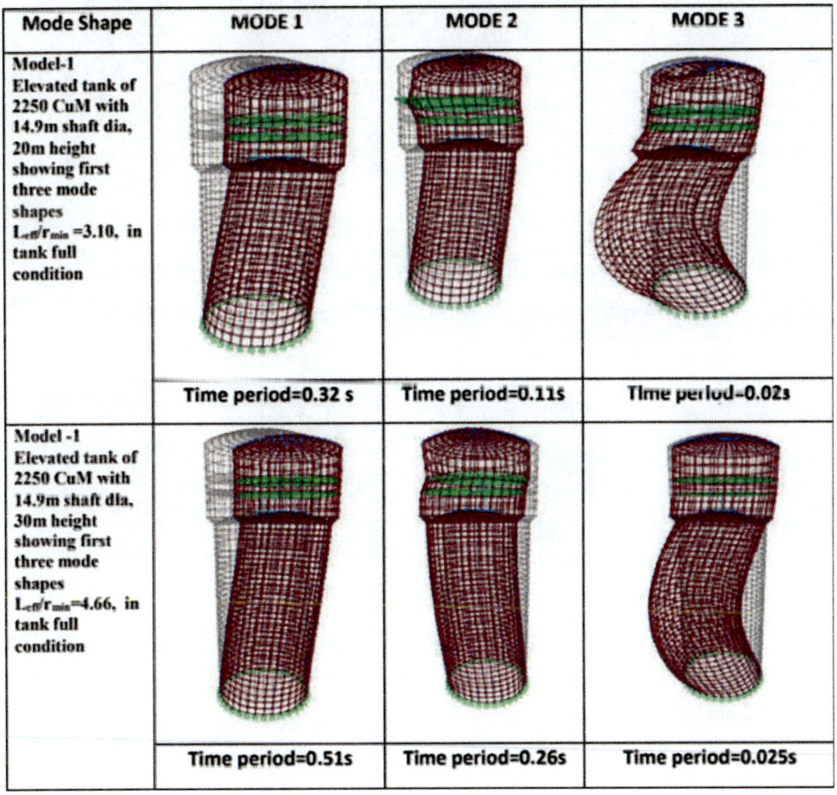


**Fig. 4** First three modes of vibration of the shaft supported tank in tank full condition, for diameters of the shaft staging of 14.9 m, 9.0 m and 6.0 m respectively in SAP 2000 software

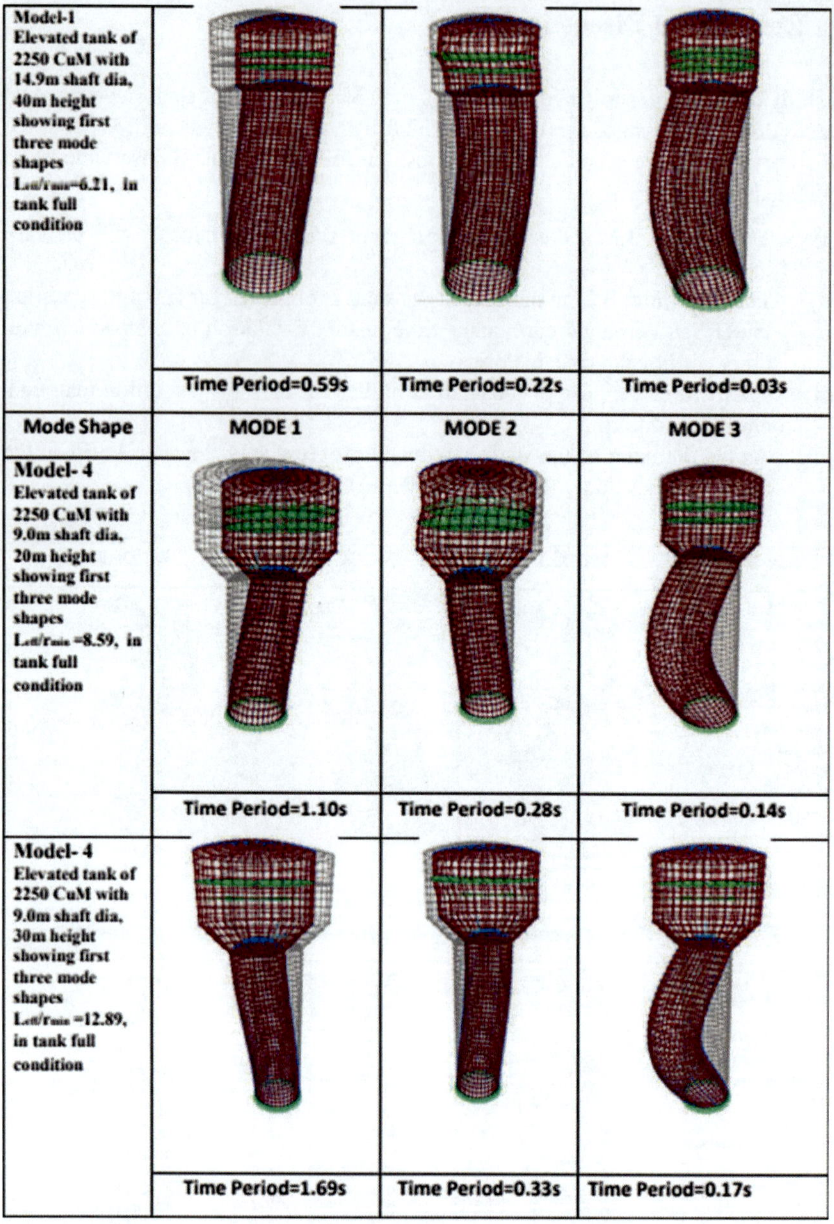


**Fig. 4** (continued)

| | | | |
|---|---|---|---|
| **Model- 4** **Elevated tank of 2250 CuM with 9.0m shaft dia, 40m height showing first three mode shapes $L_{eff}/r_{min}$ =17.19, in tank full condition** |  |  |  |
| | Time Period=2.35s | Time Period=0.40s | Time Period=0.20s |
| **Mode** | **MODE1** | **MODE 2** | **MODE 3** |
| **Model- 6** **Elevated tank of 2250 CuM with 6.0m shaft dia, 20m height showing first three mode shapes $L_{eff}/r_{min}$ =19.58, in tank full condition** |  |  |  |
| | Time Period=1.92s | Time Period=0.47s | Time Period=0.18s |
| **Model- 6** **Elevated tank of 2250 CuM with 6.0m shaft dia, 30m height showing first three mode shapes $L_{eff}/r_{min}$ =29.35, in tank full condition** | 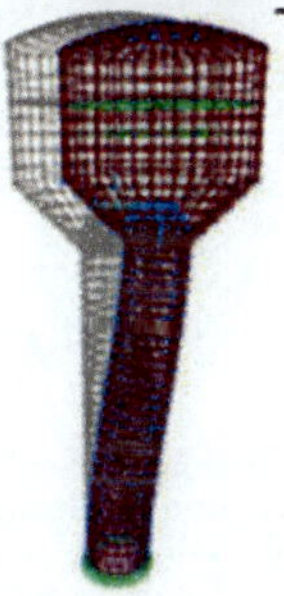 |  |  |
| | Time Period=2.94s | Time Period-0.58s | Time Period=0.25 sec |

**Fig. 4** (continued)

| | | | |
|---|---|---|---|
| **Model- 6** Elevated tank of 2250 CuM with 6.0m shaft dia, 40m height showing first three mode shapes $L_{eff}/r_{min}$ =39.17, in tank full condition | 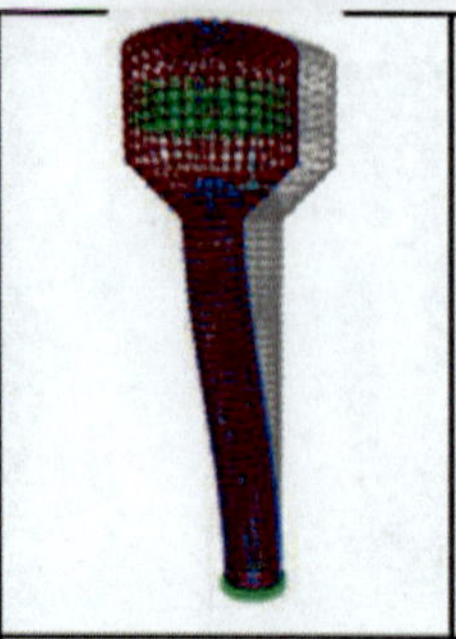 | 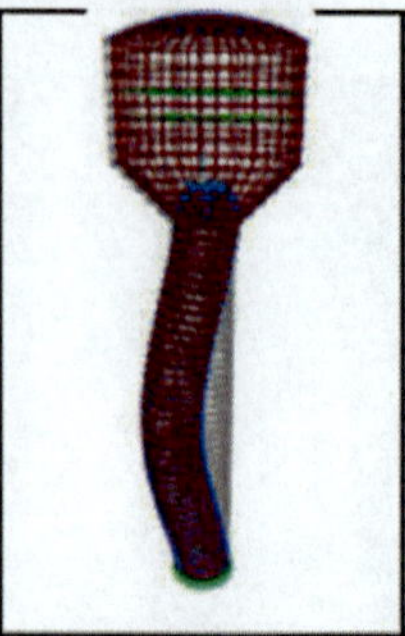 |  |
| | Time Period=3.29s | Time Period=0.66s | Time Period=0.30s |

**Fig. 4** (continued)

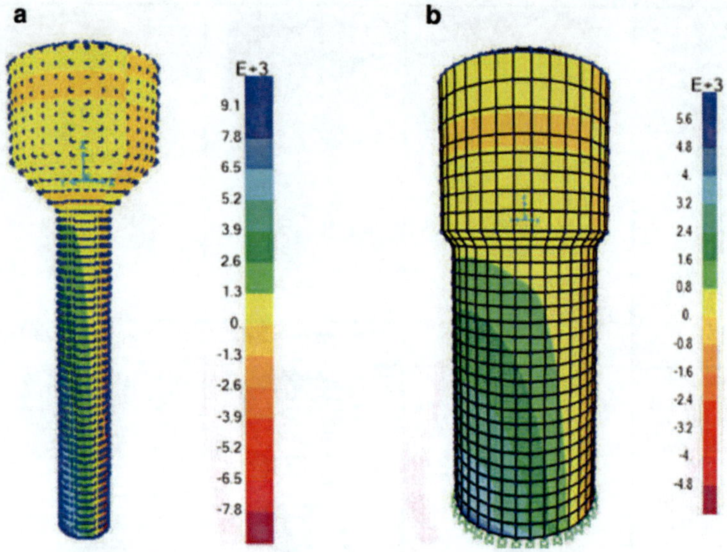


**Fig. 5** **a** Bending stress contour of the shaft staging of 40 m height and 6 m diameter of shaft (unit KN/m$^2$) under the effect of San Fernando earthquake time history. **b** Bending stress contour of the shaft staging of 20 m height and 14.9 m diameter of shaft (unit KN/m$^2$) under the effect of San Fernando earthquake time history

(v) The fundamental period of vibration of 14.9 m diameter shaft of 20 m height in tank full condition is 0.32 s whereas for the 6.0 m diameter shaft with 20 m staging height in tank full condition the fundamental period is 1.92 s under San Fernando excitation. So it is found that the lateral flexibility of the system changes remarkably with the variation in shaft diameter.

**Table 4** Lateral drift at the tank level of a 40 m high shaft supported elevated water tank under San Fernando time history

| Sl. no. | Shaft diameter (m) | Staging wall thickness (mm) | Lateral stiffness of shaft (KN/m) | Base shear (KN) f | Lateral drift at tank level (mm) |
|---|---|---|---|---|---|
| 1. | 14.9 | 150 | 112,474,473 | 10,452 | 0.09 |
| 2. | 9.0 | 150 | 24,542,234 | 6380 | 0.26 |
| 3. | 6.0 | 150 | 7,181,140 | 4263 | 0.59 |

(vi) Bending stress contour (refer Fig. 5a, b) indicates that the maximum bending stress due to lateral seismic jolt occurs approximately in the area near 1/3$^{rd}$ distance from the base of the shaft staging. But with the increase in shaft diameter the bending stress intensity reduces and the zone of stress concentration also reduces.

(vii) Lateral drift of elevated water tank on shaft for the most severe San Fernando is shown in Table 4 for three shaft diameters of 6.0 m, 9.0 m and 14.0 m respectively.

Within the limited scope of available data it is found that the seismic force induced B.M at the base of the shaft staging is maximum for the earthquake time history with greater PGA value than others (such as San Fernando time history). The BM increases with the increase in diameter of the shaft staging. With increase in the shaft diameter the lateral stiffness of the water tank structure increases attracting more seismic force inducing greater B.M in the shaft. The B.M increases gradually; however, from the graph a trend is observed that beyond the shaft diameter 0.65D the BM value increases significantly requiring thicker shaft section which makes shaft construction relatively un-economical. Normally if the height of the shaft staging increases the B.M should increase due to increase in lever arm for the lateral load induced B.M. But as the natural period of the structure increases with an increase in the shaft height and the structure becomes more flexible and attracts less seismic force, resulting in less B.M in the shaft cross-section with an increase in shaft height. As the shaft diameter decreases the lateral drift increases, which is evident from relative study of the mode shapes. Increased lateral drift induces additional B.M at the shaft aggravating overturning tendency of the elevated water tank. However from Table 4 it is found that even under the impact of severe base shear of San Fernando earthquake no significant lateral drift has been obtained.

## 6 Conclusion

From the above study, it may be concluded that if the d/D ratio of the shaft is kept low then consumption of construction material for casting the shaft is on the lesser side but it may increase lateral drift endangering the stability of the elevated water tank on shaft with respect to collapse. If d/D ratio is kept relatively large lateral stiffness of the shaft increases and lateral displacement is reduced, but material consumption for shaft construction also becomes high. A stiffer system attracts more seismic force thus B.M in the shaft base increases. Thus from experience it may be stated that optimally for construction of elevated water tank on shaft a preferable d/D ratio should be kept around 0.55–0.75 (about 55–75% of the tank diameter) for safer and economic design. Beyond these values construction material consumption tend to increase and seismic induced BM is found to be on the higher side. Construction joints at the subsequent lift of the slip formwork are the weakest points. If this joints especially those within 1/3rd height of the base of the shaft are not properly cast, they are likely to open up under flexural tension and circumferential crack may badly undermine the strength of the shaft portion near the base aggravating tendency of collapse.

**Acknowledgements** It is hereby indicated that the author is extremely grateful to the encouragement and support given by my Ph.D. thesis supervisor Dr. Partha Ghosh of the Department of Construction Engineering Jadavpur University Kolkata while preparing the paper. The undersigned also gratefully acknowledges the support obtained from Mr. Dripta Bandopadhaya a consultant structural Engineer in checking the various calculations and important discussions made regarding the various aspects of the paper which was most beneficial in developing the paper.

## References

1. Rai DC (2003) Performance of elevated water tanks in $M_W$ 7.7 Bhuj earthquake of January 26th, 2001. In: Proceedings of Indian Academy of Sciences (earth planet science), vol 112, pp 421–429
2. Bargi K (1992) Special structures, damage report of the Manjil earthquake on June 21, 1990. In: Tenth world conference on Earthquake Engineering. Rotterdam, Netherlands, pp 115–120
3. Jain SK et al (1997) Some observations on engineering aspects of Jabalpur earthquake of 22 May 1997. EERI Newslett 32(2)
4. Rai DC, Murthy CVR (2006) Effects of the 2005 Muzaffarabad (Kashmir) earthquake on built environment. Curr Sci 90(8):1066–1070
5. IS 1893-1984, Indian standard criteria for earthquake resistant design of structures, Bureau of Indian Standards, New Delhi
6. Housner GW (1963) The dynamic behaviour of water tanks. Bull Seismol Soc Am 53(2):381–387
7. IS 1893(Part-2)-2014, Indian standard criteria for earthquake resistant design of structures, Part-2, Liquid retaining tanks, Bureau of Indian Standards, New Delhi
8. Sood VK, Singh S (1983) Seismic analysis of shaft supported elevated tanks. J Inst Eng (India), Civil Eng Divis 64:143–148
9. Memari AM, Ahmadi MM (1992) Behaviour of reinforced concrete water tower during Manjil-Roudbar earthquake of June 1990. In: Tenth world conference on earthquake engineering. Rotterdam, Netherlands, pp 4953–4959
10. SAP 2000 (1998) Integrated finite element analysis and design of structures. Computers and Structures, Inc., Berkley, California
11. Muthuvijay P, Prakash A (2014) Analysis of sloshing impact on overhead liquid storage structures. Int J Res Eng Technol 2(8)
12. IS 11682-1985, Indian Standard Criteria for design of RCC staging for overhead tanks. Bureau of Indian Standards, New Delhi

# Seismic Assessment of RC Framed Staging of Elevated Water Tanks

Trishit Chandra and Saraswati Setia

## 1 Introduction

Seismic behaviour of elevated water tanks are very complex due to the coupled fluid–structure interaction. Past earthquakes have revealed the vulnerability of these structures under seismic excitation. Engineers study and learn from those disastrous experiences to increase the robustness of these complex structures. Seismic codes are evolving every day following the scientific contributions from different ideas throughout the globe. Present seismic codes of different countries including Indian Seismic Code on Liquid Retaining Tanks [1] follow the Housner model of spring-mass system [2–5]. Previous documented literatures suggest that according to Housner model if the ratio of the convective mode to the impulsive mode of time period ($T_c/T_i$) is greater than 2.5, then the coupled fluid–structure interaction can be uncoupled for simplified analysis [6–8]. Also, greater value will decrease the chance of forming coupled resonance condition between the fluid and the structure [9]. This ratio is strongly dependent on the number of panels and the height to diameter ratio of the RC framed staging ($h_s/D_s$). For some values of the staging aspect ratio ($h_s/D_s$), the value of $T_c/T_i$ can be less than 2.5. Hence, the present study focuses on finding some optimal ranges or specific values of the staging aspect to satisfy the desirable criteria. Also, the study reveals some efficient remedies for the structures to come under that provision without major negotiation to the proposed aspect ratio.

T. Chandra (✉) · S. Setia
National Institute of Technology Kurukshetra, Kurukshetra 136119, Haryana, India



S. Kolathayar and S. C. Chian (eds.), *Recent Advances in Earthquake Engineering*, Lecture Notes in Civil Engineering 175,
https://doi.org/10.1007/978-981-16-4617-1_6

## 2 Methodology

### 2.1 Tank Models

Elevated circular concrete water tank with flat roof is considered for this seismic assessment study. Two models of small (80 $m^3$) and large (500 $m^3$) capacities are taken into account with different staging aspect ratio. Analysis of elevated intze tanks can be simplified by considering an equivalent cylindrical tank container of same capacity [6, 7]. Therefore, the present study can also be used for intze type tank of same capacity. The dimensional parameters of different structural elements are listed in Table 1.

The above-mentioned dimensions of structural elements are consistent throughout the entire analysis. Bracings of staging system are comprised of horizontal circumferential tie beams only. As the aspect ratio of staging ($h_s/D_s$) varies, the aspect ratio of the tank container ($h_t/D$) and the number of panels change. In consequence of that these parameters are not listed as constant dimensions in the Table 1.

### 2.2 Analysis of Tanks

Various seismic codes of different countries follow the analysis method proposed by Malhotra et al. [10] which is a generalized and simplified extension of Housner model. The present study also takes the path shown by Malhotra to assess convective time period (Tc) of the tank models. The lateral stiffness of the staging is calculated following a method proposed by Sameer and Jain [11] to evaluate the impulsive time period (Ti) of elevated tanks. The small tank is considered to be located in a low seismic zone, whereas the large tank is situated in high seismic zone. Both the tanks are built on a soft soil.

The variable parameters in this study are the height of staging ($h_s$), diameter of staging ($D_s$), diameter of tank container (D), height of water level (h), height of sloshing ($h_{sl}$), height of tank container ($h_t$), number of columns in plan and the number of panels. Among these parameters D is taken as approximately equal to $D_s$

**Table 1** Detail of structural elements of tank models

| Structural elements | 80 $m^3$ tank | 500 $m^3$ tank |
|---|---|---|
| Thickness of roof (tr) | 0.12 m | 0.2 m |
| Thickness of wall (tw) | 0.2 m | 0.25 m |
| Thickness of base slab (tb) | 0.2 m | 0.25 m |
| Dimension of floor beam | 0.6 m × 0.25 m | 0.7 m × 0.4 m |
| Diameter of staging column | 0.45 m | 0.8 m |
| Dimension of staging brace | 0.45 m × 0.3 m | 0.7 m × 0.5 m |

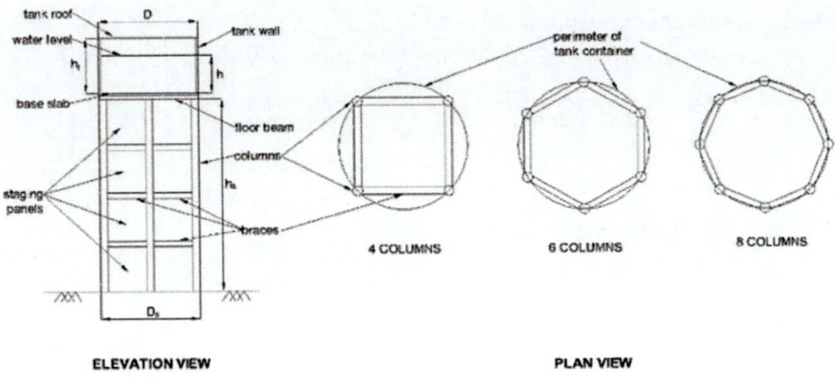


**Fig. 1** Detail of structural elements and their corresponding notations

and $h_t$ is considered as the upper limit approximate sum of h and $h_{sl}$ where $h_{sl}$ is also a dependent variable of h/D. Therefore, as long as the $D_s$ is constant the value of D, h, $h_{sl}$, $h_t$ will not vary. In case of both the tanks, variations of the results are carried out for four, six and eight numbers of columns in plan. All these notations and detail of structural elements are shown below (see Fig. 1).

The parameters $h_s$, $D_s$ and number panels are taken for three different cases. In the first case, $D_s$ and panel height are fixed with variation in $h_s$ only. In the second case, $h_s$ and $D_s$ are fixed with variation in number panels only. And in the last case, $h_s$ and panel height are fixed with variation in $D_s$. For the last case, the variation in $D_s$ causes D, h, $h_{sl}$ and $h_t$ to vary also. And for all the cases, number of columns in plan are taken as three different values of 4, 6 and 8. These three cases for both the tanks are tabulated in Tables 2 and 3, respectively.

All the three cases are performed according to the above-mentioned methods to obtain the impulsive and convective mode of time periods and their ratio. For both the tanks in case 1 and 2, the aspect ratios of the tank containers are taken by considering h/D as nearly equal to 0.5 for finding rest of the parameters accordingly.

## 3 Results

The results obtained from the entire assessment of the tank models are discussed in this section. As the number of columns in the plan increases, the stiffness of the structure increases, subsequently, the impulsive time period gets reduced and the ratio $T_c/T_i$ gets increased. This effect can be seen in all the data provided below (see Figs. 2, 3, 4, 5, 6 and 7). Our goal is to achieve the value of $T_c/T_i$ greater than 2.5 and for any scenario this value should not lie below 2.5. For all the cases of both the tanks, it can be observed that 4 columns in plan is intuitively more vulnerable to

**Table 2** Variation of parameters in 80 $m^3$ tank model

| | $D_s = D$ (m) | h (m) | $h_{sl}$ (m) | $h_t$ (m) | Panel height (m) | No. of panels | $h_s$ (m) | $h_s/D_s$ |
|---|---|---|---|---|---|---|---|---|
| Case 1 | 6 | 2.85 | 0.25 | 3.2 | 3 | 2 | 6 | 1 |
| | | | | | | 3 | 9 | 1.5 |
| | | | | | | 4 | 12 | 2 |
| | | | | | | 5 | 15 | 2.5 |
| | | | | | | 6 | 18 | 3 |
| | | | | | | 7 | 21 | 3.5 |
| | | | | | | 8 | 24 | 4 |
| Case 2 | 6 | 2.85 | 0.25 | 3.2 | 6 | 2 | 12 | 2 |
| | | | | | 4 | 3 | | |
| | | | | | 3 | 4 | | |
| | | | | | 2.4 | 5 | | |
| | | | | | 2 | 6 | | |
| Case 3 | 12 | 0.71 | 0.33 | 1.1 | 3 | 4 | 12 | 1 |
| | 8 | 1.6 | 0.23 | 2 | | | | 1.5 |
| | 6 | 2.85 | 0.25 | 3.2 | | | | 2 |
| | 4.8 | 4.45 | 0.23 | 4.7 | | | | 2.5 |
| | 4 | 6.4 | 0.21 | 6.7 | | | | 3 |
| | 3.43 | 8.7 | 0.19 | 8.9 | | | | 3.5 |
| | 3 | 11.4 | 0.18 | 11.6 | | | | 4 |

generate coupled resonance between fluid and structure compared to 6 and 8 number of columns.

From Fig. 2, it can be seen that $T_c/T_i$ value gets reduced as the number of panels as well as $h_s$ increases. For 4, 6 and 8 number of columns in plan, it is safe to limit the $h_s/D_s$ value to 1.5, 2.5 and 3.5, respectively, with 4 panels in each case. Comparing Figs. 3 and 4, it can be concluded that at least 3 panels should be provided when h/D and $h_s/D_s$ is limited to 0.5 and 1.5, respectively. Also, from economic view-point if number of panels cannot be increased above 5 then $h_t/D$ and $h_s/D_s$ should be limited to 1 and 3, respectively. Hence, for any tank of capacity less than 100 $m^3$, the maximum value of $h_s/D_s$ should be taken as 2 with 4 columns and 5 panels, 2.5 with 6 columns and 5 panels and 3 with 8 columns and 5 panels, by keeping the $h_t/D$ value less than 1. Otherwise, if these values exceed, then either the dimensions of the columns and braces must be increased or the radial braces must be used.

From Fig. 5, it can be seen that $T_c/T_i$ value gets reduced as the number of panels as well as $h_s$ increases. For 4, 6 and 8 number of columns in plan, it is safe to limit the $h_s/D_s$ value to 1.5, 2 and 2.5, respectively, with 4 panels in each case. Comparing Figs. 6 and 7, it can be concluded that at least 4 panels should be provided when h/D and $h_s/D_s$ is limited to 0.5 and 1.5, respectively. Also, from economic view-point if

**Table 3** Variation of parameters in 500 m$^3$ tank model

| | $D_s = D$ (m) | h (m) | $h_{sl}$ (m) | $h_t$ (m) | Panel height (m) | No. of panels | $h_s$ (m) | $h_s/D_s$ |
|---|---|---|---|---|---|---|---|---|
| Case 1 | 11 | 5.26 | 1.21 | 6.5 | 5.5 | 2 | 11 | 1 |
| | | | | | | 3 | 16.5 | 1.5 |
| | | | | | | 4 | 22 | 2 |
| | | | | | | 5 | 27.5 | 2.5 |
| | | | | | | 6 | 33 | 3 |
| | | | | | | 7 | 38.5 | 3.5 |
| | | | | | | 8 | 44 | 4 |
| Case 2 | 11 | 5.26 | 1.21 | 6.5 | 11 | 2 | 22 | 2 |
| | | | | | 7.33 | 3 | | |
| | | | | | 5.5 | 4 | | |
| | | | | | 4.4 | 5 | | |
| | | | | | 3.67 | 6 | | |
| | | | | | 3.14 | 7 | | |
| | | | | | 2.75 | 8 | | |
| Case 3 | 22 | 1.315 | 2.17 | 3.5 | 5.5 | 4 | 22 | 1 |
| | 14.67 | 2.96 | 1.45 | 4.5 | | | | 1.5 |
| | 11 | 5.26 | 1.21 | 6.5 | | | | 2 |
| | 8.8 | 8.22 | 1.12 | 9.4 | | | | 2.5 |
| | 7.33 | 11.84 | 1.02 | 12.9 | | | | 3 |
| | 6.28 | 16.14 | 0.94 | 17.1 | | | | 3.5 |
| | 5.5 | 21.05 | 0.89 | 22 | | | | 4 |

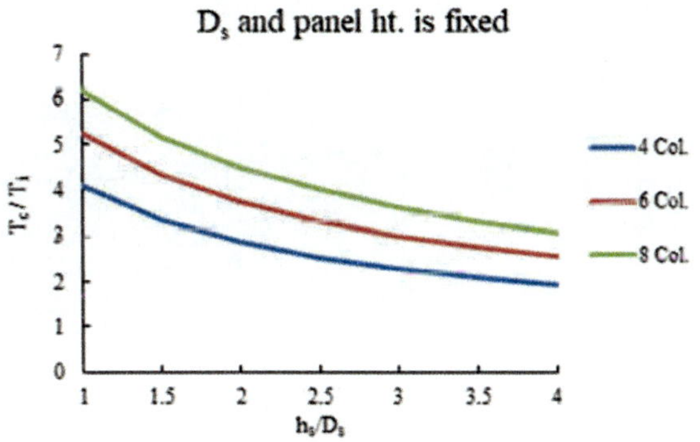


**Fig. 2** Variation of $T_c/T_i$ with respect to $h_s/D_s$ (Case 1) of 80 m$^3$ tank model

**Fig. 3** Variation of $T_c/T_i$ with respect to panel numbers (Case 2) of 80 $m^3$ tank model

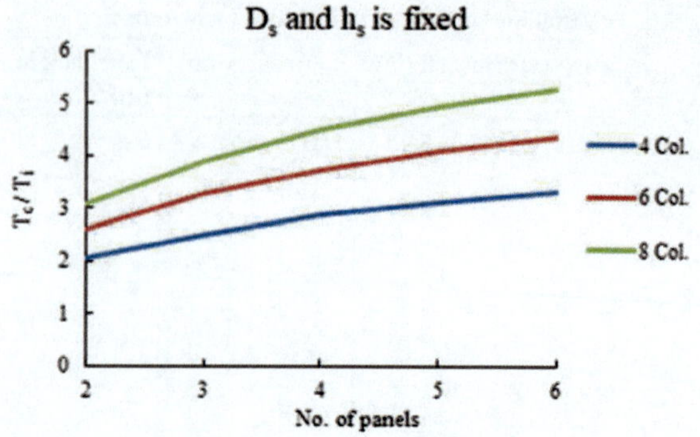


**Fig. 4** Variation of $T_c/T_i$ with respect to $h_s/D_s$ (Case 3) of 80 $m^3$ tank model

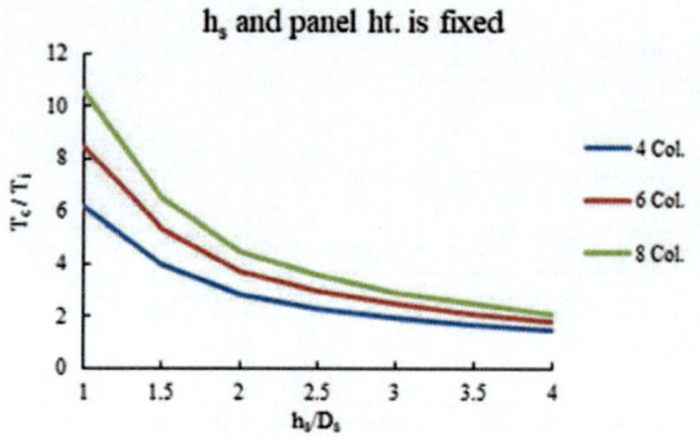


**Fig. 5** Variation of $T_c/T_i$ with respect to $h_s/D_s$ (Case 1) of 500 $m^3$ tank model

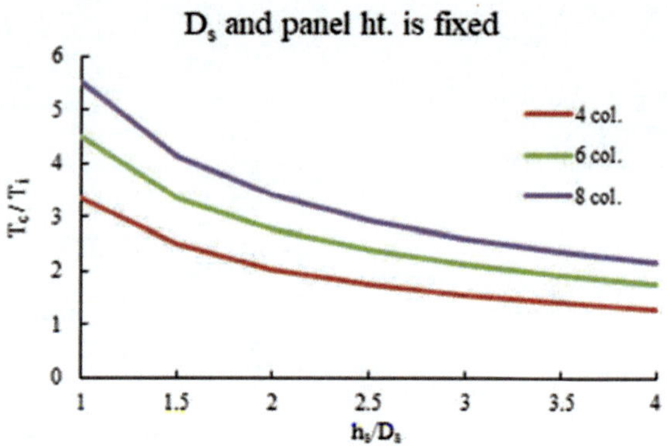


number of panels cannot be increased above 6 then $h_t/D$ and $h_s/D_s$ should be limited to 1 and 3, respectively. Hence, for any tank of capacity up to 500 $m^3$, the maximum value of $h_s/D_s$ should be taken as 1.5 with 4 columns and 6 panels, 2.5 with 6 columns and 6 panels and 3.5 with 8 columns and 7 panels, by keeping the $h_t/D$ value less than 1. Otherwise, if these values exceed, then either the dimensions of the columns

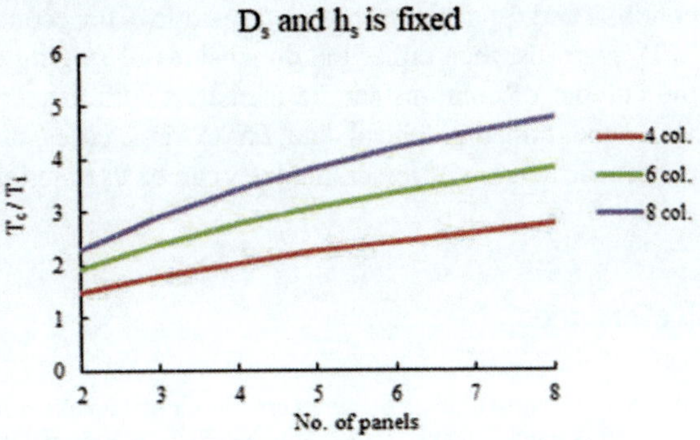


**Fig. 6** Variation of $T_c/T_i$ with respect to panel numbers (Case 2) of 500 $m^3$ tank model

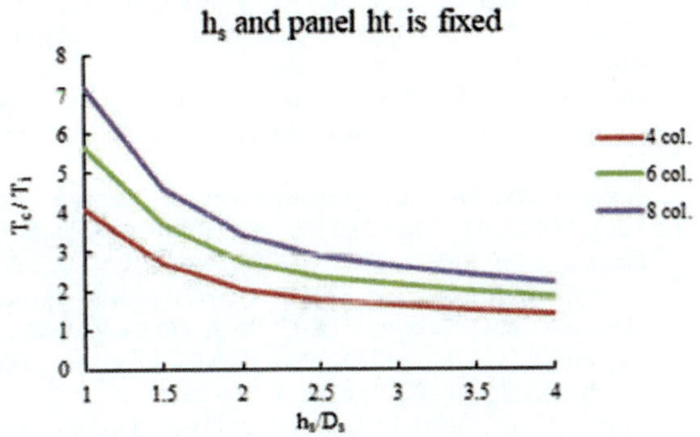


**Fig. 7** Variation of $T_c/T_i$ with respect to $h_s/D_s$ (Case 3) of 500 $m^3$ tank model

and braces must be increased or another concentric staging with lesser diameter of circumference must be used.

## 4 Conclusion

Elevated water tanks are more vulnerable under seismic excitation than ground-supported tanks because of the elevated limped mass system. When the ratio of the convective mode to the impulsive mode of time period ($T_c/T_i$) lies below 2.5, the coupled resonance between fluid and structure generates which can cause detrimental damage or collapse of the structure. This study suggests some efficiently proportioned aspect ratio of staging ($h_s/D_s$) as well as container ($h_t/D$), number of columns and panels in the staging. In case of tanks having capacity of less than 100 $m^3$, the $h_s/D_s$ value should be limited to 2 with 4 columns and 5 panels keeping $h_t/D$ within 1 for economical design. If the value of $h_s/D_s$ exceeds, then either the dimensions of staging elements can be increased or the number of columns can be increased with panel numbers by interpolating the above-mentioned graphical data. In case of tanks having capacity up to 500 $m^3$, the $h_s/D_s$ value should be limited to 3 with 8 columns and 6 panels keeping $h_t/D$ within 1 for economical design. If the value of $h_s/D_s$ exceeds, then either the dimensions of staging elements can be increased or the number of columns can be increased with panel numbers by interpolating the above-mentioned graphical data. For extreme cases, the use of radial braces or extra concentric staging of lesser diameter can be used in large tanks.

## References


1. IS 1893 (2014) Indian Standard criteria for earthquake resistant design of structures. Part 2: liquid retaining tanks. Bureau of Indian Standards, New Delhi
2. Housner GW (1954) Earthquake pressures on fluid containers. In: 8th technical report under Office of Naval Research, NR-081-095, California Institute of Technology
3. Housner GW (1957) Dynamic pressure on accelerated fluid containers. Bull Seismol Soc Am 47(1):15–35
4. Housner GW (1963) Dynamic analysis of fluids in containers subjected to acceleration. Nuclear reactors and earthquakes, report no. TID 7024, U. S. Atomic Energy Commission, Washington, DC
5. Housner GW (1963) The dynamic behavior of water tanks. Bull Seismol Soc Am 53(2):381–387
6. Jaiswal OR, Rai DC, Jain SK (2004) Codal provisions on design seismic forces for liquid storage tanks: a review. Report no. IITK-GSDMA-EQ-01-V1.0, Indian Institute of Technology, Kanpur
7. Jaiswal OR, Rai DC, Jain SK (2004) Codal provisions on seismic analysis of liquid storage tanks: a review. Report no. IITK-GSDMA-EQ-04-V1.0, Indian Institute of Technology, Kanpur
8. Jaiswal OR, Rai DC, Jain SK (2007) Review of seismic codes on liquid-containing tanks. Eng Spectra Earthq Eng Res Inst 23(1):239–260
9. Priestley MJN, Wood JH, Davidson BJ (1986) Seismic design of storage tanks. Recommendation of a study group of the New Zealand National Society for Earthquake Engineering
10. Malhotra PK, Wenk T, Wieland M (2000) Simple procedure for seismic analysis of liquid-storage tanks. Struct Eng Int 10(3):197–201
11. Sameer US, Jain SK (1994) Lateral-load analysis of frame stagings for elevated water tanks. J Struct Eng 120(5):1375–1394

# Effect of Pier Height on Behavior of Integrated Bridge Under Far and Near-Field Earthquake

**Kedar Ramesh Kumbhojkar, M. K. Shrimali, S. D. Bharti, and Vishal Kamble**

## 1 Introduction

The bridge is the part of the roadway network which demand a major chunk of fund and labor for its construction. Moreover, it should remain operational even after a major seismic event so that the rescue team can reach the affected area to provide help. Hence to prevent damage in the major seismic event of the bridge, various protection techniques and structural forms of bridges have been developed. The integrated bridge (IB) is one of the structural forms which is more earthquake resistant as compared to conventional bridge (CB). In the case of the integrated bridge, the deck of the bridge is monolithically connected with pier and/or abutment. On the other hand, in the case of the conventional bridge, the deck of the bridge is connected through bearing to the substructure. Moment sharing between the deck girder and pier is possible in the case of IB due to the monolithic connection between deck and pier which enhances the structural performance of IB. Integrated abutment bridge (IAB) is one type of IB in which only the abutment and deck of the bridge is monolithically connected. Intermediate piers of the IAB are connected to the deck through bearing. The first conceptualization of IB was done in 1930 and it was in the form of IAB. The first IAB named teen run bridge was constructed in 1938 in the USA [1]. IB is gaining popularity in the USA, UK, and Canada because of it's various advantages [2, 3]. IB's key advantages are (i) cost-effective, speedy and simple form of construction due to devoid of bearing. (ii) as joints are completely eliminated it provides vibration free and comfortable riding. (iii) lesser maintenance cost. To evaluate the behavior of IB many research works were reported on the seismic analysis of IB. Wolde-Tinsae et al. [2] presented a survey of existing IBs/IABs in the USA which reported satisfactory performance of IB. Greimann et al. [4] studied pile stress and pile–soil interaction in

K. R. Kumbhojkar (✉) · M. K. Shrimali · S. D. Bharti · V. Kamble
National Centre for Disaster Mitigation and Management, Malaviya National Institute of Technology, Jaipur 302017, India



S. Kolathayar and S. C. Chian (eds.), *Recent Advances in Earthquake Engineering*, Lecture Notes in Civil Engineering 175,
https://doi.org/10.1007/978-981-16-4617-1_7

skewed and non-skewed IAB. Reduction of the vertical load carrying capacity of piles was found in the study. Kunde and Jangid [5] compared three mathematical models of an isolated bridge considering and ignoring the flexibilities of pier and deck. The main conclusion of the study was model considering flexibilities provided maximum peak responses. Tongaonkar and Jangid [6] studied the soil-structure interaction effect on the base-isolated bridge. Rubber isolators were considered at the deck level. It was observed that the flexibility of soil beneath the foundation significantly affected the responses of the bridge especially displacement of the bearing at the abutment underestimated if the soil-structure interaction effect was neglected. Carlo et al. [7] studied the effect of soil-structure interaction on the pier of the conventional bridge. for the purpose of the study, SDOF was simulated considering lumped mass model. The study concluded that soil-structure interaction increased with an increase in the stiffness of the structure. Cai et al. [8] examined the seismic performance of the tall pier of the bridge under near-field earthquake. Concrete filled steel tubular pier with the energy dissipating mild steel plates were considered for the analysis. The study reported that the arrangement of pier considered for the tall bridge was satisfactory. Tamanani et al. [9] evaluated EC08 code criteria for irregular pier height of the bridge and proposed a method to balance the irregularity simultaneous damage in the pier. Thippeswamy et al. [10] examined the relative performance of five different structural arrangements of IAB. Analysis of all bridges was done only for the static loading either vertical or lateral load. Lateral load due to creep and shrinkage effect and thermal expansion was also considered. Support condition at deck level, footing type (spread or pile) was the parameter for variation. The study reported that the performance of IAB with a flexible pile foundation was better. Caner and Zia [11] performed experimental work on the three-span continuous deck girder for various combinations of support conditions. For the purpose of the test, both RC and steel girders were used with different combinations of roller and pin supports. It was observed in the study that the effect of type of support was negligible on the results. Kozak [12] examined the effect of soil-structural interaction on the three-span IAB. For this study, an existing IAB in Illinois (USA) was simulated in the Opensees. NLTHA and pushover analysis were performed. It was observed in the study that pile capacity was reduced in the soft soil. Mitoulis et al. [13] examined the cost-effectiveness of three different structural forms of IB with respect to the isolated deck of the bridge from the substructure using a rubber isolator. The partially integrated bridge in which elastomers were provided at the specific location was found more cost-effective than the other configuration.

From the literature review, it can be noted that many studies reported on the performance of various structural forms of IAB or IB subjected to static or dynamic loading considering flexibilities of soil (i.e., soil-structure interaction) or modeling assumption for piers but not much studies reported on the effect of height of pier on the response of IB under the different type of earthquake, namely, far-field and near-field earthquake.

In the present study, three models namely, P-10, P-20, and P-30 are simulated with pier heights 10, 20 and, 30 m for both IB and CB. Thus total six models are simulated. The seismic behavior of P-20 and P-30 with respect to P-10 is investigated under unidirectional earthquakes. Earthquakes used for the NLTHA are far-field earthquake (FFE), near-field earthquake with forward directivity effect (FDE) and, fling step effect (FSE). The earthquakes are applied in the longitudinal direction and transverse direction of the bridge separately. PGA of all earthquakes used for analysis was scaled to 0.2g, 0.4g, 0.6g, 0.8g. Major response quantities of interest are the absolute maximum deck moment (i.e., maximum moment in the deck), absolute maximum deck displacement, and pier moments. To comprehend the effect of pier height on the IB these response quantities of IB are compared with corresponding response quantities of CB.

## 2 Theoretical Background

Many structural forms of IBs came into the practice depending upon design constraints, site conditions, and traffic requirements. A semi-integrated bridge (SIB) is one type of IB in which the deck is monolithically connected with the pier bents whereas, at the abutment, the deck is connected through the bearing. Lateral movement of the deck due to lateral load is allowed at the bearing provided at the abutment location. Hence the rotation demand of foundation in the case of semi-integrated bridge drastically reduces as compared to IAB. Again, in the case of IAB, soil-structure interaction between the soil and abutment needs to be considered in the analysis which can be avoided in the semi-integrated bridge. So for the present study semi-integrated bridge is considered and for simplicity, it is called as an integrated bridge (IB). The types of earthquakes and the bridge used for the present study are described in this section.

### 2.1 *Type of Earthquake*

The type of earthquake is one parameter chosen as each type of earthquake possesses distinct characteristics. Far-field and near-field earthquake is characterized by the closest fault distance which is called as Joyner-Boore distance ($R_{jb}$). The earthquake recorded within 15–20 km distance from a fault is called a near-field earthquake [14, 15]. Near-field earthquake, ground motion contains a large amount of energy in pulse form. When the structure is confronted with this large amount of detrimental energy suddenly, it demands high energy dissipation in a short span of time. Thus, the structure goes in an inelastic range, So NTHA is essential for near-field earthquake [16]. Forward directivity effect and fling step effect are the two distinct characteristics of near-field earthquake, which exhibits strong pulses or dominant frequencies on narrowband against in case of the far-field has dominant frequency over broadband. Forward directivity effect is associated with the direction of wave propagation. Wave travel along the direction of rupture toward the site and the velocity of wave propagation is very close to shear velocity. Fling step earthquake occurred due to permeant displacement at the fault and it can be observed parallel to fault or slip direction. This effect gets reflected in the velocity and displacement time histories [14, 16–18]. Typical time histories of acceleration, velocity, and displacement are shown in Fig. 4.

## 2.2 Modeling

A three-span RC bridge is considered for the present study as shown in Fig. 1 and Table 1. The main aim of the study to examine the effect of pier height so three models of both IB and CB are modeled in CSI Bridge software [19] by varying pier height. The geometric configuration of the P-10 model is taken from the literature [3]. The moment of inertia of longitudinal and transverse girder of the deck in the case of the P-10 is nearly the same as that of the model used by the author. Also, the moment of inertial about the weak axis of the pier also approximately the same. In all six models, piers and transverse girders are modeled as line elements. The deck of bridge consists deck slab supported by the grid of longitudinal girders and cross diaphragms. The deck of the bridge is modeled as a line element of equivalent

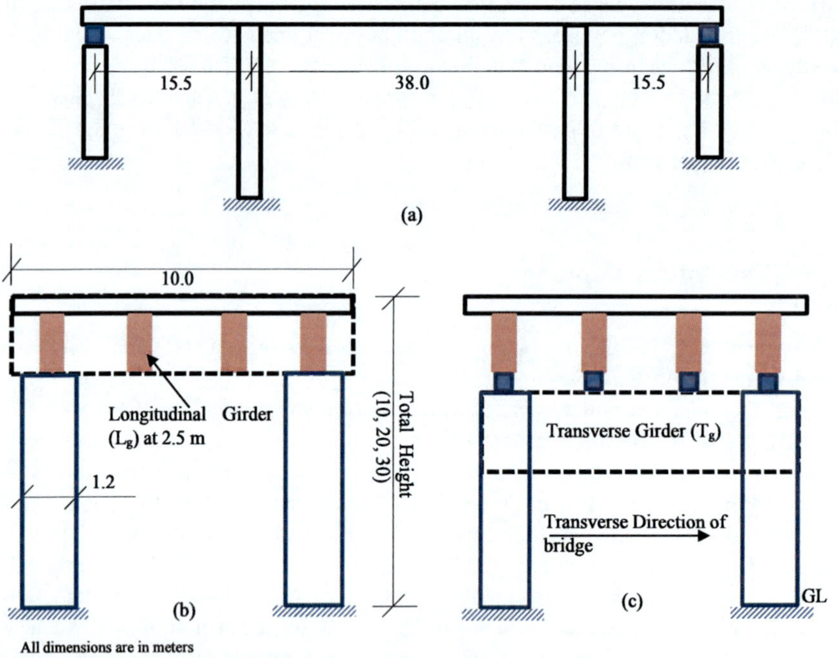


**Fig. 1** Geometric details of IB and CB **a** longitudinal section of IB **b** transverse section of IB **c** transverse section of CB

**Table 1** Material and section details of the bridge model

| Bridge type | Bridge model (bridge type-Pier-height) | Bridge height in m | Size of pier in m (% steel) | Size of $L_g$ and $T_g$ in m | Deck slab in m | Time period (s) | |
|---|---|---|---|---|---|---|---|
| | | | | | | 1st mode | 2nd mode |
| IB | IB-P-10 | 10 | 1.20 × 0.75 (1.3%) | 1.8 × 0.5 | 0.2 m thickness | 0.63 | 0.42 |
| IB | IB-P-20 | 20 | | | | 1.35 | 1.09 |
| IB | IB-P-30 | 30 | | | | 1.6 | 1.48 |
| CB | CB-P-10 | 10 | | | | 1.35 | 1.19 |
| CB | CB-P-20 | 20 | | | | 1.6 | 1.33 |
| CB | CB-P-30 | 30 | | | | 1.67 | 1.52 |

stiffness. This is called as a spine model of the bridge. Response quantities of interest are absolute maximum bending moment (M2 and M3) and deck displacement. So, spine model is used which saves computational time as compared to the area object model in which the deck is modeled as an area element for slab and line elements for longitudinal girders and cross diaphragms. Bending moment M2 and M3 for any section is the moment about minor and major axis of section as shown in Fig. 2 Elastomers are provided at the abutments to connect the deck and abutments which are modeled as link elements. The nonlinear behavior of the elastomer is shown in the form of a force–displacement curve in Fig. 3. The abutment and foundation beneath the pier are not modeled but instead, fixity is considered at the pier base and at the elastomeric base at the abutment.

**Fig. 2** Major and minor axis moments of section

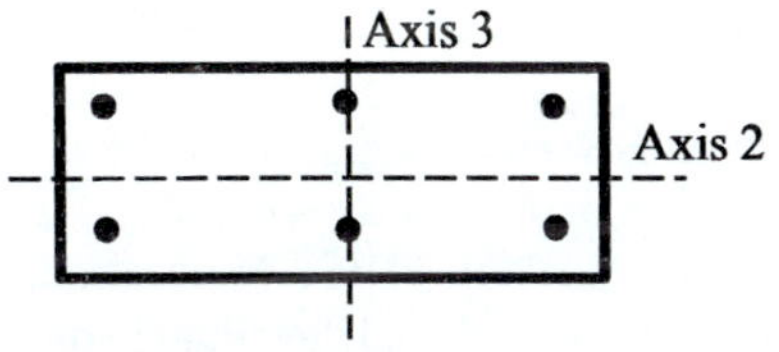


M3 and M2 are moments about the major (3-3) and minor (2-2) axis of the section

**Fig. 3** Force–displacement relation for elastomer

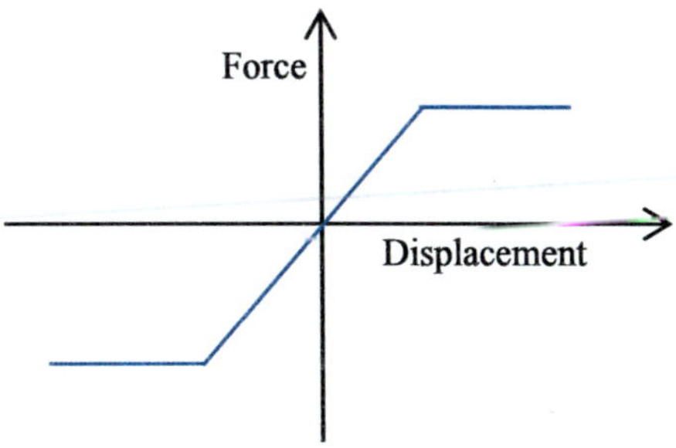

### 2.3 Analysis

3-D models of P-10, P-20, and P-30 for both IB and CB are analyzed for unidirectional earthquake loading in longitudinal and transverse directions separately to perform the parametric study. Earthquake time histories are applied at the base of piers and abutments. In order to study the effect of type of earthquake NLTH analysis is performed. To consider the nonlinearity of model default plastic hinges available in the CSI bridge according to ASCE 41-13 (Table No. 10-8) are assigned to piers. The plastic hinges are assigned at 0.1 and 0.9 relative distance of pier length (i.e., on both ends of piers) and P-M2-M3 interaction is considered for the piers. The direct integration method is used for numerical solutions to perform NLTHA. Proportional viscous damping is considered in terms of the damping ratio.

## 3 Numerical Study

A three-span RC bridge with span lengths 15.5 m–38 m–15.5 m is considered as shown in Fig. 1 and Table 1 for the present study. M-40 concrete and Fe-500 rebar steel is considered for the RC bridge. The deck width of the bridge is 10 m. The deck consists of four longitudinal girders placed at 2.5 m spacing and cross diagrams in the transverse direction at equal spacing within the span. Pier bent consists of two piers and one connecting transverse girder. To study the effect of pier flexibility, pier height varied as 10, 20, and 30 m. Elastomers are used to connect the deck with the abutment in the case of IB whereas, in the case of the CB deck is connected to substructure through elastomers. The elastomers are designed according to IRC 083-2. Its elastic perfectly plastic behavior is modeled as a multi-linear link element in the CSI bridge for which effective stiffness of elastomer and its force–displacement curve is used as an input as shown in Fig. 2. Effective lateral stiffness of elastomer ($K_h$) is considered as 2700 kN/m. Fixity is considered for vertical direction as the bearing is very stiff in the vertical direction. In order to compare results between IB and CB, all geometric dimensions, component dimensions, and pier reinforcement are kept the same. To perform NLTH analysis, three types of earthquakes are considered. These earthquakes are listed in Table 2. The list of the earthquake given in Table 2 is divided into three parts as per the type of earthquake and the type of earthquake is confirmed from the previous study [15, 18]. Five time histories of each type of earthquake are applied in longitudinal and transverse directions separately. PGA of time histories is varied 0.2g, 0.4g, 0.6g, and 0.8g by using PGA scaling. These accelerograms are obtained from the PEER NGA database [20]. Sample time histories of the earthquake (one from each type) used for the analysis are shown in Fig. 4. The fundamental periods of all six models are listed in Table 1 and it is clear from the table that the fundamental periods of all six models are in the range of 0.63–1.67 s.

Response quantities of interest are absolute maximum deck moment, absolute maximum pier moments, and absolute maximum deck displacement. The average

**Table 2** List of time histories used for NLTHA [15, 18, 20]

| Sr. No. | Earthquake name (year) | Station name | $M_w$ | Mechanism | $R_{jb}$ (km) | PGA in g (angle) |
|---|---|---|---|---|---|---|
| *List of far-field earthquake* | | | | | | |
| 1 | Superstition Hill (1987) | Brawley Airport | 6.22 | Strike-slip | 17.59 | 0.16 (90) |
| 2 | Kern County (1952) | Taft Lincoln School | 7.36 | Reverse | 38.42 | 0.18 (111) |
| 3 | Loma Prieta (1989) | Sunol—Forest Fire Station | 6.93 | Reverse oblique | 47.41 | 0.084 (90) |
| 4 | San Fernando (1971) | LA—Hollywood Stor. FF | 6.61 | Reverse | 22.77 | 0.225 (90) |
| 5 | Northridge-01(1994) | Downey—Country Maint Bldg | 6.69 | Reverse | 43.2 | 0.158 (90) |
| *List of near-field earthquake with forward directivity effect* | | | | | | |
| 1 | Kobe Japan (1995) | KJM | 6.9 | Strike slip | 0.94 | 0.62 (90) |
| 2 | Cape Mendocino (1992) | Petrolia | 7.01 | Reverse | 0 | 0.59 (0) |
| 3 | Northridge-01 (1994) | Jensen Filter Plant Admin. Building | 6.69 | Reverse | 0 | 0.41 (22) |
| 4 | Northridge-01 (1994) | LA Dam | 6.69 | Reverse | 0 | 0.43 (64) |
| 5 | Imperial Valley-06 (1979) | El Centro Differential Array | 6.53 | Strike slip | 5.09 | 0.35 (270) |
| *List of near-field earthquake with the fling step effect* | | | | | | |
| 1 | Chi-Chi Taiwan (1999) | TCU049 | 7.62 | Reverse oblique | 3.76 | 0.279 (E) |
| 2 | Chi-Chi Taiwan (1999) | TCU065 | 7.62 | Reverse oblique | 0.57 | 0.789 (E) |
| 3 | Chi-Chi Taiwan (1999) | TCU067 | 7.62 | Reverse oblique | 0.62 | 0.499 (E) |
| 4 | Chi-Chi Taiwan (1999) | TCU072 | 7.62 | Reverse oblique | 0 | 0.477 (E) |
| 5 | Chi-Chi Taiwan (1999) | TCU074 | 7.62 | Reverse oblique | 0 | 0.59 (E) |

response is considered for five earthquakes. To study the relative performance of P-20 and P-30 models, responses of these models are normalized with its corresponding responses of the P-10 model.

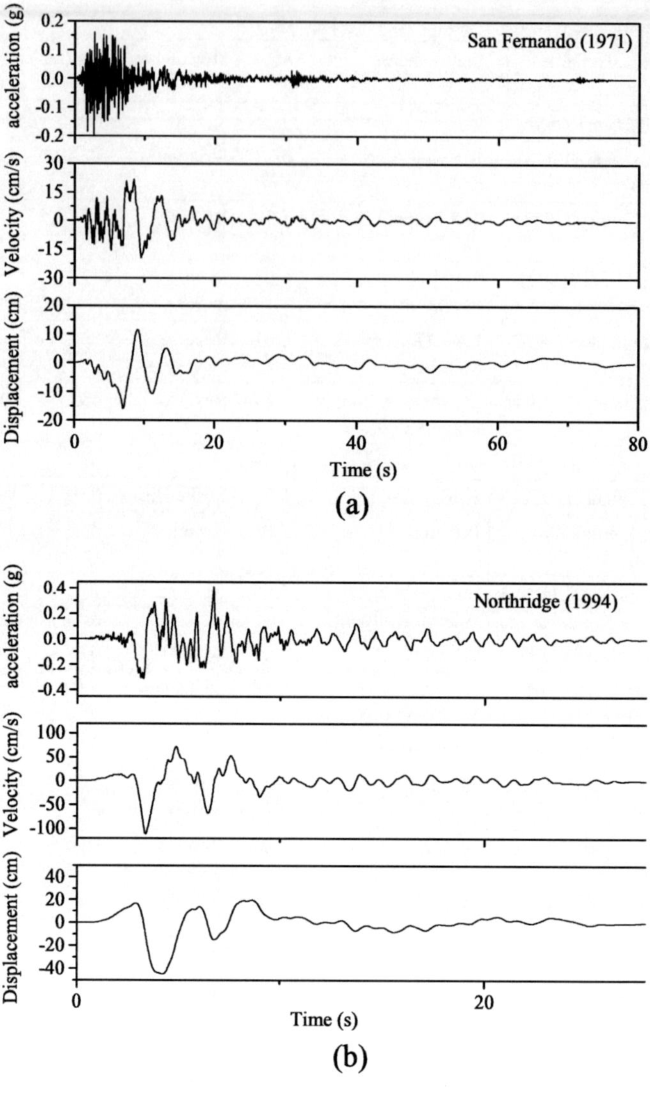


**Fig. 4** Time histories of acceleration, velocity, and displacement of (**a**) FFE, (**b**) FDE, and (**c**) FSE

## 4 Results and Discussion

Both IB and CB are analyzed as mentioned before for unidirectional earthquakes in the longitudinal and transverse direction and average response quantities from each type of earthquake are considered. Results are presented in form of normalized response quantities with respect to P-10, i.e., Response quantities of P-20 and P-30

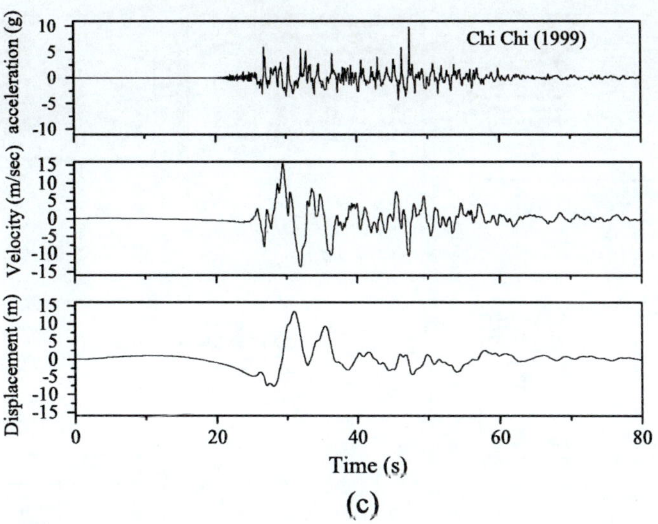


**Fig. 4** (continued)

models of both IB and CB are normalized with corresponding responses of the P-10 model. Normalized results with respect to the P-10 model (i.e., model having the least pier height) will help to comprehend the effect of pier height on the response. Thus, the ratio of response quantity of P-20 or P-30 with the P-10 model, i.e., normalized response indicates an increase or decrease in the response with respect to the P-10 model. If the ratio is less than one then it shows the reduction in the response.

Normalized deck moments, normalized pier moments, and normalized deck displacement for IB and CB are presented below for two cases (i) when the earthquake is applied in the longitudinal direction (ii) earthquake applied in the transverse direction of the bridge. Figure 2 shows major axis moments (M3) and minor axis moments (M2) of a particular section. As only unidirectional loading is considered either major or minor axis moment will occur. Thus, for the first case when the earthquake is applied in the longitudinal direction of the bridge then, deck moment M3 is considered and pier moment M2 is considered. on the other hand, when the earthquake is applied in the transverse direction of the bridge, deck moment M2, and pier moment M3 are considered.

### *4.1 Earthquake Applied in the Longitudinal Direction of the Bridge*

Figure 5(a–c) shows the variation of normalized deck moment, pier moment, and deck displacement for IB, whereas Fig. 5(d, e) shows the variation of normalized deck moment, pier moment, and deck displacement for CB.

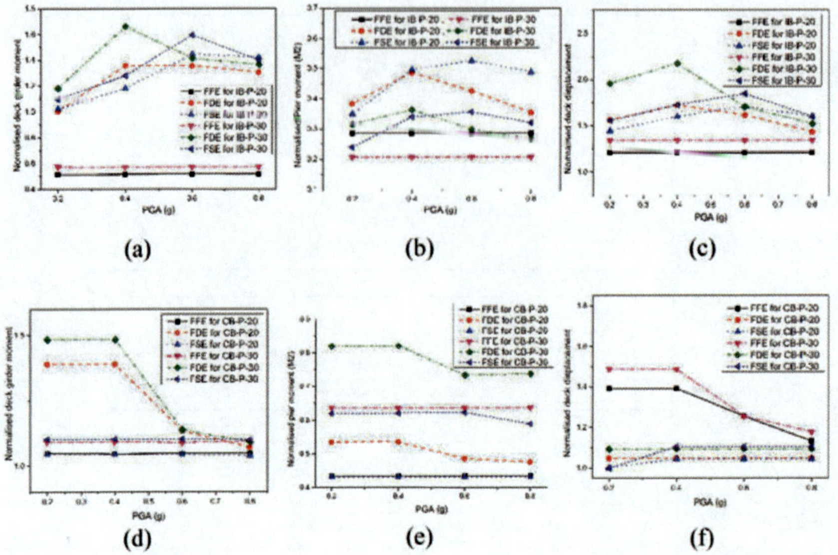


**Fig. 5** The normalized response of IB and CB when earthquake applied in the longitudinal direction of the bridge (**a**) normalized deck moment (M3) of IB (**b**) normalized pier moment (M2) of IB (**c**) normalized deck displacement in the longitudinal direction of IB (**d**) normalized deck moment (M3) of CB (**e**) normalized pier moment (M2) of CB (**f**) normalized deck displacement in the longitudinal direction of CB

It can be seen from Fig. 5(a) that deck moment (M3) decreases as pier height increases as compared to the P-10 model for the far-field earthquake on the other hand for near-field earthquakes with directivity effect and fling step effect, deck moment increases as pier height increases. The increase in the pier height increases the flexibility of the model which provides a frequency separation effect for the far-field earthquake. Whereas in the case of near-field earthquake with directivity effect and fling step effect, excessive deflection of pier increases the moment in the deck. Thus moment sharing between the pier and deck girders can be noted in this case.

In the case of pier moment (M2), it can be seen from Fig. 5(b) pier moments decrease as pier height increases but again the relative reduction in M2 moments for P-20 and P-30 model with respect to the P-10 model for near-field earthquake with both directivity and fling step effect is lesser than the far-field earthquake. Least reduction provided by the near-field earthquake with directivity effect.

Figure 5(c) shows the normalized deck displacement. The deck displacement increases as the pier height increases. It can be noted from Fig. 5(c) that for the deck displacement, far-field earthquake provides a lesser relative increase in the deck displacement of P-20 and P-30 model with respect to the P-10 model as compared to near-field earthquake with directivity effect and fling step effect. The maximum relative increase in the deck displacement is provided by the near-field earthquake with the fling step effect.

Figure 5(d, e) shows normalized response quantities for the CB. Figure 5(d) shows the deck moment M3 variation with respect to PGA. It can be noted from the figure that very lesser change in the deck moment M3 for far-field earthquake unlike for IB. Again similar to the case of IB, the maximum increase in the deck moment M3 is for near-field earthquake with a directivity effect.

Figure 5(e) shows the variation of pier moment M2 in the pier. Reduction in the pier moment for the P-20 and P-30 model is observed as compared to the P-10 model. Like IB, again reduction in the pier moment is least for the near-field earthquake with directivity effect.

Figure 5(f) shows the variation of normalized displacement in the CB. The increase in deck displacement increases as the height of the pier increases. but the relative increase in P-20 and P-30 models with respect to the P-10 model is not much significant.

It can be noted from Fig. 5, by comparing the responses of IB with respect to CB, variation in the response of CB is lesser than IB. Thus in the case of CB, reduction or increase in the response of P-20 or P-30 model with respect to corresponding P-10 model is lesser as compared to IB. This is the case because the elastomer used between the deck and substructure provides quasi isolation. So increase in height or flexibility does not affect much on the response.

In the case of IB, from Fig. 5(a–c) it can be noted that variation of normalized response quantities with respect to PGA is not significant for the far-field earthquake. This is the case again because of frequency separation between structure and the far-field earthquake. In the case of FDE and FSE, variation is maximum for PGA 0.6g. In the case of CB, it can be noted from Fig. 5(d–f), the variation of normalized response with respect to PGA is much lesser as compared to IB.

### *4.2 Earthquake Applied in the Transverse Direction of the Bridge*

Figure 6(a–f) show normalized response quantities of IB and CB when the sets of earthquakes are applied in the transverse direction of the bridge. The relative increase in the deck moment (M2) and deck displacement, as well as the reduction in the pier moment (M3), follow the same trend as that for the previous case, i.e., when sets of earthquakes are applied in the longitudinal direction. So, the trend of variation with respect to PGA, for the type of earthquake and type of bridge (IB or CB) is the same.

From Figs. 5 and 6 it is clear that the relative increase in the deck moments and deck displacement is higher when the earthquake is applied in the transverse direction. But the relative reduction in the pier moments is decreased in the case when the earthquake is applied in the transverse direction.

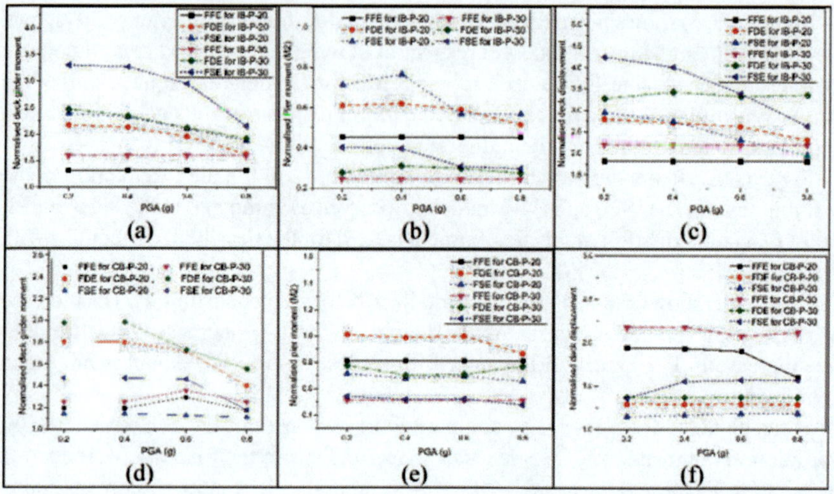


**Fig. 6** The normalized response of IB and CB when earthquake applied in the transverse direction of the bridge (**a**) normalized deck moment (M2) of IB (**b**) normalized pier moment (M3) of IB (**c**) normalized deck displacement in the transverse direction of IB (**d**) normalized deck moment (M2) of CB (**e**) normalized pier moment (M3) of CB (**f**) normalized deck displacement in the transverse direction of CB

# 5 Conclusions

Normalized response quantities obtained from the NLTH analysis are presented in form of variation of normalized response with respect to PGA in the previous section. The observation of the study leads to the following conclusions:

- The maximum relative reduction of deck moments and pier moments are obtained for the far-field earthquake. This is the case because of the frequency separation of bridge and earthquake.
- In the case of near-field earthquake with directivity effect and fling step effect, deck moment increases as the height of pier increases because of the drastic increase in lateral displacement.
- Pier moment reduces for all the cases as the height of the pier increase. The maximum reduction is for the far-field earthquake.
- An increase in the deck moments and deck displacement with respect to an increase in the pier height for the CB is lesser in comparison with IB. The relative reduction in the pier moments of the CB is also lesser.
- Variation of normalized response with respect to PGA is lesser for the CB as compared to IB as the elastomer provides quasi isolation and hence the flexibility of pier does not affect much on the response.
- The increase in the responses for the P-20 or P-30 model with respect to the P-10 model is higher when the earthquake is applied in the transverse direction.

Moreover reduction in the pier moment for the P-20 or P-30 with respect to the P-10 model is lesser when the earthquake is applied in the transverse direction.

## References


1. Martin P, Burke J (2009) Integral and semi-integral bridges, 1st edn. Wiley, Iowa
2. Wolde-Tinsae AM, Klinger JE, White EJ (1988) Performance of jointless bridges. J Perform Constr Facil 2:111–125
3. Ní Choine M, Oconnor AJ, Padgett JE (2015) Comparison between the seismic performance of integral and jointed concrete bridges. J Earthq Eng 19:172–191. https://doi.org/10.1080/13632469.2014.946163
4. Greimann LF, Yang P-S, Wolde-Tinsae AM (1986) Nonlinear analysis of integral abutment bridges. J Struct Eng 112:2263–2280
5. Kunde MC, Jangid RS (2006) Effects of Pier and Deck flexibility on the seismic response of isolated bridges. J Bridg Eng 11:109–121. https://doi.org/10.1061/(asce)1084-0702(2006)11:1(109)
6. Tongaonkar NP, Jangid RS (2003) Seismic response of isolated bridges with soil-structure interaction. Soil Dyn Earthq Eng 23:287–302. https://doi.org/10.1016/S0267-7261(03)00020-4
7. Carlo GDE, Dolce M, Liberatore D (2000) Influence of soil-structure interaction on the seismic response of bridge piers. In: 12th World conference on earthquake engineering, pp 1–8
8. Cai Z, Wang Z, Lin K, Sun Y, Zhuo W (2020) Seismic behavior of a bridge with new composite tall piers under near-fault ground motion conditions. Appl Sci 10:1–17. https://doi.org/10.3390/app10207377
9. Tamanani M, Gian Y, Ayoub A (2016) Evaluation of code criteria for bridges with unequal pier heights. Bull Earthq Eng 14:3151–3174. https://doi.org/10.1007/s10518-016-9941-4
10. Thippeswamy HK, GangaRao HVS, Franco JM (2002) Performance evaluation of jointless bridges. J Bridg Eng 7:276–289. https://doi.org/10.1061/(ASCE)1084-0702(2002)7:5(276)
11. Caner A, Zia P (1998) Behavior and design of link slabs for jointless bridge decks. PCI J 43:68–78. https://doi.org/10.15554/pcij.05011998.68.80.
12. Kozak DL, LaFave JM, Fahnestock LA (2018) Seismic modeling of integral abutment bridges in Illinois. Eng Struct 165:170–183. https://doi.org/10.1016/j.engstruct.2018.02.088
13. Mitoulis SA, Tegos IA, Stylianidis KC (2010) Cost-effectiveness related to the earthquake resisting system of multi-span bridges. Eng Struct 32:2658–2671. https://doi.org/10.1016/j.engstruct.2010.04.036
14. Somerville PG, Smith NF, Graves RW, Abrahamson NA (1997) Modification of empirical strong ground motion attenuation relations to include the amplitude and duration effects of rupture directivity. Seismol Res Lett 68:199–222. https://doi.org/10.1785/gssrl.68.1.199
15. Bhandari M, Bharti SD, Shrimali MK, Datta TK (2019) Seismic fragility analysis of base-isolated building frames excited by near- and far-field earthquakes. J Perform Constr Facil 33:04019029. https://doi.org/10.1061/(asce)cf.1943-5509.0001298
16. Mollaioli F, Bruno S, Decanini LD, Panza GF (2006) Characterization of the dynamic response of structures to damaging pulse-type near-fault ground motions. Meccanica 41:23–46. https://doi.org/10.1007/s11012-005-7965-y
17. Somerville PG (2002) Characterizing near fault ground motion for the design and evaluation of bridges. In: Proceedings of the third national seismic conference and workshop on bridges and highways, pp 137–148
18. Kalkan E, Kunnath SK (2006) Effects of fling step and forward directivity on seismic response of buildings. Earthq Spectra 22:367–390. https://doi.org/10.1193/1.2192560
19. CSI Bridge (2020) Integrated software for structural analysis and design of bridge. Computers and Structures Inc., Berkeley, California

20. PEER NGA data base. PEER (Pacific Earthquake Engineering Research) Center (2000) Strong ground motion database. Univ California, Berkeley, CA n.d.:1. https://ngawest2.berkeley.edu/users/sign_in?unauthenticated=true

# Seismic Performance Evaluation of Integral Bridges Considering Soil-Structure Interaction

**Duy-Duan Nguyen and Huu-Cong Nguyen**

## 1 Introduction

The integral bridge is the structure, which integrates the superstructure and basic substructures. This bridge is known as having more economic benefit than a conventional bridge since it is constructed without supplemental devices such as expansion joints and bridge bearings [1]. Kozak et al. [2] investigated seismic responses of integral abutment bridges in Illinois (USA), in which they focused on the steel plate and precast prestressed concrete girders were considered. Zordan et al. [3] conducted a parametric study on the variation of soil properties and then obtained the behavior of the bridge under temperature change. Erhan and Dicleli [4] showed that integral bridges own a superior seismic performance in terms of smaller inelastic structure displacement, rotations, or forces compared to conventional bridges. Additionally, this study considered the individual behavior of the pile-pile cap connection under seismic load and of the girder-abutment connection, which are sensitive since large bending moments are transferred at these locations. In the study of Far et al. [5], they pointed out that thermal and seismic loads greatly affect the response of integral abutment bridges due to the integrity of the structure and complex soil-structure-pile interactions. Choi et al. [6] performed a series of pushover analyses to evaluate the capacity curves and stiffness of an integral abutment railroad bridge in the longitudinal direction.

Numerous studies on seismic performances of integral abutment bridges were implemented using the static analysis method [2, 6–8] and the time-history analysis approach [8–12]. However, studies on the detailed plastic hinge formation and proposed damage states of the integral bridge have been very limited so far. Previous studies mostly focused on the performance of integral bridge with the individual

D.-D. Nguyen (✉) · H.-C. Nguyen
Department of Civil Engineering, Vinh University, Vinh, Vietnam
e-mail: duyduankxd@vinhuni.edu.vn



S. Kolathayar and S. C. Chian (eds.), *Recent Advances in Earthquake Engineering*, Lecture Notes in Civil Engineering 175,
https://doi.org/10.1007/978-981-16-4617-1_8

components, which may neglect some important interactions between components. Understanding the seismic behavior of integral bridges is crucial not only to implement cost-effective designs but also to properly evaluate seismic damage evolution of existing bridges. Since there is little experimental data on the seismic response of fully integral bridges, numerical models are essential for understanding structural behavior.

The purpose of this study is to perform seismic evaluation of an integral abutment bridge considering soil-structure interaction using pushover analyses. Capacity curves of the bridge are obtained for both longitudinal and transversal directions. Plastic hinge formations during pushover analyses are monitored. Based on nonlinear static analyses, a set of damage states of the integral bridge are proposed.

## 2 Description of the Investigated Bridge

The integral abutment RC bridge comprises of seven spans, and the first and the last span length is 26 m, while the length of other spans is 35 m, as shown in Fig. 1. The height of all pier columns is 15 m, and the abutment height is 10.4 m. The bridge girder is various RC sections, as also shown in Fig. 1. Detailed cross-sectional dimensions and reinforcing bar arrangements are shown in Fig. 2. Foundations of bridge piers are designed with three bored piles, while nine bored piles are used for the foundation of abutments. Further information about the bridge can be found in Choi et al. [6].

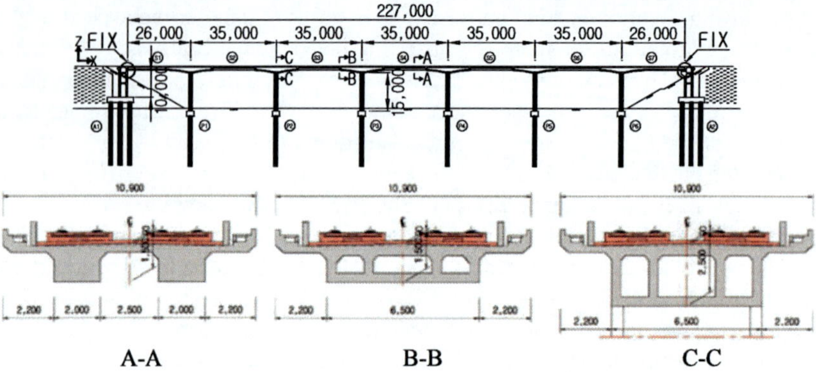


Fig. 1 Elevation view of the bridge and girder dimensions

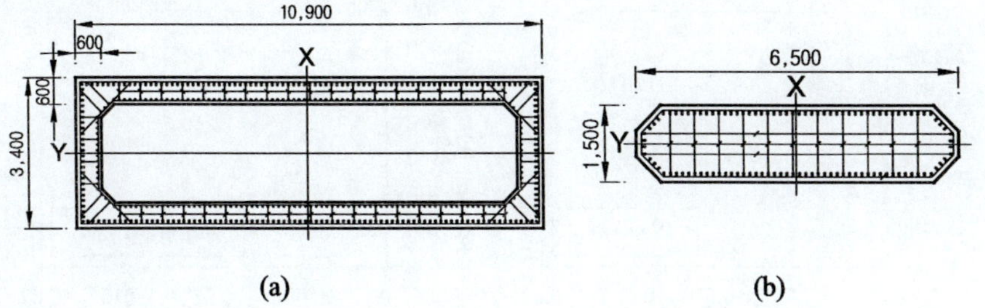


**Fig. 2** Cross section and reinforcing detail of the piers **(a)** and abutments **(b)**

# 3 Structural Modeling

The numerical model of the bridge is developed using SAP2000, the finite element analysis program [13]. The bridge girder is assumed to elastically behave during seismic excitations. Meanwhile, the bridge piers and abutments are modeled as nonlinear beam-column elements. For that, the model proposed by Mander et al. [14] is adopted for nonlinear concrete material, while the steel model presented in Park and Pauley [15] is applied for reinforcing bars. Figure 3 shows the stress–strain curves of the concrete and steel models. Figure 4 shows moment–curvature relationships of cross-sections of piers and abutments.

Moreover, to consider soil-structure interaction, the piles are also modeled in terms of elastic beam elements, in which the bi-liner soil springs are attached to the pile element nodes, as shown in Fig. 4. It should be noted that the foundations of the selected bridge are mostly embedded in the dense sand and weathered-rock layers. Figure 5 shows the modeling schemes of structural members of the bridge. The 3D finite element modeling of the bridge in SAP2000 and modal analysis results are

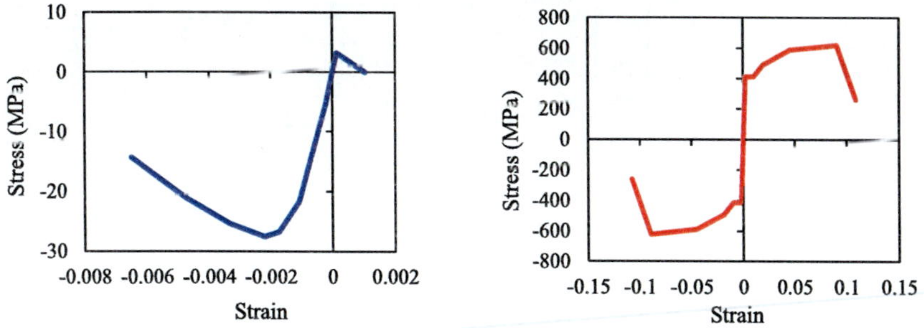


**Fig. 3** Material models for concrete and reinforcement

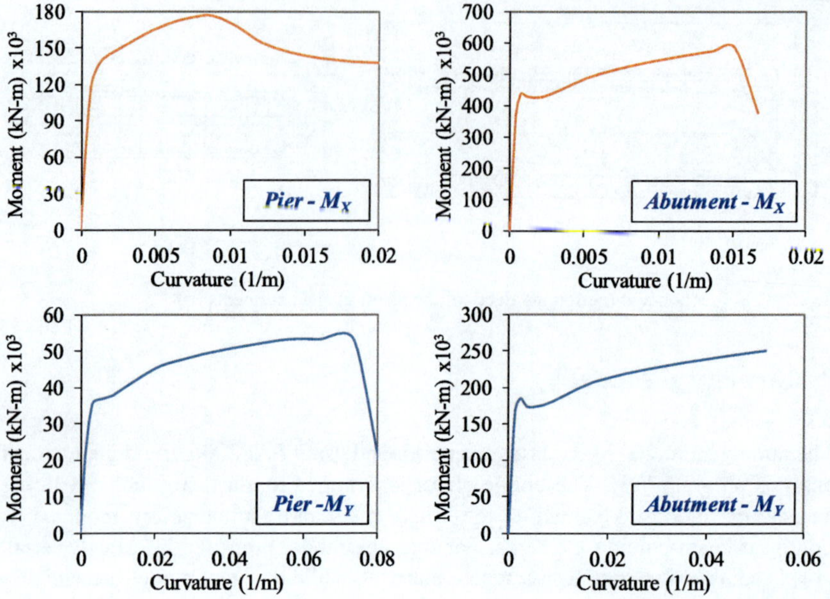


**Fig. 4** Moment–curvature relationships of pier and abutment cross-sections

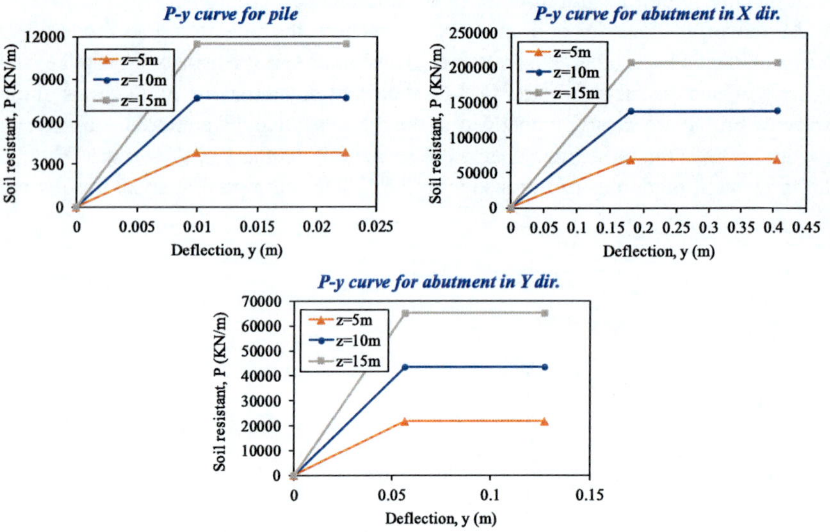


**Fig. 5** *P*-*y* curves for considering the soil-structure interaction

shown in Fig. 6 and Fig. 7, respectively. The results of eigen-value analysis is shown in Fig. 8.

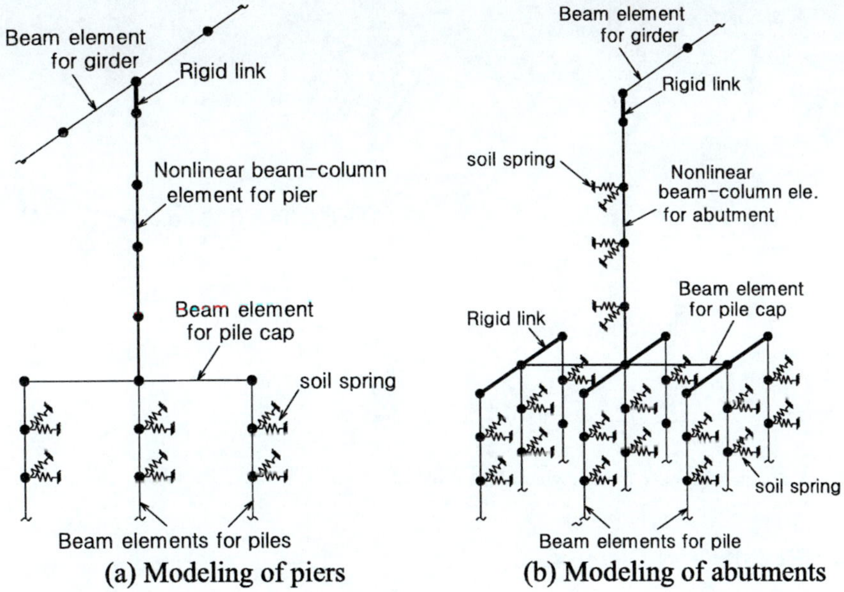


**Fig. 6** Modeling schemes of the bridge elements

**Fig. 7** 3D FEM modeling of the integral abutment bridge

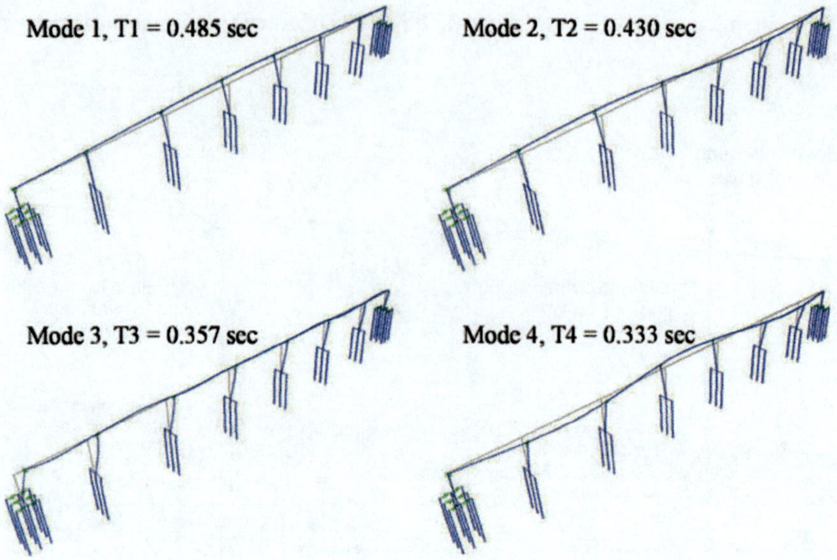


**Fig. 8** Eigen values and mode shapes of the bridge

## 4 Results and Discussion

Pushover analyses in both X- and Y-direction are performed to obtain the capacity curves as well as the plastic hinge formation of the bridge. Figure 9 shows the pushover curves of the bridge in the longitudinal and transversal direction, i.e., X- and Y-direction, respectively. It can be observed that the stiffness of the bridge in the longitudinal direction is significantly larger than that in the transversal direction. This can be attributed to the reason that the integral bridge is a multispan-framed structure, which provides a higher in-plane stiffness, even though the cross-section

**Fig. 9** Pushover curves of the bridge

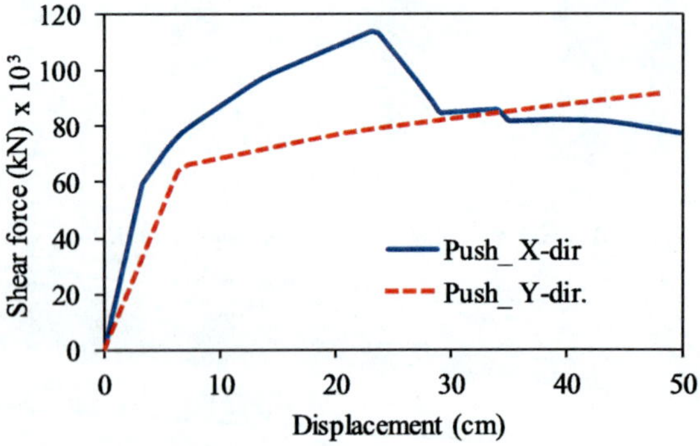

of the piers and abutments along with the X-direction are larger than those in the Y-direction.

Figure 10 shows the sequence of plastic hinges in the longitudinal direction. It can be observed that the first yielding is formed at the bottom of the left abutment according to a lateral displacement at the top pier approximately of 3.0 cm. Then, the second plastic hinge is occurred at the bottom of the right abutment. As lateral load increased, the plastic hinges spread to the top of near-abutment piers, and then expand to the top of the mid-area piers. Finally, plastic hinges form at the bottom of

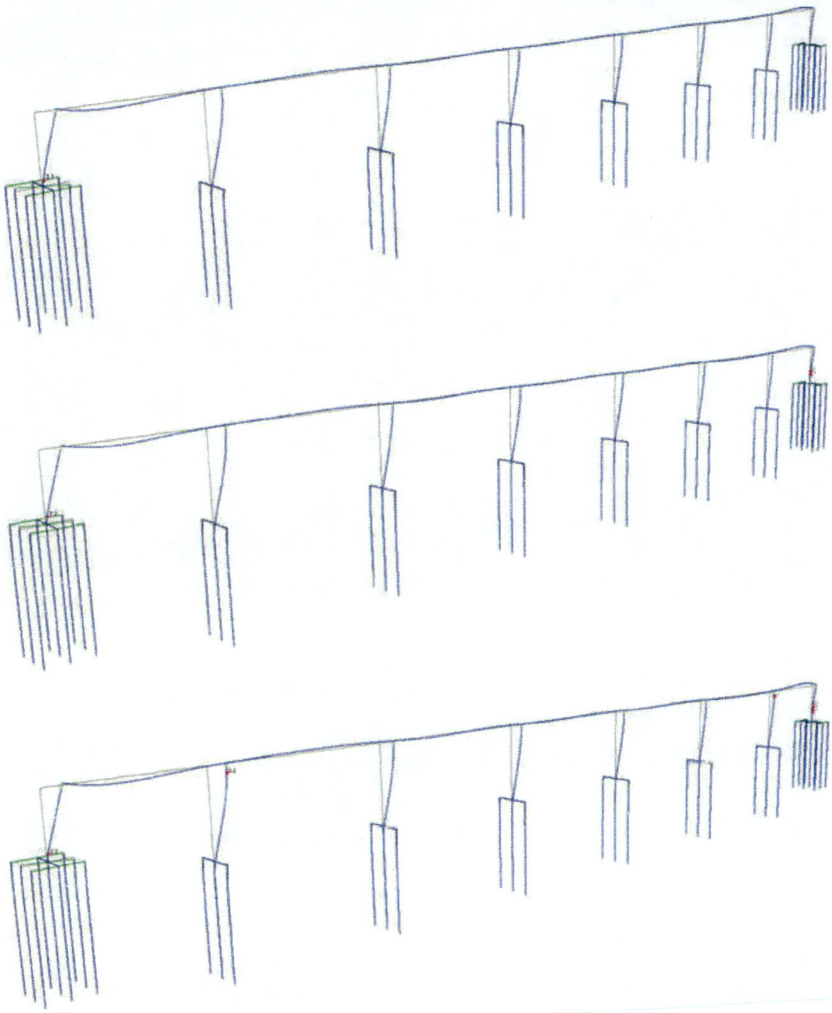

**Fig. 10** Plastic hinge evolution during pushover in X-direction (longitudinal)

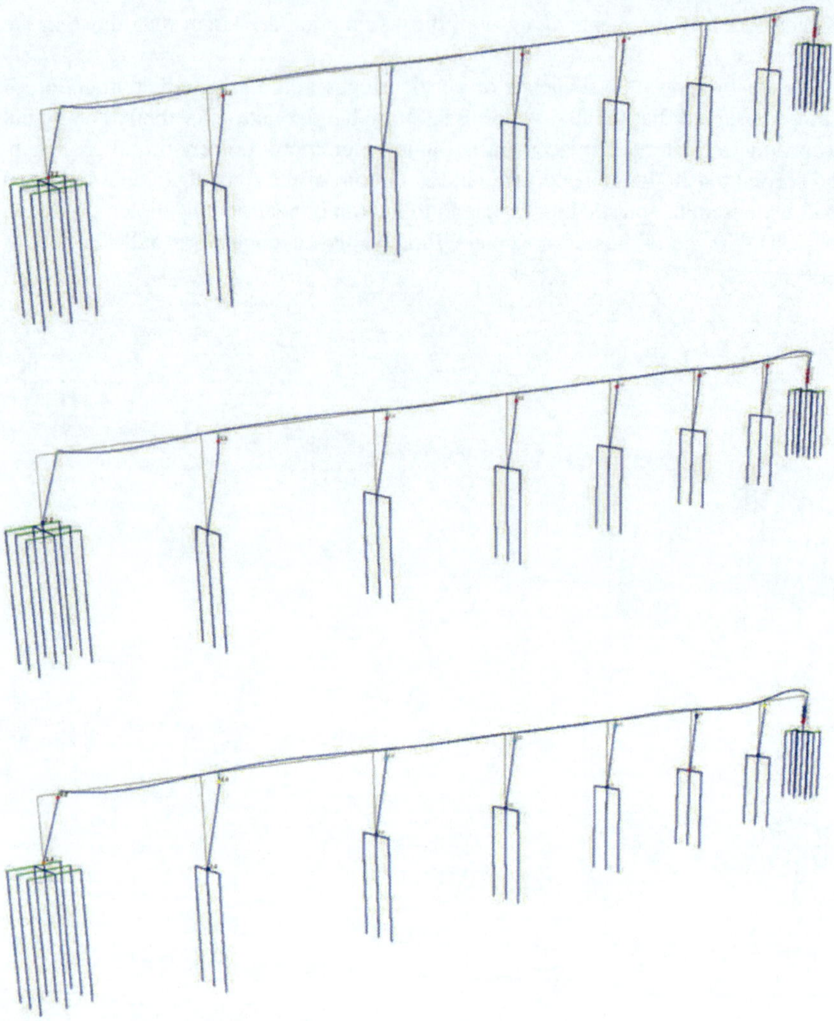

**Fig. 10** (continued)

the bridge piers. The concrete is crushed at the horizontal displacement of 22 cm. After this stage, the stiffness of the bridge is rapidly decreased.

Figure 11 shows the plastic hinges evolution of the bridge under transversal pushover analysis. The plastic hinges are firstly occurred at the bottom of the mid-area piers (at the horizontal displacement of 7.0 cm), and then formed at the bottom of other piers. Different from pushover in the longitudinal direction, the structure can reach the collapse state without plastic hinges at the abutments and top of the piers.

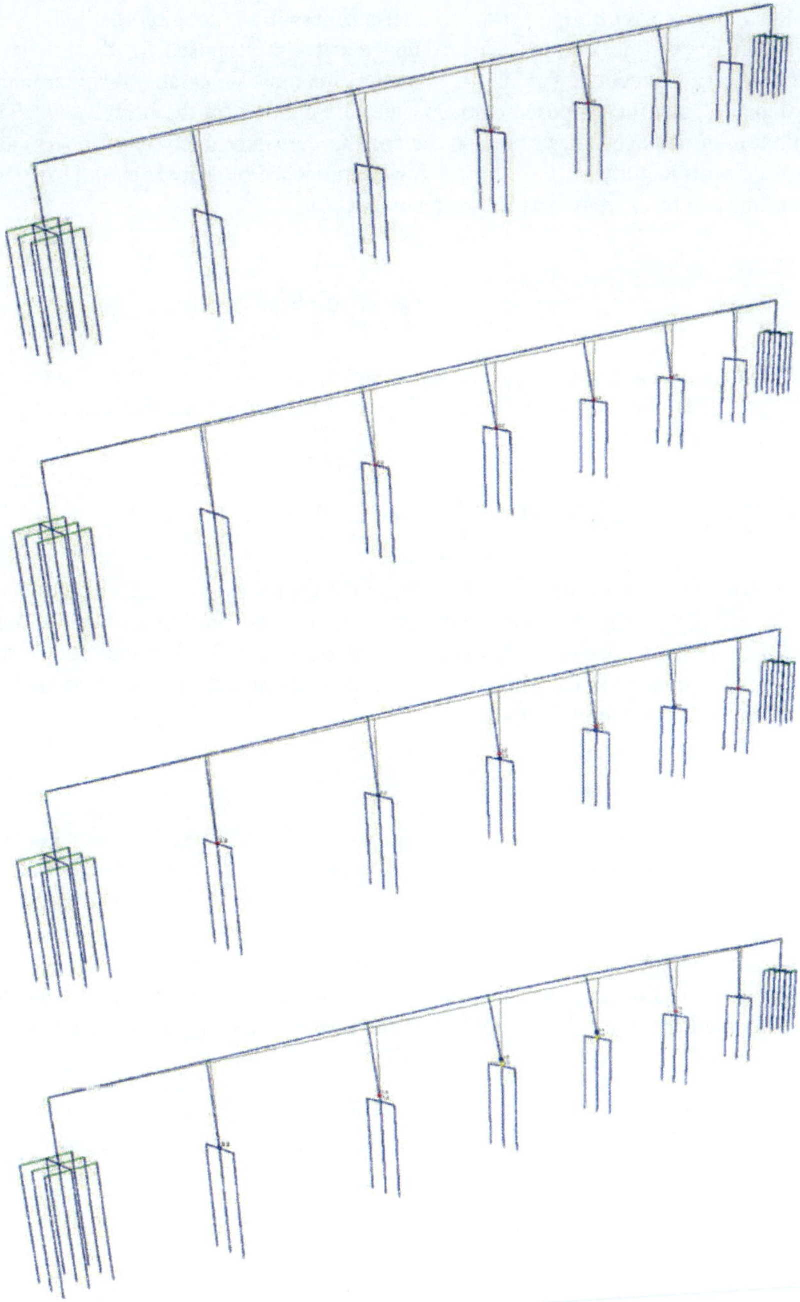

**Fig. 11** Plastic hinge evolution during pushover in Y-direction (transversal)

Since the integral bridge shows a larger stiffness in the longitudinal rather that in the transversal direction, a set of damage states is proposed for the bridge in the transversal direction. The pier is known as the most vulnerable member under earthquakes, and the proposed damage states are defined for the bridge piers. The definition of damage states based on the column curvature ductility ratio ($\mu_\phi$) and displacement ductility ratio ($\mu_\Delta$), which were proposed by Buckle et al. [16]. This definition can be expressed as following expression:

$$\mu_\varphi = 1 + \frac{\mu_\Delta - 1}{3\frac{l_p}{L}\left(1 - 0.5\frac{l_p}{L}\right)} \tag{1}$$

Due to practical reasons, the damage states related to displacements are more facilitated than the curvature ductility. Equation (2) can be rewritten in terms of displacement ductility ratio as

$$\mu_\Delta = 1 + 3\frac{l_p}{L}\left(1 - 0.5\frac{l_p}{L}\right)(\mu_\varphi - 1), \tag{2}$$

where $L$ is the length of the pier, $l_p$ is the length of the plastic hinge given by Eq. (3), and hereafter, $d_b$ is the diameter of the column longitudinal reinforcing bar. It should be noted that curvature ductility ratio ($\mu_\phi$) can be also defined as the ratio of the developing curvature in the piers due to bending moment to the curvature at the first yielding state of the reinforcement.

$$l_p = 0.08L + 9d_b \tag{3}$$

In this study, four damage states (DSs), namely, slight, moderate, extensive, and collapse are defined based on the moment–curvature relationship of the pier cross-section, as illustrated in Fig. 12. The four damage states are determined based on changing stiffness of the pier cross-section, specifically related to cracking, yielding, crack and yielding developed, and ultimate strength, respectively. The damage index is then expressed in terms of the displacement ductility ratio. This approach can be found in the literature [17–19]. Table 1 shows proposed damage states of all bridge

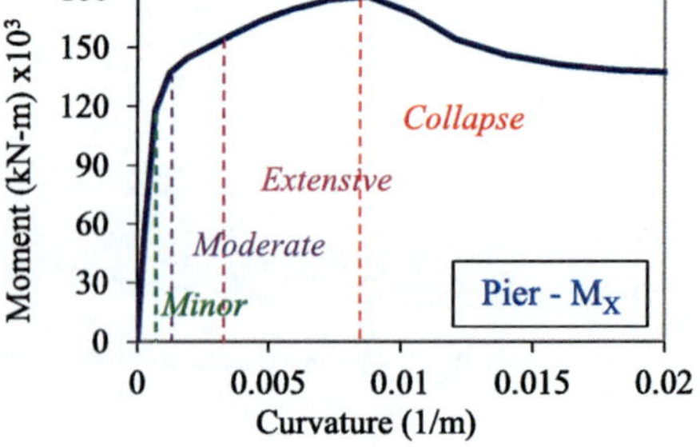


**Fig. 12** Definition of damage states for the bridge pier

**Table 1** Proposed damage states and indices of Hwang et al. [20]

| | Component | Demand parameter | Threshold value | | | |
|---|---|---|---|---|---|---|
| | | | Slight (DS1) | Moderate (DS2) | Extensive (DS3) | Collapse (DS4) |
| This study | Pier | Displacement ductility ratio | 1.0 | 1.2 | 1.8 | 3.7 |
| Hwang et al. [20] | Pier | Displacement ductility ratio | 1.0 | 1.2 | 1.76 | 4.76 |

piers, which are related to the damage indices. It should be noted that all piers have the same height of 15.0 m, hence, to be in the same value of the defined damage levels. The propose damage states can be useful for developing fragility curves of the integral bridges.

## 5 Conclusions

This paper presents seismic performances of an integral abutment bridge using pushover analyses. Fully integral abutment and soil-pile interaction are considered in the numerical model. Following conclusions are drawn based on numerical analyses.

- The stiffness of the bridge in the longitudinal direction is significantly larger than that in the transversal direction.
- The evolution of plastic hinges in the longitudinal and transversal directions is different. Plastic hinges are formed at the bottom and top of the piers and abutments subjected to longitudinal load, meanwhile, the integral bridge can reach the collapse state under transversal load without plastic hinges at the abutments and top of the piers.
- A set of damage states is proposed to assess the seismic performance of the bridge. It can be useful in conducting fragility analyses of the bridge.

## References

1. Greimann LF, Yang P-S, Wolde-Tinsae AM (1986) Nonlinear analysis of integral abutment bridges. J Struct Eng 112:2263–2280
2. Kozak D, LaFave J, Fahnestock L (2018) Seismic modeling of integral abutment bridges in Illinois. Eng Struct 165:170–183
3. Zordan T, Briseghella B, Lan C (2011) Parametric and pushover analyses on integral abutment bridge. Eng Struct 33:502–515
4. Erhan S, Dicleli M (2015) Comparative assessment of the seismic performance of integral and conventional bridges with respect to the differences at the abutments. Bull Earthq Eng 13:653–677

5. Far NE, Maleki S, Barghian M (2015) Design of integral abutment bridges for combined thermal and seismic loads. Earthq Struct 9:415–430
6. Choi BH, Moreno LB, Lim C-S, Nguyen D-D, Lee T-H (2019) Seismic performance evaluation of a fully integral concrete bridge with end-restraining abutments. Adv Civ Eng
7. Kim DS, Kim UJ (2020) Evaluation of passive soil stiffness for the development of integral abutments for railways. J Korean Soc Hazard Mitig 20:13–19
8. Itani AM, Pekcan G (2011) Seismic performance of steel plate girder bridges with integral abutments. United States. Federal Highway Administration. Office of Bridge Technology
9. Kotsoglou A, Pantazopoulou SJ (2009) Assessment and modeling of embankment participation in the seismic response of integral abutment bridges. Bull Earthq Eng 7:343
10. Dicleli M, Erhan S (2011) Effect of foundation soil stiffness on the seismic performance of integral bridges. Struct Eng Int 21:162–168
11. Masrilayanti (2014) The behaviour of integral bridges under vertical and horizontal earthquake ground motion. University of Salford
12. Franchin P, Pinto PE (2014) Performance-based seismic design of integral abutment bridges. Bull Earthq Eng 12:939–960
13. SAP2000 (2015) Linear and nonlinear static and dynamic analysis and design of three-dimensional structures. Computers and Structures, Inc., Berkeley, California, USA
14. Mander JB, Priestley MJ, Park R (1988) Theoretical stress-strain model for confined concrete. J Struct Eng 114:1804–1826
15. Park R, Paulay T (1975) Reinforced concrete structures. Wiley
16. Buckle IG, Friedland I, Mander J, Martin G, Nutt R, Power M (2006) Seismic retrofitting manual for highway structures. Part 1, Bridges. Turner-Fairbank Highway Research Center
17. Choi E, DesRoches R, Nielson B (2004) Seismic fragility of typical bridges in moderate seismic zones. Eng Struct 26:187–199
18. Lee T-H, Nguyen D-D (2018) Seismic vulnerability assessment of a continuous steel box girder bridge considering influence of LRB properties. Sādhanā 43:14
19. Tran N-L, Nguyen T-H, Phan V-T, Nguyen D-D (2021) Seismic fragility analysis of reinforced concrete piers of steel box girder bridges: a parametric study. Mater Today: Proc 38:2310–2315
20. Hwang H, Liu JB, Chiu Y-H (2001) Seismic fragility analysis of highway bridges. Mid-America Earthquake Center CD Release 01–06

# Non Linear Seismic Analysis of RC Framed Structure by Extended N2 Method

**T. Sakthi and V. Vasugi**

## 1 Introduction

The structure of a seismic overall performance is strongly dependent on the damage of the building and the suffered by it under the design earthquake [1]. The asymmetric setback is one of the major reasons for harm in a multi-storey building. The past earthquake has shown that building with Asymmetric Setback causes more seismic force and leads to structural failure [2]. The structure may have irregularities in both plan and elevation. Irregularities in structures cause more damage during an earthquake because more seismic force occur on the building [3]. The regular structures are required to evaluate the seismic parameters as per existing codes. Since an irregular structure is not available to evaluate the seismic parameters as per existing codes [4].

The simplified N2 method is a simple non-linear seismic evaluation method of structure. It synthesizes with pushover method (MDOF) and linear dynamic response spectrum analysis method (SDOF) [5]. The pushover curve is base shear versus roof level displacement. The pushover curve characterizes the highest values at the base of the structure which indicates that maximum base shear [6]. Response spectrum analysis (RSA) is a linear dynamic analysis method that measures the vibration of each mode to indicate the maximum seismic response in the structure [7].

N2 method is used in the single mode of building vibrates predominately. This method is constantly fulfilled, in the case of multi-storey building and irregularity building. The development of the N2 method is known as the Extended N2 method is relevant to structure in higher mode [8].

The application of the extended N2 method examines with real behaviour of the structure all through an earthquake by non-linear dynamic time-history analysis method (THAM) [9]. In the present study, the basic ideas of the N2 method are

T. Sakthi · V. Vasugi (✉)
School of Civil Engineering, Vellore Institute of Technology, Chennai 600127, India
e-mail: vasugi.v@vit.ac.in



S. Kolathayar and S. C. Chian (eds.), *Recent Advances in Earthquake Engineering*, Lecture Notes in Civil Engineering 175,
https://doi.org/10.1007/978-981-16-4617-1_9

**Fig. 1** Plan

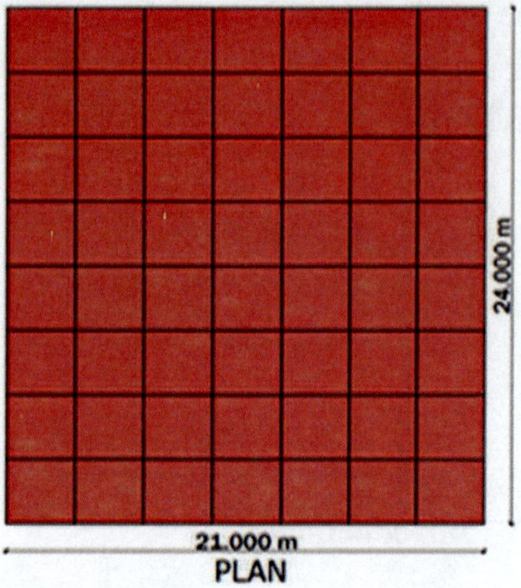


**Fig. 2** Section

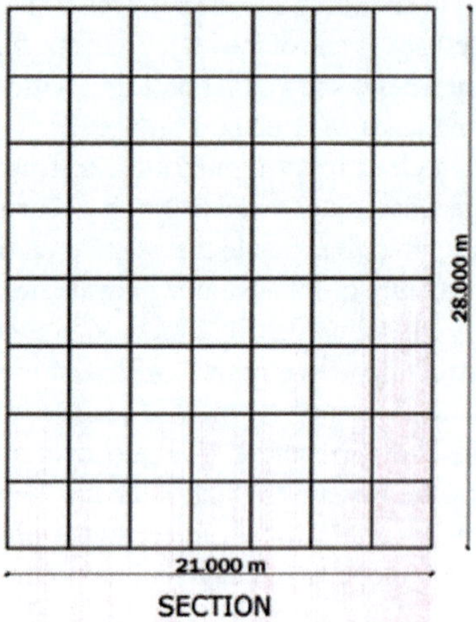


explained followed by an extended N2 method as per IS 1893-2016 and to evaluate the seismic parameters [10].

## 2 Plan

In the present study, Seven Stories of Reinforced Cement Concrete Framed Structure with Vertical Irregularity is selected as shown in Figs. 1, 2 and 3. Plan area 24 m × 21 m structure has been adopted.

Fig. 3 Elevation

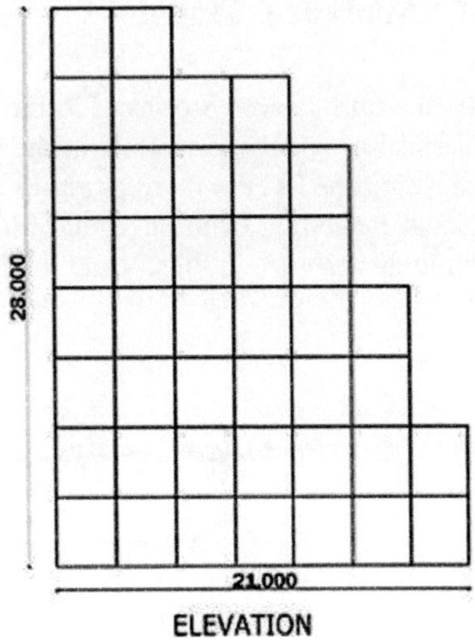


Length of each bay = 3 m
Width of each bay = 3 m
Total Length = 24 m
Total width = 21 m
Total height = 28 m
No of X-direction bays = 7
No of Y-direction bays = 8.

## 3 Extended N2 Method Procedure

- Create the model in the SAP 2000.
- The material properties section should be assigned.
- Assign the loads such dead load as 100%, live load as 50% and earthquake load as 100% in both x and y direction.
- Set the all-load cases as not run and model load case as run.
- After run the analysis, time period (T) and fundamental frequencies for every mode are determined.
- And the displacement and storey drift for each floor determined by using equivalent static analysis method.
- The same procedure followed for creating model for push over analysis
- In pushover analysis, the assign gravity and horizontal load cases.
- In extended N2 method, the correction factor applied to N2 method.
- Set all the load cases are not run and pushover load as run.
- After run the analysis, displacement and storey drift are determined.

# 4 Analytical Model

In this study, Seven Stories of Reinforced Cement Concrete Framed Structure with Vertical Irregularity is selected. The structure is located on the medium soil type in seismic zone III ($z = 0.16$) as per IS 1893-2016. Using SAP 2000 software, Dimensional modelling of the irregular buildings is generated. 3D view of the irregular building is shown in Figs. 4 and 5. The description of the structure is given in Table 1.

## *4.1 Elastic Modal Analysis*

The elastic modal analysis is done by considering 100% dead load and 50% live load taken for analysis as per IS 1893-2016. The N2 method is assumed to vibrate predominantly in the fundamental modes. In the SAP 2000 software. This method is done, and the values are accompanied. Storey masses [M] = [304.92, 283.34, 243.70, 223.41, 213.47, 198.94, 100.74, 99.56] KN. The fundamental time period T = 1.059 s. The fundamental frequency = 0.944 Cps, Circular frequency = 5.93 rad/s and the fundamental mode shape $\{\Phi\} = \{1, 0.76, 0.72, 0.63, 0.62, 0.57, 0.53, 0.46\}$.

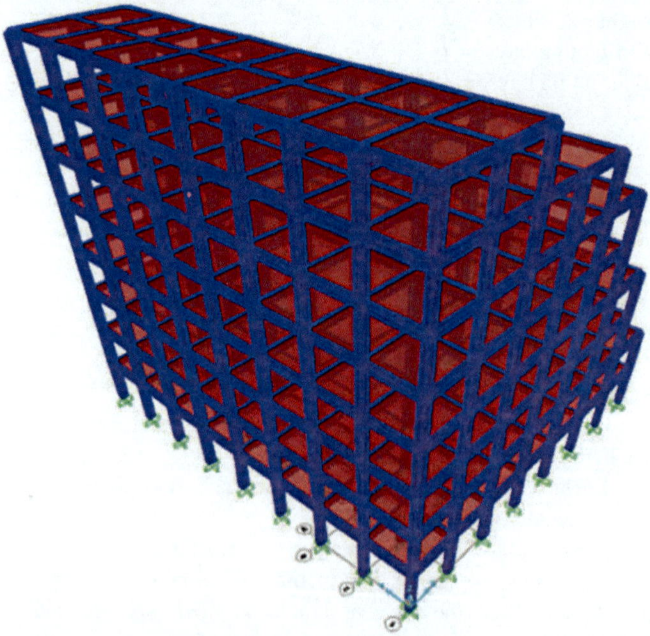

**Fig. 4.** 3D view of irregular building

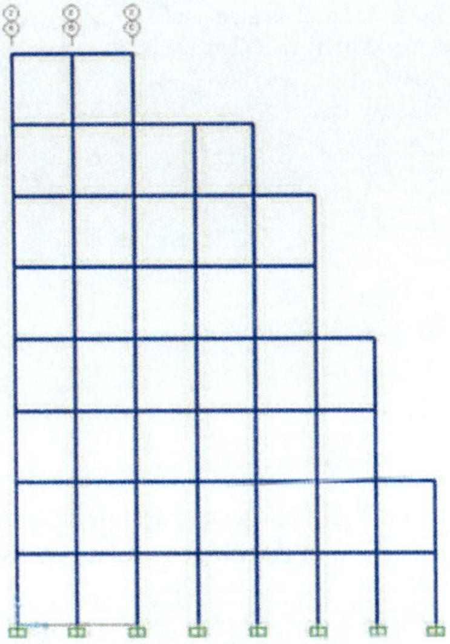

Fig. 5 XZ view with support

Table 1 Description of the structure

| Type of structure | Framed structure |
|---|---|
| Building material | RCC |
| No of stories | G + 7 |
| Beam size | 0.3 m × 0.4 m |
| Column size | 0.3 m × 0.45 m |
| Slab thickness | 125 mm |
| Floor load | 3 $kN/m^2$ |
| Floor finish | 1.5 $kN/m^2$ |
| Total load | 4.5 $kN/m^2$ |
| Compressive strength (fck) | M20 |
| Yield strength (fy) | Fe415 |
| % of imposed load to be considered for self-weight | 50% |

The Displacement and Storey drifts are obtained in the Elastic Modal Analysis are presented in Table 2.

### 4.2 Non-linear Static Analysis of MDOF System

Non-linear static analysis of MDOF system is one of the seismic evaluation methods. In this analysis method, load cases are assigned as gravity and pushover cases in each

**Table 2** Displacement and storey drift by modal analysis

| Storey level | Displacement (m) | Storey drift (m) |
|---|---|---|
| 8 | 0.0329 | 0.0073 |
| 7 | 0.0256 | 0.0046 |
| 6 | 0.021 | 0.0024 |
| 5 | 0.0186 | 0.0054 |
| 4 | 0.0132 | 0.0032 |
| 3 | 0.0101 | 0.0046 |
| 2 | 0.0054 | 0.0029 |
| 1 | 0.0025 | 0.0025 |
| Base | 0 | 0 |

X and Y directions and assign the hinges in column and beam. After run the analysis, formation of the hinge results is obtained.

### *4.3 Transformation of MDOF System to SDOF System*

Equivalent mass of SDOF (m*) = 132,709.305 kg.

$\sum m^* \Phi^2 = 89578.821$ kg. From Eq. 1, transformation factor is

Transformation factor $\Gamma = m^*$

$$\sum m * \Phi^2 = 1.48 \tag{1}$$

### *4.4 Response Spectrum Analysis of SDOF System*

Response spectrum analysis is the linear dynamic analysis method is done by considering the seismic zone factor is 0.16 in seismic zone III, and soil type is considered as medium type soil II. As per IS 1893-2016, Importance Factor, I = 1 and Response Reduction Factor, R = 5 are taken. From demand curve, maximum acceleration spectrum is 0.16 m/s$^2$ and maximum time period is 2.3 s. The Displacements and Storey drifts are accompanied from simplified N2 Method are presented in Table 3.

## 5 Extended N2 Method

The advancement procedure of N2 method is called as extended N2 method. The extended N2 method is used in higher mode. The application of extended N2 method

**Table 3** Displacement and storey drift by N2 method

| Storey level | Displacement (m) | Storey drift (m) |
|---|---|---|
| 8 | 0.0193 | 0.0008 |
| 7 | 0.0185 | 0.0013 |
| 6 | 0.0172 | 0.0017 |
| 5 | 0.0155 | 0.0024 |
| 4 | 0.0131 | 0.0029 |
| 3 | 0.0102 | 0.0035 |
| 2 | 0.0067 | 0.0037 |
| 1 | 0.003 | 0.003 |
| Base | 0 | 0 |

**Table 4** Correction factor

| Storey level | Correction factor (m) |
|---|---|
| 8 | 1.7046 |
| 7 | 1.3837 |
| 6 | 1.2209 |
| 5 | 1.2 |
| 4 | 1.4198 |
| 3 | 0.9901 |
| 2 | 0.8059 |
| 1 | 0.8333 |
| Base | 0 |

is suitable for vertical irregularity building. The extended N2 method is carried out the correction factor to N2 method. The correction factor is defined as the ratio of the displacement of modal analysis to the displacement of non-linear static analysis. The correction factor is presented in Table 4, and the extended N2 method are presented Table 5.

## 6 Non-linear Dynamic Time History Analysis

Non-linear dynamic time history method is done by considering El Centeno earthquake database. The time history method is used to show the real behaviour of the structure all through an earthquake. The displacements and the storey drifts obtained from non-linear dynamic Time History Analysis are presented in Table 6.

**Table 5** Displacement and storey drift by Extended N2 method

| Storey level | Displacement (m) | Storey drift (m) |
|---|---|---|
| 8 | 0.0328 | 0.0008 |
| 7 | 0.0255 | 0.0013 |
| 6 | 0.0209 | 0.0017 |
| 5 | 0.0186 | 0.0024 |
| 4 | 0.0185 | 0.0029 |
| 3 | 0.01 | 0.0035 |
| 2 | 0.0054 | 0.0037 |
| 1 | 0.003 | 0.003 |
| Base | 0 | 0 |

**Table 6** Displacement and storey drift by non-linear dynamic time history analysis

| Storey level | Displacement (m) | Storey drift (m) |
|---|---|---|
| 8 | 0.0154 | 0.0033 |
| 7 | 0.0121 | 0.0092 |
| 6 | 0.0029 | 0.0058 |
| 5 | 0.0087 | 0.0024 |
| 4 | 0.0063 | 0.0015 |
| 3 | 0.0048 | 0.0023 |
| 2 | 0.0025 | 0.0013 |
| 1 | 0.0012 | 0.0012 |
| Base | 0 | 0 |

## 7 Results and Discussion

G + 7 storey building with Asymmetric Setback is considered for this study. The structural behaviour of the N2 method, extended N2 method and Time History method due to displacement a storey drift is studied and the following results are obtained the SDOF system. The comparison of the Displacements and Storey drifts examined from the different methods are presented in Tables 7 and 8.

### *7.1 Comparison of Displacement*

The comparison of different seismic methods is mentioned in Table 7 and is shown in Fig. 6. Table 7 shows that displacement increases for the upper storey compared to the lower storey. N2 method having more displacement values compared to another seismic method. The extended N2 method and modal analysis method displacement shows slightly accurate values are obtained (Fig. 7).

**Table 7** Comparison of displacement

| Displacement (m) | | | | |
|---|---|---|---|---|
| Storey level | Modal analysis | N2 method | Extended N2 method | Time history analysis |
| 8 | 0.0329 | 0.0193 | 0.0328 | 0.0033 |
| 7 | 0.0256 | 0.0185 | 0.0255 | 0.0092 |
| 6 | 0.021 | 0.0172 | 0.0209 | 0.0058 |
| 5 | 0.0186 | 0.0155 | 0.0186 | 0.0024 |
| 4 | 0.0132 | 0.0131 | 0.0185 | 0.0015 |
| 3 | 0.0101 | 0.0102 | 0.01 | 0.0023 |
| 2 | 0.0054 | 0.0067 | 0.0054 | 0.0013 |
| 1 | 0.0025 | 0.003 | 0.003 | 0.0012 |
| Base | 0 | 0 | 0 | 0 |

**Table 8** Comparison of storey drift

| Storey drift (m) | | | | |
|---|---|---|---|---|
| Storey level | Modal analysis | N2 method | Extended N2 method | Time history analysis |
| 8 | 0.0073 | 0.0008 | 0.0073 | 0.0033 |
| 7 | 0.0046 | 0.0013 | 0.0046 | 0.0092 |
| 6 | 0.0024 | 0.0017 | 0.0023 | 0.0058 |
| 5 | 0.0054 | 0.0024 | 0.0001 | 0.0024 |
| 4 | 0.0032 | 0.0029 | 0.0085 | 0.0015 |
| 3 | 0.0046 | 0.0035 | 0.0046 | 0.0023 |
| 2 | 0.0029 | 0.0037 | 0.0024 | 0.0013 |
| 1 | 0.0025 | 0.003 | 0.003 | 0.0012 |
| Base | 0 | 0 | 0 | 0 |

**Fig. 6** Hinges

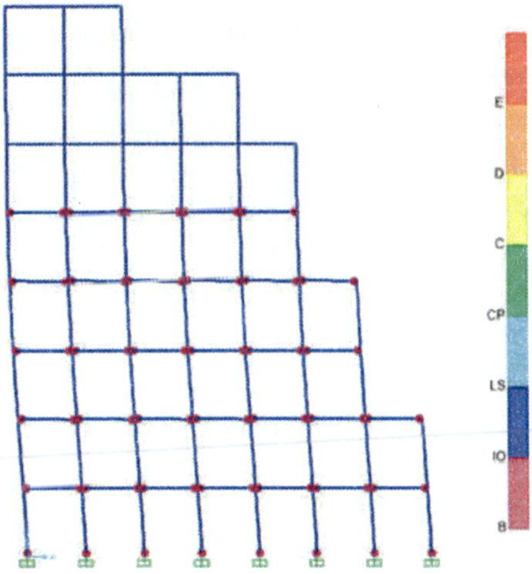

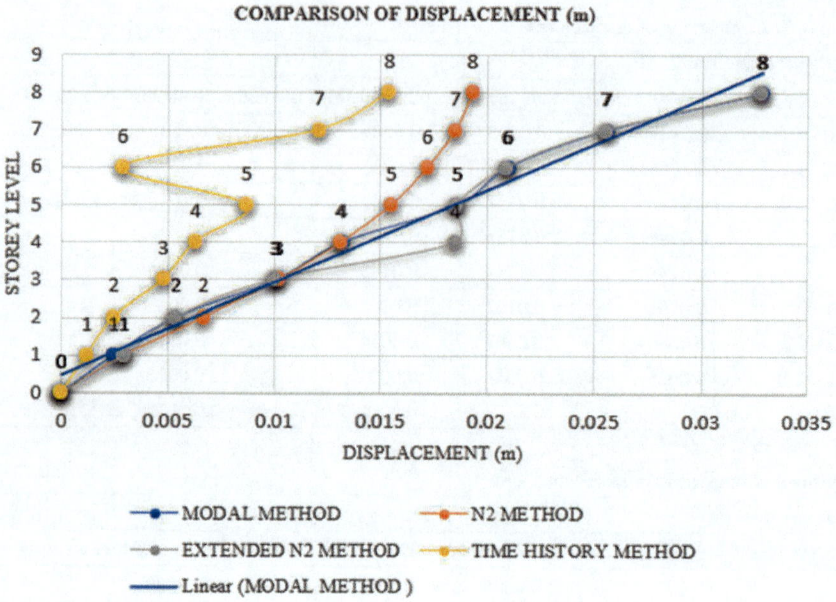


**Fig. 7** Displacement

## *7.2 Comparison of Store Drift*

The comparison of the storey drifts is mentioned in Table 8 and is shown in Fig. 8. Table 8 shows that the storey drift of each floor varies based on displacement. The storey drift of different seismic methods shows that displacement varies for each floor and storey drifts also varied. The modal and ex-tended N2 method having slightly accurate storey drift values compared to the other seismic methods. The N2 method and Extended N2 method storey drift for lower storey having similar values compared to other methods.

# 8 Conclusion

In this paper, Reinforced Cement Concrete Framed Structure with Asymmetric Setback by using the different seismic analysis methods is studied. The followings were concluded:

- From the study, it is observed that the pushover method is simple tools to determine the strength of the structures for seismic analysis.
- From the Extended N2 method (EN2), it observed that the displacements and storey drifts increase when compared to the simplified N2 method.

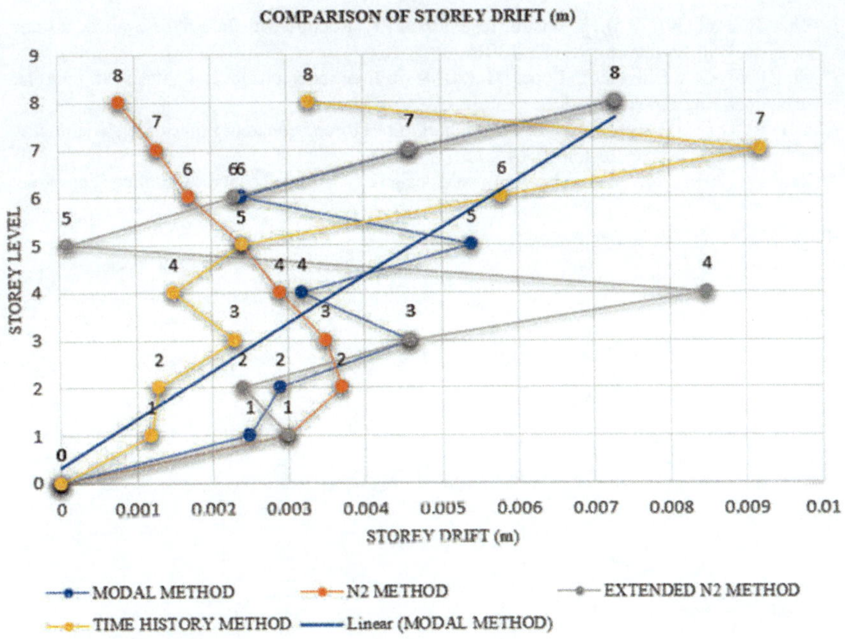


**Fig. 8** Comparison of storey drift

- The displacements and storey drifts of the extended N2 Method (EN2) results that more displacement 0.0328 m for the upper storey and less displacement 0.003 m for the lower storey.
- The extended N2 method (EN2) was found to result the same displacement and storey drift results for the upper storey when compared to the Time history analysis method (NTHAM).

## References


1. Paulo B (2020) Pushover analysis of unreinforced irregular masonry buildings: lessons from different modeling approaches. Mater Today: Proc 218:116–219
2. Prajwal TP, Imtiaz A (2020) Nonlinear analysis of irregular buildings considering the direction of seismic waves. Eng Struct 4:9828–9832
3. Khanal B (2019) Seismic elastic performance of L-shaped building frames through plan irregularities. Structures 27:22–36
4. Domenico G, Falsone G (2018) Improved response-spectrum analysis of base-isolated buildings: a substructure-based response spectrum method. Eng Struct 162:198–212
5. Ramadevi K (2018) Nonlinear seismic response of structural systems having vertical irregularities due to discontinuities in columns. Int J Innov Technol Explor Eng 90:2190–3075
6. Swathi T, Manjula S (2017) Usage of N2 approach for the overall performance of plan uneven structure. Comput Struct 8:2396–2398

7. Bhatt C, Bento K (2016) The seismic response of RC structure with the aid the equivalent static analysis. Bull Earthq Eng 9:1183–1201
8. Cimellaro GK, Giovine J, Lopez G (2014) Bidirectional pushover analysis of irregular structures. Eng Struct 4:1–13
9. Devassy D (2014) Application of N2 method to reinforced concrete frames with asymmetric setback. Int J Civ Eng Technol 5:143–154
10. Rehan A (2014) Performance based seismic design of reinforced concrete building. Eng Struct 80:2483–2493

# Assessment of Seismic Performance and Vulnerability of a 20-Storied Reinforced Concrete Moment-Resisting Frame Designed by Direct Displacement-Based Design and Force-Based Design

Sumit Ganvit, Sejal Purvang Dalal, Ronak Motiani, and Purvang Dalal

## 1 Introduction

The force-based design is used in India traditionally. It is known to be an iterative process involving trials until the required strength and serviceability requirements are met as prescribed in code. This method works efficiently when the linear elastic analysis is performed but has a limitation of addressing the inelastic behavior of the structure under seismic loads. One of the reasons for this is that the FBD estimates the response using the initial stiffness and elastic damping of the structure. This leads to miscalculation of the structural responses under seismic loads.

Over the last decade, the importance of displacement, rather than forces, has become better appreciated, and because of this, displacement-based design procedures are continuously being developed. Since the damages are related directly to displacements, therefore, the direct displacement-based design (DDBD) method for the design of new structures and evaluation and retrofitting of existing structures has attracted many professionals and researchers recently. This method includes the inelastic response of the structure, which is reflected during the whole design process especially in the calculation and distribution of the design seismic loads. In case of

S. Ganvit
Bardoli Municipal Corporation, Bardoli, Gujarat, India

S. P. Dalal (✉)
SVIT, Vasad, Gujarat, India

R. Motiani
Civil Engineering Department, Pandit Deendayal Energy University, Gandhinagar, Gujarat, India

P. Dalal
Faculty of Technology, Dharmsinh Desai University, Nadiad, Gujarat, India
e-mail: pur_dalal.ec@ddu.ac.in



S. Kolathayar and S. C. Chian (eds.), *Recent Advances in Earthquake Engineering*,
Lecture Notes in Civil Engineering 175,
https://doi.org/10.1007/978-981-16-4617-1_10

framed structures, the required design forces and moments in the beams and columns are calculated for potential hinge locations.

In this study, the force-based design of a 20-storied RC study frame was done by traditional methods currently in practice as per Indian Standards [1–4]. The displacement-based design of the study frames was done as per the suggestions of [5, 6]. The seismic performance of these frames was compared in terms of strength and deformation by performing the nonlinear static pushover analysis method. Subsequently, the vulnerability assessment was done as per the HAZUS [7] methodology based on the responses obtained from the nonlinear static pushover analysis.

## 2 The Study Frame

A 20-storied RC building with a special moment-resisting frame structure having rectangular plan, which has symmetry about both X and Y axes, was selected for the study. The detailed geometric, material, and site-specific parameters are listed in Table 1 and Fig. 1. Due to its symmetry, the building frame was analyzed by considering two different plane frames in each of the XX and YY directions. The specifications as per Indian Standards [1–4] were followed for loading, analysis, and design.

**Table 1** The details of the study frames

| Parameter | Details |
|---|---|
| Dimensions in the X direction | 8 bays at 5 m each |
| Dimensions in the Y direction | 5 bays at 5 m each |
| Story height (m) | 3.3 |
| Height of the roof from base, $h_n$ (m) | 66 |
| Thickness of slab at roof (mm) | 150 |
| Thickness of slab at all floors (mm) | 175 |
| Natural period, T ($0.075h^{0.75}$) (sec) | 1.74 |
| Soil site | Medium |
| Zone factor (Z) | 0.36 |
| Importance factor (I) | 1 |
| Yield stress of the reinforcing steel rebars and stirrups (N/mm$^2$) | 415 |
| Yield strain of the reinforcing steel rebars and stirrups | 0.002 |
| Modulus of elasticity of the reinforcing steel rebars and stirrups (N/mm$^2$) | 200,000 |
| Concrete grade | M25 |
| Imposed load at roof (kN/m$^2$) | 1.5 |
| Imposed load at roof (kN/m$^2$) | 3.5 |

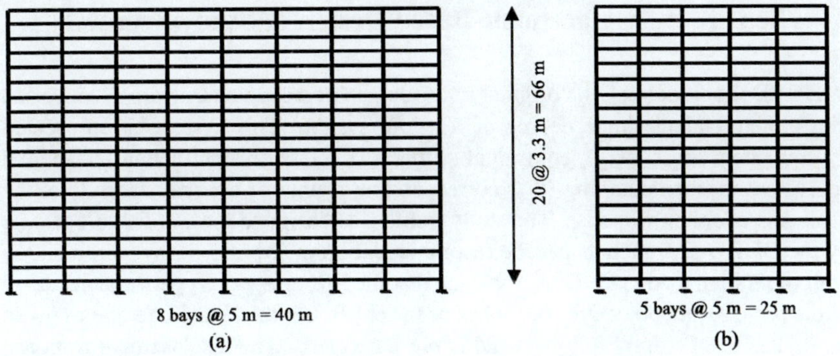


**Fig. 1** **a** Front view of XX direction frame. **b** Front view of YY direction frame

## 3 The Force-Based Design

The study frame was analyzed for seismic loading as special moment resisting frame, which has been assigned a response reduction factor ($R$) as 5 as per [2]. The design base shear $V_B$ was calculated by the response spectrum method using the design acceleration spectrum specified in [2] and then distributed along the height using the complete quadratic combination method. Undamped free vibration analysis was performed to obtain the natural periods ($T_k$) and mode shapes ($\varphi_k$) of those of its $N_m$ modes of oscillation [$k \in (1, N_m)$] that need to be considered as per [2]. The ETABS software [8] was used to find these seismic forces. The seismic parameters of the study frames are presented in Table 2.

The study frame was then analyzed in ETABS for lateral (seismic) and vertical loads as per the [2] load combinations while also considering the vertical earthquake effects and secondary (P-delta) effects. The values of maximum forces and moments, thus, obtained were used for the limit state design of beams and columns as per [3, 4].

**Table 2** The seismic parameters of the study frames for the FBD method

| Parameter | XX frame | YY frame |
|---|---|---|
| Seismic weight of roof floor (kN) | 480 | 300 |
| Seismic weight of all other floors (kN) | 842 | 534 |
| Total seismic weight, W (kN) | 16,525 | 10,486 |
| Design base shear, $V_B$ (kN) | 465.8 | 295.6 |
| $V_B$/W (percent of seismic weight) | 2.8 | 2.8 |

# 4 The Direct Displacement-Based Design of Study Frames

In DDBD methodology, the original multi-degree of freedom structure is converted into the equivalent single degree of freedom system. This system is represented by equivalent mass ($m_{eff}$), equivalent stiffness ($k_{eff}$), equivalent height ($H_{eff}$), and equivalent viscous damping ($\xi_{eq}$) corresponding to that of the real structure at the peak displacement response. The study frame was designed for a drift limit of 0.02 by the DDBD approach as per the recommendation of [5] after its synchronization with the Indian Code [2]. Thus, it means that the life safety (LS) performance level of the basic safety objective (BSO) has been set [9]. For this, the procedure as given by [6] has been strictly followed and is presented next. The displacement response spectra for the respective equivalent viscous damping required in this study have been procured from the acceleration design spectrum of [2] for the medium soil site as per procedure given in [6]. The design base shear is calculated as Eq. (1). It is distributed along the height and then further analysis and calculation of design loads and moments in individual structural members is done as suggested by [5]. The whole analysis process can be done manually and requires no software. The design of structural members is done the same as the FBD method as per [3, 4]. The detailed seismic parameters thus calculated for the DDBD method are shown in Table 3.

$$V_B = k_{eff}\Delta_d \tag{1}$$

The critical story displacement $\Delta_c$ is equal to the height of lower story times the drift limit. For drift limit of 0.02, the $\Delta_c$ works out to be 0.066 m in both the XX

**Table 3** The seismic parameters of the study frames for the DDBD method

| Parameter | XX frame | YY frame |
|---|---|---|
| Seismic mass of roof floor (tons) | 53.52 | 34.4 |
| Seismic mass of all other floors (tons) | 85.84 | 54.45 |
| Total seismic mass, M (tons) | 1684.5 | 1068 |
| Critical normalized mode shape, $\delta_c$ | 0.066 | 0.066 |
| Drift reduction factor, $\omega_\theta$ | 0.93 | 0.93 |
| Design displacement, $\Delta_d$ (m) | 0.66 | 0.66 |
| Yield displacement, $\Delta_y$ (m) | 0.41 | 0.43 |
| Displacement ductility, $\mu$ | 1.6 | 1.53 |
| Equivalent viscous damping, $\xi_{eq}$ | 0.117 | 0.112 |
| Effective height, $H_{eff}$ (m) | 43.2 | 43.2 |
| Effective mass, $m_{eff}$ (t) | 1357 | 861.02 |
| Effective stiffness, $k_{eff}$ | 954 | 615 |
| Effective time period, $T_{eff}$ (sec) | 7.49 | 7.43 |
| Design base shear, $V_B$ (kN) | 628.9 | 405.7 |
| $V_B$/W (percent of seismic weight) | 3.8 | 3.8 |

frame and YY frame because the height of the lower story is 3.3 m. The normalized inelastic mode shape $\delta_i$ (i.e. the mode shape of the inelastic system) of the frame is defined in [5] and should be obtained as Eq. (2)

$$\delta_i = \frac{4}{3}.\left(\frac{h_i}{h_n}\right).\left(1 - \frac{h_i}{4h_n}\right) \tag{2}$$

where, $h_i$ (and $h_n$) = height of level i (and roof) measured from base.

The $\delta_i$ as defined in Eq. (2) is scaled with respect to the critical story displacement $\Delta_c$ and to the corresponding inelastic mode shape at the critical story level $\delta_c$ by using Eq. (3). The lower story is the critical story that implies that $\delta_c$ is equal to the $\delta_i$ of lowest story. The design story displacement $\Delta_i$ of the individual masses is calculated as:

$$\Delta_i = \omega_\theta.\delta_i.\left(\frac{\Delta_c}{\delta_c}\right) \tag{3}$$

where $\omega_\theta$ (Eq. 4) is the drift reduction factor to consider the higher mode effect as follows

$$\omega_\theta = 1.15 - 0.0034h_n \leq 1.0 \tag{4}$$

The design displacement, $\Delta_d$ (Eq. 5), is given by the following equation as per [5]

$$\Delta_d = \frac{\sum_{i=1}^{n}\left(m_i\Delta_i^2\right)}{\sum_{i=1}^{n}\left(m_i\Delta_i\right)} \tag{5}$$

where $m_i$ is the mass at level $i$. The design displacement ductility $\mu$ of the frames is then calculated as the standard relationship.

$$\mu = \frac{\Delta_d}{\Delta_y} \tag{6}$$

The yield displacement $\Delta_y$ (Eq. 7) in the above Eq. (6) depends on the yield drift of inner bay $\theta_{y1}$ and outer bays $\theta_{y2}$ (Eq. 9) as well as the equivalent height $H_{eff}$ (Eq. 8) and is obtained by the standard procedure as per [5] for beams with constant depths as following:

$$\Delta_y = \left(\frac{2\theta_{y1} + \theta_{y2}}{3}\right) * H_{eff} \tag{7}$$

$$H_{eff} = \frac{\sum m_i\Delta_i h_i}{\sum m_i\Delta_i} \tag{8}$$

$$\theta_{y1}(or\ \theta_{y2}) = 0.5 * \varepsilon_y \frac{Length\ of\ the\ beam\ of\ bay\ 1\ (or\ bay\ 2)}{Depth\ of\ the\ beam\ of\ bay 2\ (or\ bay\ 2)} \tag{9}$$

$\varepsilon_y$ is the yield strain of the material (0.002). Initially, the depth of the beams is assumed as 600 mm, and length of beams can be seen from Fig. 1. Based on the relationship as presented in Fig. 2 by [5], the equivalent viscous damping $\xi_{eq}$ is worked out. Figure 3 presents the displacement response spectrum for the equivalent viscous damping ($\xi_{eq}$) values, which has been procured from the elastic design spectrum of [2] for both XX and YY frames as per the procedure given in [6]. The values of $T_{eff}$ for the corresponding $\Delta_d$ are marked in Fig. 3.

The effective mass and stiffness are calculated as in Eqs. (10) and (11), respectively:

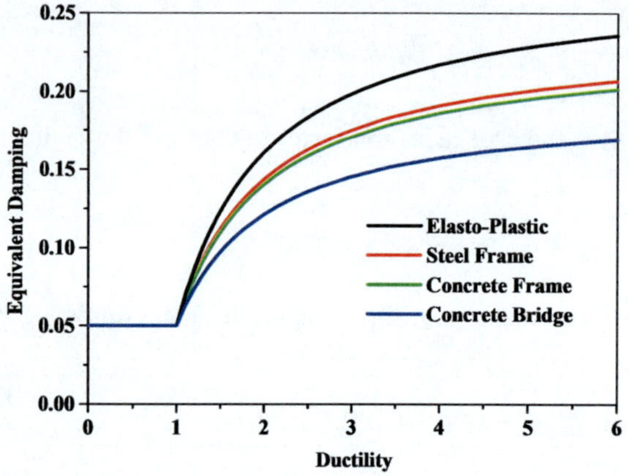


**Fig. 2** Equivalent damping versus ductility

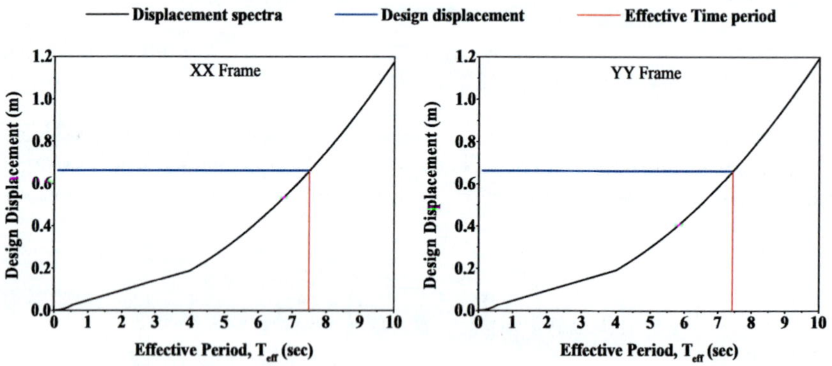


**Fig. 3** The displacement response spectra for the study frames

$$k_{\text{eff}} = \frac{4\pi^2 m_{eff}}{T_{eff}^2} \tag{10}$$

$$m_{eff} = \frac{\sum_{i=1}^{n}(m_i \Delta_i)}{\Delta_d} \tag{11}$$

The lateral load at a particular story level, $Q_i$ (Eq. 12) is calculated by distributing the design base shear as [5]:

$$Q_i = Q_{top} + 0.9V_B\left(\frac{(m_i \Delta_i)}{\sum_{i=1}^{n}(m_i \Delta_i)}\right) \tag{12}$$

where, $Q_{top} = 0.1V_B$ at roof level and $Q_{top} = 0$, at all other story level.

## 5 Assessment of Seismic Performance

Nonlinear static analysis of study frames was done to check and compare the seismic performance in terms of strength and deformation using the ETABS as per [6] pertaining to [9–11]. The frames were modeled in ETABS as suggested in [12]. Lumped plasticity models with the hinge properties by considering only the flexural inelastic mechanisms as defined in [10] are used. Conforming reinforcements have been used to assign the plastic rotations for the beams and columns. The performance point (or target displacement) is obtained considering bilinear force–displacement relationship by the coefficient method. The first yield has been taken as the yield displacement. The three performance levels of basic safety objective, namely, immediate occupancy (IO), life safety (LS), and collapse prevention (CP) for which the permissible target drift is 0.01, 0.02, and 0.04, respectively [9] are shown in the Pushover curves (Fig. 4) of the study frames as obtained from the software.

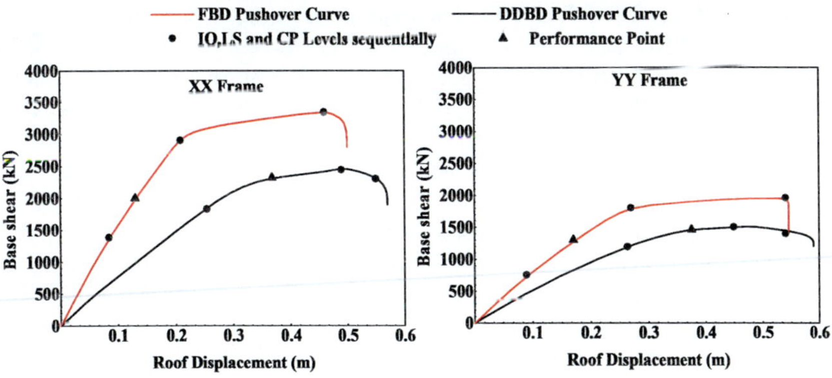


Fig. 4 Pushover curves of the study frames

**Table 4** Strength and ductility of the study frames

| Parameter | XX frames | | YY frames | |
|---|---|---|---|---|
| | FBD | DDBD | FBD | DDBD |
| Limiting value of response, $V_o$ (kN) | 2001 | 2320 | 1300 | 1458 |
| Design base shear, $V_B$ (kN) | 489 | 629 | 311 | 406 |
| Overstrength factor, $R_s$ ($V_o/V_B$) | 4.09 | 3.68 | 4.18 | 3.59 |
| Yield displacement, $d_y$ (m) | 0.09 | 0.25 | 0.09 | 0.26 |
| Performance point, $d_t$ (m) | 0.13 | 0.36 | 0.17 | 0.37 |
| Ductility factor, $R_\mu$ ($d_t/d_y$) | 1.44 | 1.44 | 1.89 | 1.42 |
| Response reduction factor, R ($R_\mu * R_s$) | 5.9 | 5.29 | 7.9 | 5.09 |
| Ultimate displacement, $d_u$ (m) | 0.46 | 0.57 | 0.54 | 0.55 |

Figure 4 shows that the performance point of all the frames is achieved between IO and LS levels. Thus, the life safety performance level is met for all the frames. The FBD frames have more reported higher strength and stiffness whereas DDBD frames are more flexible. The first yield is formed in all the frames at a base shear value higher than the corresponding design base shear. The response reduction factor (Eq. 13) that represents the reserved strength and ductility in a structure was calculated at performance points and is presented in Table 4 as per the procedure given in [11].

$$R = R_s R_\mu R_R \tag{13}$$

The overstrength factor, $R_S$ is calculated as given in [11]. As the time period of the study frame is more than 1 s, the period dependent ductility factor, $R_\mu$ is equal to the ductility factor $\mu$ ($d_t/d_y$). The redundancy factor, $R_R$ is taken as unity [11]. It is evident from Table 4 that in DDBD frames, the overstrength factor is lesser than the FBD frames even though they were designed for a higher value of design base shear compared with the FBD frames. It means that the DDBD method gives better section utilization compared with FBD method, which ultimately leads to economy. The lesser values of $R$ of the DDBD frames attribute an economical design.

As per [2], FBD frames are supposed to satisfy the permissible drift criteria of 0.004 at design loads. It is worth to mention here that it already reduces the lateral loads by applying the ($R$) value as 5 in the beginning of the calculation of seismic forces. Thus, indirectly, it can be pointed out that the design drift, in this case, is 5 times 0.004, which is 0.02 [6]. The maximum story drift in the FBD frames was higher than the permissible value of 0.004 particularly at lower stories as seen in Fig. 5. As mentioned earlier, the DDBD frames are directly designed for a story drift of 0.02 and hence in the absence of code guidelines for this method, the permissible drift may be taken as 0.02. The story drift in the DDBD frames was below its design drift of 0.02.

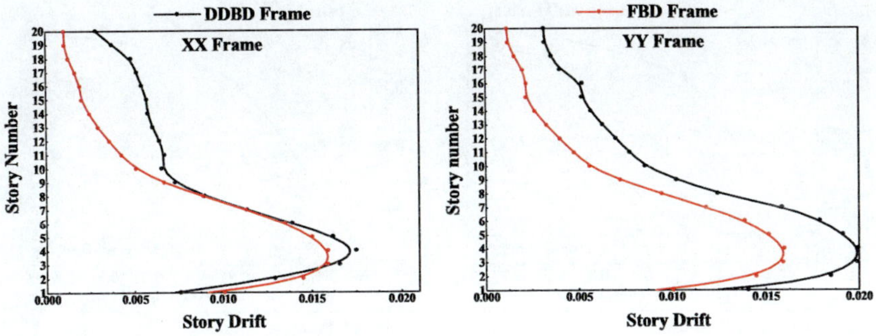


**Fig. 5** Story drift of the study frames

## 6 Assessment of Vulnerability

The vulnerability assessment of a structure can be evaluated by developing fragility curves. These curves show the probability of the structure, which exceeds the particular performance limit of the given damage state during strong ground motions. These curves can be developed for different parameters like spectral displacement (Sd), spectral acceleration (Sa), peak ground acceleration (PGA), interstory drift ratio (IDR), etc. The fragility curves in this study give the probability that the expected global damage of the study frames exceeds a given damage state as a function of spectral displacement (Sd). The damage states defined as per [7] are slight damage, moderate damage, extensive damage, and complete damage. The damage states thresholds as per [13] were used to define the damage states in the study (Table 5). The fragility curves are developed based on the pushover analysis results of Table 4 and are presented in Figs. 6 and 7.

For each damage state, the corresponding fragility curve is defined by plotting the probability of exceedance $P[d \geq d_{si}]$ as ordinate and the spectral displacement (Sd) (of the performance point) as abscissa. The fragility curves follow a lognormal probability distribution and are defined in terms of probability density function as in Eq. (14).

$$P\left[\frac{ds_i}{Sd}\right] = \phi\left[\frac{1}{\beta_{dsi}}\ln\left(\frac{Sd}{\overline{Sd}_{dsi}}\right)\right] \quad (14)$$

**Table 5** The damage states thresholds

| Sr. No. | Damage state | Damage state thresholds ($\overline{Sd}_{si}$) | $\beta_{dsi}$ |
|---|---|---|---|
| 1 | $Sd_1$ (slight) | $0.7\ d_y$ | 0.63 |
| 2 | $Sd_2$ (moderate) | $d_y$ | 0.63 |
| 3 | $Sd_3$ (severe/extensive) | $0.75d_y + 0.25\ d_u$ | 0.63 |
| 4 | $Sd_4$ (complete) | $d_u$ | 0.69 |

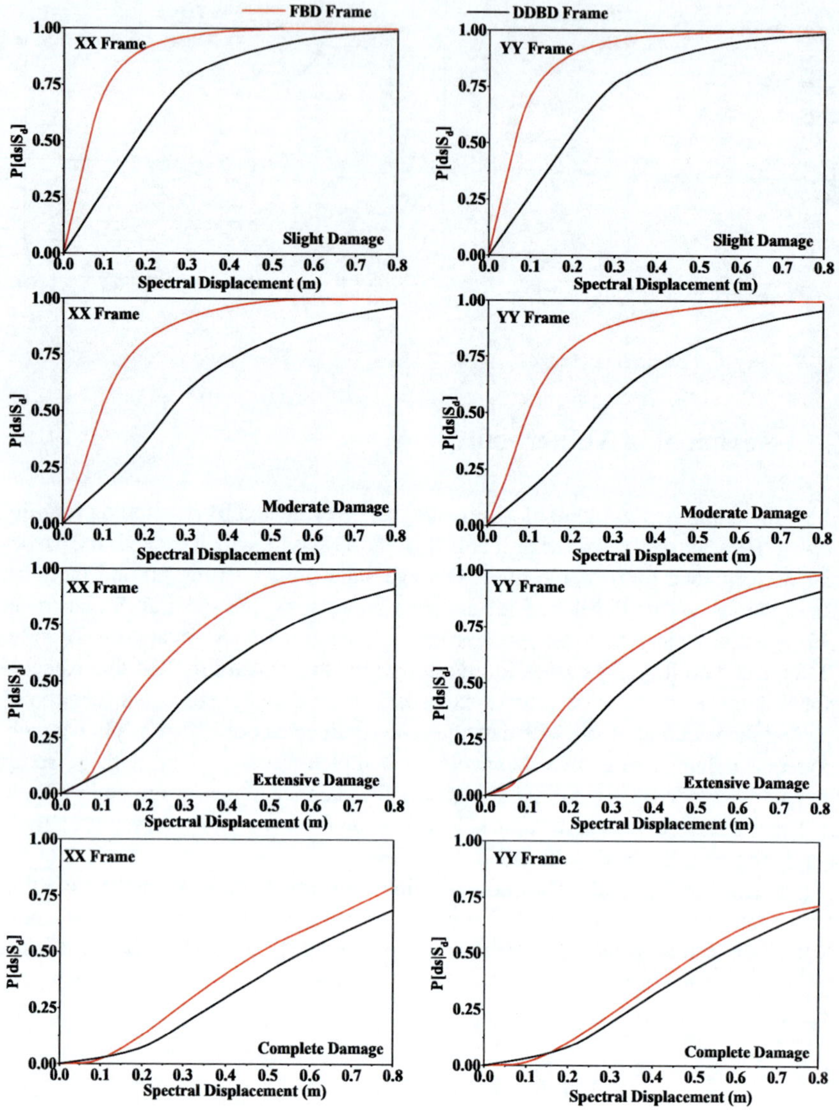


**Fig. 6** Comparison of probability of exceedance of slight, moderate, extensive, and complete damage of study frames

where

$d_{si}$ = given damage state.

$\overline{Sd}_{si}$ = threshold spectral displacement.

$\beta_{dsi}$ = standard deviation of the natural logarithm of the displacement threshold as per [7].

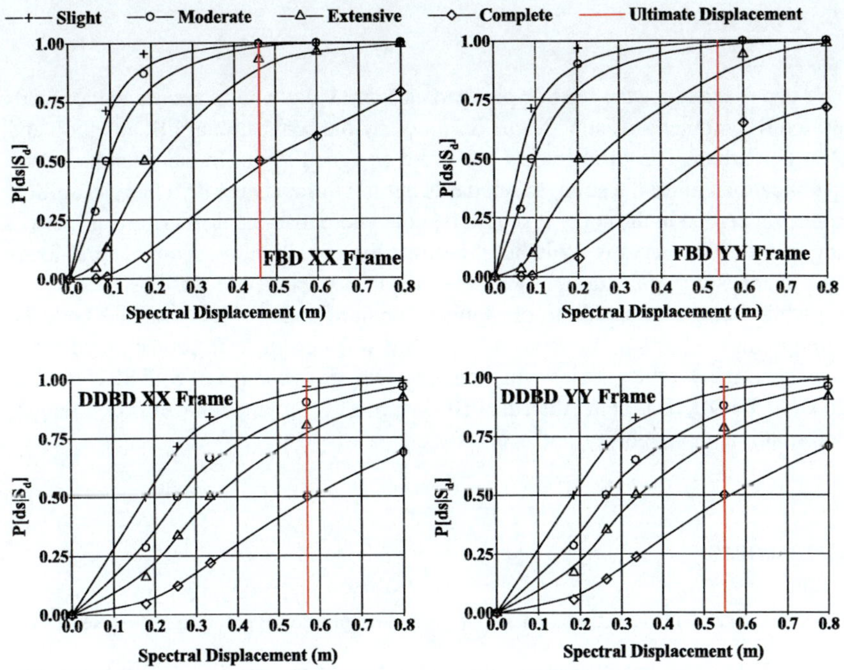


**Fig. 7** Fragility curves of the study frames

$\Phi$ = standard normal cumulative distribution function.

It is seen in Fig. 6 that the probability of exceedance for all damage states is higher in case of FBD frames compared with DDBD frames. As seen from the fragility curves of Fig. 7, the FBD XX frames are 99%, 98%, 93%, and 50% vulnerable to slight, moderate, extensive, and complete damage, respectively, at the time it reaches its maximum displacement just before failure. For the FBD YY frames, there are 99%, 98%, 91%, and 52% chances of slight, moderate, extensive, and complete damage, respectively, at ultimate displacement. The DDBD XX frames are 96%, 89%, 80%, and 50% vulnerable to slight, moderate, extensive, and complete damage, respectively, at maximum ultimate displacement whereas in the DDBD YY frames, these values are 95%, 87%, 78%, and 50%.

If we emphasize the permissible lateral displacement limit of 0.264 m, (0.004$h$) as per [2], it is observed that the FBD XX frame is 99%, 96%, 73%, and 21% and FBD YY frame is 98%, 95%, 76%, and 14% vulnerable to slight, moderate, extensive, and complete damage, respectively, when it reaches this limit. On the other hand, DDBD XX frame has 73%, 52%, 35%, and 13% and DDBD YY frame has 71%, 50%, 35%, and 14% probabilities of slight, moderate, extensive, and complete damage, respectively, at lateral displacement of 0.264 m.

## 7 Conclusion

This paper examines the seismic performance and vulnerability assessment of a 20-storied RC moment-resisting frame designed by the conventional FBD method and DDBD method. The FBD frames that were designed as per Indian Standard code specifications fulfill the strength standards but fail to satisfy the drift (serviceability) criteria specified in the code. It should be noted here that, the drift criteria as per [2] are on a very conservative side but it permits the reduction of seismic lateral loads by a factor 5 during the design process. The DDBD frame that was designed as per [5] while being compatible to the Indian Standard code [6] has satisfied both the strength and deformation criteria for which it was designed. In probabilistic terms also, the DDBD frames are less vulnerable to damage as compared with FBD frames. Overall, the DDBD is better than the FBD in terms of computational efforts, strength, deformation, vulnerability, and economy.

## References

1. BIS:875 (2003) Indian standard design loads-code of practice. Bur. Indian Stand., New Delhi, Parts I-II
2. BIS:1893 (2016) Indian standard criteria for earthquake resistant design of structures. Bur. Indian Stand., New Delhi, Part I
3. BIS:456 (2000) Indian standard plain and reinforced concrete—code of practice. Bur. Indian Stand., New Delhi
4. BIS:13920 (2003) Indian standard ductile detailing of reinforced concrete structures subjected to seismic forces–code of practice. Bur. Indian Stand. New Delhi
5. Pristley M, Calvi MJN, Lowalsky GM (2007) Displacement based seismic design of structures. IUSS Press, Pavia
6. Qammer SS, Dalal SP, Dalal P (2019) Displacement-based design of RC frames using design spectra of indian code and its seismic performance evaluation. J Inst Eng Ser A 100:367–379. https://doi.org/10.1007/s40030-019-00373-z
7. HAZUS (2003) HAZUS-MH MR4 multi-hazard loss estimation methodology—earthquake model: technical manual. Department of Homeland Security, Fed. Emerg. Manag. Agency
8. ETABS Version 17 (2016) Structural and earthquake engineering software. Computers and Structures, America
9. FEMA 356 (2000) Prestandard and commentary for the seismic rehabilitation of buildings. Fed. Emerg. Manag. Agency
10. ASCE/SEI 41-17 (2017) Seismic evaluation and retrofit of existing buildings. https://doi.org/10.1061/9780784414859
11. ATC 19 (1995) Structural response modification factors. Appl Technol Counc 1–69
12. Dalal SP, Dalal P (2021) Strength deformation and fragility assessment of reinforced concrete moment resisting frame designed by force based design and the performance based plastic design method for seismic loads. Structures 29:1154–1164. https://doi.org/10.1016/j.istruc.2020.11.029
13. Barbat AH, Pujades LG, Lantada N (2008) Seismic damage evaluation in urban areas using the capacity spectrum method: application to Barcelona. Soil Dyn Earthq Eng 28:851–865. https://doi.org/10.1016/j.soildyn.2007.10.006

# Comparative Study on Seismic Behaviour of G+41 Storey Frame with X, K and V Bracings

**Rosemary K. Thomas, Swetha Elizabeth Philip, and V. Vasugi**

## 1 Introduction

Construction materials, machines, technology and knowledge as well as the lifestyle of people are improving and so is the need of the hour to construct high rise buildings. However, construction of tall structures has its own challenges and problems. Vertical loading increases as height of the building increases. Also there is a large effect from horizontal wind load on the building. Unexpected seismic load also can affect the building stability [1, 2]. Structures designed to resist earthquakes must have excellent stiffness and strength to control deflection and hence prevent any possible damage or collapse [3].

One of the main challenges while designing tall structures is its ability to absorb the horizontal forces and to transfer the resulting moment into the foundation. Bracings play an important role in seismic resistance of structures [2, 4–6]. They contribute much to the resistance of lateral loads acting on high rise buildings. Seismic analysis of these buildings are mainly done to obtain various parameters like storey shear, storey displacement, base shear, joint displacement and so on, that defines the seismic behaviour of buildings. In this study we analysed braced steel frame with concrete slab for base shear and storey displacement by using equivalent static and response spectrum method of analysis.

R. K. Thomas · S. E. Philip
Vellore Institute of Technology, Chennai, India

V. Vasugi (✉)
School of Civil Engineering, Vellore Institute of Technology, Chennai, India
e-mail: vasugi.v@vit.ac.in



S. Kolathayar and S. C. Chian (eds.), *Recent Advances in Earthquake Engineering*, Lecture Notes in Civil Engineering 175,
https://doi.org/10.1007/978-981-16-4617-1_11

## 2 Bracing System

The seismic performance depends upon the lateral force resisting system provided in the structures. Incorporating bracings into the system can give high seismic resistance to the structure. As per studies, braced structures have less drift compared to unbraced structures which shows that the performance of structures can be increased by providing different bracings [7]. On the basis of the reaction mechanism or structural behaviour for resisting the lateral loads, there are various structural systems for steel structures. There are different arrangements in bracing a structure in which the mainly used as well as the conventional method is placing the braces in a vertical line, in story height and bay width dimensions. The other arrangements have been found by tests and optimizations. Researches are still going on in the field [8].

Braced frames have excellent stiffness due to limited ductility. There are several ways of providing braces to increase the seismic resistance of buildings [6]. They are mainly classified as:

1. Vertical bracing system: Bracing in vertical planes (between lines to ground level and provide lateral stability of columns) provides load paths to transfer horizontal forces.
2. Horizontal bracing system: The bracing at each floor provides load paths for the transference of horizontal forces to the planes of vertical bracing. Horizontal bracing is needed at each floor level; however, the floor system itself may provide sufficient resistance.

Different types of bracings used in buildings are given below:

- Single diagonal bracing
- X bracing or cross bracing
- K Bracing
- V Bracing
- Inverted V Bracing
- Eccentric bracing

Bracings are usually provided all throughout the height of the building and it forms a vertical truss structure. These structures hence increase the strength of the building and hence provide resistance to the effect of lateral loads like wind and seismic loads and this in turn increases the stability of the whole building [9]. The performance of bracings vary according to its type and load acting on them.

Steel bracings are also used in retrofitting of RC and steel structures affected by earthquake. The ease of construction and the relatively low cost make steel bracings more effective in comparison to other conventional upgrading techniques such as adding concrete or masonry shear walls or base isolation systems. Adding bracings enhances the global capacity of the buildings with respect to strength, deformation and ductility compared to the case with no bracings [10]. In case of high rise buildings, stiffness plays an important role. Moment resisting frames and braced frames have been commonly used as lateral load resisting structural elements in steel buildings.

Moment resisting frames provide ductility through yielding, but due to their flexibility they do not satisfy the criteria for stiffness. However, excellent stiffness is provided by concentric braced frames due to their limited ductility [11].

A braced frame is an earthquake-resistant design system of steel-frame structures, and due to its high rigidity, strength and cost effectiveness, it is widely used in areas where frequent earthquakes occur [12]. Braces of concentrically braced frames (CBFs) are subjected to deformations caused by repeated tension and compression after buckling, and through this behaviour, seismic energy is dissipated. Sufficient stiffness and strength should be there for structures constructed in seismically active regions to minimize deflections. In an economically effective design, we cannot permit the structure to remain in elastic region under severe earthquakes. The inherent damping of yielding structural elements provides ductility against sudden failure, and this can be utilized to lower the strength requirement, leading to a more economical design. Always it is desirable to design a structural system that combines stiffness and ductility, in cost effective manner [3].

## 3 Problem Statement

G+41 frame is chosen for the work. It is a steel frame with concrete slab. Three building models are considered with X, K and V bracings at right end, left end and mid portion of the building. Both response spectrum and equivalent static methods are taken into account for the seismic analyses.

The details of the structure are given below:

- G+41 steel frame with concrete slab is considered for the work.
- Number of bays = 9 × 7 (5 m each)
- Plan dimension = 45 × 35 m
- Height of storey = 3.5 m, height of elevated portion above lift and stair area = 2.5 m
- Thickness of slab = 150 mm
- Column ISMB550
- Beam ISMB500
- Zone V
- Response Reduction Factor = 4
- Damping = 5%
- Soil type = Medium soil

As per IS 1893, the following combinations were used:

- 1.7 (D.L + L.L)
- 1.7 (D.L + EX
- 1.7 (D.L + EZ)
- 1.7 (D.L – EX)
- 1.7 (D.L – EZ)

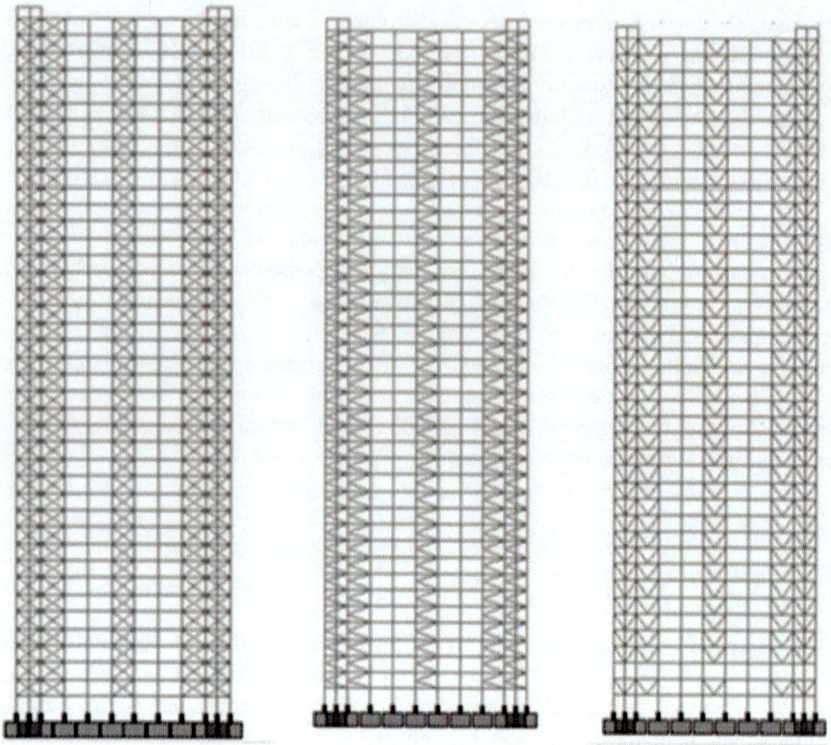

Fig. 1 X, K and V braced frames

- 1.3 (D.L + L.L + EX)
- 1.3 (D.L + L.L + EZ)
- 1.3 (D.L + L.L – EX)
- 1.3 (D.L + L.L – EZ)

Modelling and analysis of structures were done in STAAD pro V8i software. The building models with X, K and V bracing are shown in the following Fig. 1.

## 4 Results and Discussions

All the structures were modelled and analysed in Staad Pro V8i software by equivalent static and response spectrum methods. Following graphs represent the comparison of seismic behaviour of different bracing systems after equivalent static and response spectrum analysis. Figures 2 and 3 represent the storey displacement for frames with the three bracing systems. Storey displacement increased from storey 1–42 for all three cases. The highest displacement was for X bracing in case of both equivalent

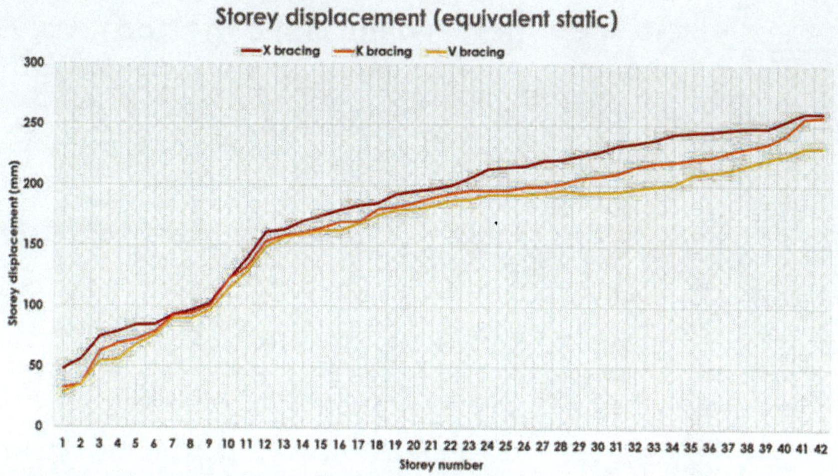


**Fig. 2** Storey displacement after equivalent static analysis

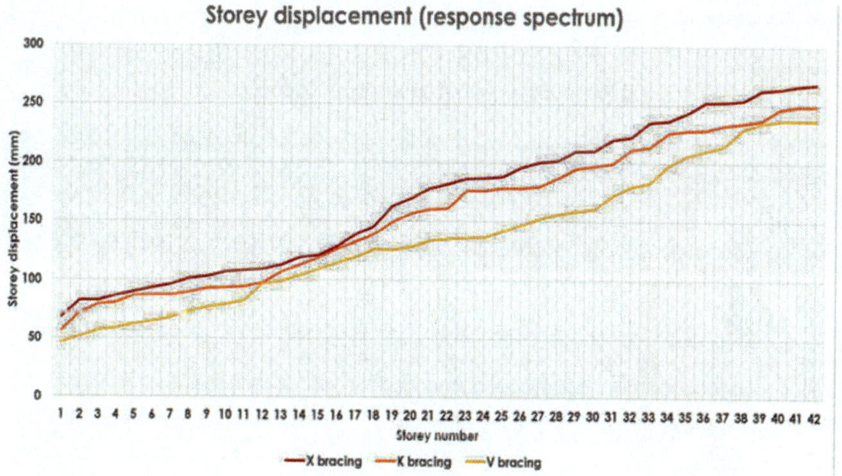


**Fig. 3** Storey displacement after response spectrum analysis

static and from response spectrum analysis. The lowest was for V braced frame in both the analyses.

Figures 4 and 5 represent the base shear for different braced systems from equivalent static and response analysis. V bracing showed highest base shear while K bracing showed the lowest in both the methods.

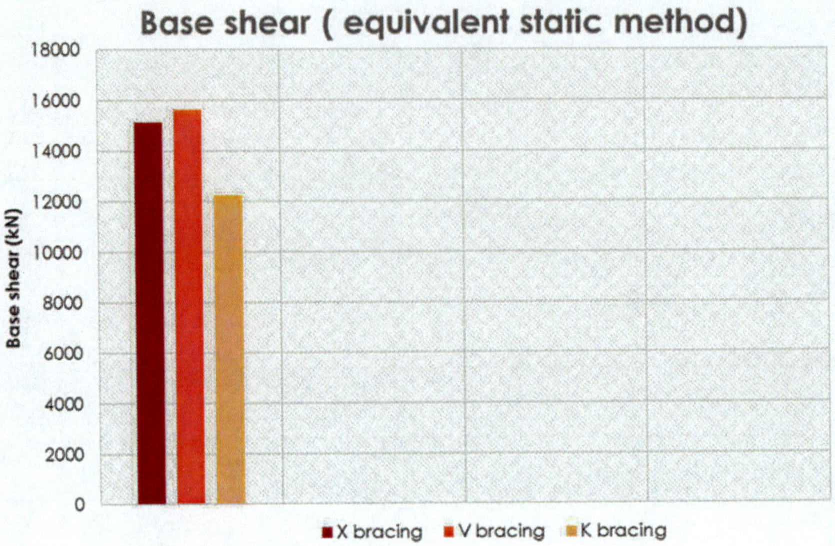


**Fig. 4** Base shear after equivalent static analysis

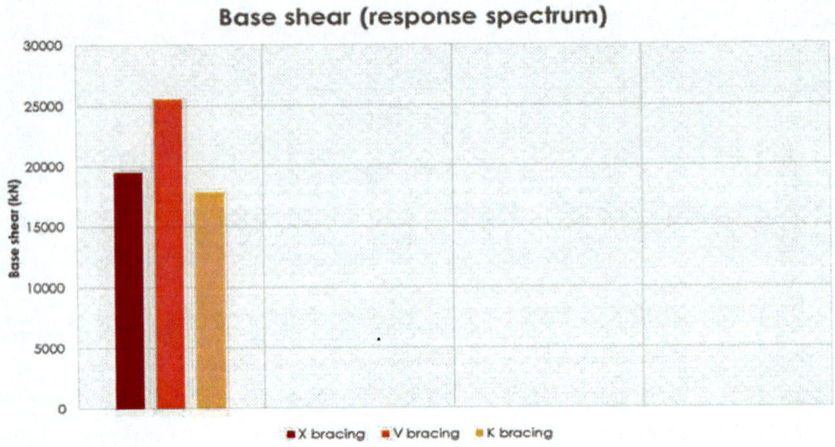


**Fig. 5** Base shear after response spectrum analysis

## 5 Conclusions

Response spectrum and equivalent static analysis were done for frames with X, K and V bracing and compared for base shear and storey displacement. The followings were observed.

- Storey displacement for X braced frame was 4.9% higher than the lower value of V braced frame in equivalent static method and 16.5% higher in response spectrum method.
- Base shear was higher for V braced frame, around 42.3% higher in equivalent static analysis and 49.6% higher in case of response spectrum analysis than the K braced frame which exhibited low base shear.
- As per the analyses done, the seismic behaviour of V braced frame was more efficient with respect to the other bracing types used for the G+41 steel frame in zone V considered in this study.
- K braced framed frame performed second best in terms of displacement when compared to X braced frame.

## References

1. Alghuff AY, Shihada SM, Tayeb BA (2019) Comparative study of static and response spectrum methods for seismic analysis of regular RC buildings. J Appl Sci 19:495–503
2. Abhishek KK, Rajeeva SV (2017) Effect of bracing systems on seismic behavior of typical RC tall building. Int J Res Sci Innov 04(09)
3. Sheidaii MR, Tahamouli Roudsari M, Gordini M (2016) Investigation of the nonlinear seismic behavior of knee braced frames using the incremental dynamic analysis method. Int J Adv Struct Eng 8 151–159
4. Ravali B, Poluraju P (2019) Seismic analysis of industrial structure using bracings and dampers. Int J Recent Technol Eng 7
5. Patel B, Mali R, Jadhav PR, Mohan Ganesh G (2017) Seismic behavior of different bracing systems in high rise RCC buildings. Int J Civ Eng Technol 8(3):973–981
6. Siddiqi ZA, Hameed R, Akmal U (2014) Comparison of different bracing systems for tall buildings. Pak J Eng Appl Sci 14:17–26
7. Nassani DE, Hussein AK, Mohammed AH (2017) Comparative response assessment of steel frames with different bracing systems under seismic effect. Structures 11:229–242
8. Moosa M, Mohammadreza G, Mohammad G (2019) Evaluating inelastic performance of mega-scale bracing systems in low- and medium-rise structures. Asian J Civ Eng 20:383–393
9. Khaleel MT, Kumar D (2016) Seismic analysis of steel frames with different bracings using ETABS software. Int Res J Eng Technol 3(08)
10. Kadid A, Yahiaoui D (2011) Seismic assessment of braced RC frames. Procedia Eng 14:2899–2905
11. Patil DM, Sangle KK (2015) Seismic behaviour of different bracing systems in high rise 2-D steel buildings. Structures (Elsevier) 3
12. Yeom H-J, Yoo J-H (2018) Analytical investigation on seismic behavior of inverted V-braced frames. Int J Steel Struct 18:189–198

# Influence of Mass Irregularity on the Response of RC Frame with Stiffness Irregularity by Non-linear Time History Analysis

Athar Tauheed, Mehtab Alam, and T. K. Datta

## 1 Introduction

From the history of the earthquakes, it is inferred that the majority of the earthquakes are not a single event but followed by a number of subsequent events of sequential ground motions. The mainshock is followed by numerous aftershocks in succession as could be seen in Christchurch earthquake of 2010. It was succeeded by thousands of earthquakes in the succeeding months till February 2011 [1]. Similarly, the Nepal earthquake of 2015 (Gorkha earthquake) hits near the city of Kathmandu and was a very powerful one [2]. It is seen that there is a tremendous loss of strength and stiffness of the reinforced concrete structures in cases of repetitive ground motions, but this is ignored by the current design codes and regulations, which only talks about one single event that is mainshock. Damage might not be severe during mainshock but lack of repair due to time constraints or some other reasons, the structure may suffer severe damage even collapse in the succeeding aftershocks. A vertically irregular structure is described as one having non-uniform distribution of stiffness, mass, and strength along with the height of the structure. In the urban areas, there are several multi-storeyed structures with vertical and mass irregularities owing to the high demands of parking, showrooms, banquet halls, service floors etc. These types of structures are exposed to enhanced structural forces as compared with the designed ones, hence become highly susceptible to repeated ground motions.

A. Tauheed (✉) · M. Alam
Department of Civil Engineering, Jamia Millia Islamia, Jamia Nagar, New Delhi 110025, India

T. K. Datta
Department of Civil Engineering, Indian Institute of Technology Delhi, Hauz Khas, New Delhi 110067, India



S. Kolathayar and S. C. Chian (eds.), *Recent Advances in Earthquake Engineering*, Lecture Notes in Civil Engineering 175,
https://doi.org/10.1007/978-981-16-4617-1_12

Numerous researchers had tried to study the effects of sequential earthquakes with respect to SDOF and MDOF. Loulelis et al. carried out an extensive analysis to investigate the seismic behaviour of SMRF when exposed to sequential ground motions. It was concluded that the seismic damages as well as the ductility demands of the structures are higher for multiple earthquakes as against a single episode of the earthquake. Along with this, the sequential earthquakes also require incremented displacement demands. Faisal et al. [3] carried out an extensive study on the ductility demands of three-dimensional inelastic concrete frames of 3-, 6-, 9-, 12-storeys taking into consideration the effects of multiple ground motions and concluded that it increases remarkably in the case of repetitive earthquakes with the help of empirical relationships. Hatzivassiliou and Hatzigeorgiou carried out a comprehensive study on four structures (2-regular and 2-irregular) keeping in mind the inelastic response of structures to multiple earthquakes as against a single event. He deduced that the sequential earthquakes led to an increment in the seismic demands, which, in turn, affect the ductility demand parameter of the buildings. Raghunandan et al. investigated the capacity of collapse of RCC frames using incremental dynamic analysis and concluded that the building would survive if the damage resulting due to mainshock was not severe. Hatzigeorgiou et al. came up with a new method to study the ductility demands of the SDOF systems under the influence of sequential ground motions. The force reduction factor was considered as a parameter to achieve both cumulative damages as well as ductility demands of the structure. It was concluded that the design earthquake led to less conservative force reduction factors as against a series of earthquakes. Goda [4] studied the ductility demand of inelastic SDOF systems and examined the consequences of aftershocks with the help of real and artificial seismic sequences. A similarity was observed in the ductility demands of both real and artificial sequences, which implied that the latter could be used as a substitute for the real ones. Hosseinpour et al. had studied both regular and irregular eight storeyed structures to investigate the fragility curves for them under the effect of multiple earthquakes. A conclusion was drawn that the regular building portrayed lesser vulnerability under the influence of aftershock polarity, direction of aftershocks as compared with the irregular ones.

This study examines the seismic behaviour of a regular reinforced concrete two-dimensional frame having 11 storeys that have been designed as per IS code (Type 1). The frames with stiffness irregularity only in the first storey (Type 2), with mass (200%) and stiffness irregularity in the first storey (Type 3), with mass (300%) and stiffness irregularity in the first storey (Type 4), with stiffness irregularity in the sixth storey (Type 5), with mass (200%) and stiffness irregularity in the sixth storey (Type 6), with mass (300%) and stiffness irregularity in the sixth storey (Type 7), with stiffness irregularity in the top storey (Type 8), with mass (200%) and stiffness irregularity in the top storey (Type 9), with mass (300%) and stiffness irregularity in the top storey (Type 10). These frames are subjected to synthetically generated sequences of mainshock and aftershock events compatible to IS (1893:2016) response spectrum. An elaborative comparison is done to study the performances of regular as well as irregular frames occurring at first, fifth, and top storey. A similar trend as mentioned above for the storey height 4.5 m and mass irregularity 200% and 300% mass has been followed to frames Type 11 to Type 19.

## 2 Building Frame Model

An 11-storey 2D-RC frame symmetrical in plan with 4 bays, each 5 m width in both directions has been considered. The frame that is taken up for study is the middle frame of the building in seismic zone V, and the soil type considered is medium as per the IS code 1893:2016 [5–7]. Nineteen models of the frames are taken up for study in this paper, out of which only one is regular. The storey height is taken to be constant for each of the storeys (Type 1) as shown in Fig. 1a and changed by incrementing height (3.2–3.85 m/4.5 m) in first/sixth/top storeys (Type 2 to Type 10) as shown in Fig. 1b–d leading to stiffness irregularity. Mass irregularity is also introduced at the floor of the storey with stiffness irregularity by increasing the live load by 200 and 300% in two different cases. Columns of size 450 mm × 450 mm and beams of size 230 mm × 450 mm are taken from the foundation to terrace of the frames. M25 grade of concrete and Fe500 steel are used. The dead load (DL) is taken as 12.5 kN/m and the live load (LL) is taken to be 5 kN/m. The dead load and live load taken for the roof are 15 kN/m and 3.75 kN/m, respectively, and for the partition wall load is 7.5 kN/m. Non-linear time history analyses of the building frames are carried out for scaled ground acceleration time histories, which are compatible with IS1893:2016 corresponding to 0.36g and 0.45g PGAs as mainshocks by making use of software SAP2000. FEMA guidelines are utilized to define the default frame hinges for the columns (P-M2-M3 hinges) as well as the end beams (M3 hinges) taking into account the provision of default hinges present in SAP2000.

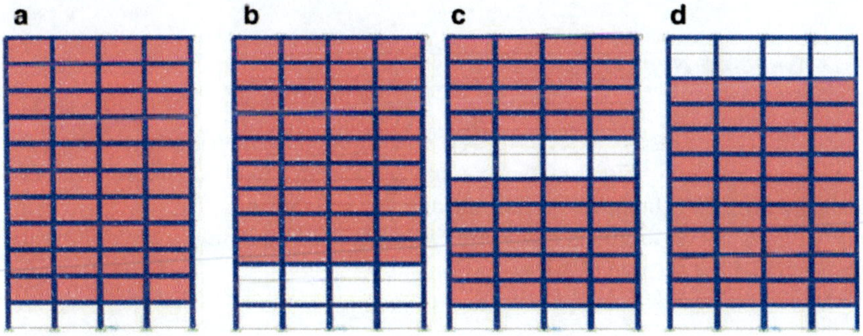


**Fig. 1** Elevation detail of 11-storey RC **a** (regular frame), **b** (stiffness irregularity at the first storey), **c** (stiffness irregularity at the sixth storey), **d** (stiffness irregularity at the top storey)

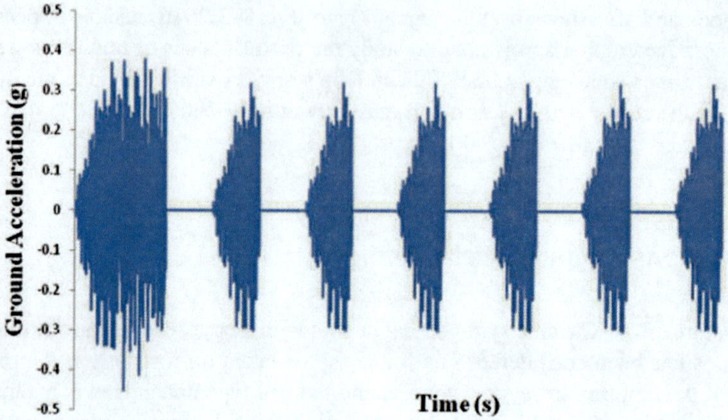


**Fig. 2** Time history compatible to IS1893:2016

## 3 Sequential Ground Motion

SeismoSignal software is used for scale artificial time history generation compatible to response spectrum (IS1893:2016) as mainshock and repetitive aftershocks. Each sequential earthquake episode is a combination of mainshock (MS) followed by six aftershocks (first AS, second AS, third AS, fourth AS, fifth AS, and sixth AS). The aftershock is calculated as two-third of PGA of mainshock. The PGAs are scaled to 0.36g and 0.45g. Time duration for which the mainshock lasts is 30 s whereas, for each of the aftershocks, it is 15 s. Between any two consecutive events of earthquake, the time duration is 100 s. An acceleration of magnitude zero is induced to bring the structure to rest as shown in Fig. 2.

## 4 Result and Discussion

### *4.1 Seismic Behaviour*

A non-linear time history analysis (NDA) is being conducted on the regular/irregular frames as depicted in Fig. 1 on being subjected to sequential earthquakes with two mainshocks (PGAs 0.36, 0.45g) accompanied by six aftershocks as explained using Fig. 2. Parameters discussed below are related to the response of regular/irregular frames under the seismic events.

## 4.2 Maximum Horizontal Displacements

The general trend observed from the analyses carried out is that the horizontal transient displacement decreases slightly during the first aftershock but increases gradually in the succeeding aftershocks. The regular frame (Type 1) sustains the mainshock and aftershocks of 0.36 PGA but in the case of 0.45 PGA, it fails during the last aftershock. For the irregular frames (Type 2 to Type 10), the structures survive the mainshock of 0.36g PGA and its aftershocks in both the cases (height increases from 3.2 to 3.85 m) and (height increases from 3.2 to 4.5 m). A common observation for both cases is that the value of transient horizontal displacement is always less in case of regular frames as compared with the irregular frames during the mainshock as well as its aftershocks.

For Case 1, as per Table 1, the Type 2 (stiffness irregularity at first storey), frames fail during the fourth aftershock of 0.45 PGA. The Type 3 (mass increased by 200% along with stiffness irregularity at first storey) and Type 4 (mass increased by 300% along with stiffness irregularity at first storey)) fail during the fifth aftershock. At lower PGAs, the decrease in horizontal transient displacement is minimal with an increase in mass, but on the contrary at higher PGAs, the decrease is highly significant. The Type 5 (stiffness irregularity at sixth storey), Type 6 (mass increased by 200% along with stiffness irregularity at sixth storey), Type 7 (mass increased by 200% along with stiffness irregularity at sixth storey), Type 8 (stiffness irregularity at top storey), Type 9 (mass increased by 200% along with stiffness irregularity at top storey), and Type 10 (increased by 300% along with stiffness irregularity at top storey) frames survive the mainshock and the first four aftershocks but fail during the fifth aftershock. For Types 5–7, it is observed that when mass irregularity is introduced with stiffness irregularity at lower PGAs, the increase in the horizontal transient displacement is minimal with an increase in mass, but on the contrary at higher PGAs, the increase is remarkable. For Types 8–10, it is observed for 0.36 PGA, there is a slow increase in the horizontal transient displacement with the increase in mass whereas on the other hand, for 0.45g PGA, the value initially decreases with an increase in mass, but increases later.

Considering Case 2, as per Table 2, the Type 11 (stiffness irregularity at first storey), Type 12 (increased by 200% along with stiffness irregularity at first storey), and Type 12 (increased by 300% along with stiffness irregularity at first storey) frames fail during the fourth aftershock of 0.45 PGA. At lower as well as higher PGAs, no impact in horizontal transient displacement is seen with an increase in mass. The values are almost constant. The Type 13 (stiffness irregularity at sixth storey), Type 14 (increased by 200% along with stiffness irregularity at sixth storey), and Type 15 (increased by 300% along with stiffness irregularity at sixth storey) frames survive the mainshock and the first four aftershocks but fail during the fifth aftershock. For

**Table 1** Maximum transient top floor displacement (mm) for case 1

| | Mainshock | | First aftershock | | Second aftershock | | Third aftershock | | Fourth aftershock | | Fifth aftershock | | Sixth aftershock | |
|---|---|---|---|---|---|---|---|---|---|---|---|---|---|---|
| | 0.36g | 0.45g | 0.24g | 0.30g | 0.24g | 0.30g | 0.24g | 0.30g | 0.24g | 0.30g | 0.24g | 0.30g | 0.24g | 0.30g |
| Type 1 | 249.8 | 307.4 | 246.6 | 296.1 | 250.8 | 358.1 | 258.9 | 418.8 | 266.6 | 495.1 | 273.7 | 601.5 | 280.5 | Exceeds limit |
| Type 2 | 262.7 | 337.6 | 267.3 | 343.5 | 294.1 | 458.9 | 318.1 | 597.8 | 340.8 | Exceeds limit | 364.8 | Exceeds limit | 391.7 | Exceeds limit |
| Type 3 | 263.2 | 335 | 267.8 | 335.3 | 294.8 | 431.4 | 319.1 | 550.4 | 342.1 | 694.4 | 366.5 | Exceeds limit | 394 | Exceeds limit |
| Type 4 | 263.1 | 330.5 | 267.1 | 326 | 293.6 | 418.6 | 317.5 | 535.9 | 340.1 | 679.3 | 363.8 | Exceeds limit | 390.4 | Exceeds limit |
| Type 5 | 258.6 | 301.6 | 254 | 281.8 | 270.3 | 355.5 | 282.3 | 428 | 293.7 | 512.7 | 304.6 | Exceeds limit | 315.6 | Exceeds limit |
| Type 6 | 258.2 | 303 | 253.6 | 292.3 | 270.8 | 376.5 | 284.1 | 463.1 | 296.6 | 582.5 | 308.6 | Exceeds limit | 320.8 | Exceeds limit |
| Type 7 | 258.1 | 303.8 | 253.3 | 293.4 | 271.6 | 379.2 | 286.2 | 463.3 | 299.9 | 582.2 | 313 | Exceeds limit | 326.5 | Exceeds limit |
| Type 8 | 259.3 | 316.4 | 253.9 | 304 | 259.5 | 371.6 | 270.3 | 441.3 | 281.5 | 536 | 292.4 | Exceeds limit | 303.1 | Exceeds limit |
| Type 9 | 260.8 | 315.1 | 253.2 | 302.2 | 260 | 373.3 | 272.2 | 446.3 | 284.6 | 545.7 | 296.7 | Exceeds limit | 308.6 | Exceeds limit |
| Type 10 | 262.7 | 314.5 | 256.9 | 301.3 | 265.5 | 375.9 | 279.6 | 453 | 294 | 558.5 | 308.1 | Exceeds limit | 322.1 | Exceeds limit |

**Table 2** Maximum transient top floor displacement (mm) for case 2

| | Mainshock | | First aftershock | | Second aftershock | | Third aftershock | | Fourth aftershock | | Fifth aftershock | | Sixth aftershock | |
|---|---|---|---|---|---|---|---|---|---|---|---|---|---|---|
| | 0.36g | 0.45g | 0.24g | 0.30g | 0.24g | 0.30g | 0.24g | 0.30g | 0.24g | 0.30g | 0.24g | 0.30g | 0.24g | 0.30g |
| Type 1 | 249.8 | 307.4 | 246.6 | 296.1 | 250.8 | 358.1 | 258.9 | 418.8 | 266.6 | 495.1 | 273.7 | 601.5 | 280.5 | Exceeds limit |
| Type 11 | 272.3 | 355.5 | 305.8 | 353.2 | 348.7 | 442.5 | 388.8 | 545.7 | 431.4 | Exceeds limit | 479.4 | Exceeds limit | 533.7 | Exceeds limit |
| Type 12 | 272.4 | 355.7 | 306.1 | 347.7 | 349.2 | 433.9 | 389.5 | 533.6 | 432.3 | Exceeds limit | 480.5 | Exceeds limit | 535 | Exceeds limit |
| Type 13 | 272.4 | 356 | 306.1 | 351.3 | 349.5 | 439.6 | 389.9 | 542 | 433 | Exceeds limit | 481.6 | Exceeds limit | 536.4 | Exceeds limit |
| Type 14 | 257 | 301.4 | 244.7 | 281 | 276.6 | 368.9 | 300.9 | 453.4 | 322.6 | 548.8 | 343.5 | Exceeds limit | 364.7 | Exceeds limit |
| Type 15 | 256.4 | 301.2 | 243.4 | 278.9 | 276 | 365.6 | 300.8 | 449.3 | 322.8 | 544.3 | 344.2 | Exceeds limit | 366.2 | Exceeds limit |
| Type 16 | 255.5 | 297.8 | 241.7 | 268.1 | 274.9 | 349.8 | 299.9 | 429 | 322.2 | 519.8 | 343.8 | Exceeds limit | 366.2 | Exceeds limit |
| Type 17 | 281.8 | 343.4 | 269 | 326.3 | 279.2 | 413.3 | 300.7 | 507.9 | 323.3 | Exceeds limit | 346.3 | Exceeds limit | 369.8 | Exceeds limit |
| Type 18 | 283.6 | 341 | 271.6 | 324 | 283.9 | 413.7 | 305.9 | 515.5 | 329.7 | 640 | 353.9 | Exceeds limit | 378.4 | Exceeds limit |
| Type 19 | 284.9 | 337 | 274.9 | 319.2 | 289.4 | 408.5 | 311.5 | 497.7 | 335.7 | 616.3 | 360.6 | Exceeds limit | 386.1 | Exceeds limit |

Types 14–16, it is observed that when mass irregularity is introduced with stiffness irregularity at lower PGAs, the decrease in the horizontal transient displacement is minimal. At higher PGAs, a gradual decrease is seen when mass is increased. Type 17 (stiffness irregularity at top storey) fails during the fourth aftershock whereas Type 18 (increased by 200% along with stiffness irregularity at top storey), and Type 19 (increased by 300% along with stiffness irregularity at top storey) fail during the fifth aftershock. It is observed that for 0.36 PGA, there is a slow increase in the horizontal transient displacement with the increase in mass whereas on the other hand, for 0.45g PGA, the value decreases for mainshock, and its first aftershock, and increases for the rest.

## 4.3 *Maximum Horizontal Permanent Displacement*

The frame exposed to repetitive ground motions experiences residual displacement, which goes on incrementing with every subsequent aftershock, that adds up to increase the horizontal top displacement and may even exceed the limit in some cases as exhibited in Fig. 3 and Tables 2 and 3. For Case 1, the value of residual floor displacement of regular buildings is lesser than the irregular buildings except for the mainshock of Types 2–7 in case of lower PGA, and the mainshock of Types 5–7 of the higher PGAs. For Case 2, the value is lesser except for the first two events on Types 5–7 of both the PGAs.

For Case 1, in Types 2–4, it is seen that to lower PGAs, the value of residual displacement gradually decreases with the increase in mass. But the decrease is

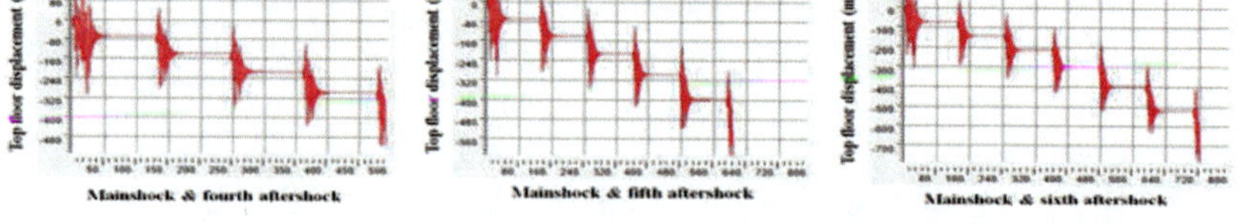

**Fig. 3** Typical time history of horizontal top displacement (Type 2) under repeated ground motion with PGA as 0.36g

Table 3 Maximum residual floor displacement (mm) for case 1

| | Mainshock | | First aftershock | | Second aftershock | | Third aftershock | | Fourth aftershock | | Fifth aftershock | |
|---|---|---|---|---|---|---|---|---|---|---|---|---|
| | 0.36g | 0.45g | 0.24g | 0.30g | 0.24g | 0.30g | 0.24g | 0.30g | 0.24g | 0.30g | 0.24g | 0.30g |
| Type 1 | 100.7 | 100.9 | 102.4 | 162.9 | 110.5 | 223.8 | 118.3 | 294.7 | 125.4 | 380.9 | 132.2 | 487.30 |
| Type 2 | 100.2 | 121.8 | 127 | 230.5 | 151.2 | 346.1 | 174.1 | 485.4 | 198 | Exceeds limit | 225 | Exceeds limit |
| Type 3 | 100.6 | 114.7 | 127.5 | 216.3 | 152.1 | 316.4 | 175.2 | 436.9 | 199.7 | 581.5 | 227.1 | Exceeds limit |
| Type 4 | 99.75 | 106.7 | 125.2 | 203.3 | 150.4 | 303.4 | 173.1 | 422.6 | 196.8 | 566.7 | 223.4 | Exceeds limit |
| Type 5 | 92.45 | 67.76 | 107.9 | 139.7 | 119.8 | 212.3 | 131.2 | 296.2 | 142.2 | 396.7 | 153.3 | Exceeds limit |
| Type 6 | 90.47 | 74.05 | 107 | 157.8 | 120.2 | 243.9 | 132.8 | 345.5 | 144.9 | 469 | 157.1 | Exceeds limit |
| Type 7 | 88.63 | 72.73 | 106.4 | 158.7 | 121 | 243.2 | 134.7 | 345.1 | 147.9 | 469.3 | 161.5 | Exceeds limit |
| Type 8 | 104.5 | 103.6 | 108 | 170.9 | 118.8 | 240.9 | 130 | 323.3 | 140.9 | 422.9 | 151.7 | Exceeds limit |
| Type 9 | 100.8 | 98.13 | 105.7 | 169.[illegible] | 117.8 | 242.4 | 130.4 | 328.6 | 142.5 | 432.2 | 154.5 | Exceeds limit |
| Type 10 | 101.6 | 93.66 | 108.3 | 168.3 | 122.5 | 245.6 | 137 | 336.4 | 151.1 | 444.8 | 165.3 | Exceeds limit |

remarkable at higher PGAs for these types. The Type 2 frame exceeds the permissible limits during the fourth episode of aftershock while in Type 3 and Type 4, it collapses during the fifth aftershock. In Types 5–7, the value of the maximum permanent displacement decreases during the first few aftershocks but increases slightly in the later aftershocks pertaining to the increase in mass. At higher PGAs, the increment is very slow during the first few aftershocks but a steep rise is seen in the later aftershocks. These three frames exceed their permissible limits during the fifth aftershock. Frames of Types 8–10 respond with lower permanent residual displacement for both PGAs during the mainshock. But approximately equals during the first aftershock and then gradually keeps on increasing with the increase in mass in the remaining aftershocks.

For Case 2 of Types 11–13, it is seen that in lower as well as higher PGAs, the change in the value of residual displacement is negligible with the increase in mass. All the three frames collapse during the fourth aftershock. In Types 14–16, the value of the maximum permanent displacement as seen in lower as well as higher PGAs keep on decreasing throughout the mainshock and its aftershocks, pertaining to the increase in mass. These three frames exceed their permissible limits during the fifth aftershock. Now taking frames 17–19 into consideration, it is observed that for the lower PGA, the value is increasing gradually throughout but for higher PGAs, the value decreases with the increase in mass. Type 17 collapses during the fourth episode of earthquake while Types 18 and 19 collapse during the fifth aftershock (Table 4).

## 4.4 *Maximum Transient Storey Drift*

Storey drift is explained on the basis of maximum difference prevalent between the lateral displacements of two consecutive storeys. The storey drift of the regular buildings is lesser as compared with the irregular buildings in both Cases 1 and 2 except for the mainshock of Types 2–4 in Case 1.

In Case 1, for Types 2–4, at lower PGAs, the impact on storey drift is not very significant but at higher PGAs, there is a decrease when the mass is increased. At higher PGAs, Type 2 exceeds the permissible limits (FEMA 356 guidelines) during the fourth episode of earthquake whereas Types 3 and 4 do the same during the fifth episode. For Types 5–7 at lower PGAs, the value decreases slightly during the mainshock and the first aftershock but increases thereafter with the increase in mass. Now at higher PGAs, the value decreases slightly during the mainshock but increases thereafter with the increase in mass. The frame collapses during the fifth aftershock at higher PGAs for all three types. For Types 8–10 at both the PGAs, the value increases slightly during the mainshock and its aftershocks with the increase in mass. The frame fails during the fifth aftershock in higher PGAs (Table 5).

**Table 4** Maximum residual top storey displacement (mm) for case 2

| | Mainshock | | First aftershock | | Second aftershock | | Third aftershock | | Fourth aftershock | | Fifth aftershock | |
|---|---|---|---|---|---|---|---|---|---|---|---|---|
| | 0.36g | 0.45g | 0.24g | 0.30g | 0.24g | 0.30g | 0.24g | 0.30g | 0.24g | 0.30g | 0.24g | 0.30g |
| Type 1 | 100.7 | 100.9 | 102.4 | 162.9 | 110.5 | 223.8 | 118.3 | 294.7 | 125.4 | 380.9 | 132.2 | 487.30 |
| Type 11 | 121.5 | 116.3 | 164.7 | 213.9 | 204.9 | 314.6 | 247.2 | 434.8 | 295 | Exceeds limit | 349.2 | Exceeds limit |
| Type 12 | 121.7 | 112.1 | 165.1 | 205.7 | 205.4 | 303 | 247.9 | 419.6 | 295.9 | Exceeds limit | 350.3 | Exceeds limit |
| Type 13 | 121.7 | 114.6 | 165.2 | 210.8 | 205.8 | 310.8 | 248.5 | 430.3 | 296.8 | Exceeds limit | 351.6 | Exceeds limit |
| Type 14 | 65.82 | 41.52 | 98.07 | 129.7 | 122.5 | 214.3 | 144.3 | 309.4 | 165.3 | 422.7 | 186.7 | Exceeds limit |
| Type 15 | 63.23 | 37.97 | 96.24 | 125.3 | 121.1 | 208.9 | 143.3 | 303.2 | 164.7 | 416.3 | 186.8 | Exceeds limit |
| Type 16 | 60.1 | 26.15 | 93.77 | 108.6 | 118.8 | 187.7 | 141.3 | 277.8 | 163.1 | 386 | 185.5 | Exceeds limit |
| Type 17 | 114.5 | 119.7 | 123.2 | 203.9 | 143 | 295.6 | 164.9 | 394.2 | 187.5 | Exceeds limit | 210.8 | Exceeds limit |
| Type 18 | 114.1 | 113.5 | 125 | 201.1 | 146.3 | 296.1 | 169.5 | 401.8 | 193.4 | 527 | 217.6 | Exceeds limit |
| Type 19 | 114.6 | 105.3 | 127.8 | 192.9 | 149.9 | 282.2 | 174.1 | 333.7 | 198.6 | 502.8 | 223.7 | Exceeds limit |

**Table 5** Maximum transient storey drift (mm) for case 1

| | Mainshock | | First aftershock | | Second aftershock | | Third aftershock | | Fourth aftershock | | Fifth aftershock | | Sixth aftershock | |
|---|---|---|---|---|---|---|---|---|---|---|---|---|---|---|
| | 0.36g | 0.45g | 0.24g | 0.30g | 0.24g | 0.30g | 0.24g | 0.30g | 0.24g | 0.30g | 0.24g | 0.30g | 0.24g | 0.30g |
| Type 1 | 31.7 | 38 | 32.32 | 41.67 | 32.75 | 49.1 | 33.46 | 55.8 | 34.59 | 63.8 | 35.69 | 73.7 | 36.65 | Exceeds limit |
| Type 2 | 31.25 | 50.33 | 33.89 | 54.83 | 37.62 | 68.59 | 41.1 | 87.21 | 44.2 | Exceeds limit | 47.2 | Exceeds limit | 50.3 | Exceeds limit |
| Type 3 | 31.26 | 49.32 | 34 | 52.08 | 37.77 | 63.14 | 41.2 | 76.54 | 44.3 | 95.37 | 47.3 | Exceeds limit | 50.6 | Exceeds limit |
| Type 4 | 31.25 | 47.73 | 33.91 | 49.51 | 37.53 | 59.88 | 40.74 | 74.1 | 44.1 | 93.38 | 46.2 | Exceeds limit | 50.2 | Exceeds limit |
| Type 5 | 40.80 | 43.7 | 35.3 | 38.64 | 40.4 | 48.4 | 43 | 60.1 | 45.1 | 73.1 | 47.1 | Exceeds limit | 49.1 | Exceeds limit |
| Type 6 | 39.3 | 42.4 | 35.1 | 40.07 | 40.5 | 50.3 | 43.3 | 61.7 | 45.6 | 76.2 | 47.8 | Exceeds limit | 50.1 | Exceeds limit |
| Type 7 | 38.2 | 42.2 | 35.0 | 40.38 | 40.6 | 50.9 | 43.6 | 62 | 46.1 | 76.5 | 48.4 | Exceeds limit | 51 | Exceeds limit |
| Type 8 | 34.1 | 38.3 | 32.63 | 41.87 | 32.84 | 49.8 | 34.28 | 57.5 | 35.84 | 66.5 | 37.33 | Exceeds limit | 38.7 | Exceeds limit |
| Type 9 | 35.1 | 38.3 | 32.83 | 41.73 | 32.96 | 50.1 | 34.7 | 58.3 | 36.46 | 67.8 | 38.14 | Exceeds limit | 39.7 | Exceeds limit |
| Type 10 | 35.6 | 38.3 | 33.19 | 41.62 | 33.9 | 50.5 | 35.88 | 59.1 | 37.88 | 69.2 | 39.73 | Exceeds limit | 41.7 | Exceeds limit |

In Case 2, there is no significant impact of mass on the values of storey drift at lower PGAs but during higher PGAs in Types 11–13, there is a slight decrease in the value of storey drift. These frames collapse during the fourth aftershock. Negligible decrease is seen in Types 14–16 for both PGAs and all the three frames fail during the fifth aftershock. In Types 17–19, the value of storey drift slightly increases for lower PGA but the reverse happens in case of higher PGA. Type 17 collapses during the fourth episode whereas Types 18 and 19 collapse during the fifth aftershock (Table 6).

## 4.5 Maximum Permanent Storey Drift

It is a critical parameter that helps us calculate the potential damages incurred on a structure due to its exposure to a seismic load. For both regular as well as irregular frames at lower and higher PGAs, the value of maximum residual storey drift increases gradually after each subsequent shock. For Case 1, under lower PGAs all the frames (Type 1 to Type 10) survive the mainshock and aftershocks, but collapse occur during fifth aftershock except in Type 2 frame which collapses during fourth aftershock at higher PGAs. For Case 2, at lower PGAs, all the types (Type 1 to Type 10), the frames survive the mainshock and its aftershocks but at higher PGAs Types 5–10 collapse during the fifth aftershock except Types 2–4, and 8, which collapse during the fourth aftershock (Tables 7 and 8).

## 4.6 Formation of Plastic Hinges

It is seen that the location of formation of hinges and their state define the damage as well as its degree in the structure. All the types of hinges that are formed during the repetitive ground motions of PGAs 0.36g and 0.45g are exhibited in Tables 10, 11 and 12 and Fig. 4. Tables 10, 11 and 12 depict the total count of hinges formed and also their transformation from one state to another during the mainshock and its aftershocks. The plastic rotation that takes place during the repetitive earthquakes leads to storey drift and permanent floor displacement (Table 9).

**Table 6** Maximum transient storey drift (mm) for case 2

| | Mainshock | | First aftershock | | Second aftershock | | Third aftershock | | Fourth aftershock | | Fifth aftershock | | Sixth aftershock | |
|---|---|---|---|---|---|---|---|---|---|---|---|---|---|---|
| | 0.36g | 0.45g | 0.24g | 0.30g | 0.24g | 0.30g | 0.24g | 0.30g | 0.24g | 0.30g | 0.24g | 0.30g | 0.24g | 0.30g |
| Type 1 | 31.7 | 38 | 32.32 | 41.67 | 32.75 | 49.1 | 33.46 | 55.8 | 34.59 | 63.8 | 35.69 | 73.7 | 36.66 | Exceeds limit |
| Type 11 | 50.41 | 71.4 | 52.15 | 67.68 | 59.04 | 83.57 | 65.37 | 103.25 | 73.4 | Exceeds limit | 82.41 | Exceeds limit | 92.66 | Exceeds limit |
| Type 12 | 50.39 | 71.52 | 52.22 | 64.71 | 59.11 | 80.1 | 65.49 | 99.36 | 73.53 | Exceeds limit | 82.64 | Exceeds limit | 92.79 | Exceeds limit |
| Type 13 | 50.37 | 71.6 | 52.25 | 66.58 | 59.16 | 82.29 | 65.57 | 101.98 | 73.62 | Exceeds limit | 82.8 | Exceeds limit | 93.11 | Exceeds limit |
| Type 14 | 36.3 | 54.8 | 46.6 | 46.6 | 54.8 | 63.7 | 60 | 79.8 | 64.6 | 97.8 | 69.2 | Exceeds limit | 74 | Exceeds limit |
| Type 15 | 36.7 | 54.8 | 46.3 | 46.6 | 54.6 | 63.4 | 59.9 | 79.6 | 64.6 | 97.7 | 69.3 | Exceeds limit | 74.1 | Exceeds limit |
| Type 16 | 37 | 54.7 | 45.8 | 46.4 | 54.3 | 61.7 | 59.4 | 76.5 | 64.3 | 93.5 | 69.1 | Exceeds limit | 74 | Exceeds limit |
| Type 17 | 35.3 | 38.97 | 32.37 | 51.64 | 32.74 | 51.1 | 35.2 | 60.5 | 37.9 | Exceeds limit | 40.7 | Exceeds limit | 43.5 | Exceeds limit |
| Type 18 | 35.4 | 39.22 | 32.6 | 41.48 | 33.45 | 51.5 | 36.1 | 61.4 | 39.2 | 72.1 | 42.1 | Exceeds limit | 45.1 | Exceeds limit |
| Type 19 | 35.6 | 39.29 | 33.07 | 41.15 | 34.32 | 51.2 | 37 | 61.1 | 40.2 | 71.8 | 43.3 | Exceeds limit | 46.3 | Exceeds limit |

**Table 7** Maximum residual storey drift (mm) for case 1

| | Mainshock | | First aftershock | | Second aftershock | | Third aftershock | | Fourth aftershock | | Fifth aftershock | |
|---|---|---|---|---|---|---|---|---|---|---|---|---|
| | 0.36g | 0.45g | 0.24g | 0.30g | 0.24g | 0.30g | 0.24g | 0.30g | 0.24g | 0.30g | 0.24g | 0.30g |
| Type 1 | 14.76 | 14.04 | 17.76 | 22.5 | 19.37 | 31.1 | 20.64 | 40.3 | 21.74 | 50.7 | 22.78 | 63 |
| Type 2 | 15.56 | 22.32 | 20.42 | 31.96 | 23.98 | 47.33 | 27.21 | 68.28 | 30.3 | Exceeds limit | 33.8 | Exceeds limit |
| Type 3 | 15.62 | 19.85 | 20.49 | 27.85 | 24.1 | 39.23 | 27.36 | 55.24 | 30.6 | 76.71 | 34 | Exceeds limit |
| Type 4 | 15.66 | 17.4 | 20.49 | 24.86 | 24.03 | 36.54 | 27.22 | 52.88 | 30.4 | 74.69 | 33.8 | Exceeds limit |
| Type 5 | 15.89 | 10.71 | 23.26 | 25.13 | 26.02 | 37.3 | 28.17 | 50.8 | 30.15 | 66.3 | 32.2 | Exceeds limit |
| Type 6 | 15.64 | 10.23 | 23.18 | 24.98 | 26.12 | 30.2 | 28.44 | 54.3 | 30.64 | 72.2 | 32.93 | Exceeds limit |
| Type 7 | 15.42 | 10.29 | 23.11 | 24.78 | 26.28 | 39.1 | 28.78 | 54.5 | 31.19 | 72.3 | 33.74 | Exceeds limit |
| Type 8 | 15.1 | 13.15 | 18.48 | 22.54 | 20.66 | 32.2 | 22.45 | 42.5 | 24.12 | 54.3 | 25.74 | Exceeds limit |
| Type 9 | 15.12 | 12.68 | 18.54 | 22.26 | 20.89 | 32.3 | 22.87 | 43.1 | 24.71 | 55.2 | 26.52 | Exceeds limit |
| Type 10 | 15.33 | 12.45 | 18.45 | 22.23 | 21.73 | 32.7 | 22.8 | 43.8 | 54.78 | 56.4 | 28.26 | Exceeds limit |

**Table 8** Maximum residual storey drift (mm) for case 2

| | Mainshock | | First aftershock | | Second aftershock | | Third aftershock | | Fourth aftershock | | Fifth aftershock | |
|---|---|---|---|---|---|---|---|---|---|---|---|---|
| | 0.36g | 0.45g | 0.24g | 0.30g | 0.24g | 0.30g | 0.24g | 0.30g | 0.24g | 0.30g | 0.24g | 0.30g |
| Type 1 | 14.76 | 14.04 | 17.76 | 22.5 | 19.37 | 31.1 | 20.64 | 40.3 | 21.74 | 50.7 | 22.78 | 63 |
| Type 11 | 19.29 | 23.9 | 22.03 | 37.32 | 27.87 | 54.95 | 35.17 | 77.55 | 43.99 | Exceeds limit | 54.08 | Exceeds limit |
| Type 12 | 19.33 | 20.98 | 22.1 | 33.93 | 27.96 | 51.05 | 35.29 | 73.12 | 44.11 | Exceeds limit | 54.21 | Exceeds limit |
| Type 13 | 19.31 | 22.65 | 22.13 | 35.96 | 28.02 | 53.57 | 35.4 | 76.11 | 44.29 | Exceeds limit | 54.44 | Exceeds limit |
| Type 14 | 16.82 | 8.31 | 27.04 | 26.36 | 32.28 | 42.3 | 36.87 | 61.2 | 41.43 | 83.4 | 46.15 | Exceeds limit |
| Type 15 | 16.23 | 8.18 | 26.6 | 25.86 | 31.92 | 41.6 | 36.58 | 60.5 | 41.24 | 82.9 | 46.11 | Exceeds limit |
| Type 16 | 15.58 | 7.94 | 26.09 | 24.4 | 31.39 | 38.5 | 36.1 | 56.1 | 40.77 | 77.3 | 45.76 | Exceeds limit |
| Type 17 | 15.32 | 14.37 | 18.84 | 23.5 | 22.06 | 34.9 | 25.03 | 46.8 | 27.95 | Exceeds limit | 30.95 | Exceeds limit |
| Type 18 | 15.39 | 14.14 | 19.2 | 23.34 | 22.65 | 35 | 25.82 | 47.6 | 28.92 | 61.3 | 32.01 | Exceeds limit |
| Type 19 | 15.53 | 13.66 | 19.64 | 22.86 | 23.23 | 34.5 | 26.53 | 46.9 | 29.78 | 60.3 | 33.06 | Exceeds limit |

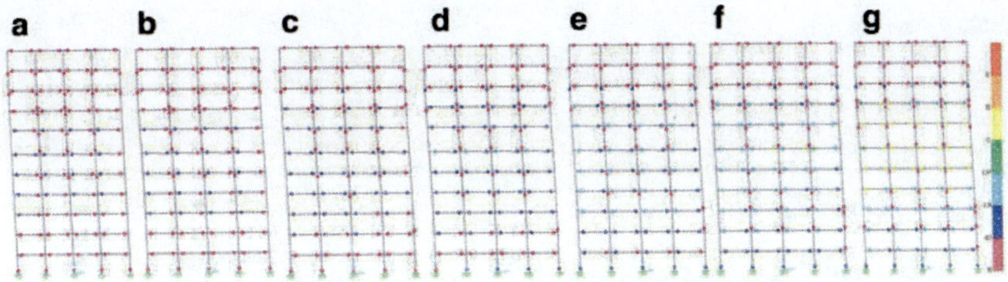


**Fig. 4** Typical Hinges pattern for Type 2 under a sequential ground motion for **a** (Mainshock), **b** (Mainshock and first aftershock), **c** (Mainshock and second aftershock), **d** (Mainshock and third aftershock), **e** (Mainshock and fourth aftershock), **f** (Mainshock and fifth aftershock) and **g** (Mainshock and sixth aftershock)

**Table 9** Total number of hinges pattern for 0.36g as mainshock (Case 1)

| Case 1 | Mainshock | Aftershocks | | | | | |
|---|---|---|---|---|---|---|---|
| | | 1 | 2 | 3 | 4 | 5 | 6 |
| *Type 1 (Regular building)* | | | | | | | |
| B-IO | 101 | 101 | 101 | 97 | 93 | 93 | 89 |
| IO-LS | – | – | – | 4 | 8 | 8 | 12 |
| Total hinge | 101 | 101 | 101 | 101 | 101 | 101 | 101 |
| *Type 2 (Stiffness irregularity at first storey)* | | | | | | | |
| B-IO | 93 | 93 | 84 | 81 | 74 | 65 | 53 |
| IO-LS | 8 | 8 | 17 | 21 | 29 | 39 | 51 |
| LS-CP | – | – | – | – | – | – | – |
| Total hinge | 101 | 101 | 101 | 102 | 103 | 104 | 104 |
| *Type-3 (Stiffness and mass (200%) irregularity at first storey)* | | | | | | | |
| B-IO | 93 | 93 | 84 | 81 | 73 | 65 | 53 |
| IO-LS | 8 | 8 | 17 | 21 | 30 | 39 | 51 |
| LS-CP | – | – | | | – | – | – |
| Total hinge | 101 | 101 | 101 | 102 | 103 | 104 | 104 |
| *Type 4 (Stiffness and mass (300%) irregularity at first storey)* | | | | | | | |
| B-IO | 93 | 93 | 85 | 81 | 73 | 65 | 54 |
| IO LS | 8 | 8 | 16 | 21 | 30 | 39 | 50 |
| LS-CP | – | – | | – | – | – | – |
| Total hinge | 101 | 101 | 101 | 102 | 103 | 104 | 104 |
| *Type 5 (Stiffness irregularity at sixth storey)* | | | | | | | |
| B-IO | 91 | 91 | 92 | 93 | 87 | 87 | 85 |
| IO-LS | 12 | 12 | 12 | 12 | 18 | 21 | 24 |
| Total hinge | 103 | 103 | 104 | 105 | 105 | 108 | 109 |
| *Type 6 (Stiffness and mass (200%) irregularity at sixth storey)* | | | | | | | |
| B-IO | 92 | 92 | 92 | 93 | 87 | 87 | 82 |
| IO-LS | 12 | 12 | 12 | 14 | 19 | 22 | 27 |
| Total hinge | 104 | 104 | 104 | 107 | 106 | 109 | 109 |

(continued)

**Table 9** (continued)

| Case 1 | Mainshock | Aftershocks | | | | | |
|---|---|---|---|---|---|---|---|
| | | 1 | 2 | 3 | 4 | 5 | 6 |
| *Type 7 (Stiffness and mass (300%) irregularity at sixth storey)* | | | | | | | |
| B-IO | 93 | 93 | 93 | 91 | 87 | 85 | 83 |
| IO-LS | 12 | 12 | 12 | 15 | 20 | 24 | 27 |
| Total hinge | 105 | 105 | 105 | 106 | 107 | 109 | 110 |
| *Type 8 (Stiffness irregularity at top storey)* | | | | | | | |
| B-IO | 101 | 101 | 101 | 95 | 94 | 92 | 86 |
| IO-LS | 5 | 5 | 5 | 11 | 12 | 15 | 21 |
| Total hinge | 106 | 106 | 106 | 106 | 106 | 107 | 107 |
| *Type 9 (Stiffness and mass (200%) irregularity at top storey)* | | | | | | | |
| B-IO | 100 | 100 | 100 | 95 | 93 | 87 | 87 |
| IO-LS | 6 | 6 | 6 | 11 | 13 | 20 | 21 |
| Total hinge | 106 | 106 | 106 | 106 | 106 | 107 | 108 |
| *Type 10 (Stiffness and mass (300%) irregularity at top storey)* | | | | | | | |
| B-IO | 99 | 99 | 96 | 95 | 88 | 87 | 84 |
| IO-LS | 8 | 8 | 11 | 12 | 19 | 22 | 25 |
| Total hinge | 107 | 107 | 107 | 107 | 107 | 109 | 109 |

**Table 10** Total number of hinges pattern for 0.45g as mainshock (Case 1)

| Case 1 | Mainshock | Aftershocks | | | | | |
|---|---|---|---|---|---|---|---|
| | | 1 | 2 | 3 | 4 | 5 | 6 |
| *Type 1 (Regular building)* | | | | | | | |
| B-IO | 85 | 83 | 64 | 51 | 43 | 49 | 53 |
| IO-LS | 22 | 24 | 45 | 61 | 47 | 41 | 18 |
| LS-CP | – | – | – | – | 23 | 33 | 34 |
| CP-C | – | – | – | – | – | – | 9 |
| C-D | – | – | – | – | – | – | 13 |
| Total hinge | 107 | 107 | 109 | 112 | 113 | 123 | 127 |
| *Type 2 (Stiffness irregularity at first storey)* | | | | | | | |
| B-IO | 74 | 65 | 41 | 42 | 42 | 53 | 47 |
| IO-LS | 29 | 38 | 57 | 45 | 31 | 7 | 18 |
| LS-CP | – | – | 8 | 28 | 14 | 18 | 10 |
| CP-C | – | – | – | – | – | – | – |
| C-D | – | – | – | – | 29 | 49 | 57 |
| Total hinge | 103 | 103 | 106 | 115 | 116 | 127 | 132 |
| *Type 3 (Stiffness and mass (200%) irregularity at first storey)* | | | | | | | |
| B-IO | 73 | 65 | 41 | 37 | 42 | 44 | 54 |
| IO-LS | 30 | 38 | 61 | 53 | 21 | 5 | 7 |
| LS-CP | – | – | 4 | 20 | 50 | 19 | 3 |
| CP-C | – | – | – | – | 3 | 1 | 1 |

(continued)

**Table 10** (continued)

| Case 1 | Mainshock | Aftershocks | | | | | |
|---|---|---|---|---|---|---|---|
| | | 1 | 2 | 3 | 4 | 5 | 6 |
| C-D | – | – | – | – | – | 49 | 65 |
| Total hinge | 103 | 103 | 106 | 110 | 116 | 118 | 130 |
| *Type 4 (Stiffness and mass (300%) irregularity at first storey)* | | | | | | | |
| B-IO | 74 | 70 | 47 | 36 | 42 | 44 | 53 |
| IO-LS | 29 | 33 | 58 | 53 | 25 | 6 | 7 |
| LS-CP | – | – | 1 | 20 | 49 | 19 | 3 |
| CP-C | – | – | – | – | – | 1 | 1 |
| C-D | – | – | – | – | – | 47 | 65 |
| Total hinge | 103 | 103 | 106 | 109 | 116 | 117 | 129 |
| *Type 5 (Stiffness irregularity at sixth storey)* | | | | | | | |
| B-IO | 99 | 96 | 77 | 64 | 55 | 60 | 64 |
| IO-LS | 18 | 21 | 44 | 56 | 47 | 33 | 17 |
| LS-CP | – | – | – | 4 | 23 | 38 | 26 |
| CP-C | – | | – | – | – | 1 | 2 |
| C-D | – | – | – | – | – | 2 | 29 |
| Total hinge | 117 | 117 | 121 | 124 | 125 | 134 | 138 |
| *Type 6 (Stiffness and mass (200%) irregularity at sixth storey)* | | | | | | | |
| B-IO | 95 | 95 | 65 | 51 | 48 | 49 | 54 |
| IO-LS | 21 | 21 | 53 | 58 | 45 | 16 | 2 |
| LS-CP | – | – | – | 11 | 29 | 42 | 17 |
| CP-C | – | – | – | – | – | 6 | 2 |
| C-D | – | – | – | – | – | 10 | 53 |
| Total hinge | 116 | 116 | 118 | 120 | 122 | 123 | 128 |
| *Type 7 (Stiffness and mass (300%) irregularity at sixth storey)* | | | | | | | |
| B-IO | 97 | 96 | 63 | 55 | 48 | 49 | 55 |
| IO-LS | 20 | 21 | 55 | 50 | 45 | 14 | 2 |
| LS-CP | – | – | – | 15 | 29 | 44 | 16 |
| CP-C | – | – | – | – | – | 6 | 2 |
| C-D | – | – | – | – | – | 9 | 53 |
| Total hinge | 117 | 117 | 118 | 120 | 122 | 122 | 130 |
| *Type 8 (Stiffness irregularity at top storey)* | | | | | | | |
| B-IO | 90 | 87 | 61 | 50 | 51 | 51 | 54 |
| IO-LS | 23 | 26 | 52 | 63 | 46 | 33 | 15 |
| LS-CP | – | – | – | 4 | 24 | 42 | 29 |
| CP-C | – | – | – | – | – | – | 5 |
| C-D | – | – | – | – | – | 1 | 30 |
| Total hinge | 113 | 113 | 113 | 117 | 121 | 127 | 133 |

(continued)

**Table 10** (continued)

| Case 1 | Mainshock | Aftershocks | | | | | |
|---|---|---|---|---|---|---|---|
| | | 1 | 2 | 3 | 4 | 5 | 6 |
| *Type 9 (Stiffness and mass (200%) irregularity at top storey)* | | | | | | | |
| B-IO | 89 | 87 | 64 | 50 | 48 | 49 | 53 |
| IO-LS | 24 | 26 | 52 | 62 | 49 | 31 | 15 |
| LS-CP | | | | 6 | 24 | 44 | 24 |
| CP-C | | | | | | - | 5 |
| C-D | | | | | | 3 | 36 |
| Total hinge | 113 | 113 | 116 | 118 | 121 | 127 | 133 |
| *Type 10 (Stiffness and mass (300%) irregularity at top storey)* | | | | | | | |
| B-IO | 89 | 88 | 63 | 49 | 42 | 45 | 48 |
| IO-LS | 24 | 25 | 52 | 60 | 52 | 29 | 17 |
| LS-CP | | | | 8 | 24 | 41 | 19 |
| CP-C | | | | | | 3 | 5 |
| C-D | | | | | | 7 | 41 |
| Total hinge | 113 | 113 | 115 | 117 | 118 | 125 | 130 |

**Table 11** Total number of hinges pattern for 0.36g as mainshock (Case 2)

| Case 2 | Mainshock | Aftershocks | | | | | |
|---|---|---|---|---|---|---|---|
| | | 1 | 2 | 3 | 4 | 5 | 6 |
| *Type 1 (Regular building)* | | | | | | | |
| B-IO | 101 | 101 | 101 | 97 | 93 | 93 | 89 |
| IO-LS | – | – | – | 4 | 8 | 8 | 12 |
| Total hinge | 101 | 101 | 101 | 101 | 101 | 101 | 101 |
| *Type 2 (Stiffness irregularity at first storey)* | | | | | | | |
| B-IO | 95 | 87 | 69 | 57 | 50 | 37 | 34 |
| IO-LS | 4 | 12 | 32 | 44 | 48 | 52 | 51 |
| LS-CP | – | – | – | – | 3 | 12 | 17 |
| Total hinge | 99 | 99 | 101 | 101 | 101 | 101 | 102 |
| *Type 3 (Stiffness and mass (200%) irregularity at first storey)* | | | | | | | |
| B-IO | 95 | 87 | 68 | 57 | 50 | 37 | 34 |
| IO-LS | 4 | 12 | 33 | 44 | 48 | 52 | 49 |
| LS-CP | – | – | – | – | 3 | 12 | 19 |
| Total hinge | 99 | 99 | 101 | 101 | 101 | 101 | 102 |
| *Type 4 (Stiffness and mass (300%) irregularity at first storey)* | | | | | | | |
| B-IO | 95 | 87 | 77 | 56 | 50 | 52 | 33 |
| IO-LS | 4 | 12 | 34 | 45 | 48 | 37 | 49 |
| LS-CP | | | | | 3 | 12 | 20 |
| Total hinge | 99 | 99 | 101 | 101 | 101 | 101 | 102 |

(continued)

**Table 11** (continued)

| Case 2 | Mainshock | Aftershocks | | | | | |
|---|---|---|---|---|---|---|---|
| | | 1 | 2 | 3 | 4 | 5 | 6 |
| *Type 5 (Stiffness irregularity at sixth storey)* | | | | | | | |
| B-IO | 108 | 108 | 109 | 102 | 91 | 90 | 89 |
| IO-LS | 8 | 8 | 8 | 16 | 27 | 30 | 37 |
| Total hinge | 116 | 116 | 117 | 118 | 118 | 120 | 126 |
| *Type 6 (Stiffness and mass (200%) irregularity at sixth storey)* | | | | | | | |
| B-IO | 108 | 108 | 108 | 101 | 91 | 89 | 89 |
| IO-LS | 8 | 8 | 9 | 17 | 27 | 31 | 37 |
| Total hinge | 116 | 116 | 117 | 118 | 118 | 120 | 126 |
| *Type 7 (Stiffness and mass (300%) irregularity at sixth storey)* | | | | | | | |
| B-IO | 106 | 106 | 106 | 100 | 91 | 89 | 84 |
| IO-LS | 8 | 8 | 9 | 16 | 25 | 31 | 40 |
| Total hinge | 114 | 114 | 115 | 116 | 116 | 120 | 124 |
| *Type 8 (Stiffness irregularity at top storey)* | | | | | | | |
| B-IO | 98 | 98 | 94 | 90 | 89 | 78 | 73 |
| IO-LS | 7 | 7 | 11 | 16 | 19 | 31 | 38 |
| Total hinge | 105 | 105 | 105 | 106 | 108 | 109 | 111 |
| *Type 9 (Stiffness and mass (200%) irregularity at top storey)* | | | | | | | |
| B-IO | 97 | 97 | 93 | 90 | 82 | 75 | 73 |
| IO-LS | 8 | 8 | 12 | 16 | 25 | 35 | 39 |
| Total hinge | 105 | 105 | 105 | 106 | 107 | 110 | 112 |
| *Type 10 (Stiffness and mass (300%) irregularity at top storey)* | | | | | | | |
| B-IO | 97 | 97 | 93 | 88 | 79 | 75 | 71 |
| IO-LS | 8 | 8 | 13 | 18 | 27 | 35 | 43 |
| Total hinge | 105 | 105 | 105 | 106 | 106 | 110 | 114 |

**Table 12** Total number of hinges pattern for 0.45g as mainshock (Case 2)

| Case 2 | Mainshock | Aftershocks | | | | | |
|---|---|---|---|---|---|---|---|
| | | 1 | 2 | 3 | 4 | 5 | 6 |
| *Type 1 (Regular building)* | | | | | | | |
| B-IO | 85 | 83 | 64 | 51 | 43 | 49 | 53 |
| IO-LS | 22 | 24 | 45 | 61 | 47 | 41 | 18 |
| LS-CP | – | – | – | – | 23 | 33 | 34 |
| CP-C | – | – | – | – | – | – | 9 |
| C-D | – | – | – | – | – | – | 13 |
| Total hinge | 107 | 107 | 109 | 112 | 113 | 123 | 127 |
| *Type 2 (Stiffness irregularity at first storey)* | | | | | | | |
| B-IO | 75 | 71 | 56 | 52 | 52 | 53 | 45 |
| IO-LS | 33 | 37 | 49 | 44 | 27 | 16 | 23 |
| LS-CP | – | – | 11 | 22 | 28 | 20 | 20 |

(continued)

**Table 12** (continued)

| Case 2 | Mainshock | Aftershocks | | | | | |
|---|---|---|---|---|---|---|---|
| | | 1 | 2 | 3 | 4 | 5 | 6 |
| CP-C | – | – | – | – | 1 | – | – |
| C-D | – | – | – | – | 16 | 38 | 42 |
| Total hinge | 108 | 108 | 116 | 118 | 124 | 127 | 130 |
| *Type 3 (Stiffness and mass (200%) irregularity at first storey)* | | | | | | | |
| B-IO | 78 | 74 | 58 | 50 | 51 | 53 | 49 |
| IO-LS | 31 | 35 | 46 | 48 | 33 | 16 | 18 |
| LS-CP | – | – | 9 | 18 | 30 | 21 | 20 |
| CP-C | – | – | – | – | – | – | – |
| C-D | – | – | – | – | 9 | 37 | 42 |
| Total hinge | 109 | 109 | 113 | 116 | 123 | 127 | 129 |
| *Type 4 (Stiffness and mass (300%) irregularity at first storey)* | | | | | | | |
| B-IO | 77 | 73 | 57 | 51 | 53 | 54 | 43 |
| IO-LS | 32 | 36 | 44 | 42 | 29 | 16 | 25 |
| LS-CP | – | – | 11 | 19 | 32 | 20 | 20 |
| CP-C | – | – | – | – | – | – | – |
| C-D | – | – | – | – | 11 | 38 | 42 |
| Total hinge | 109 | 109 | 112 | 112 | 125 | 128 | 130 |
| *Type 5 (Stiffness irregularity at sixth storey)* | | | | | | | |
| B-IO | 100 | 100 | 77 | 64 | 47 | 51 | 60 |
| IO-LS | 18 | 18 | 41 | 46 | 58 | 40 | 20 |
| LS-CP | – | – | – | 12 | 18 | 23 | 29 |
| CP-C | – | – | – | – | – | 6 | 3 |
| C-D | – | – | – | – | – | 10 | 27 |
| Total hinge | 118 | 118 | 118 | 122 | 123 | 130 | 139 |
| *Type 6 (Stiffness and mass (200%) irregularity at sixth storey)* | | | | | | | |
| B-IO | 100 | 100 | 77 | 64 | 51 | 51 | 58 |
| IO-LS | 18 | 18 | 41 | 46 | 55 | 40 | 22 |
| LS-CP | – | – | – | 12 | 18 | 21 | 29 |
| CP-C | – | – | – | – | – | 10 | 3 |
| C-D | – | – | – | – | – | 8 | 27 |
| Total hinge | 118 | 118 | 118 | 122 | 124 | 130 | 139 |
| *Type 7 (Stiffness and mass (300%) irregularity at sixth storey)* | | | | | | | |
| B-IO | 98 | 105 | 85 | 70 | 55 | 52 | 55 |
| IO-LS | 14 | 14 | 34 | 41 | 53 | 43 | 26 |
| LS-CP | – | – | – | 10 | 16 | 27 | 25 |
| CP-C | – | – | – | – | – | 3 | 3 |
| C-D | – | – | – | – | – | 6 | 25 |
| Total hinge | 112 | 119 | 119 | 121 | 124 | 131 | 135 |

(continued)

**Table 12** (continued)

| Case 2 | Mainshock | Aftershocks | | | | | |
|---|---|---|---|---|---|---|---|
| | | 1 | 2 | 3 | 4 | 5 | 6 |
| *Type 8 (Stiffness irregularity at top storey)* | | | | | | | |
| B-IO | 71 | 71 | 68 | 68 | 70 | 73 | 76 |
| IO-LS | 55 | 57 | 55 | 42 | 11 | 7 | 13 |
| LS-CP | – | – | 14 | 32 | 21 | 4 | 4 |
| CP-C | – | – | – | – | 2 | 1 | – |
| C-D | – | – | – | – | 42 | 68 | 69 |
| Total hinge | 126 | 128 | 137 | 142 | 146 | 153 | 162 |
| *Type 9 (Stiffness and mass (200%) irregularity at top storey)* | | | | | | | |
| B-IO | 88 | 81 | 50 | 40 | 36 | 36 | 40 |
| IO-LS | 23 | 30 | 61 | 55 | 60 | 37 | 25 |
| LS-CP | | | | 19 | 27 | 31 | 21 |
| CP-C | | | | | | 9 | 6 |
| C-D | | | | | | 15 | 42 |
| Total hinge | 111 | 111 | 111 | 114 | 123 | 128 | 134 |
| *Type 10 (Stiffness and mass (300%) irregularity at top storey)* | | | | | | | |
| B-IO | 89 | 84 | 51 | 44 | 44 | 38 | 42 |
| IO-LS | 22 | 27 | 60 | 55 | 57 | 40 | 31 |
| LS-CP | | | | 18 | 25 | 29 | 16 |
| CP-C | | | | | | 12 | 8 |
| C-D | | | | | | 10 | 39 |
| Total hinge | 111 | 111 | 111 | 117 | 126 | 129 | 136 |

## 5 Conclusion

The conclusions obtained after the analyses of 4 bays/11 storey frames with considered stiffness irregularity, and combined stiffness–mass irregularity together at first, sixth, and top storey level under the influence of sequential ground motions having PGAs 0.36g and 0.45g are:

1. In the case of regular buildings for lower or higher PGAs, the severity of response of the frame is less severe than the irregular frames. At higher PGAs, the irregular frames collapse during earlier aftershock, and the regular frame collapses under last aftershocks.
2. In case of stiffness irregularity alone or stiffness irregularity combined with mass irregularity, the responses are higher for the first storey as compared with the sixth and the top storey.
3. When the stiffness decreases by 43%:

   - In the first storey, the stiffness irregularity when combined with increasing mass has a positive response on the frame and it delays the collapse of the structure in higher PGAs.

- On the top storey as well as the sixth storey when stiffness irregularity is introduced with increasing mass, there is an increment in the severity of state of damage to the structures.
- The response of the frame to stiffness irregularity at the first storey is more severe and collapse occurs under earlier aftershock as compared with the frames with similar irregularity in sixth and top storey.

4. When the stiffness decreases by 65%:

   - In the first storey, the stiffness irregularity when combined with increasing mass irregularity has a negligible effect on the response of the frame.
   - The stiffness irregularity along with increasing mass irregularity in the sixth story reduces the severity of response of the frame at the mid-level.
   - In the top storey, the stiffness irregularity when combined with increasing mass irregularity produces a favourable effect on the response of the frame and delays the collapse of the structure under higher PGAs.

## References

1. Christchurch earthquake of 2010–2011. https://www.britannica.com/event/Christchurch-earthquakes-of-2010-2011
2. Nepal earthquake of 2015. https://www.britannica.com/topic/Nepal-earthquake-of-2015
3. Loulelis D, Hatzigeorgiou G, Beskos D (2012) Moment resisting steel frames under repeated earthquakes. Earthq Struct3(3–4):231–248; Faisal A, Majid TA, Hatzigeorgiou GD (2013) Investigation of story ductility demands of inelastic concrete frames subjected to repeated earthquakes. Soil Dyn Earthq Eng 44:42–53
4. Goda K (2012) Nonlinear response potential of mainshock–aftershock sequences from Japanese earthquakes. Bull Seismol Soc Am 102(5):2139–2156
5. Indian Standard, IS 1893 (2016) Criteria for earthquake resistance design of structures, Part-I. Bureau of Indian Standards, New Delhi
6. Indian Standard, IS 456 (2000) Indian standard plain and reinforced concrete—code of practice. Bureau of Indian Standards, New Delhi, India
7. SAP2000, C.S.I. (2016) Analysis reference manual. Computer and Structures Inc., Berkeley
8. Hatzivassiliou M, Hatzigeorgiou GD (2015) Seismic sequence effects on three-dimensional reinforced concrete buildings. Soil Dyn Earthq Eng 72:77–88
9. Raghunandan M, Liel AB, Luco N (2015) Aftershock collapse vulnerability assessment of reinforced concrete frame structures. Earthq Eng Struct Dyn44(3):419–439(4)
10. Hatzigeorgiou G (2010) Ductility demands control under repeated earthquakes using appropriate force reduction factors. J Earthq Tsunami 4(03):231–250
11. Hosseinpour F, Abdelnaby A (2017) Fragility curves for RC frames under multiple earthquakes. Soil Dyn Earthq Eng 98(7):222–234
12. FEMA-356 (2000) Prestandard and commentary for the seismic rehabilitation of buildings. Federal Emergency Management Agency, Washington DC
13. Hatzigeorgiou GD, Beskos DE (2009) Inelastic displacement ratios for SDOF structures subjected to repeated earthquakes. Eng Struct 31(11):2744–2755
14. Amadio C, Fragiacomo M, Rajgelj S (2003) The effects of repeated earthquake ground motions on the non-linear response of SDOF systems. Earthq Eng Struct Dynam 32(2):291–308
15. Hatzigeorgiou G (2010) Behavior factors for nonlinear structures subjected to multiple near-fault earthquakes. Comput Struct 88(5–6):309–321

16. Hatzigeorgiou GD (2010) Ductility demand spectra for multiple near-and far-fault earthquakes. Soil Dyn Earthq Eng 30(4):170–183
17. Moustafa A, Takewaki I (2011) Response of nonlinear single-degree-of-freedom structures to random acceleration sequences. Eng Struct 33(4):1251–1258
18. Zhai C-H, Wen W-P, Chen Z, Li S, Xie L-L (2013) Damage spectra for the mainshock–aftershock sequence-type ground motions. Soil Dyn Earthq Eng 45:1–12
19. Zhai C-H, Wen W-P, Li S, Chen Z, Chang Z, Xie L-L (2014) The damage investigation of inelastic SDOF structure under the mainshock–aftershock sequence-type ground motions. Soil Dyn Earthq Eng 59:30–41
20. Zhang Y, Chen J, Sun C (2017) Damage-based strength reduction factor for nonlinear structures subjected to sequence-type ground motions. Soil Dyn Earthq Eng 92:298–311
21. Hatzigeorgiou GD, Liolios AA (2010) Nonlinear behaviour of RC frames under repeated strong ground motions. Soil Dyn Earthq Eng 30(10):1010–1025

# Comparative Study on Static and Dynamic Analysis of RC Buildings of Different Heights in Different Seismic Zones

**Rajdeep Saharia, Debashis Bhuyan, Saunak Chowdhury, and Karabi Bharadwaj**

## 1 Introduction

An earthquake is a random and unpredictable event. It can cause significant damage to structure as well as human life which makes it one of the most destructive natural calamities. With the increase in total world population the number of high-rise building is increasing exponentially in the modern time. In a country like India which has been hit by several high magnitude earthquakes many times in the past, the seismic analysis and earthquake resistant design of structures have become prime challenges for all the civil engineers. Seismic analysis is mainly concerned with the behaviour of a structure under the action of earthquake loads. Two types of methods are used for analysis of earthquake loads, namely, static analysis method and dynamic analysis method. Static method, which includes Equivalent Static Method, does not take the dynamic behaviour of the loads into consideration and assumes that during an earthquake the building responds in its fundamental mode. On the contrary, dynamic analysis method, which includes Response Spectrum Method, considers multiple mode of response and takes the dynamic aspect of the loads acting on a building into consideration.

In this study, buildings of different heights located in different seismic zones are analysed using both static and dynamic analysis methods in STAAD-Pro software and the responses of various members of the buildings are compared for both the methods.

R. Saharia (✉) · D. Bhuyan · S. Chowdhury · K. Bharadwaj
Department of Civil Engineering, Tezpur University, Sonitpur, Assam, India

K. Bharadwaj
e-mail: karabi@tezu.ernet.in



S. Kolathayar and S. C. Chian (eds.), *Recent Advances in Earthquake Engineering*, Lecture Notes in Civil Engineering 175,
https://doi.org/10.1007/978-981-16-4617-1_13

## 2 Literature Review

Gottala and Yajdhani [1] carried out static and dynamic analysis of G + 9 building in STAAD-Pro as per the IS 1893(Part-1): 2002. Comparing the values of the responses obtained from both the methods, the authors found that the moments obtained from dynamic analysis are 35–45% higher than those obtained from static analysis. It was also observed that nodal displacements values are 50% higher for dynamic analysis than those for static analysis.

Sharma and Maru [2] performed static and dynamic analysis for regular G + 30 buildings situated in zone-II and zone-III. The structure had a plan area of 25 m × 45 m with a storey height of 3.6 m each. The authors concluded that the values of moments and displacements obtained from dynamic analysis were 10–15% and 17–28% higher, respectively, than those obtained from static analysis.

## 3 Objective

The IS 1893(Part 1): 2016 recommends the use of Equivalent Static Method only for regular buildings of height less than 15 m situated in Seismic Zone II and Response Spectrum Method for all the buildings other than regular building of height less than 15 m located in Zone II [3]. However, in common practice it is observed that the Equivalent Static Method is also used for analysis of buildings other than those suggested by the code. The use of inappropriate method may significantly affect the performance of the building during an earthquake.

The objective of this study is to make a comparative analysis between responses obtained from Equivalent Static Method and Response Spectrum Method. For this purpose, G + 3, G + 5 and G + 8 buildings situated in Seismic Zone II and Seismic Zone V are modelled in STAAD-Pro software. The analyses of these buildings are carried out in the software by both the methods as per Indian Standard code and the responses like axial force, shear force, bending moment and nodal displacement are compared.

## 4 Methods of Seismic Analysis

### *4.1 Equivalent Static Method (ESM)*

The dynamic nature of the loads acting on the building must be taken into account for earthquake resistant design of buildings. However, in this simplified technique, the effect of earthquake force is substituted by a static force that is distributed laterally on a structure by using formulas given in the code [4]. The basic principle behind this method is the approximation of a MDOF system to a SDOF system that responds in its fundamental mode. In most of the codes of practice, this method is permitted for regular, low-to-medium rise buildings in lower seismic zones.

### *4.2 Response Spectrum Method (RSM)*

Response spectra curves represent the maximum response of an idealized SDOF system subjected to a particular earthquake ground motion corresponding to its natural time period. A building possesses multiple modes of vibration during earthquake shaking. All the modes of responses are taken into account in response spectrum method. Depending on the modal frequency and modal mass, a response is obtained from the design spectrum corresponding to each mode. In the end, an estimate of the total response of the structure is found out by combining all the modes [4]. In most of the codes of practice, this method is permitted for irregular, high-rise buildings in higher seismic zones.

## 5 Building Configuration

In our study, we have considered three models of G + 3, G + 5 and G + 8 buildings of equal plan area in both seismic zones II and V. Some general specifications related to building models and soil data are listed below [5].

| | |
|---|---|
| No. of bays | 5 in both x and y-direction |
| Bay length | 3 m in both x and y-direction |
| Storey height | 3 m |
| Plan dimension | 15 m × 15 m |
| Size of beam | 300 mm × 450 mm |
| Size of column: | |
| For 12 m high (G + 3) building | 300 mm × 300 mm |
| For 18 m high (G + 5) building | 350 mm × 400 mm |
| For 27 m high (G + 8) building | 450 mm × 400 mm |
| Wall thickness | 150 mm |
| Thickness of slab | 100 mm |
| Type of soil | Type-II, medium soil as per IS 1893 (Part 1): 2016 |
| Grade of concrete | M25 |
| Grade of reinforcement | Fe415 |
| Specific weight of concrete | 25 $kNm^{-3}$ |
| Specific weight of infill | 20 $kNm^{-3}$ |

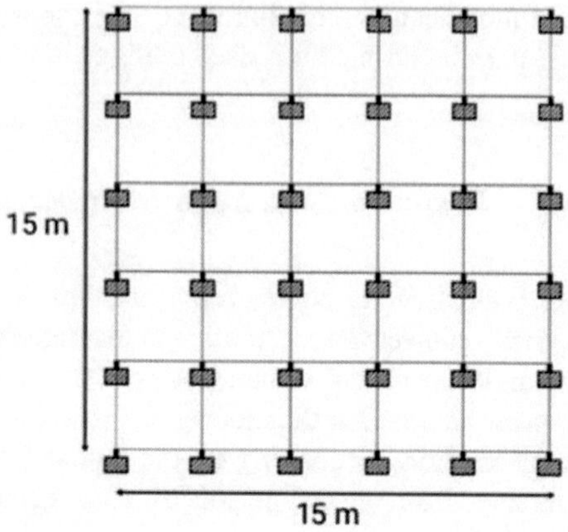


Fig. 1 Plan view of building models

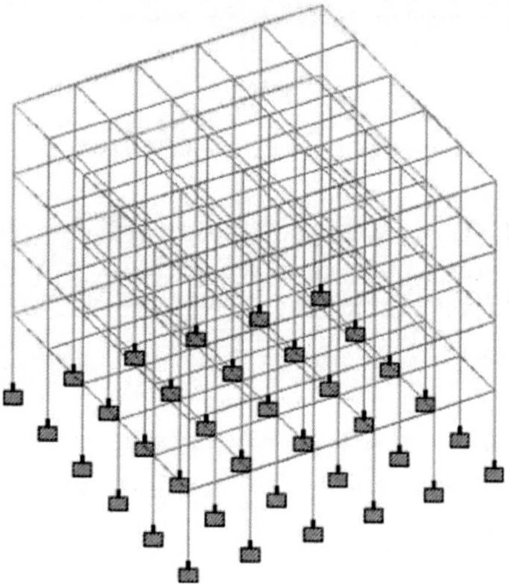

Fig. 2 3D view of 12 m high building

## 6 Results and Observations

Analyses of the buildings of various heights situated in seismic zone II and V were carried out and the responses were compared.

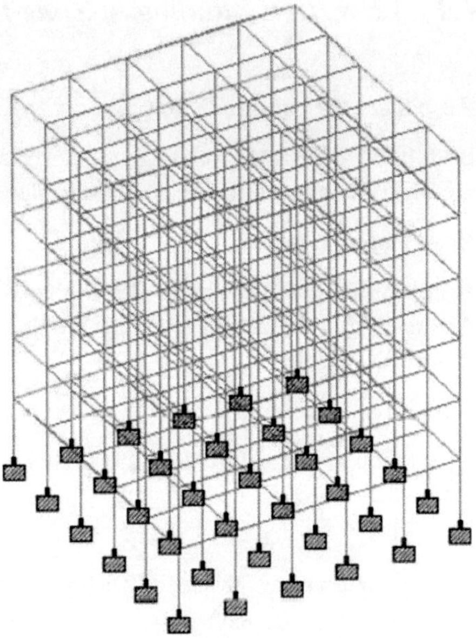

**Fig. 3** 3D view of 18 m high building

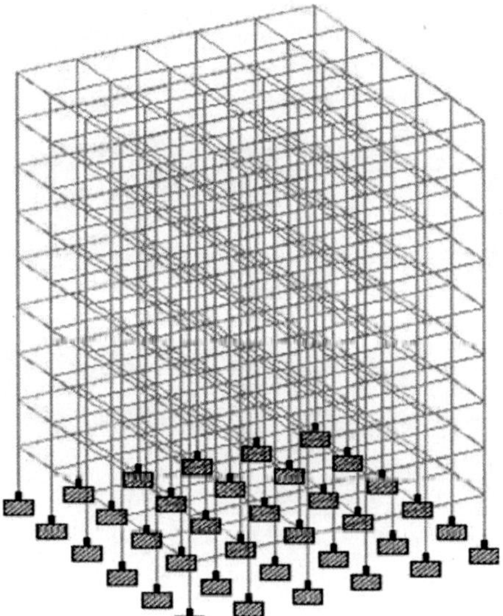

**Fig. 4** 3D view of 27 m high building

## *6.1 12 m High Building in Zone II*

**Responses for columns**

Both ESM and RSM gives same values of axial forces. Values of bending moments are 7–35% higher for RSM than the values obtained from ESM (Table 1).

**Responses for beams**

Values of shear forces are 4–14% higher for RSM than those obtained from ESM. Again, values of bending moments are 9–26% higher for RSM than those obtained from ESM (Table 2).

**Nodal displacements**

Nodal displacements are 20–32% higher for RSM than those obtained from ESM (Table 3).

**Table. 1** Responses in column for 12 m high building in zone II

| Column No. | Location | Axial Force (kN) | | Bending Moment (kNm) | |
|---|---|---|---|---|---|
| | | ESM | RSM | ESM | RSM |
| 199 | Ground level | 702.3 | 702.3 | −23.3 | −35.5 |
| 205 | 1st floor | 518.9 | 518.9 | 22.7 | 31.5 |
| 211 | 2nd floor | 338.7 | 338.7 | 20.3 | 24.4 |
| 217 | 3rd floor | 160.4 | 160.4 | 13.2 | 14.1 |

**Table. 2** Responses in beam for 12 m high building in zone II

| Beam No. | Location | Shear Force (kN) | | Bending Moment (kNm) | |
|---|---|---|---|---|---|
| | | ESM | RSM | ESM | RSM |
| 177 | 1st floor | 53.9 | 62.7 | 42.8 | 58.1 |
| 182 | 2nd floor | 50.4 | 55.9 | 42.2 | 51.6 |
| 187 | 3rd floor | 44.7 | 46.7 | 35.5 | 39.1 |
| 192 | 4th floor | 38.7 | 38.6 | −19.7 | −19.9 |

**Table. 3** Nodal displacements for 12 m high building in zone II

| Node No. | Location | Nodal displacement (mm) | |
|---|---|---|---|
| | | ESM | RSM |
| 127 | 1st floor | 2.75 | 4.07 |
| 133 | 2nd floor | 5.89 | 8.28 |
| 139 | 3rd floor | 8.66 | 11.431 |
| 145 | 4th floor | 10.45 | 13.16 |

## 6.2 12 m High Building in Zone V

**Responses for columns**

Both ESM and RSM gives the same values of axial forces. Values of bending moments are 9–35% higher for RSM than those obtained from ESM (Table 4).

**Responses for beams**

Shear forces are 16–24% higher for RSM than the values obtained from ESM. Again, bending moments are 20–30% higher for RSM than those obtained from ESM (Table 5).

**Nodal displacements**

Nodal displacements are 20–32% higher for RSM than those obtained from ESM (Table 6).

**Table. 4** Responses in column for 12 m high building in zone V

| Column No. | Location | Axial Force (kN) | | Bending Moment (kNm) | |
|---|---|---|---|---|---|
| | | ESM | RSM | ESM | RSM |
| 199 | Ground level | 702.3 | 702.3 | 82.8 | −127.9 |
| 205 | 1st floor | 518.9 | 518.9 | −80.6 | −113.0 |
| 211 | 2nd floor | 338.7 | 338.7 | 70.1 | 87.4 |
| 217 | 3rd floor | 160.4 | 160.4 | 45.8 | 50.3 |

**Table. 5** Responses in beam for 12 m high building in zone V

| Beam No. | Location | Shear Force (kN) | | Bending Moment (kNm) | |
|---|---|---|---|---|---|
| | | ESM | RSM | ESM | RSM |
| 177 | 1st floor | 102.5 | 134.4 | 126.7 | 181.6 |
| 182 | 2nd floor | 83.1 | 105.8 | 94.34 | −129.4 |
| 187 | 3rd floor | 94.2 | 113.9 | 117.5 | 151.4 |
| 192 | 4th floor | 82.4 | 98.1 | −92.3 | −116.1 |

**Table. 6** Nodal displacements for 12 m high building in zone V

| Node No. | Location | Nodal displacement (mm) | |
|---|---|---|---|
| | | ESM | RSM |
| 127 | 1st floor | 10 | 14.64 |
| 133 | 2nd floor | 21.22 | 29.8 |
| 139 | 3rd floor | 31.16 | 41.16 |
| 145 | 4th floor | 37.61 | 47.32 |

### *6.3 18 m High Building in Zone II*

**Responses for columns**

Both ESM and RSM gives the same values of axial forces. Again, values of bending moments are 4–12% higher for ESM than those obtained from RSM (Table 7).

**Responses for beams**

Values of shear forces are 10–14% higher in case of ESM than those obtained from RSM. Again, values of bending moments are 10–24% higher in case of ESM than those obtained from RSM (Table 8).

**Nodal displacements**

Nodal displacements are 20–32% higher for ESM than those obtained from RSM (Table 9).

**Table. 7** Responses in column for 18 m high building in zone II

| Column No. | Location | Axial force (kN) | | Bending moment (kNm) | |
|---|---|---|---|---|---|
| | | ESM | RSM | ESM | RSM |
| 295 | Ground level | 872.6 | 870.4 | –43.7 | 33.9 |
| 301 | 1st floor | 736.1 | 736.1 | –47.2 | –40.0 |
| 307 | 2nd floor | 594.1 | 594.1 | –49.8 | –40.4 |
| 313 | 3rd floor | 446.5 | 446.5 | –49.4 | –38.6 |
| 319 | 4th floor | 294.9 | 294.9 | –44.5 | –34.3 |
| 325 | 5th floor | 140.2 | 140.2 | –39.8 | –32.1 |

**Table. 8** Responses in beam for 18 m high building in zone II

| Beam No. | Location | Shear force (kN) | | Bending moment (kNm) | |
|---|---|---|---|---|---|
| | | ESM | RSM | ESM | RSM |
| 265 | 1st floor | 61.5 | 55.6 | –54.2 | 48.3 |
| 270 | 2nd floor | 59.8 | 52.7 | 59.8 | 48.1 |
| 275 | 3rd floor | 57.7 | 50.0 | 56.6 | 43.8 |
| 280 | 4th floor | 52.6 | 46.9 | 49.9 | 38.0 |
| 285 | 5th floor | 47.5 | 46.8 | 39.1 | 31.7 |
| 290 | 6th floor | 39.2 | 39.2 | 21.6 | 18.3 |

**Table. 9** Nodal displacements for 18 m high building in zone II

| Node No. | Location | Nodal displacement (mm) | |
|---|---|---|---|
| | | ESM | RSM |
| 175 | 1st floor | 2.31 | 1.84 |
| 181 | 2nd floor | 5.31 | 4.1 |
| 187 | 3rd floor | 8.23 | 6.13 |
| 193 | 4th floor | 10.84 | 7.80 |
| 199 | 5th floor | 12.90 | 9 |
| 205 | 6th floor | 14.15 | 9.67 |

## *6.4 18 m High Building in Zone V*

**Responses for columns**

Values of axial forces are 4–12% higher in case of ESM than those obtained from RSM. Again, values of bending moments are 15–32% higher for ESM than those obtained from RSM (Table 10).

**Responses for beams**

**Table. 10** Responses in column for 18 m high building in zone V

| Column No. | Location | Axial force (kN) | | Bending moment (kNm) | |
|---|---|---|---|---|---|
| | | ESM | RSM | ESM | RSM |
| 295 | Ground level | 1174.8 | 1041.3 | −151.9 | −128.6 |
| 301 | 1st floor | 961.6 | 852.7 | −148.8 | −120.7 |
| 307 | 2nd floor | 742.3 | 662.2 | −146.6 | −112.5 |
| 313 | 3rd floor | 527.1 | 477.1 | −137.7 | −98.8 |
| 319 | 4th floor | 324.8 | 301.8 | −116.1 | −79.2 |
| 325 | 5th floor | 145.6 | 140.7 | −84.3 | −57.5 |

**Table. 11** Responses in beam for 18 m high building in zone V

| Beam No. | Location | Shear force (kN) | | Bending moment (kNm) | |
|---|---|---|---|---|---|
| | | ESM | RSM | ESM | RSM |
| 265 | 1st floor | 132.2 | 111.0 | 169.8 | 139.9 |
| 270 | 2nd floor | 129.9 | 104.3 | 174.9 | 126.5 |
| 275 | 3rd floor | 120.9 | 92.9 | 160.9 | 114.7 |
| 280 | 4th floor | 104.8 | 78.9 | 128.1 | 92.0 |
| 285 | 5th floor | 76.2 | 61.3 | 93.8 | 63.2 |
| 290 | 6th floor | 48.8 | 41.9 | 42.8 | 30.0 |

**Table. 12** Nodal displacements for 18 m high building in zone V

| Node No. | Location | Nodal displacement (mm) | |
|---|---|---|---|
| | | ESM | RSM |
| 176 | 1st floor | 8.35 | 6.63 |
| 182 | 2nd floor | 19.13 | 14.77 |
| 188 | 3rd floor | 29.61 | 22.07 |
| 194 | 4th floor | 39.01 | 28.04 |
| 200 | 5th floor | 46.42 | 32.36 |
| 206 | 6th floor | 50.87 | 34.75 |

Values of shear forces are found to be 14–25% higher for ESM than those obtained from RSM. Again, values of bending moments are 18–32% higher for ESM than those obtained from RSM (Table 11).

**Nodal displacements**

Nodal displacements are 20–32% higher for ESM than those obtained from RSM (Table 12).

## *6.5 27 m High Building in Zone II*

**Responses for column**

Compared to RSM, the ESM gives 15–20% and 25–35% higher values of axial forces and bending moment, respectively (Table 13).

**Responses for beams**

**Table. 13** Responses in column for 27 m high building in zone II

| Column No. | Location | Axial force (kN) | | Bending moment (kNm) | |
|---|---|---|---|---|---|
| | | ESM | RSM | ESM | RSM |
| 145 | Ground level | 1742.4 | 1403.4 | 59 | 41.6 |
| 151 | 1st floor | 1556.5 | 1264.3 | 61 | 44.1 |
| 157 | 2nd floor | 1369.8 | 1117.6 | 63 | 46.8 |
| 163 | 3rd floor | 1182.6 | 964.9 | 67 | 45.7 |
| 169 | 4th floor | 990.8 | 807.6 | 59 | 47.6 |
| 175 | 5th floor | 795.5 | 646.8 | 63 | 46.9 |
| 181 | 6th floor | 597.2 | 483.4 | 61 | 44.9 |
| 187 | 7th floor | 397.9 | 318.2 | 53 | 40.3 |
| 193 | 8th floor | 185.7 | 151.0 | 51 | 42.4 |

**Table. 14** Responses in beam for 27 m high building in zone II

| Beam No. | Location | Shear force (kN) | | Bending moment (kNm) | |
|---|---|---|---|---|---|
| | | ESM | RSM | ESM | RSM |
| 101 | 1st floor | 63 | 54.1 | 62 | −49.45 |
| 106 | 2nd floor | 67 | 56.4 | 67 | −52.67 |
| 111 | 3rd floor | 67 | 55.4 | 67 | −51.10 |
| 116 | 4th floor | 66 | 53.8 | 67 | −48.56 |
| 121 | 5th floor | 64 | 51.8 | 63 | −45.55 |
| 126 | 6th floor | 60 | 49.4 | 58 | −41.87 |
| 131 | 7th floor | 55 | 46.2 | 44 | −37.06 |
| 136 | 8th floor | 50 | 42.4 | 41 | −31.39 |
| 141 | 9th floor | 39 | 38.6 | 28 | −24.16 |

**Table. 15** Nodal displacements for 27 m high building in zone II

| Node No. | Location | Nodal displacement (mm) | |
|---|---|---|---|
| | | ESM | RSM |
| 68 | 1st floor | 1.8 | 1.35 |
| 74 | 2nd floor | 4.69 | 3.30 |
| 80 | 3rd floor | 7.59 | 5.23 |
| 86 | 4th floor | 10.46 | 7.00 |
| 92 | 5th floor | 13.18 | 8.59 |
| 98 | 6th floor | 15.66 | 9.96 |
| 104 | 7th floor | 17.78 | 11.08 |
| 110 | 8th floor | 19.40 | 11.90 |
| 116 | 9th floor | 20.43 | 12.40 |

The values of shear forces and bending moments obtained by ESM were 3–20% and 3–30% higher, respectively (Table 14)

**Nodal displacements**

The nodal displacement values were found to be 25–39% higher in case of ESM (Table 15).

## 6.6 *27 m High Building in Zone V*

**Responses for columns**

The values of axial forces and bending moments obtained by ESM were 3–24% and 22–39% higher, respectively, than those obtained from RSM (Table 16).

**Table. 16** Responses in column for 27 m high building in zone V

| Column No. | Location | Axial force (kN) | | Bending moment (kNm) | |
|---|---|---|---|---|---|
| | | ESM | RSM | ESM | RSM |
| 145 | Ground level | 2248.2 | 1692.2 | 185.7 | 143.9 |
| 151 | 1st floor | 1551.6 | 1495.1 | 187.8 | 135.1 |
| 157 | 2nd floor | 1520.8 | 1289.3 | 190.7 | 131.2 |
| 163 | 3rd floor | 1457.2 | 1084.8 | 195.4 | 124.7 |
| 169 | 4th floor | 1121.4 | 883.5 | 180.2 | 116.8 |
| 175 | 5th floor | 795.8 | 687.0 | 176.1 | 107.8 |
| 181 | 6th floor | 597.7 | 497.4 | 155.2 | 95.8 |
| 187 | 7th floor | 399.2 | 318.5 | 124.5 | 77.7 |
| 193 | 8th floor | 190.1 | 151.2 | 91.7 | 62.7 |

**Table. 17** Responses in beam for 27 m high building in zone V

| Beam No. | Location | Shear force (kN) | | Bending moment (kNm) | |
|---|---|---|---|---|---|
| | | ESM | RSM | ESM | RSM |
| 101 | 1st floor | 140.5 | 109.0 | 172 | −132.1 |
| 106 | 2nd floor | 155.4 | 115.6 | 200 | −142.8 |
| 111 | 3rd floor | 155.9 | 112.6 | 200 | −136.6 |
| 116 | 4th floor | 151.7 | 106.4 | 194 | −126.9 |
| 121 | 5th floor | 143.7 | 99.0 | 182 | −115.7 |
| 126 | 6th floor | 130.5 | 90.0 | 162 | −102.1 |
| 131 | 7th floor | 111.9 | 78.3 | 134 | −84.6 |
| 136 | 8th floor | 88.6 | 64.4 | 99 | −63.5 |
| 141 | 9th floor | 60.4 | 48.2 | 52 | 39.9 |

**Responses for beams**

In case of beam, the values of shear force and bending moments obtained by ESM were 20–31% and 23–37% higher, respectively (Table 17).

**Nodal displacements**

The nodal displacement values obtained by ESM were 28–40% higher than those obtained by RSM (Table 18).

**Table. 18** Nodal displacements for 27 m high building in zone V

| Node No. | Location | Nodal displacement (mm) | |
|---|---|---|---|
| | | ESM | RSM |
| 68 | 1st floor | 6.79 | 4.85 |
| 74 | 2nd floor | 16.88 | 11.89 |
| 80 | 3rd floor | 27.35 | 18.82 |
| 86 | 4th floor | 37.65 | 25.23 |
| 92 | 5th floor | 47.46 | 30.94 |
| 98 | 6th floor | 56.39 | 35.87 |
| 104 | 7th floor | 64.02 | 39.89 |
| 110 | 8th floor | 69.86 | 42.85 |
| 116 | 9th floor | 73.52 | 44.68 |

### 6.7 *Observations from Manual Calculations*

We considered a G + 4 building situated in seismic zone V and applied ESM and RSM to it separately. In the comparative analysis of member responses of a critical frame, the following results were observed:

1. The values of axial forces in columns are 5–12% higher, and bending moments are 15–30% higher for ESM than those obtained from RSM.
2. The shear forces in beams are 6–12% higher and bending moments are 18–25% higher in case of ESM than those obtained from RSM.
3. Nodal displacements are found to be 20–30% higher in case of ESM than those obtained from RSM.

## 7 Conclusion

- For 12 m high building, in Seismic Zone II and V, both ESM and RSM give almost the same values of axial forces in columns. However, RSM gives higher values of bending moments in columns than ESM in both the zones. Similarly, shear forces and bending moments in beams and nodal displacements are found to be higher in case of RSM in both the zones.
- For 18 m high building, in Seismic Zone II, no change is observed in the values of axial forces for columns obtained from both the methods. But, in Seismic Zone V, the axial forces in columns are found to be marginally higher in case of ESM than those obtained from RSM. Moreover, bending moments in columns and beams, shear forces in beams and nodal displacements are found to be higher from ESM than those obtained from RSM in both the Seismic Zones.
- For 27 m high building, all the responses of beams and columns and nodal displacements are found to be higher in case ESM in both the Seismic Zones.

- From our comparative analysis it can be concluded that RSM gives higher values of responses for low rise buildings, while ESM gives higher value of responses for high rise buildings.
- Hence, ESM is more economical in case of low-rise buildings and for high rise buildings RSM is found to be more cost effective.

## References

1. Gottala A, Yajdhani S (2015) Comparative study of static and dynamic seismic analysis of multistoried building. IJSTE Int J Sci Technol Eng 2
2. Sharma A, Maru S (2014) Dynamic analysis of multistoried regular building. IOSR J Mech Civil Eng (IOSR-JMCE) 11(1). e-ISSN: 2278–1684, p-ISSN: 2320–334X, Ver. II
3. IS 1893 (2016) Indian standard criteria for earthquake resistant design of structures-part I: General provisions and buildings (Sixth Revision). Bureau of Indian Standards New Delhi
4. Agarwal P, Shrikhande M (2014) Earthquake resistant design of structures, 13th edn. Asoke K, Ghosh (eds) PHI Learning Private Limited, Rimjhim House, 111, Patparganj Industrial Estate, Delhi-110092
5. IS 456 (2000) Indian standard criteria for plain and reinforced concrete-code of practice (Fourth Revision). Bureau of Indian Standards New Delhi

# Effect of Dog-Legged Staircase on the Seismic Response of Hill Buildings

**Ahmed Bilal, Zaid Mohammad, and Abdul Baqi**

## 1 Introduction

The RC framed buildings built on hill slopes show entirely different dynamic response as compared to those resting on the level ground under earthquake loads [1]. Two building configurations, viz. stepback and setback-stepback are the most common building types that are being constructed on steep slopes. Due to the unsymmetrical structural configuration in the hill buildings, the centre of stiffness and centre of mass vary along the height and increase shear force moment values in structural members when subjected to the lateral forces. Also, it was observed that the columns on the uphill side of the building have larger stiffness and showed higher shear forces in comparison to the columns in downhill side, proving to be highly vulnerable under seismic loads [2–9].

Staircase is one of the most important of the building systems due to its functional importance. Due to its complex modelling, staircase is designed separately from the building as a non-structural element. However, previous studies [10–12, 15] have shown that the staircase elements significantly alter the seismic response of the structure. Thus, the influence of the staircase in the dynamic analysis of RC framed buildings could not be ignored. Lavado and Gonzalez [10] investigated RC frame structures with and without staircase elements and observed a local rigidity effect in the columns surrounding stairwell, hence enhancing the axial stresses in connecting structural frame members and their affecting strength and ductility demands. Previous studies [11, 12] also suggested that the staircase increases structural strength and lateral stiffness of the framed building resulting into reduction of modal time periods.

A. Bilal
Civil Engineering Section, University Polytechnic, A. M. U., Aligarh, Uttar Pradesh, India

Z. Mohammad (✉) · A. Baqi
Department of Civil Engineering, Z.H.C.E.T., A. M. U., Aligarh, Uttar Pradesh, India
e-mail: zmohammad_co@myamu.ac.in



S. Kolathayar and S. C. Chian (eds.), *Recent Advances in Earthquake Engineering*, Lecture Notes in Civil Engineering 175,
https://doi.org/10.1007/978-981-16-4617-1_14

This particular behaviour also attracts seismic shear forces due to which connected columns in staircase frame could fail in short column effect. Qiwang [13] reported that the Quang staircase arrangement may affect the torsional mode of the building and location of the staircase must be analysed for dynamic effects. Zhang and Yuan [14] and Mohammad et al. [15] investigated the influence of dog-legged staircase on the seismic performance of RC framed structure and found that the going members contribute stiffness as K-type bracings that induce additional deformations in the adjoining frame members. Pendhari and Sangai presented an elaborative study on the staircase modelling and its detrimental effects of stairs on the seismic behaviour of overall structure to avoid serious damages to the stair components and surrounding RC frame element [16]. Kumbhar et al. [17] presented a numerical study on the influence of location of stairs on the dynamic performance RC framed buildings using non-linear static analysis. It was found that ignorance of staircase in modelling led to unsafe design of structure. Aghajani-Delavar et al. [18] studied the effect of location of stairs on the seismic behaviour of a 6-storey building. It was concluded that the staircase should be considered in the design. Karaaslan and Avsar [19] presented a parametric study in which Modal as well as bi-directional Non-Linear Time History analysis were conducted and seismic parameters compared and a different damage pattern in the structural members was observed due to introduction of staircase.

The state-of-the-art studies conducted on hill buildings [1–9] discussed the structural performance of hill buildings. The influence of staircase structure on the dynamic performance of RC framed buildings built on plain ground has been explored [10–19]. However, none of the investigations were conducted on the behaviour of hill buildings with the consideration of staircase systems under earthquake loads. Moreover, as per codal provisions stated in IS 1893 (Part 1), it was recommended to perform modal analysis of the buildings that have geometrical, mass and stiffness irregularities [20]. Also, it was suggested that the non-linear behaviour of hill buildings should be investigated in order to ascertain the actual response of the structure. Thus, the present study explores the influence of most commonly constructed dog-legged staircase system on the seismic performance of two hill building configurations, viz. stepback and setback-stepback. The three-dimensional models were created using a finite element software, and examined by employing Equivalent Static and Response Spectrum Method of analysis. The seismic parameters obtained from the numerical study were discussed as variations in base shear, time period, storey drift values and shear forces at foundation levels. At last, the vulnerability and suitability of the different configurations against seismic excitations was compared.

## 2 Materials and Methods

This study evaluates the structural behaviour of two hill building configurations constructed on steep hill slope, viz. stepback and setback-stepback, under earthquake loads. The influence of dog-legged staircase system on the seismic performance of the considered configurations was analysed and location of the stairs was also varied to observe the dynamic behaviour of hill buildings. Both Equivalent Static and Response Spectrum methods were used to determine the seismic behaviour of the hill buildings. The seismic parameters obtained from the numerical study were discussed as variations in base shear, time period, storey drift values and shear forces at foundation levels in along as well as across hill slope direction. The material parameters have been taken from previously published study [1, 2]. Rigid frame diaphragm was considered in floor systems, and foundation supports were modelled to be fixed.

Six models of stepback and setback-stepback configurations were analysed with and without dog-legged staircase. The buildings were modelled with 5 bays in along as well as across slope directions. In all the models, the length of each bay was taken as 7 and 5 m in along and across the slope, respectively. The inclination of ground was assumed to be 27° [1, 2]. The staircase systems were introduced at two locations, viz. at the side bay and in the middle bay of the stepback building. Also, in case of stepback-setback configuration, the stairs were given at two places at different stories to provide the similar number of flights as in stepback buildings (see Fig. 1). The load due to the masonry infills has been considered at periphery of the building frames. The various geometric and seismic parameters were considered in the analysis have been mentioned in Table 1.

## 3 Results and Discussion

Both hill building configurations with and without staircase system were analysed for the seismic loads in along as well as across slope directions including the effect of accidental eccentricity as per codal provisions [20]. The three-dimensional models were analysed using Equivalent Static and Response Spectrum Method of analysis. The results obtained from the analyses were discussed in terms of time period for the first three modes, storey drift and lateral shear force at foundation level and compared within the considered configurations.

The modal time periods ascertained from the numerical analyses of the building models have been described in Table 2. It could be clearly observed that the modal time periods of the hill configurations were significantly decreased after the inclusion of staircase elements in the modelling and analysis. It can be observed that the time period value in Mode I was reduced to 85.3 and 84.4%, respectively, for stepback buildings with staircase situated at end and in the middle bay, in comparison to that of bare frame. Similarily, for setback-stepback configurations, buildings with staircase at end and middle bay showed 13.1 and 14.1% reduction in the time period as compared with that of bare frame building. Further, setback-stepback type buildings showed approximately 20% less fundamental time period than stepback type buildings with as well as without staircase elements.

The storey drift variation of both stepback and setback-stepback buillding configurations was evaluated in the direction of hill slope (see Figs. 2 and 3). It was observed that the staircase elements considerably affect the deformation behaviour

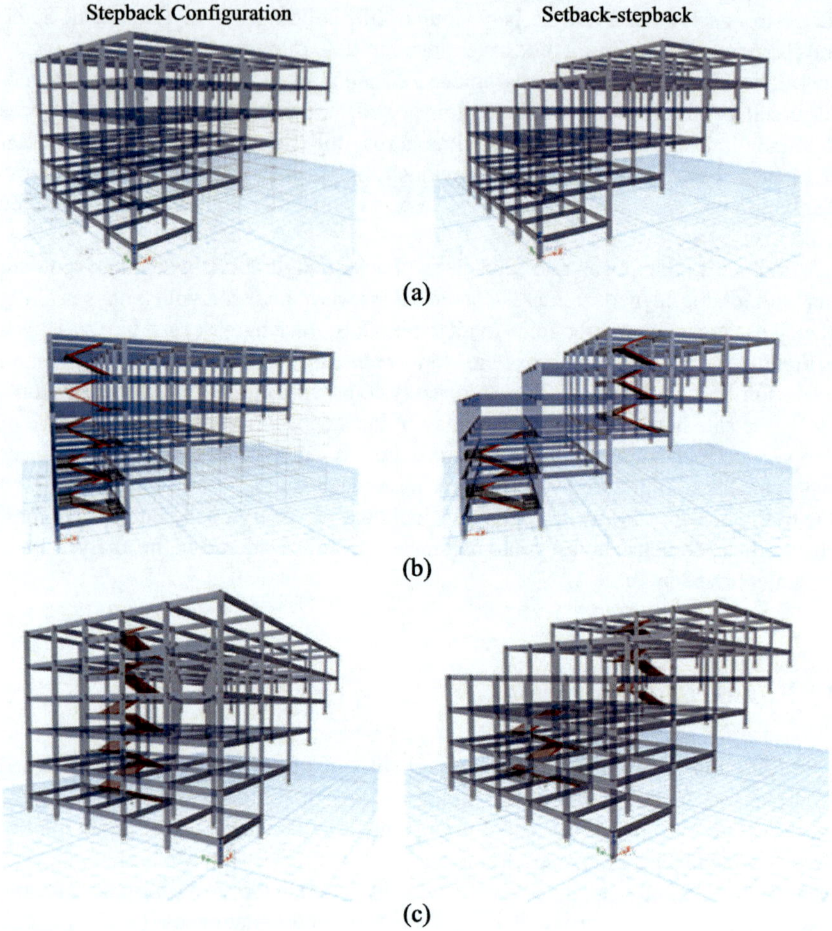


**Fig. 1** Different hill building configurations: **a** bare RC frame without staircase, **b** with staircase at side bay and **c** with staircase at middle bay

**Table 1** Parameters considered in the seismic analysis of hill building configurations [1, 2]

| Geometric parameters | Seismic parameters |
|---|---|
| Thickness of slab: 0.150 m | Zone: V |
| Floor height: 3.5 m | I: 1.5 |
| Depth of foundation: 1.75 m | R: 5 |
| Column size: 0.23 × 0.50 m | Soil type: I (Rock) |
| Beam size: 0.23 × 0.50 m | Live load: 3 kN/m$^2$ |
| Stair width: 2.5 m | Floor finish: 1.5 kN/m$^2$ |
| Landing width: 1.5 m | |

**Table 2** Modal time periods in different building configurations

| Configuration | Stepback | | | Setback-stepback | | |
|---|---|---|---|---|---|---|
| | Mode I | Mode II | Mode III | Mode I | Mode II | Mode III |
| Bare frame | 0.818 | 0.452 | 0.416 | 0.639 | 0.422 | 0.371 |
| With staircase at end bay | 0.698 | 0.438 | 0.386 | 0.555 | 0.393 | 0.337 |
| With staircase at middle bay | 0.690 | 0.459 | 0.395 | 0.549 | 0.397 | 0.338 |

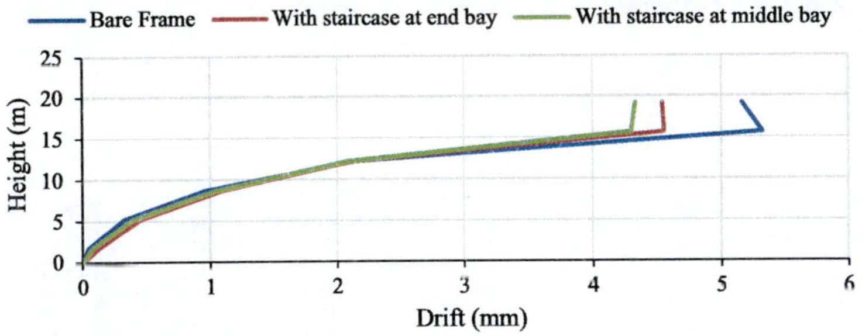


**Fig. 2** Storey drift variation in stepback building configuration subjected to earthquake loading in the direction of hill slope

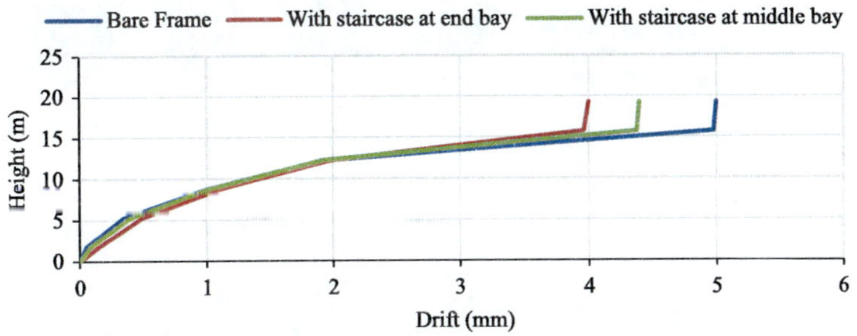


**Fig. 3** Storey drift variation in setback-stepback building configuration subjected to earthquake loading in the direction of hill slope

of both hill building configurations and effectively reduced the drift due to bracing effect provided by the dog-legged staircase structure. In case of stepback buildings, the storey drift was found to be reduced to 85.5 and 80.1% of that of bare frame, after the inclusion of staircase at end and middle bays, respectively. Whereas, the 79.9 and 87.9% drift value of that of bare frame setback-stepback building was onserved in buildings with staircase at end and middle bays, respectively. Further, setback-stepback configurations showed overall 9% less storey drift in comparison to stepback buildings, showing less vulnerability against lateral loads due to distributed structural weight across the entire structure.

The total shear force variation at the foundation of individual frames in along as well as across hill slope directions in stepback as well as setback-stepback building configurations have been observed (see Figs. 4 and 5). It could be observed that the columns in the upper hill side attract large amount of shear force (approximately ten fold) than the columns at down hill side when subjected to earthquake forces along the hill slope. Whereas, similar amount of shear forces were observed in the frames when subjected to seismic load in across slope direction. Further, it was concluded that both stepback and setback-stepback buildings have attracted more shear force after including the staircase in the analysis. Moreover, hill buildings with staircase

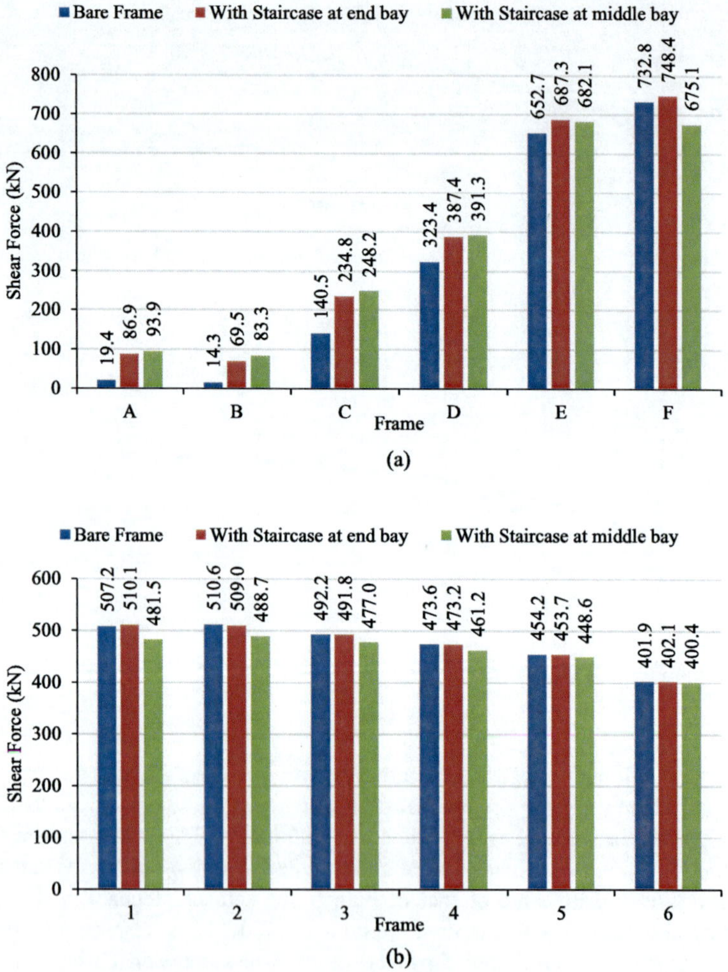


**Fig. 4** Base shear at foundation in various frames in stepback building configuration in **a** along and **b** across hill slope directions

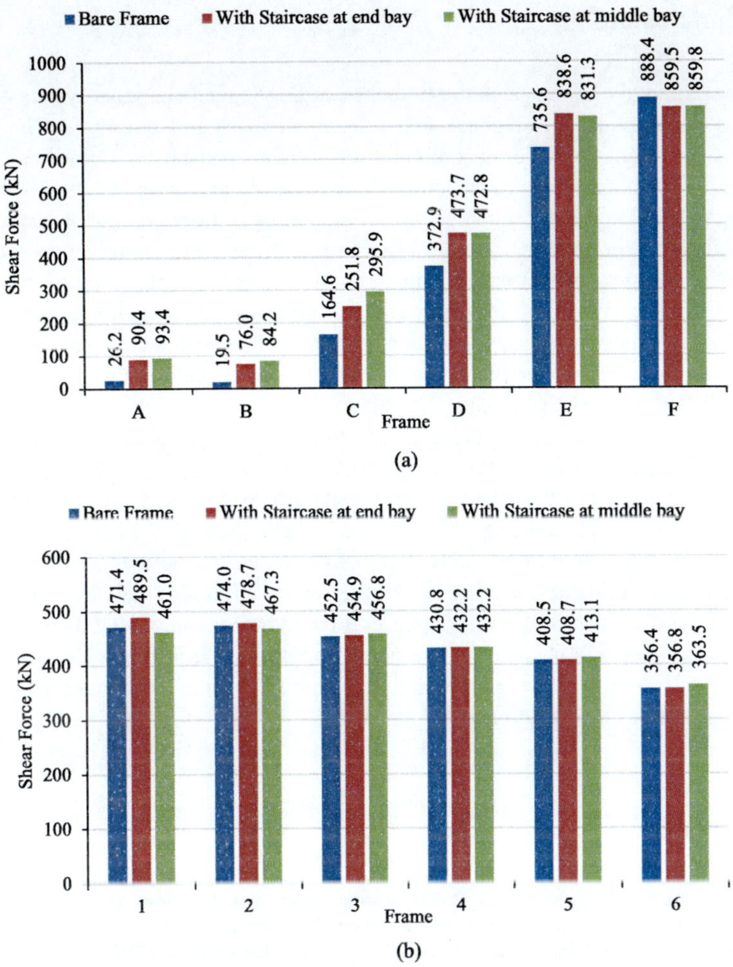


**Fig. 5** Base shear at foundation in various frames in setback-stepback building configuration in **a** along and **b** across hill slope directions

structure located at middle bay displayed less shear force than in buildings with staircase at end bay. Thus, make it most effective location for the staircase. It could be concluded that the modelling of staircase elements in the structure plays important role in the behaviour of the building and should not be ignored in the seismic analysis of earthquake vulnerable buildings such as hill buildings.

## 4 Conclusion

The present study investigates the influence of dog-legged staircase system on the seismic performance of stepback and setback-stepback hill buildings. The finite element models of the building configurations were analysed using Equivalent Static as well as Response Spectrum method. The dynamic properties of the hill buildings were evaluated and compared within the configurations. It was observed that the modal time periods of the both hill configurations were significantly reduced after the inclusion of staircase elements in the modelling and analysis. Further, setback-stepback type buildings showed approximately 20% less fundamental time period value than stepback type buildings with as well as without staircase elements. Moreover, the staircase elements considerably affect the deformation behaviour of both hill building configurations and effectively reduced the drift due to bracing effect provided by the dog-legged staircase structure. Also, setback-stepback configurations showed overall 9% less storey drift in comparison to stepback buildings, showing less vulnerability against lateral loads. It was observed that both stepback and setback-stepback buildings attracted more shear force after including the staircase in the analysis. Moreover, hill buildings with staircase structure located at middle bay displayed less shear force than in buildings with staircase at end bay. Thus, make it most effective location for the staircase. It could be concluded that the modelling of staircase elements in the structure plays important role in the behaviour of the building and should not be ignored in the seismic analysis of earthquake vulnerable buildings such as hill buildings.

## References


1. Mohammad Z, Baqi A, Arif M (2017) Seismic response of RC framed buildings resting on hill slopes. In: 11th international symposium on plasticity and impact mechanics (IMPLAST 2016). Procedia Engineering, vol 173. Elsevier, New Delhi, pp 1792–1799. https://doi.org/10.1016/j.proeng.2016.12.221
2. Mohammad Z, Razi MA, Baqi A (2021) Influence of masonry infill panels on the seismic performance of irregular buildings. In: Kumar SS, Raman SN, Bhattacharjee B, Bhattacharjee J (eds) Advances in geotechnics and structural engineering, vol 143. Lecture notes in civil engineering. Springer, Singapore, pp 1–11
3. Kumar S, Paul DK (1998) A simplified method for elastic seismic analysis. J Earthq Eng 2(2):241–266. https://doi.org/10.1080/13632469809350321
4. Birajdar BG, Nalawade SS (2004) Seismic analysis of buildings resting on sloping ground. In: 13th world conference on earthquake engineering (13WCEE). Paper no. 1472, Vancouver, B.C., Canada. https://www.iitk.ac.in/nicee/wcee/article/13_1472.pdf
5. Kadid A, Boumrkik A (2008) Pushover analysis of reinforced concrete frame structures. Asian J Civil Eng (Building & Housing) 75–83. https://doi.org/10.1007/s11803-013-0179-8
6. Kaushik HB, Rai DC, Jain SK (2008) A rational approach to analytical modelling of masonry infills in reinforced concrete frame buildings. In: 14th world conference on earthquake engineering (14WCEE). Corpus ID: 165155308, Beijing, China. https://www.iitk.ac.in/nicee/wcee/article/14_05-01-0317.PDF

7. Davis R, Krishnan P, Menon D, Prasad AM (2004) Effect of infill stiffness on seismic performance of multi-storey RC framed buildings in India. In: 13th world conference on earthquake engineering (13WCEE). Paper no. 1198, Vancouver, B.C., Canada. https://www.iitk.ac.in/nicee/wcee/article/13_1198.pdf
8. Murty CVR, Jain SK (2000) Beneficial influence of masonry infill walls on seismic performance of RC frame buildings. In: 12th world conference on earthquake engineering (12WCEE). Paper no. 1790, Auckland, New Zealand. https://www.iitk.ac.in/nicee/wcee/article/1790.pdf
9. Mohammad Z (2019) Effect of unreinforced masonry infills on seismic performance of hill buildings. VW Appl Sci 1(1):37–47. https://doi.org/10.36297/vw.applsci.v1i1.29
10. Lavado J, Gonzalez ML (2004) Influence of stair slabs in reinforced concrete buildings under seismic loads. In: Proceedings of the seventh international conference on computational structures technology, Stirlingshire, UK, 273
11. Cosenza E, Vederame GM, Zambrano A (2008) Seismic performance of stairs in the existing reinforced concrete building. In: 14th WCEE, Beijing, China
12. Zhu JJ, Wang CX, Lee Y (2011) Analysis of impacts of staircase upon seismic responses of structure. Appl Mech Mater 94–96(2011):469–475
13. Qiwang Su (2010) Finite element analysis of staircase under earthquake action. Adv Mater Res 163–167(2011):2964–2968
14. Zhang PC, Yuan Q (2011) The seismic design of staircase in frame structure. Adv Mater Res 368–373(2012):865–868
15. Mohammad Z, Danish M, Shariq M, Masood A, Baqi A (2013) Effect of staircase on RC frame structure under seismic load. In: Proceedings of international conference on trends and challenges in concrete structures, Ghaziabad, UP, India, pp 497–509
16. Pendhari AR, Sangai BD (2015) Modeling of staircase and its effect on the seismic performance of RCC building. Int J of Sci Technol 3(3):7–11
17. Kumbhar OG, Kumar R, Adhikary S (2015) Effect of staircase on seismic performance of RC frame building. Earthq Struct 9(2):375–390. https://doi.org/10.12989/eas.2015.9.2.375
18. Aghjani-Delavar M, Varnosfaderani MP, Bargi K (2017) Effects of staircase on the seismic performance of reinforced concrete frame buildings considering the position of the staircase. In: Proceedings of international conference on contemporary Iran on civil engineering architecture and urban development, Iran
19. Karaaslan A, Avsar O (2019) Seismic response of substandard RC frame buildings in consideration of staircases. Earthq Struct 17(3):283–295. https://doi.org/10.12989/eas.2019.17.3.283
20. IS 1893 (Part 1): 2016 (2016) Criteria for earthquake resistant design of structures. BIS, New Delhi

# Effect of Supporting Structure's Torsion on Floor Acceleration Demands in Buildings on Slopes

**Ankur Jain and Mitesh Surana**

## 1 Introduction

Components and systems that are neither a part of the gravity nor a part of the primary lateral load-resisting system but offer functionality to a building structure are termed as secondary systems or non-structural components (NSCs). The seismic response of these NSCs is either sensitive to the inertia forces or to the inter-story drifts. Thus, based on the sensitivity of their response, these NSCs can be subdivided under three main categories: (i) acceleration-sensitive NSCs, (ii) drift-sensitive NSCs, and (iii) combined acceleration- and drift-sensitive NSCs. The relative share of these NSCs to the total cost of the building can vary between 70–80% in the case of commercial buildings [1]. Specifically, in hospital buildings, the in-operability of these NSCs could hamper the post-earthquake relief operations, whereas, in the case of office buildings, it may lead to business interruption and downtime losses. Therefore, developing adequate seismic design provisions for ensuring seismic safety of NSCs on buildings got attention in the recent past.

Several studies have already been conducted in the past addressing seismic design of acceleration-sensitive NSCs [2]; however, the studies which especially focussed on incorporating the effect of supporting structures torsion on seismic response of NSCs are very limited. Yang and Huang [3] studied the behaviour of the NSCs mounted on a multi-story building, exposed to significant torsional deformations, under seismic actions. The amplification of equipment's acceleration due to the presence of torsion was observed to be dependent on the modal participation factor for the torsional modes of vibration. Agrawal and Datta [4] observed that yielding of the torsionally coupled primary system has a significant effect on the response of the secondary structure, which was observed to be reduced under the tuned condition and amplified

A. Jain · M. Surana (✉)
Indian Institute of Technology Ropar, Rupnagar, Punjab 140001, India
e-mail: msurana@iitrpr.ac.in



S. Kolathayar and S. C. Chian (eds.), *Recent Advances in Earthquake Engineering*, Lecture Notes in Civil Engineering 175,
https://doi.org/10.1007/978-981-16-4617-1_15

under the non-tuned condition. Agrawal and Datta [5] studied the response of the NSCs, mounted on the torsionally coupled primary structure, under bi-directional seismic ground motions and reported that under tuned conditions, with an increase in the eccentricity of the primary system, the response of the secondary system increases.

Aldeka et al. [6] studied seismic behaviour of the NSCs attached to irregular reinforced-concrete (RC) buildings. They found that the acceleration of NSCs increases with an increase in the input PGA, provided the primary structure responds elastically. They highlighted that NSCs attached to the flexible edge (FE) of high-rise buildings get more influenced by the torsional behaviour as compared to the low-rise buildings. Aldeka et al. [7] observed that the eccentricity ratio of the primary structure has an insignificant effect on the behaviour of the NSCs attached to the centre of rigidity (CR) of the primary structure. Surana et al. [8] investigated the floor acceleration demands in multi-story hillside step-back (SB) and split–foundation (SF) RC buildings. They found that both FEMA P750 and Eurocode 8 models undervalue the peak floor acceleration demands in hillside buildings. Vijayanarayanan and Goswami [9] conducted the linear and nonlinear time history analyses of some typical RC buildings resting on hill slopes. They observed the floor acceleration response along the FE to be higher than as that of the stiff edge (SE).

The past studies [2–9] reported that the dynamic characteristics of the primary structure plays a crucial role in seismic design of NSCs. So far, the various seismic design provisions which are in existence in different codes for seismic design of NSCs have been developed for regular buildings. In hilly terrain, the buildings are usually constructed following the land's slope to suit its geometry [8, 9]. These buildings on hill slopes pose both the plan and elevation irregularities due to the differences in the column heights within the same storey, and differences in the strength, mass, and stiffness of the successive stories, resting on a slope. As a consequence of the presence of plan irregularity, these buildings exhibit torsional response while subjected to excitation in across-slope direction. Thus, this study makes an attempt to study the effect of torsion on floor response of buildings resting on slope.

Accordingly, in this study, flat ground (FG) and SB buildings with two different heights (i.e., 2- and 4-storey) have been analysed to study their floor acceleration response. Floor response spectra (FRS) corresponding to NSC damping ratio of 5% are evaluated at two different floor levels, i.e., at the floor level immediately above the topmost foundation level and at the roof level, at different locations on the respective floors (i.e., at the FE, the CR and the SE). Torsional amplification factor (TAF), defined as the ratio between the floor spectral ordinate at the FE or SE to the floor spectral ordinate at the CR for any spectral period of interest, is computed in across-slope direction. A correlation of the TAF with the various parameters described in building codes to quantify the torsional irregularity in the buildings has been studied.

## 2 Numerical Study

### 2.1 Building and Modelling Details

A group of RC frame buildings with FG and SB configuration, having an identical plan shape as shown in Fig. 1, with two different building heights, i.e., 2- and 4-storey, are analysed in the current study. For SB buildings, the height above the topmost foundation level has been assumed to determine the number of storeys [8], whereas, the storeys below the topmost foundation level are obtained considering a slope angle of ~27°. The height of the storey is taken as 3.3 m and set constant for all the storeys. In SB buildings, the short columns are assigned a height of 1.1 and 2.75 m, in successive stories resting on slope. Three-dimensional structural models of the FG and SB buildings with 2- and 4-storey (above the topmost foundation level), considered in this study, are created in OpenSees [10]. The beams and columns are modelled using *ForceBeamColumn* elements, while the slab is modelled as a rigid diaphragm. The effective cracked moment of inertia for beams and columns is considered as 30% of the gross moment of inertia as per guidelines given in ASCE 41-17 [11]. Dead loads and live loads are considered as per recommendations of Indian standards, IS 875 Part 1 [12] and IS 875 Part 2 [13], respectively. The buildings considered are designed as special moment-resisting frames (SMRF) for Seismic Zone V on soil type I (rock site), following the provisions of the relevant Indian standards, IS 456 [14], IS 1893 Part 1 [15], and IS 13920 [16]. The typical beams and columns are proportioned to result longitudinal reinforcements between 0.75–1.5% and 2–4%, respectively.

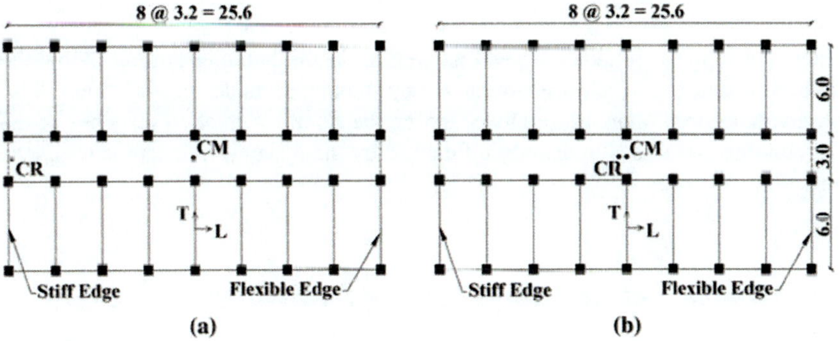


**Fig. 1** Typical plan layouts: **a** floor plan of SB building, at the floor immediately above the topmost foundation level, and **b** floor plan of the SB building, at the roof level. The dotted lines determine the boundaries of the floor slab (CM and CR represents the locations of the centre of mass and centre of rigidity in the floor plan, Longitudinal (L) direction is considered to be along-slope, whereas, transverse (T) direction is considered to be across-slope)

**Table 1** Dynamic characteristics of the building models considered

| Building model | | Period of vibration (s) | | Period ratio | Modal mass participation ratio (%) | |
|---|---|---|---|---|---|---|
| *N* | Direction | First mode ($T_1$) | Second mode ($T_2$) | $T_2/T_1$ | $\alpha_{m1}$ | $\alpha_{m2}$ |
| 2FG | Along-slope | 0.63 | 0.21 | 0.34 | 65.00 | 05.65 |
| | Across-slope | 0.85 | 0.26 | 0.30 | 64.00 | 08.77 |
| 4FG | Along-slope | 1.23 | 0.40 | 0.33 | 72.00 | 08.07 |
| | Across-slope | 1.74 | 0.53 | 0.31 | 71.00 | 09.28 |
| 2SB | Along-slope | 0.62 | 0.21 | 0.34 | 38.66 | 03.78 |
| | Across-slope | 0.88 | 0.31 | 0.35 | 39.43 | 36.24 |
| 4SB | Along-slope | 1.21 | 0.39 | 0.33 | 51.68 | 05.70 |
| | Across-slope | 1.74 | 0.53 | 0.31 | 50.93 | 07.12 |

N–no. of floors above the topmost foundation level, FG–building on the flat ground, SB -step-back building, $\alpha_{m1}$ and $\alpha_{m2}$ are the modal mass participation ratios corresponding to the fundamental and second modes of vibration in a given direction of excitation

## 2.2 *Dynamic Characteristics of the Considered Buildings*

The dynamic characteristics of the buildings investigated in the present study are summarized in Table 1. Consistent with the observations of the previous studies [8], in case of SB buildings, the period corresponding to the fundamental mode of vibration ($T_1$), is controlled by the number of storeys above the topmost foundation level, in both the directions of excitations (i.e., along and across slope). The structural response in low-rise SB buildings is dominated by the fundamental mode in the building portion above the topmost foundation levels, whereas, at the storey just above the topmost foundation level as well as in the building portion below the topmost foundation level, it is dominated by the higher modes of vibration. These observations are important in understanding the floor acceleration response of the SB buildings, as it is significantly influenced by the dynamic characteristics of the structure.

## 2.3 *Ground Motion Records Selected for Analysis*

In order to study the behaviour of the considered buildings, bidirectional linear time-history analyses are conducted by implementing the suite of 22 pairs of far-field recorded earthquake ground motions as given in the FEMA P695 [17]. The moment magnitude of these ground motions ranges from $M_w$ 6.5 to $M_w$ 7.6 with an average magnitude of $M_w$ 7.0. The time-history analyses are performed by applying both the horizontal components of a seismic ground-motion records at the same time, along the two orthogonal directions (i.e., along the slope and across the slope). For each of the

FG and SB buildings, a total of 44 linear time history analyses have been conducted by interchanging the two orthogonal components of seismic ground-motions along two principal axes of the buildings. The viscous damping effects in the time history analyses are considered by defining a Rayleigh damping of 2.5% [18] at periods equal to 1.5 times the fundamental mode period, and the period resulting in 95% cumulative mass participation in both the directions.

## 3 Torsional Irregularities in Step-Back Buildings

Torsional irregularity is one of the most prominent type of the structural irregularity which is often observed in the existing housing stocks. There are several factors which in turn induce torsional irregularity in buildings, some of which include the use of unequal column heights within the same storey, non-uniform distribution of stiffness of the lateral load resisting elements in the building plan, asymmetric placement of infill walls in the building. The presence of any of the aforementioned factors results in a net difference in the location of the CM and CR, often referred as the eccentricity in the floor plan. The presence of eccentricity in any floor induces torsional effects in the building, usually causing an amplification in the seismic response at the FE.

According to ASCE 7-16 [18] provisions, if the ratio between the maximum horizontal displacement ($\Delta_{max}$) at one end and the average horizontal displacement ($\Delta_{avg}$) at the two ends of the structure is greater than 1.20, then the structure is considered as torsionally irregular. Further, when this ratio exceeds a value of 1.40, the structure is said to have an extreme torsional irregularity. Similarly, according to IS1893:2016 [15], a building is said to have a torsional irregularity, when the ratio between the maximum horizontal displacement ($\Delta_{max}$) at one end, and the minimum horizontal displacement at the other end ($\Delta_{min}$) exceeds a value of 1.50. Contrarily, a different sort of definition of torsional (plan) irregularity is defined in Eurocode 8 [19], and the torsional irregularity is said to exist, when the normalized eccentricity ratio ($e/r$, where $e$ is the eccentricity between the CM and CR, and $r$ is the torsional radius) surpasses a value of 0.30. In the present study, three different indices are used to assess the extent of torsional irregularity in the investigated structures.

Figure 2 presents the variation of these three indices, i.e., normalized eccentricity ratio $\Delta_{max}/\Delta_{avg}$ and $\Delta_{max}/\Delta_{min}$ obtained from a 3D linear dynamic analysis along the height of 2- and 4-storey SB buildings. The maximum torsional effects are observed to exist at the storey immediately above the topmost foundation level. This observation is consistent among all the three indices considered herein to quantify the effect of torsion. Further, at the roof level, the extent of torsion present in the 2-storey SB building is higher than as compared to the 4-storey SB building, implying reduction in the torsional effects at the roof level, with an increase in the number of storeys.

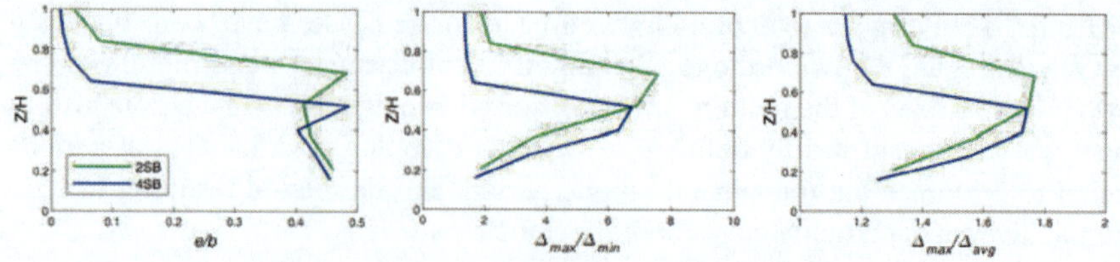


**Fig. 2** Variations of the torsional irregularity indices for SB buildings considered in this study. $Z$ is the height of the floor measured from the bottommost foundation level, and $H$ is the total height of the building measured from the bottommost foundation level. Here, $b$ is the width of the floor plan, in the direction perpendicular to the excitation

# 4 Results and Discussion

## 4.1 Spectral Amplification Function

In the present study, the bi-directional linear dynamic analyses have been conducted and FRS corresponding to NSC damping ratio of 5% are evaluated at the CR, FE, and SE, at two different floor levels, i.e., at the floor level where the effect of torsion is maximum (i.e., the storey immediately above the topmost foundation level), and at the roof level, in both the directions. The floor response is presented in the form of spectral amplification factors (SAF), defined as the ratio between the 5%-damped FRS to 5%-damped ground response spectra, in the direction under consideration.

Figures 3 and 4 present a comparison of the obtained SAFs as a function of tuning ratio ($T_s/T_1$), for the investigated 2- and 4-storey buildings, in the direction along and across the slope. Sharp peaks in the spectral amplification functions can be observed corresponding to the different contributing modes of vibration, at the floor level under consideration.

It can be observed that the median floor response of both FG and SB buildings are almost identical, especially at the CR, in along slope direction (Figs. 3a, c and 4a, c). Further, there are minor differences in the median SAFs in along slope direction, at the storey immediately above the topmost foundation level, especially at the FE and the SE (Figs. 3c and 4c). On the other hand, significant differences exist in the floor response at the CR, when compared with the floor response obtained at the FE or the SE, in across-slope direction (Figs. 3b, d, and 4b, d). In general, the floor spectral accelerations are observed to be more at the FE and less at the SE, when compared with the respective values at the CR, in across-slope direction. Further, this difference at the FE is significantly higher at the storey immediately above the topmost foundation level (Figs. 3d and 4d). This observation can be explained with the fact that the storey immediately above the topmost foundation level has the highest torsional irregularity (Fig. 2). The effect of torsion in floor response is observed to be maximum under the tuned response of the NSC, i.e., at $T_s/T_1$~0.30–0.40 and at $T_s/T_1$~1.00 (Figs. 3b, d and 4b, d). Further, the observed amplification in acceleration demands due to torsion is limited to the influence zones of the different contributing

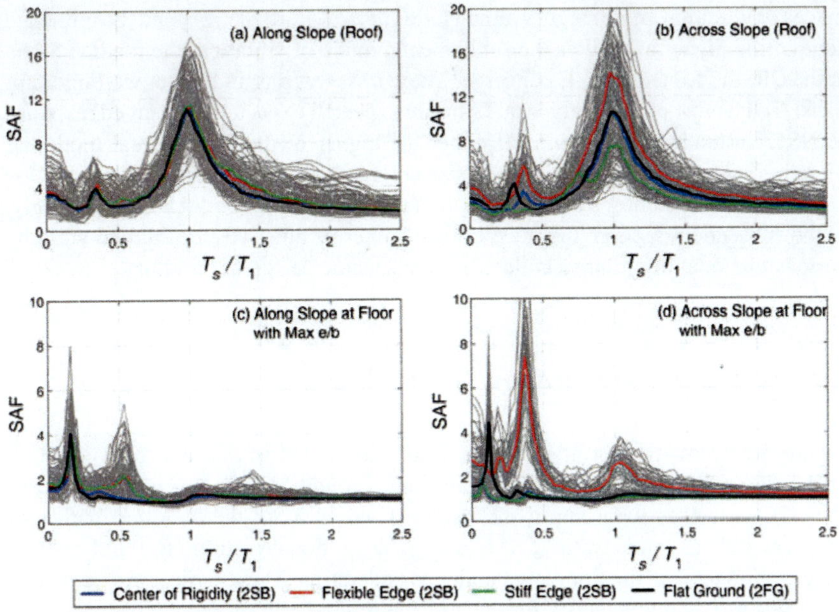


**Fig. 3** Comparison of the SAF obtained for the 2FG building with those obtained for 2SB building at the CR, at the FE, and at the SE. Different coloured lines represent the respective median values

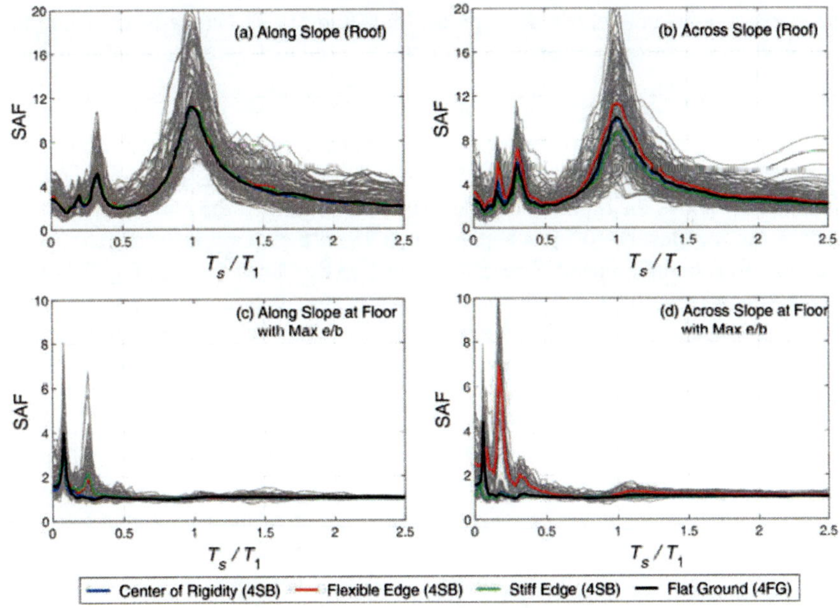


**Fig. 4** Comparison of the SAF obtained for the 4FG building with those obtained for 4SB building, at the CR, at the FE, and at the SE. Different coloured lines represent the respective median values

modes of vibration, and at higher tuning ratios ($T_s/T_1 \geq 2.0$), i.e., beyond the influence zone of the higher as well as the fundamental mode of vibration, the median SAFs at the CR, the FE and the SE converge. These observations underline two important facts: (i) the most critical values of TAFs are expected to occur under tuned response of NSC, in the influence zones of either the higher or the fundamental modes of vibration, and (ii) for non-tuned response corresponding to the higher tuning ratios ($T_s/T_1 \geq 2.0$), a building's torsion has an insignificant effect on SAFs. Accordingly, in the subsequent section, the most critical values of the TAFs are studied with the various torsional irregularity indices used in seismic design of buildings.

## 4.2 Torsional Amplification Factor (TAF)

Figure 5 presents the variation of the TAFs (computed as the ratio between SAF at the FE to SAF at the CR, under the tuning condition, i.e., when the period of the NSC coincides with one of the modes of vibration of the building) with four different indices, namely, $\Delta_{max}/\Delta_{min}$, $\Delta_{max}/\Delta_{avg}$, floor rotation ($F_\theta$) and angular acceleration ($\alpha_\theta$), at the floor level under consideration. These indices are computed from linear dynamic analysis for each of the ground-motion records separately. It can be observed that usually, at the FE, torsional amplification occurs, whereas, at the SE, a torsional de-amplification occurs. The magnitude of the TAF is higher for the higher modes of vibration as compared to the fundamental mode of vibration, at the storey immediately above the topmost foundation level. On the other hand, the torsional amplification is comparable, for NSCs tuned to the fundamental or higher modes of vibration.

Table 2 reports the median values of TAFs obtained at the FE, and the SE, of the SB buildings, investigated in the present study. It is to be noticed that the maximum value of the median TAF at the roof level is 1.67, whereas it is 6.17 at the storey immediately above the topmost foundation level. Contrarily, the minimum value of TAF at the roof level is 0.56, whereas it is 0.96, at the storey immediately above the topmost foundation level. These values can be explained through Fig. 2, which showed the presence of severe torsional effects, in the storey immediately above the topmost foundation level, in SB buildings. Further, a TAF value close to unity (at the SE), at the storey immediately above the topmost foundation level can be attributed to closer proximity of the CR and the SE (Fig. 1a).

Table 3 reports the correlation coefficients of different torsional irregularity indices investigated in the present study, with the estimated TAFs, at the FE. In addition, the coefficient of variations (CoVs) in estimating the considered torsional irregularity indices, from the linear time history analyses, using natural earthquake ground motions, are also presented. For the investigated SB buildings, TAFs are observed to be best correlated with the parameter $\Delta_{max}/\Delta_{min}$. The correlation of TAFs with $\Delta_{max}/\Delta_{avg}$ is observed to be slightly lesser than as compared to $\Delta_{max}/\Delta_{min}$ (Table 3). Further, the correlation of TAFs with the floor rotation ($F_\theta$) and angular acceleration ($\alpha_\theta$) is observed to reduce significantly. On the other hand, the CoVs in estimation

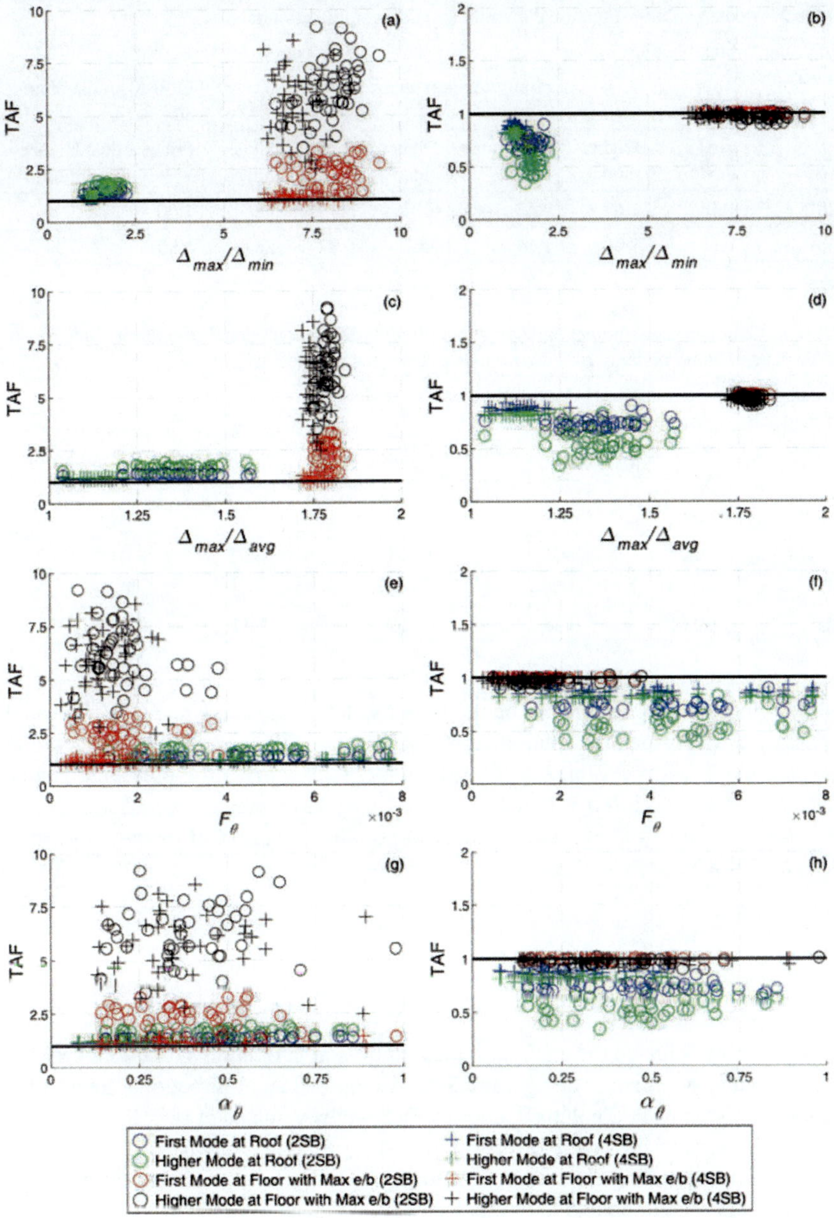


**Fig. 5** Variation of the TAF at different floors of the investigated SB buildings with different torsional irregularity indices considered in the present study. Plots on the left column reports TAF at the FE, whereas, plots on the right column reports TAF at the SE. The horizontal black colour line is drawn corresponding to a TAF equal to unity to distinguish the cases of torsional amplification and de-amplification, respectively

**Table 2** Median values of torsional amplification/de-amplification factors

| Building model | Flexible edge | | | | Stiff edge | | | |
|---|---|---|---|---|---|---|---|---|
| | Roof level | | Floor with maximum *e*/*b* | | Roof level | | Floor with maximum *e*/*b* | |
| | First mode | Higher mode | First mode | Higher mode | First mode | Higher mode | First mode | Higher mode |
| 2SB | 1.36 | 1.67 | 2.47 | 6.17 | 0.72 | 0.56 | 1.00 | 0.96 |
| 4SB | 1.15 | 1.20 | 1.14 | 5.67 | 0.87 | 0.82 | 1.00 | 0.97 |

**Table 3** Correlation coefficient of TAFs at the FE with different torsional irregularity indices and CoVs in estimating torsional irregularity indices from the time history analyses

| Torsional irregularity index | Correlation coefficient | Coefficient of variation | | | |
|---|---|---|---|---|---|
| | | 2SB (Floor with maximum *e*/*b*) | 4SB (Floor with maximum *e*/*b*) | 2SB (Roof level) | 4SB (Roof level) |
| $\Delta_{max}/\Delta_{min}$ | 0.61 | 0.08 | 0.08 | 0.14 | 0.09 |
| $\Delta_{max}/\Delta_{avg}$ | 0.58 | 0.01 | 0.01 | 0.07 | 0.04 |
| $F_\theta$ | -0.39 | 0.46 | 0.49 | 0.51 | 0.53 |
| $\alpha_\theta$ | 0.06 | 0.48 | 0.50 | 0.42 | 0.46 |

of these indices are observed to be the least for the parameter $\Delta_{max}/\Delta_{avg}$, followed by $\Delta_{max}/\Delta_{min}$. Further, the estimation of floor rotation ($F_\theta$) and angular acceleration ($\alpha_\theta$) are observed to exhibit significantly higher CoVs. From the presented results and discussions, it is inferred that the torsional irregularity indices, e.g., $\Delta_{max}/\Delta_{min}$ or $\Delta_{max}/\Delta_{avg}$, offers a superior choice for incorporating the effect of torsion in seismic design of NSCs.

## 5 Conclusions

This paper investigated the effects of inherent torsion in buildings on floor acceleration demands, for seismic design of NSCs. A total of 176 bi-directional linear time history analyses were conducted, and the FRS were evaluated at two different floor levels. The derived FRS were further used to study the TAFs and its correlation with the indices representing the extent of the torsional irregularity present in the building. The following major conclusions are drawn from this study:

- In SB buildings, the effect of torsion is maximum at the storey immediately above the topmost foundation level, in the direction across the slope. These torsional effects are observed to reduce along the height of the building, above the topmost foundation level.

- Torsion of the buildings is observed to have a higher impact on the design floor acceleration demands, when the NSC is tuned with one of the modes of vibration of the building. For non-tuned response, outside the influence zone of the different modes of vibration of the building, the effect of torsion is observed to be insignificant.
- For the elastic response of the building, torsional amplification occurs at the FE, whereas a de-amplification occurs at the SE. The median values of the TAFs are observed to be 6.17 and 1.67, at the storey immediately above the topmost foundation level, and at the roof level, respectively.
- TAFs are observed to be better correlated to the torsional irregularity indices, $\Delta_{max}/\Delta_{min}$ or $\Delta_{max}/\Delta_{avg}$ as compared to other indices such as the floor rotations and angular accelerations. Further, these torsional irregularity indices, $\Delta_{max}/\Delta_{min}$ or $\Delta_{max}/\Delta_{avg}$, are also observed to exhibit low CoVs, when obtained from time history analyses, as compared to floor rotations and angular accelerations.

The present study was conducted for low-rise buildings with FG and SB configuration exhibiting linear elastic response, and it needs to be further extended for taller buildings, considering their inelastic response.

**Acknowledgements** The first author's fellowship is granted by the Ministry of Education (MoE), Government of India (GoI). The investigations reported in this paper were conducted under the ISIRD project grant (F.No. 9-385/2019/IITRPR/3220) of the second author. The financial support received from both MoE and ISIRD Project Grant of IIT Ropar is thankfully acknowledged. The seismic ground-motion records used in this study were obtained from PEER NGA WEST 2 database.

## References

1. Taghavi S, Miranda E (2003) Response assessment of non structural building elements. PEER 2003/05 report. University of California at Berkeley, Berkeley, CA, Pacific Earthquake Engineering Research Center
2. Villaverde R (1997) Seismic design of secondary structures: state of the art. J Struct Eng 123(8):1011–1019
3. Yang Y, Huang W (1993) Seismic response of light equipment in torsional buildings. Earthq Eng Sruct Dynam 22(2):113–128
4. Agrawal AK, Datta TK (1998) Seismic response of a secondary system mounted on a torsionally coupled non-linear primary system. J Earthq Eng 2(3):339–356
5. Agarwal AK, Datta TK (1999) Seismic response of a secondary system attached to a torsionally coupled primary system under bi-directional ground motion. ISET J Earthq Technol 36(1):27–42
6. Aldeka AB, Chan AHC, Dirar S (2014) Response of non-structural components mounted on irregular RC buildings: comparison between FE and EC8 predictions. Earthq Struct 6(4):351–373
7. Aldeka AB, Dirar S, Chan AHC, Martinez-Vazquez P (2015) Seismic response of non-structural components attached to reinforced concrete structures with different eccentricity ratios. Earthq Struct 8(5):1069–1089
8. Surana M, Singh Y, Lang DH (2018) Effect of irregular structural configuration on floor acceleration demand in hillside buildings. Earthq Eng Struct Dynam 47(10):2032–2054

9. Vijayanarayanan AR, Goswami R (2020) Floor acceleration spectrum for RC moment frame buildings on slopes. In: 17th world conference on earthquake engineering, 17WCEE Sendai, Japan
10. Pacific Earthquake Engineering Research Center (PEER) (2009) OpenSees (Open system for earthquake engineering simulation), developed by PEER, University of California, Berkeley, California. http://opensees.berkeley.edu/
11. ASCE 41–17 (2017) Seismic evaluation and retrofit of existing buildings. American Society of Civil Engineers, Reston
12. IS 875-Part 1 (1987) Indian standard–code of practice for design loads for buildings and structures (Dead loads). Bureau of Indian Standards, New Delhi
13. IS 875-Part 2 (1987) Indian standard–code of practice for design loads for buildings and structures (Live loads). Bureau of Indian Standards, New Delhi
14. IS 456 (2000) Indian standard—plain and reinforced concrete, code of practice. Bureau of Indian Standards. New Delhi
15. IS 1893-Part 1 (2016) Indian standard—criteria for earthquake resistant design of structures, part 1-general provisions and buildings. Bureau of Indian Standards, New Delhi
16. IS 13920 (2016) Indian standard—ductile design and detailing of reinforced concrete structures subjected to seismic forces—code of practice. New Delhi, India
17. Federal Emergency Management Agency (FEMA) (2009) Quantification of seismic performance factors. FEMA P695, Washington, DC
18. ASCE (2017) Minimum design loads and associated criteria for buildings and other structures, ASCE 7–16. American Society of Civil Engineers, Reston, Virginia
19. CEN (2004) Eurocode 8: design of structures for earthquake resistance—part 1: general rules, seismic actions, and rules for buildings. EN 1998–1, Brussels, Belgium

# Seismic Safety of Buildings on Indian Hill Slopes—A Review

**Prateek Roshan and Shilpa Pal**

## 1 Introduction

The entire Himalayan belt extending from Shimla to Gangtok and the North-East Indian region lies in high and very high seismic risk zones as per IS 1893 (Part 1) [1]. It may be attributed to the northwards upthrusting of the Indian plate beneath the Eurasian plate at a rate of about 50 mm/year [2], making it one of the world's severest seismic prone areas.

In the recent past, the Indian Himalayan region has experienced several moderate to high magnitude earthquakes namely Kashmir earthquake [3], Sikkim earthquake [4, 5] and Nepal earthquake [6, 7]. These devastating earthquakes have indeed demonstrated the vulnerability of the structures located in hill slopes of North to North-East Indian Himalayan region and its significant impact on life safety.

Buildings in hilly areas are often augmented by various site constraints and challenges that undermine their structural performance during service life and hazards like earthquakes and landslides. Researchers have tried to assess and explore the differences that exaggerate the failure of buildings located on hill slopes in the past. Experts have tried to determine the vulnerability through proper Rapid Visual Screening (RVS) and other suitable tools. Research has also been carried out earlier to assess the relevance of slope stability and topographic amplification and its impact on the structure and the underlying foundation and slope during a seismic event.

P. Roshan (✉) · S. Pal
Delhi Technological University, Delhi 110042, India
e-mail: prateek_2k19phdce02@dtu.ac.in

S. Pal
e-mail: shilpapal@dtu.ac.in



S. Kolathayar and S. C. Chian (eds.), *Recent Advances in Earthquake Engineering*, Lecture Notes in Civil Engineering 175,
https://doi.org/10.1007/978-981-16-4617-1_16

## 2 Literature Review

National Building Code of India [8] defines any area whose height from mean sea level exceeds 600 m or any area having an average slope of 30° as a hilly area. Due to population outbursts and rapid urbanisation, there is an ever-increasing demand for construction. It has led to unplanned growth and rapid coming up of houses and buildings. A similar trend can be seen in some of the habitable areas along hill slopes as well.

Lack of flat land availability in hilly areas often results in constructing buildings on sloping grounds, causing the structural foundation to follow the natural slope. To accommodate the profile of the sloping ground, columns at foundation level have varying heights. These modifications make these buildings highly irregular in the vertical direction. It leads to considerable stiffness irregularities and non-uniform distribution of mass throughout the height of the structure. Due to irregularity in both planar directions, the dynamic characteristics of these buildings are significantly different from the ones resting on plain terrain. Thus, the seismic performance of these buildings during earthquakes is influenced by stiffness irregularity and torsional effects. Bilham and Szeliga [9] have stated that a large part of central Himalaya constitutes a region of seismic gap. The northward upthrusting of the Indian plate accumulates strain energy progressively and makes it more vulnerable to large magnitude future earthquakes.

### 2.1 *Review of Hill Buildings in India Through Site Survey and RVS*

Kumar et al. [10] presented an integrated approach to perform Rapid Visual Screening (RVS) for general building typologies in Himachal Pradesh state. The author calculated the RVS scores for 9099 buildings. The authors calculated the associated risk by assessing the structural vulnerability of different building typologies. They applied a statistical approach to obtained RVS scores to understand the damage state of building. They also reported that major buildings in the region were in the middle range of damage index so drawing meaningful conclusion was difficult. The author also proposed a modified format to carry RVS. Sarmah and Das [11] carried out a systematic vulnerability analysis of Guwahati city and developed a ward-level hazard map. The authors selected hundred residential, commercial, mixed-use buildings through random sampling technique and later scored and ranked these buildings based on their seismic vulnerability. They also assessed the likelihood of these buildings getting affected by landslides. Rautela et al. [12] evaluated the seismic risk of the Nainital township of Uttarakhand state of India. Authors surveyed around 2860 buildings in the study area using RVS technique and utilised GIS and remote sensing techniques to assess the seismic vulnerability. The study revealed that the building stock was old and deprived of seismic safety and the seismically induced losses

**Table 1** Structural configuration of some building typology in Hilly areas [13]

| Type | Structural configuration | Type | Structural configuration |
|---|---|---|---|
| | 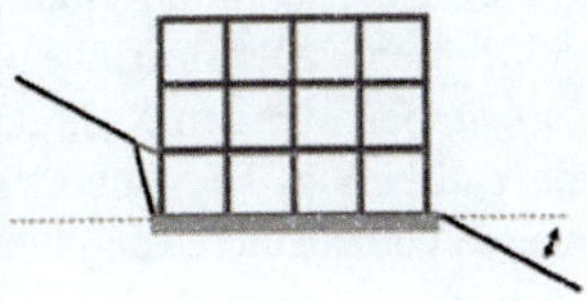 | | 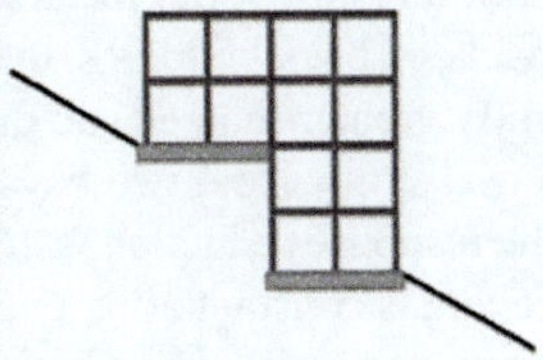 |
| (a) | Observed on low slopes | (b) | Observed on very steep slopes (split-foundation building) |
| | 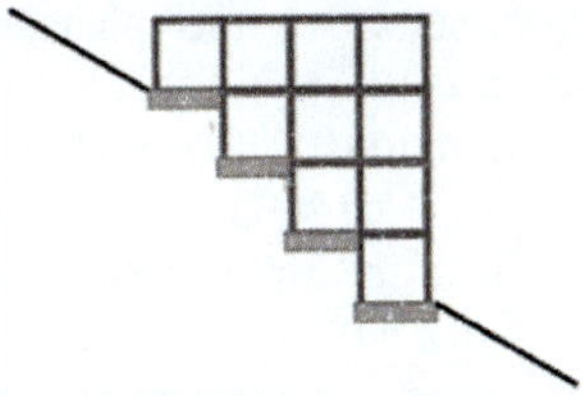 | | 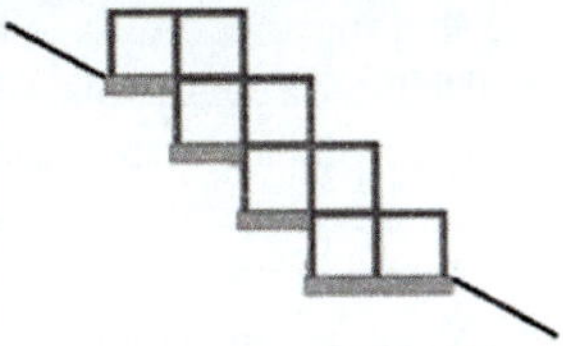 |
| (c) | Observed on low to moderate slopes (stepback building) | (d) | Addition of setbacks in case (c) (stepback setback building) |

would be more than 200 US million dollars if a major earthquake strikes the area. Surana et al. [13] carried out field surveys conducted in two hill towns Nainital and Mussoorie of India and prepared a list of the existing prevalent building stock. The authors identified different building configurations based upon the structural arrangement and relative levels of foundations some of which are shown in Table 1. The authors observed that hill building configurations were more susceptible to damage than regular buildings on flat terrain. They have reported that hill buildings' increased fragility was due to the torsional irregularity and shear failure of short columns.

### 2.2 *Review of Structural Modelling and Analysis of Building Configuration on Indian Hill Slopes*

Singh et al. [14] examined the dynamic response of hill buildings and compared the results with regular buildings on flat ground. According to authors, due to irregular configurations, the mass participation in the fundamental mode of hill buildings was much lower than the buildings with regular configurations on flat terrain. Under cross-slope earthquake motion the hill buildings are exposed to significant torsional effects. Under along-slope excitation, the varying column heights cause stiffness irregularity, and the short columns being stiffer attracts large amount of storey shear. They stated that the storey at road level was most susceptible to damage in downhill buildings. Pandey et al. [15] studied the behaviour of buildings on hill slope through a 3D static pushover and response spectrum analysis by varying support conditions using the SAP 2000 software [16]. They analysed for different soil conditions (hard, medium, and soft soils) idealised by equivalent springs. The authors stated that in step back-set back buildings, the value of the displacement at performance point was generally greater than elastic displacement for all support types in X-direction and up to two storeys and two bays for hard soil for Y-direction. The authors reported that the response reduction factor generally decreases with an increasing time period and becomes constant after a value of the time period.

Narayanan et al. [17] performed nonlinear analyses performed on reinforced concrete buildings on steep hill slopes. Based on the results, the buildings with large plan size, either in the valley or road direction, were more vulnerable to strong seismic shaking. According to their findings, buildings with a small plan along both road and valley directions had seismically performed well. Singh et al. [18] studied the seismic performance of the various step-back configuration building models using bi-directional incremental dynamic analysis (IDA) using SAP 2000 Nonlinear software [16]. IDA was carried in both orthogonal directions of the building models to evaluate the sustained intensity of ground motion before the collapse. They observed that buildings' performance was abysmal in across-slope direction due to torsional irregularities, failing at lower seismic intensities. They reported that brittle shear failure of the short columns at the ground level was one of the main reasons for the structural failure, both along and across slope earthquake excitations. Daniel and Sivakamasundari [19] evaluated the dynamic response of building on hills and compared it with the respective regular building on flat ground by equating the regular building's seismic weight on flat ground to that of the hill building. The authors concluded that regular building's energy dissipation capacity on flat ground is higher than the respective hill building. They have also reported that regular building's flexibility endures larger displacement than the respective hill building. Ghosh and Debbarma [20] analysed the seismic performance of setback structures resting on the flat ground and a hill's slope, with soft storey configuration. The authors reported that the orthogonal movement under unidirectional force was noted for the setback buildings as the taller side moves more than shorter side along the direction of the force. According to the authors, torsion is caused due to variation of mass, stiffness, and setback geometry, as was visible in the analysis results. Li et al. [21] have studied the influence of earthing tie beams on the seismic fragility of Reinforced concrete frames on steep slopes (RCFSS) by developing fragility curves using nonlinear time history analysis. The authors reported that RCFSS with an earthing tie beam experienced less fragility at each damage state (like slight, moderate, severe based on maximum inter-storey drift). They had also concluded that earthing tie beam's influence on the seismic fragility of RCFSS was more significant when it was excited by far-field ground motions than by near-field ground motions.

## 2.3 Review of Analysis of Buildings on Slopes Considering Soil-Structure Interaction

Ghosh and Debbarma [22] conducted the seismic examination of the buildings laying on varying slope angles. According to their findings, the sloping ground structures were more vulnerable, and the degree of vulnerability increases with the increase in the value of the slope angle. They observed stiffness concentration on the shorter side of the structure on the higher level of the slopes. They reported that the structures modelled without considering soil-structure interaction tends to overestimate the forces like, base shear and bending moment, and it also underestimates the response parameters such as time period, displacement and effect of torsion. Liu et al. [23] developed a model of identical inputting wave and a soil-structure interaction model suitable for evaluating the seismic response of step back-set back buildings. They reported that the shear in the ground column was considerably higher, highlighting the need for its strengthening during the design process. The author stated that the step back-set back configuration has an adequate dissipation capacity and good seismic performance based on the comparison of the distribution of plastic hinges formed. Farghaly [24] evaluated the seismic performance of hillside buildings and its foundation's system constructed on the rocky hillside slope of Doronka city. The authors suggested that the stepped isolated foundation represents the best solution for both dynamic performances and stability of buildings constructed on the hill slope. They also observed that the forces in columns for isolated footing decreased by 1.33 times than the raft foundation and also the base shear value decreased by about 1.45 times. Nayak et al. [25] carried out the seismic hazard analysis of Uttarakhand state using Probabilistic Seismic Hazard Analysis (PSHA) method. The authors reported that Peak Horizontal Acceleration (PHA) values at surface level increase considerably compared to those at bedrock level. They also stated that the soil properties and the local geology play an essential role in modifying ground motion at bedrock level, highlighting the influence of site characterisation.

## 2.4 Review of Hill Buildings Considering Slope Stability and Topographic Amplification

A slope's stability is quite a natural phenomenon depending upon soil properties, building loads transferred to the soil and slope, type and location of loads, climatic and drainage conditions of the area [26]. The stability of the slope governs the overall safety of the buildings on the hill slope. A properly designed and constructed structure may collapse if the slope is not stable. Paul and Kumar [26] studied the slope stability analysis with different configurations of buildings. They concluded that for both static and seismic conditions, the value of factor of safety was higher for stepback building than in stepback setback building. As per their findings, the stepback building transferred heavier load at the slope's downhill edge and gave a stabilising effect. They concluded that the step back building gives better stability than the stepback setback building on sloping ground. Das and Maheshwari [27] presented a summary of available literature on the analysis of shallow foundation on the sloping ground presented. They reported that generally, slopes are more unstable in the presence of pore water pressure, and during an earthquake, the excess pore water may further reduce the bearing capacity of foundations, or the slope may deform. The author stated that the limit analysis is a reliable method of analysis for the foundation on slopes as it considers the actual stress–strain behaviour of the materials. The authors concluded that the FEM (finite element based) numerical method is efficient, accurate, and sufficient to analyse foundations on slopes under seismic excitations. Kumar and Bhargava [28] analysed the slope stability for soil, soft rock and hard rock with due seismic considerations. They observed that with an increase in the degree of saturation, the factor of safety decreases. The authors stated that slope stability decreases with an increase in slope for both static and pseudo-static analyses.

Site amplification may be characterised as the increase in the shaking experienced at the ground surface in contrast at the bedrock. A couple of design codes have dealt with the topographic amplification by adding a period-independent amplification factor while developing the elastic design response spectrum. This factor is primarily connected with geometry of the topographic features and given as a function of the topographic feature's shape ratio.

Modha et al. [29] conducted a numerical study on 2D regular and symmetric triangular hills of different shape ratios (ratio of height to half of the base width of the hill) subjected to different ground motion records. They stated that the highest amplification of incident waves observed at the sharp peak of hills due to trapping of the energy, de-amplification at the toe of the hill due to scattering of waves. Amplification tends to increase with the increasing slope inclination, whereas almost equal values of amplification for hills having different base widths but the same inclination.

Researchers during their field visit in Nepal-Gorkha earthquake of 2015 have reported that most damaged buildings situated on slopes beneath the top of the ridges [7]. Ridges of hills has behaviour to focus seismic waves, causing much larger ground acceleration at the peak [7]. Singh et al. [30] calculated topographic amplification factors using the Code of Italy [31], French code [32], and Eurocode 8 [33, 34] and compared the values. The authors stated that the Italian code [31] predicts the largest amplification factors at the ridge among the three considered codes. The authors concluded that the hill building configurations could sustain much lower PGA, than their counterparts on flat terrain, designed for the same seismic hazard, using the Indian codes.

## 3 Discussion and Summary

Reinforced concrete construction has expanded drastically in the recent times to fulfil the rapid settlement expansion. Lack of earthquake awareness, inadequate follow-up of the building design codes, and poor construction practices are unfortunate reasons for such a massive number of casualties. Buildings constructed before the implementation of seismic design byelaws normally exhibit low concrete quality and poor workmanship, often with inadequate sizes of beam and columns, large stirrup spacing, inadequate reinforcement and fragile column-beam joints. These deficiencies further increase the likelihood of the brittle failure of structural members.

Some of the common inappropriate structural alterations found in buildings reported by past researchers are discussed below.

Presence of soft-stories on developments utilized for private and commercial intentions is quite possibly the most well-known construction issues. Commercial and office buildings generally lack the presence of infill at ground level causing significantly lower ground storey stiffness and storey strength. Most of the time, owners neglect attention to required measures for seismic resistance. Such houses or buildings are often constructed following the owner's requirements and convenience sans consulting engineers or architects. Consequently, these non-engineered constructions exhibit below-par performance during any seismic event suffering severe damage sometimes even leading to collapse.

The lack of satisfactory concrete cover of structural members may cause exposure of the structural reinforcing bars and may eventually cause corrosion later implying ill anchored reinforcements and a weak connection at the beam-column joints.

Unsymmetrical columns to suit the structural irregularity in the vertical direction are common in buildings located on sloping terrains. Larger differences in storey-height in residential houses should be avoided as much as possible if the occupancy usage is the same for a set of storeys. Houses having insufficient gaps in between them are a prevalent thing in almost all the reinforced concrete buildings as stated by past researchers. It poses a severe threat of pounding action in case of earthquakes. During their survey, past researchers observed several structural defects that interfere and alter the load path, short column effect at the staircase landing, a large area of window and door opening in walls is a cause of concern as it decreases the stiffness and strength of masonry walls. Often the asymmetric position of these openings induces stress concentrations and can cause a localised failure resulting in increased vulnerability of the complete structure. Non-separation of the staircase from the main structure is also a common observation in most residential construction, which is a cause of concern. In brick masonry construction, the absence of bands at different levels like plinth level and sill and level exists in most of the existing stock.

Most of the constructed buildings are very close to each other, which could suffer damages due to pounding in earthquakes. A more devastating case be when a building is very close to an adjacent and unsafe construction whose collapse can damage the building with ease. Also, soft storey, heavy overhangs, heavy rooftops and irregular plan shapes are familiar hints in buildings that make them seismically more vulnerable. Even engineered constructions are not well maintained. Structures built on hill slopes or located on unstable sites are also vulnerable to falling debris from the hill slopes. Past researchers [35] have reported that a vast number of buildings have also been constructed on steep to very steep sloped terrains.

Table 2 summarises the constructional limitations, site constraints and deviations from the standard construction practices on hill slopes. These constraints and deviations are the majority of the causes for severe damage and failures of buildings on slopes.

The shaking often damages buildings during the seismic event or due to the soil strata's settlement to a different level than was before the earthquake. Earthquake

**Table 2** Summary of constructional limitations, site constraints and structural deviations found in buildings in hilly areas

| Constructional limitations & site constraints | Deviations from standard and recommended practices |
|---|---|
| Lack of earthquake awareness | Existence of soft-stories |
| Inadequate follow-up of the building design codes | Houses having insufficient gaps in between them |
| Non-implementation of seismic detailing norms | Larger differences in storey-height |
| Safety hazard for construction site | Unsymmetrical columns |
| Poor workmanship and construction practices | Asymmetric position of window and door openings |
| Improper compaction | Large stirrup spacing |
| Ignorance of site investigation measures | Short column effect |
| Inadequate cover in beams and columns leading to exposure of reinforcement eventually causing its corrosion | Non-separation of the staircase from the main structure |
| Cutting of slopes | Irregularity in elevation |
| Preservation of natural features | Weak column-beam joints |
| Negligence of geotechnical investigations | Large area of window and door opening in walls |
| Buildings located on unstable and steep to very steep sloped terrains | Insufficient longitudinal reinforcement |
| Constraint for sewage disposal | Heavy rooftops & large projection area in a building |
| Drainage of site and maintenance of drainage pattern | Irregular plan shapes |
| Fire Safety | Heavy overhangs |
| Low concrete quality | Inadequate beam and column sizes |
| Chances of Ground settlement | Disproportionate height of building |
| Secondary hazards like landslides, mudslides etc. | Absence of bands at different levels in brick masonry construction |

vibrations result from the long-distance propagation of body waves across solid rock in-depth and solid deposits at surface. Determination of surface and sub-surface soil conditions, groundwater condition, and features in the proposed construction area are crucial as it may influence the design and address the expected post-construction problems. Site investigation is a vital step not only for new construction but for retrofitting of older structure too.

Without proper geotechnical investigations, building constructions are detrimental and a threat to structural performance and safety of occupants. IS code 14243 Part 2 [36] recommends to carry out proper field survey and stability analysis of hill slopes for cuttings and buildings should be located on stable hill slopes. Also, proper reinforced concrete design, implementation of seismic design and ductile detailing as recommended by BIS codes [1, 37, 38] should be ensured.

Some of the possible considerations that may help mitigate the failure of buildings on hill slopes, as suggested by past studies and researchers, are given below.

- *Inclusion of brick walls.* Incorporating brick walls improve the seismic resistance of buildings with shorter height [39].
- *Reduction of the span between columns.* Reduction of span length between columns may improve structural performance in the absence of additional walls [39].
- *Usage of higher-grade concrete.* Using higher-grade concrete can significantly increase the resilience of the structural framework. This aspect can be more efficient in buildings without walls [39].
- *Vulnerability assessment of critical lifeline structures.* Any post-disaster activity and relief operation dependent on critical lifeline structures like hospitals, transportation system and infrastructure, water supply system, structures with huge spaces like school buildings, auditoriums, etc. Detailed seismic vulnerability assessment of these structures in hazard-prone areas must be carried as their damage and collapse may affect post-disaster relief activities [12].
- *Geotechnical considerations.* The involvement of geotechnical engineer and related inputs at the planning stage, design and construction phase, and infrastructure maintenance is crucial for safe and cost-effective development of hilly areas. Also, proper design of cut and fill of slopes may limit haphazard slope failures to some extent [40].
- *Timely inspection of slide prone areas.* Timely inspection of areas prone to sliding must be ensured to check for the presence of tension distress and cracks, and the same, if present, shall be sealed [41].

Some of the possible solutions which may enhance the building resilience are as follows [42]:

- Suitable design considerations for buildings and their surrounding built-environment
- Use of proper conforming materials and building components
- Relevant and sound construction practices
- Periodic maintenance activities for preventive and predictive measures

- Appropriate and adequate monitoring through proper guidelines.

Incorporating the basic earthquake resistant design features like.

- strong column-weak beam design approach,
- continuous load transfer mechanism,
- simple and symmetric plan layout,
- ductile detailing

can significantly enhance structural performance and buildings may suffer less damage in case of an earthquake event and deliver a better structural performance [43–46]. This may prove very effective in increasing structural efficiency and mitigating failure to a considerable extent.

Due to natural disasters, the damages in fully and partially engineered buildings are generally not severe compared to the damages observed in marginally or non-engineered constructions [47].

An essential key to reducing seismic risk is by decreasing structural vulnerability, i.e., if buildings are less damaged, they will cause lesser losses. Vulnerability assessment for the non-structural system is as vital as for structural components of a structure. By identifying these deficient buildings and carrying suitable retrofit measures, its likely performance may be enhanced.

There is an urgent need to understand these housing risk to minimise future life losses and property losses. Observations of significant losses in past moderate magnitude earthquakes signalled a very high risk of the existing built environment. The observations and the impacts of past earthquakes conclude that the buildings in hilly areas are in a fragile state where their survivability and safety of residing people in a future earthquake is a serious concern for us all.

## References


1. BIS (Bureau of Indian Standards) (2016) Indian standard criteria for earthquake resistant design of structures. Part 1: general provisions and buildings (sixth revision). IS 1893 Part 1, New Delhi, India
2. Kumar P, Yuan X, Kumar MR et al (2007) The rapid drift of the Indian tectonic plate. Nature 449(7164):894–897. https://doi.org/10.1038/nature06214 PMID: 17943128
3. Rehman K, Qadri SMT, Ali A et al (2016) Analysis of the devastating Kashmir earthquake 2005 aftershocks. Arab J Geosci 9:379. https://doi.org/10.1007/s12517-016-2418-6
4. Prajapati SK, Kumar A, Chopra S et al (2013) Intensity map of Mw 6.9.2011 Sikkim–Nepal border earthquake and its relationships with PGA: distance and magnitude. Nat Hazards 69:1781–1801. https://doi.org/10.1007/s11069-013-0776-x
5. EERI (2012) The Mw 6.9 Sikkim-Nepal border earthquake of September 18, 2011: learning from earthquakes, special earthquake report. Earthquake Engineering Research Institute, Oakland, CA
6. Rupakhety R, Olafsson S, Halldorsson B (2017) The 2015 Mw 7.8 Gorkha earthquake in Nepal and its aftershocks: analysis of strong ground motion. Bull Earthq Eng 15:2587–2616. https://doi.org/10.1007/s10518-017-0084-z

7. ISTRUCTE (2019) The Mw 7.8 Gorkha, Nepal earthquake of the 25th April 2015, a field report by earthquake engineering field investigation team (EEFIT). The Institution of Structural Engineers, London, U.K.
8. Bureau of Indian Standards (2016) National building code of India 2016 volume 1, Bureau of Indian Standards New Delhi
9. Bilham R, Szeliga W (2008) Interaction between the Himalaya and the flexed Indian plate-spatial fluctuations in seismic hazard in India in the past millennium. In: Proceedings of the American institute of physics conference. https://doi.org/10.1063/1.2963839
10. Kumar AS, Rajaram C, Mishra S et al (2017) Rapid visual screening of different housing typologies in Himachal Pradesh. India Nat Hazards 85:1851–1875. https://doi.org/10.1007/s11069-016-2668-3
11. Sarmah T, Das S (2018) Earthquake vulnerability assessment for RCC buildings of Guwahati city using rapid visual screening. Procedia Eng 212:214–221. https://doi.org/10.1016/j.proeng.2018.01.028
12. Rautela P, Joshi GC et al (2019) Earthquake risk assessment around Nainital in Uttarakhand Himalaya, India. J Geogr Nat Disast 9(1):236. https://doi.org/10.35248/2167-0587.19.9.236
13. Surana M, Singh Y, Lang DH (2018) Seismic characterization and vulnerability of building stock in hilly regions. Nat Hazards Rev 19(1). https://doi.org/10.1061/(asce)nh.1527-6996.0000275
14. Singh Y, Gade P, Lang DH, Erduran E (2012) Seismic behaviour of buildings located on slopes—an analytical study and some observations from Sikkim earthquake of 18th September 2011. In: Proceedings of 15th world conference on earthquake engineering, Lisbon, Portugal
15. Pandey AD, Kumar P, Sharma S (2011) Seismic soil structure interaction of buildings on hill slopes. Int J Comput Civ Struct Eng 2(2):544–555
16. SAP2000 (2010) CSI analysis reference manual for SAP2000. Computers and Structures Inc., Berkeley, California
17. Narayanan AR, Goswami R, Murty CVR (2012) Performance of RC buildings along hill slopes of Himalayas during 2011 Sikkim earthquake. In: Proceedings of 15th world conference on earthquake engineering, Lisbon, Portugal
18. Singh Y, Yeluguri VR, Lang DH (2014) Seismic response of hill buildings subjected to bi-directional excitation. In: Proceedings of 10th U.S. national conference on earthquake engineering, Alaska
19. Daniel AJ, Sivakamasundari S (2016) Seismic vulnerability of building on hill slope. Int J Earth Sci Eng 9(5):1892–1899
20. Ghosh R, Debbarma R (2017) Performance evaluation of setback buildings with open ground storey on plain and sloping ground under earthquake loadings and mitigation of failure. Int J Adv Struct Eng 9:97–110. https://doi.org/10.1007/s40091-017-0151-3
21. Li Y, Xu J, Wang G (2018) Seismic fragility analysis of RC frames on steep slopes under near-fault and far-field ground motions. In: Proceedings of 16th European conference on earthquake engineering, Thessaloniki, Greece, 2018
22. Ghosh R, Debbarma R (2019) Effect of slope angle variation on the structures resting on hilly region considering soil–structure interaction. Int J Adv Struct Eng 11:67–77. https://doi.org/10.1007/s40091-019-0219-3
23. Liu L, Wu C, Li Y, Zheng N, Xie Q (2017) Seismic response of stepped building supported by stepped foundation on hillside. In: Proceedings of 16th world conference on earthquake engineering, Chile
24. Farghaly AA (2015) Evaluation of seismic performance of buildings constructed on hill-side slope of Doronka village–egypt. Int J Geotech Eng 9(2):176–189, Hindawi Publishing Corporation. https://doi.org/10.1179/1939787914Y.0000000053
25. Nayak M, Sitharam TG, Kolathayar S (2015) A revisit to seismic hazard at Uttarakhand. Int J Geotech Earthq Eng (IJGEE) 6(2):56–73. https://doi.org/10.4018/IJGEE
26. Paul DK, Kumar S (1997) Stability analysis of slope with building loads. Soil Dyn Earthq Eng 16(6):395–405. https://doi.org/10.1016/S0267-7261(97)00008-0

27. Das S, Maheshwari BK (2018) Review of seismic design of shallow foundations on hill slopes. In: Proceedings of the 16th symposium on earthquake engineering, IIT Roorkee, India, 20–22 December 2018
28. Kumar R, Bhargava K (2018) Analysis of slope stability of hilly terrain with seismic considerations. In: Proceedings of the 16th symposium on earthquake engineering, IIT Roorkee, India, 20–22 December 2018
29. Modha KG, Raj D, Singh Y (2018) Topographic amplification for triangular hill geometry. In: Proceedings of the 16th symposium on earthquake engineering, IIT Roorkee, India, 20–22 December 2018
30. Singh Y, Lang DH, Narasimha DS (2015) Seismic risk assessment in hilly areas: case study of two cities in Indian Himalayas. In: SECED 2015 conference: earthquake risk and engineering towards a resilient world, Cambridge, U.K. 9–10 July 2015
31. ICMS (2008) Indirizzi e criteri per la microzonacione sismica. grupo di lavoro "ICMS". Conferenza delle regioni e delle province autonome–dipartimento della protezione civile, Roma, vol 3 e Dvd
32. Association Française du Génie Parasismique (1990) AFPS 90, recommendations for the redaction of rules relative to the structures and installations built in regions prone to earthquakes, French Association for Earthquake Engineering
33. CEN. EN 1998–1 (2004) Eurocode 8—design of structures for earthquake resistance, part 1: general rules, seismic actions and rules for buildings. European Committee for Standardization, Brussels
34. CEN. EN 1998–5 (2014) Eurocode 8—design of structures for earthquake resistance, part 5: foundations, retaining structures and geotechnical aspects. European Committee for Standardization, Brussels
35. Ramancharla P, Goud S, Bhalkikar A et al (2019) Earthquake disaster risk index report 50 towns & 1 district in seismic zones III, IV and V. National Disaster Management Authority, Government of India, New Delhi
36. Bureau of Indian Standards (1995) I.S. 14243 Part 2, Guidelines for selection and development of hill sites, BIS, New Delhi
37. BIS (Bureau of Indian Standards) (2000) Indian standard code of practice on plain and reinforced concrete (fourth revision). IS 456:2000, New Delhi, India
38. BIS (Bureau of Indian Standards) (2016) Indian standard code of practice on ductile design and detailing of reinforced concrete structures subjected to seismic forces (first revision). IS 13920:2016, New Delhi, India
39. Dominguez-Santos D, Ballesteros-Perez P, Mora-Melia D (2017) Structural resistance of reinforced concrete buildings in areas of moderate seismicity and assessment of strategies for structural improvement. Buildings 7:89. https://doi.org/10.3390/buildings7040089
40. Sew GS, Chin TY (2000) Hill-Site development-planning, design, construction and maintenance considerations. In: IEM seminar on geotechnical engineering, Penang, September 2000. [Online], Available at: http://gnpgroup.com.my/wp-content/uploads/2017/03/2000_04.pdf, last accessed 2021/03/14
41. Panigrahi et al (2011) Investigation and design for restoration of hill slope in Mizoram. Indian Geotech J 41(4):215–225
42. Pham et al (2018) Knowing maintenance vulnerabilities to enhance building resilience. In: Proceedings of 7th international conference on building resilience, Thailand, Science Direct, Procedia Eng 212:1273–1278
43. Murty CVR, Goswami R et al (2012) Some concepts in earthquake behaviour of buildings. Gujarat State Disaster Management Authority, India. Available online https://www.iitk.ac.in/nicee/IITK-GSDMA/EBB_001_30May2013.pdf
44. Duggal SK (2013) Earthquake-Resistant design of structures, 2nd edn. Oxford University Press, New Delhi, India
45. Roshan P, Pal S, Kumar R (2018) Seismic damage assessment index of buildings using Fuzzy-AHP approach. Int J Tech Innov Modern Eng Sci 4(08):1011–1019

46. Roshan P, Pal S, Kumar R (2020) Performance assessment indexing of buildings through fuzzy AHP methodology. In: Ahmed S, Abbas S, Zia H (eds) Smart cities—opportunities and challenges. Lecture notes in civil engineering, vol 58. Springer, Singapore. https://doi.org/10.1007/978-981-15-2545-2_42
47. Bureau of Indian Standards (2000) I.S. 14804, Siting, Design and selection of materials for residential buildings in hilly areas-guidelines, BIS New Delhi

# Seismic Mainshock—Aftershock (MS-AS) Vulnerability Assessment of Reinforced Concrete Bridge Exposed to Flood Induced Scour

K. K. Jithiya, Muhamed Safeer Pandikkadavath, Sujith Mangalathu, and Praveen Nagarajan

## 1 Introduction

Bridges placed in flood-prone and seismically active locations have a higher vulnerability to earthquakes due to their susceptibility to substructure scouring. The scouring may cause loss of supports (all-around) at the foundation level, and hence there will be increased structural flexibility in the lateral directions. This in turn escalates the damage vulnerability of bridges, under the anticipated level of seismic activity [1–5]. Seismic Mainshock (MS) may follow several Aftershocks (AS) with smaller magnitude and occasionally with higher magnitude compared to the MS. AS that occurs within a relatively short extent of time after MS may have different intensity, duration, spectral shape, and energy content compared to the preceding seismic events. Often it strikes on structures that are already damaged/yielded under MS. This event further increases the bridge failure risk even during the moderate level of seismic disturbances [6–12]. Though independent investigation reports on bridges under flood induced scour followed by MS (only) and MS-AS effects on the performance of bridge structures without scouring effect are available [1–12]; the combined effect of flood induced scouring followed by MS-AS on the bridge responses is very scarce. The present study aims to assess the seismic response of a flood induced scour inflicted two span Reinforced Concrete (RC) bridge under MS-AS action. For the investigation, Sacramento County in California, a seismically active—flood-prone area is selected. The flood hazard curve for the chosen region is derived from the flood frequency analysis. The flood frequency analysis can yield the 100-year flood event design/peak discharge; successively the scour depth that has been resulted from

K. K. Jithiya · M. S. Pandikkadavath (✉) · P. Nagarajan
National Institute of Technology Calicut, Kozhikode, India
e-mail: msafeerpk@nitc.ac.in

S. Mangalathu
Data Analytics Division, Equifax Inc., Atlanta, GA, USA



S. Kolathayar and S. C. Chian (eds.), *Recent Advances in Earthquake Engineering*, Lecture Notes in Civil Engineering 175,
https://doi.org/10.1007/978-981-16-4617-1_17

this flood event is calculated by incorporating applicable bridge pier geometry, soil profile, and stream data. For the MS only and MS-AS nonlinear time history analysis, a set of 51 earthquakes that comes under various seismic hazard levels are selected [1–5]. Subsequently, the seismic response evaluations are carried out in terms of displacement ductility of the two-span RC bridge under MS only and MS-AS cases with and without considering the flood induced scour.

## 2 Flood and Seismic Hazards

Flood hazard curves using flood frequency analysis that gives the exceedance probability of annual peak discharge is a useful resource to express the flood hazard for any given region [13]. To develop the flood hazard curve of the selected study area, annual peak discharge data from 1907 to 2019 (last 113 years) is collected from United States Geological Survey (USGS) National Water Information System. These data are ranked in respective order (magnitude) to obtain the flood hazard curve in the log-normal scale as shown in Fig. 1 [14]. The resulting plot explicitly gives the relation between flood hazard level (in terms of annual probability of exceedance) and measurable flood characteristics (in terms of peak discharge). The so obtained 100-year flood event (1% annual probability of exceedance) discharge induced scour is considered at the foundation region for the seismic performance

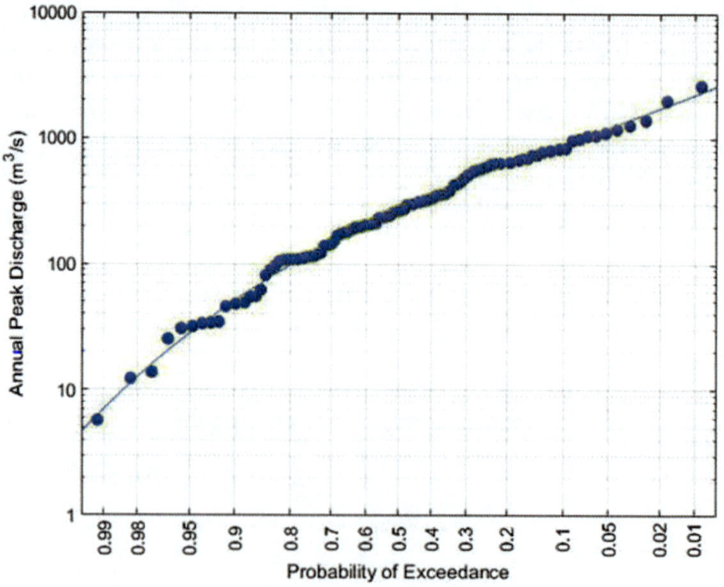


Fig. 1 Flood hazard curve

assessment of the selected two-span RC bridge [1, 2]. The developed flood hazard curve gives a 100-year maximum design flood event discharge as 2200 $m^3/s$. The scour depth corresponding to this flood event can be estimated by taking relevant details on the bridge geometry, streamflow, and soil nature of the selected location.

For the nonlinear seismic time history analysis, a suite of 51 far-field ground motion records that come under different hazard levels are selected (for both MS and AS cases). It includes 20 numbers of ground motions that come under the Maximum Considered Earthquake (MCE) hazard level with 2% exceedance in 50 years (with a return period of 2475 years) [15] and another 31 numbers as per the Federal Emergency Management Agency (FEMA) recommendations [16]. The Peak Ground Acceleration (PGA) values and magnitude of the selected ground motion records vary between 0.15 g and 1.33 g (where g is the acceleration due to gravity) and 6–7.5 in standard scale respectively. AS ground motion sequences are selected randomly from the considered set of ground motion records by keeping the PGA values in the range of 0.70–0.90 times the respective MS PGA [17].

## 3 Simplified Description of Study Bridge

As briefed earlier, for the study a two-span reinforced concrete bridge with an equal span of 39.6 m used in prior reports is adopted [1, 2]. The bridge is designed as per the California Department of Transportation bridge design norms. The bridge cross-section (hollow multiple box) has a depth of 2.1 m and a width of 12.9 m. The bridges have single column bents with span equal to 19.8 m and a diameter of 2.4 m respectively. The foundation consists of a group of 40 piles with a uniform diameter of 0.38 m and a depth of 18.3 m. A detailed description of this bridge structure can be found in the references [1, 2]. The schematic representation of relevant details of study bridges is shown in Fig. 2. For the analysis, the foundation consisting of 40 piles has been replaced with a single equivalent pile that is having the same bending stiffness as that of the pile group [1, 2, 18]. The equivalent bending stiffness $EI_{Eq}$ of the pile group can be calculated by the following formulas:

$$EI_{Eq} = EI_{Group} = n_p EI_p \tag{1}$$

$$EI_{Eq} = EI_{Group} = \sum_{i=1}^{n_p} \left\{ EI_p + EA_p (x_{pi} - x_0)^2 \right\}. \tag{2}$$

Equation 1 is applicable for sway type motion, and Eq. 2 is applicable for rocking type motion. Here $E$ represents modulus of elasticity of the foundation (pile) material, $I_{Group}$ and $I_{Eq}$ is the second moment of area (moment of inertia) of entire pile group and equivalent pile cross-section respectively; $I_p$ is the moment of inertia of individual pile; $n_p$ is the number of piles in the pile group and $(x_{pi} - x_0)$ denotes the relative distance of the individual pile from the centroid of the pile group and

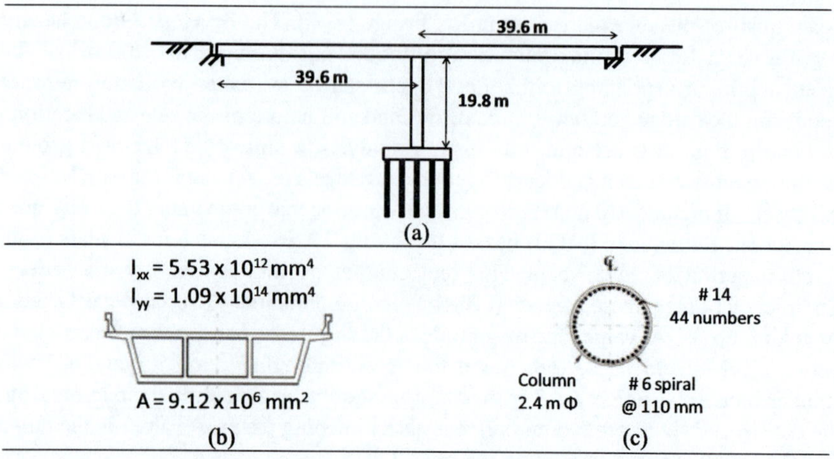


**Fig. 2** a Side view, **b** deck and **c** pier cross section details of study bridge [1, 2]

$A_p$ denotes the cross-sectional area of individual piles. Using the above-mentioned method the equivalent diameter of the single pile is calculated. For sway type motion the diameter of equivalent foundation (pile), $d_{eq}$ is obtained as 0.97 m and for the rocking condition, $d_{eq}$ is obtained as 4.2 m. In both cases, the length of the equivalent pile is the same as that of individual piles (18.3 m). The lower bound value of $d_{eq} =$ 0.97 m is considered for the investigation [1, 2].

The soil–foundation (pile) interaction between the equivalent single pile and its surroundings are modelled by incorporating suitable spring models. The lateral load resistance offered by soil (*p*) and corresponding deflection, *y* of cohesion-less soil at any given depth *H* (*p-y* curve) can be obtained using the following formulae [19].

$$p = AP_u \tanh\left[\frac{kH}{AP_u}y\right], \tag{3}$$

where factor *A* accounts for loading condition and for reverses cyclic loading conditions *A* can be taken as 0.9; $P_u$ is the ultimate lateral capacity (resistance) at a given depth *H*; *k* is the initial modulus of sub-grade reaction (foundation level), and the value of *k* is considered as 1220 ton/m$^3$ for the chosen condition [1, 2]. For finding $P_u$, the following equations can be used:

$$P_u = (C_1H + C_2d)\gamma H \tag{4}$$

$$P_u = (C_3d)\gamma H \tag{5}$$

Equation 4 is applicable for shallow foundations, and Eq. 5 can be used for deep foundations respectively. The coefficients $C_1$, $C_2$, $C_3$ are functions of friction angle and determined as per American Petroleum Institute recommendations [19]. The foundation is surrounded by three types of soils along with the depth (by taking the effective soil weight, $\gamma$ as 0.92 ton/m$^3$). That is, the soil profile consists of a silty sand layer for the top 11.7 m depth, followed by a 5.5 m deep silt layer for the intermediate depth, and a sand layer for the remaining depth of the pile length (a total length of 18 m). For calculating the $P_u$ coefficients, the average friction angle of the respective layers is considered [1, 2]. The obtained *p-y* springs are provided at 0.3 m interval downward along the length of the foundation for the study and a hinge support condition is assumed at the bottom tip of the equivalent pile. Similarly, to model the loss of lateral support due to flood induced scour, soil springs are taken out up to the depth $Y_s$ (scour depth) measured from the top of foundation ground level [1–5].

## 4 Flood Induced Scour Estimation

From the previously developed flood hazard curve, the 100 year (1% exceedance probability case) peak/desired discharge can be obtained. Taking the peak/desired discharge value, the velocity of flow, *V*, and the depth of flow $y_1$ can be estimated with the help of the following equations [13].

$$Q = bhy_1 \tag{6}$$

$$V = \frac{1}{n}\left[\frac{by_1}{b + 2y_1}\right]^{2/3} S^{1/2} \tag{7}$$

where $Q$ is the annual peak discharge, $n$ is Manning's constant for roughness and $S$ is the slope of the bed stream. The width of the stream $b$ can be assumed as the length of the bridge, and the local scour is assumed to be uniform around all the piers. Using Eqs. 6 and 7 and expressions from Hydraulic Engineering Circular, HEC-18 guidelines the scour depths $Y_s$ can be calculated for different flood hazard levels [20]. The expression for $Y_s$ is given by

$$Y_s = 2y_1 K_1 K_2 K_3 \left(\frac{a}{y_1}\right)^{0.65} Fr_1^{0.43} \tag{8}$$

where $y_1$ is the depth of flow immediately upstream to the bridge pier; $K_1$,$K_2$, and $K_3$ are correction factors that account for pier nose shape, angle of attack of flow, and bed condition, respectively, and $a$ is the width of the pier. $Fr_1$ is the Froude number at upstream of the pier, and it is defined as $V'/(g\ y_1)^{0.5}$, where $g$ is the acceleration due to gravity and $V'$ is the mean velocity of flow upstream of the pier. Since the pier is circular, the $K_1$ value can be taken as 1.0; similarly considering the angle of

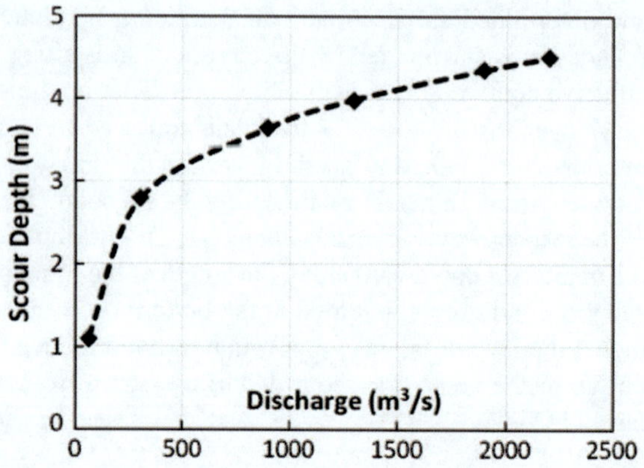


**Fig. 3** Calculated scour depth values with respect to discharge values

attack of flow as zero and assuming clear water condition, $K_2$ and $K_3$ can be taken as 1 and 1.1, respectively. For $y_1/a < 0.8$, another correction factor $K_w$ also needs to consider [20]. Further, Manning's coefficient of roughness, $n = 0.08$, and slope of the bed's stream, $S = 0.001$ are adopted for the analysis [1–3]. The estimated values of scour depths for different discharge levels are provided in Fig. 3. Initially (up to a discharge in the level of 300 $m^3/s$), the scour depth value increases sharply (1–3 m range), and then onwards it has a relatively mild increase in nature with respect to the increase of discharge value.

## 5 Bridge Modelling and Analysis

Non-linear time history analyses (MS only and MS-AS sequence) of the selected two-span bridge in the absence and absence of scour are performed in SAP2000 [21]. The scour depth value of two-span bridges ranges from 1 to 4.5 m. One representative value of scour depth, namely, 3 m is considered for the analysis. It is worth noting that the effects of loads like wind loads, vehicle loads, thermal loads, or any other secondary loads are not considered for the analysis. As (only) two-dimensional modelling and analyses are carried out; the earthquake loading and the resulting responses are considered only along the longitudinal direction of the bridge. Similarly for the analyses, the bridge deck is assumed to behave as elastic during the entire seismic activity and to capture the inelastic response of the bridge structure, relevant plastic hinges provided at critical locations on the bridge pier (top and bottom) with yield moment and yield rotation as 36,493 kNm and 0.003793 rad respectively [1–5, 22]. The corresponding elastic stiffness ($9.78 \times 10^6$ kNm) value is also incorporated in the analysis model. Additionally, soil-foundation interaction is incorporated as discussed earlier and a constant damping (Rayleigh) value of 5% is used for all the modes of the bridge structure.

From the modal analyses, it is observed that the fundamental time-period increases with deepening of the flood-induced bridge scour depth (2.29 s for no scour case and 3.45 s for 3 m scour case), indicating the increased flexibility (reduction in lateral stiffness) of bridges due to the absence of all-around lateral support at foundations. The time histories of rotations at the bottom and top of the piers and horizontal deflection at pier top are considered for the response evaluation. Measured displacements are then transformed to the displacement ductility, $\mu_{\Delta}$, and the same is used to assess (mainly) the seismic performance of the structure. Displacement ductility can be defined as the ratio of displacement (lateral) of the bridge pier to its corresponding yield displacement. Hence for finding out the yield displacement, nonlinear static analysis (push-over) is carried out and the corresponding yield displacement is obtained as 0.198 m. Displacements at top of the bridge pier can be the same as the displacement of their immediate nodes (above) on the girder due to the monolithic girder-pier connection of the structure for the modelling and analysis [1–5].

## 6 Analysis Results Discussions

For the response evaluation four damage states; namely, minor ($\mu_{\Delta} = 2.25$ to 2.90), moderate ($\mu_{\Delta} = 2.9$ to 4.60) and major ($\mu_{\Delta} = 4.60$ to 5.00) are considered. Displacement ductility above five is taken as the collapse of the structure [1–5]. On this background, the nonlinear time history of longitudinal displacement at the top pier (superstructure level) is estimated for each ground motion. Figure 4a, b shows the a representative displacement ductility time history of two-span RC bridge subject to LA31 MS (MCE; PGA = 1.30 g; duration = 30 s) followed by LA30 AS (MCE; PGA = 0.99 g; duration = 50 s) for with (3 m) and without scour cases respectively.

**Fig. 4** Representative time history response of the structure under MS-AS actions with and without flood induced scouring

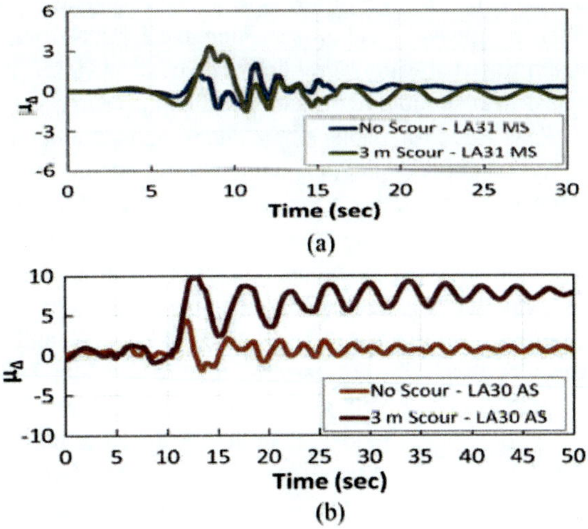

The result showed that the bridge experienced/reached a minor damage state under MS (alone) for no scour condition. But for the same MS, the structure experienced a moderate level of damage with a 3 m scour. Similarly, when the AS applied (followed by MS), the bridge without scour in a minor damage state under MS (alone) changed to moderate damage state response level, and for the case with 3 m scour lead to the collapse (peak ductility >5) of the structure. A similar trend is observed for almost all other time history analysis cases.

The time history analysis (dynamic) results further extended to the probabilistic seismic analysis domain. Probabilistic seismic analysis of structures comprises the development of seismic demand models; often serve as probabilistic models of structural response conditioned on a seismic intensity measure, *IM*. The conditional seismic demand—*IM* relationship, referred to as the Probabilistic Seismic Demand Models (PSDM). The probability that the seismic demand (related to the structural response), *D* placed on a component exceeds the capacity; *C* (related to strength) conditioned on a selected intensity measure can be evaluated by a fragility function as given in Eq. 9. This expression assumes that both demand and capacity follow lognormal distribution at the chosen intensity measure [17].

$$P[D > C|IM] = \Phi\left[\frac{\ln\left({}^{S_d}/_{S_c}\right)}{\sqrt{((\beta_{d|IM}) + (\beta_C)^2)}}\right] \tag{9}$$

$$\beta_{d|IM} = \sqrt{\sum_{i=1}^{N} \frac{(\ln(di) - \ln(S_d))^2}{N-2}} \tag{10}$$

where $\Phi(\bullet)$ is normal cumulative distribution function (standard); $S_d$ is the median estimate of demand; $S_c$ is the median value of capacity and $\beta_c$ is the dispersion of the capacity; $\beta_{d|IM}$ is referred to as the dispersion of the demand conditioned over intensity measure and $d_i$ is the demand for *i*th ground motion..$\beta_{d|IM}$ and $S_d$ can be determined using the regression of demand and intensity measures in log-transformed space with suitable regression coefficients (*a* and *b*).

Peak values of displacement ductility, $\mu_\Delta$ obtained for each analysis is considered as the seismic demand, *D*. and peak ground acceleration, PGA is considered as the intensity measure, *IM*. Three damage states before the complete collapse are considered (minor, moderate, and major) for the vulnerability assessment. The $S_c$ values are selected as per the recommendations from the past research reports [1–5]. Using the peak ductility demand obtained for different ground motion analyses, regression analysis is done among the MS, MS-AS, PGA, and displacement ductility parameters. Sequentially, regression-lines are developed for two-span RC bridge model subject to (1) No scour case (2) 3 m scour case. The slope of regression lines is more for scoured models than that of no scour models. Figure 5 shows the PSDM developed for these conditions subject to MS only and MS-AS cases respectively. Using these regression relations, Fragility curves for all the considered cases subject

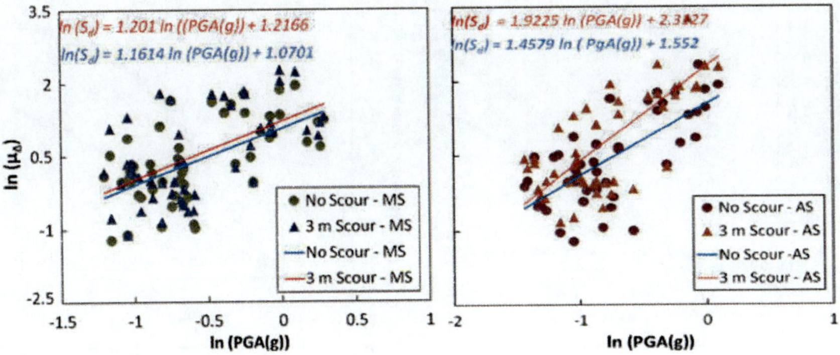


**Fig. 5** Probabilistic seismic demand models for MS and AS cases with and without considering flood induced scour effect

to various flood hazard levels are developed corresponding minor, moderate, major damage states as shown in Fig. 6. It is observed that as the scour depth increases, the exceeding probability/seismic vulnerability for each damage state increases. An escalation of exceedance probability (vulnerability to seismic events) is observed under MS-AS case compared to that of MS alone case. Corresponding to PGA of 1.0 g, the moderate damage state exceedance probability for the no-scour MS case is ~0.45 and no scour MS-AS case is 0.75 respectively. Similarly, corresponding to PGA of 1.0 g, moderate damage state exceedance probability for 3 m scour MS case is ~0.55 and 3 m scour MS-AS case is ~0.90 respectively. A comparable trend is observed for both minor and major damage states. These results indicate that sequential MS-AS action on flood induced scour affected bridges has a higher vulnerability to any of the considered damage states.

## 7 Conclusions from the Investigation

Seismic performance assessment of two-span reinforced concrete bridge in the absence and presence of flood-induced scour is investigated under seismic Mainshock alone and Mainshock followed by Aftershock cases. For the present investigation, Sacramento County in California is chosen as the bridge location area which is a seismically active and flood-prone locality. Seismic and flood hazards of the selected region are estimated and the flood-induced bridge scour caused due to various flood occurrences are calculated. The foundation consisting of 40 piles below bridge piers are modelled as a single equivalent pile that serves the same bending stiffness values as that of the entire foundation pile group under lateral movement (lower bound case). Likewise, the soil–equivalent foundation pile interaction is modelled using simplified spring models along with the depth. A finite element model of the example bridges with a scour depth of 3 m and without scouring is developed. Nonlinear time-history

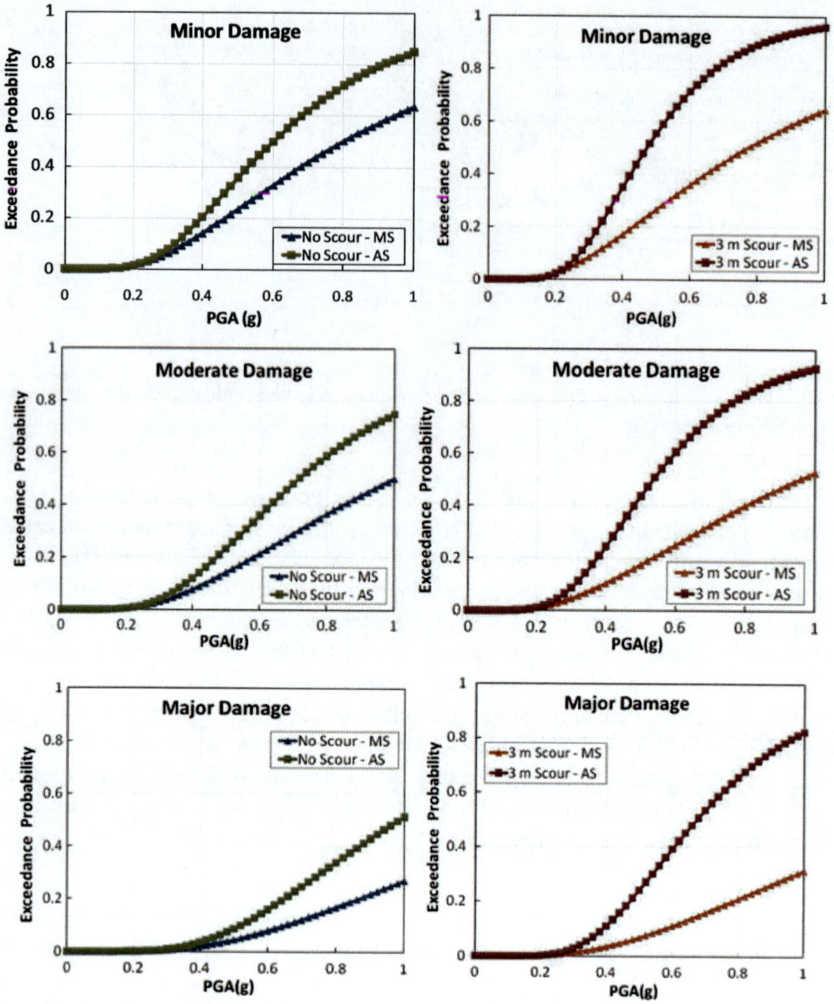


**Fig. 6** Fragility curves for two-span RC bridge under MS alone and MS-AS cases with and without scouring under different damage states

(dynamic) analyses are carried out under a selected set of ground motion records and the bridge responses are evaluated in terms of displacement at the top of the pier, which is further transformed to displacement ductility. Sequentially, Probabilistic Seismic Demand Models corresponding to major, moderate, and minor damage cases and resulting fragility curves are developed. It is found that the seismic vulnerability of bridges increases with scouring. In the event of Mainshock followed by Aftershock with scouring this trend become even worse.

## References


1. Banerjee S, Prasad GG (2013) Seismic risk assessment of reinforced concrete bridges in flood-prone regions. Struct Infrastruct Eng 9(9):952–968
2. Prasad GG, Banerjee S (2013) The impact of flood induced scour on seismic fragility characteristics of bridges. J Earthquake Eng 17:803–828
3. Yilmaz T, Banerjee S, Johnson PA (2016) Performance of two real-life California bridges under regional natural hazards. J Bridge Eng ASCE 21(3):04015063
4. Yilmaz T, Banerjee S, Johnson PA (2018) Uncertainly in risk of highway bridges assessed for integrated seismic and flood hazards. Struct Infrastruct Eng 14(9):1182–1196
5. Yilmaz T, Banerjee S (2018) Impact spectrum of flood hazard on seismic vulnerability of bridges. Struct Eng Mech 66(4):515–529
6. Alessandri S, Giannini R, Paolacci F (2013) Aftershock risk assessment and the decision to open traffic on bridges. Earthq Eng Struct Dyn 42:2255–2275
7. Kumar R, Gardoni P (2014) Effect of seismic degradation on the fragility of reinforced concrete bridges. Eng Struct 79:267–275
8. Dong Y, Frangopol DM (2015) Risk and resilience assessment of bridges under mainshock and aftershock incorporating uncertainties. Eng Struct 83:198–208
9. Ghosh J, Padget JE, Silva MS (2015) Seismic damage accumulation in high bridges in earthquake prone regions. Earthq Spectra 31(1):115–135
10. Omranian E, Abdelnaby AE, Abdollazadeh G (2018) Seismic vulnerability assessment of RC skew bridges subjected to mainshock-aftershock sequence. Soil Dyn Earthq Eng 114:186–197
11. Gidaris I, Padgett JE, Misra S (2020) Probabilistic fragility and resilience assessment and sensitivity analysis of bridges incorporating aftershock effects. Sustain Resilient Infrastruct. https://doi.org/10.1080/23789689.2019.1708174
12. Guo X, Zhang Z, Chen Z-Q (2020) Mainshock-integrated aftershock vulnerability assessment of bridge structures. Appl Sci 10:6843
13. Gupta R (2008) Hydrology and hydraulic systems. Waveland Press Inc., Illinois
14. USGS National Water Information System. http://nwis.waterdata.usgs.gov/nwis. Accessed 30 Oct 2020
15. Somerville P, Smith N, Punyamurthala S, Sun J (1997) Development of ground motion time histories for phase 2 of the FEAM/SAC steel project. SAC joint venture report no: SAC/BD-97/04
16. FEMA (2009) Quantification of building seismic performance factors. FEMA P695. Federal Emergency Management Agency, Washington DC
17. Mangalathu S, Shokrabadi M, Burton HV (2019) Aftershock seismic vulnerability and time-dependent risk assessment of bridges. PEER report no: 2019/04, Pacific Earthquake Engineering Research Centre, University of California, Berkeley
18. Yin Y, Konagai K (2001) A simplified method for expression of the dynamic stiffness of the large-scaled grouped piles in sway and rocking motions. J applied mechanics 4:415–422
19. API (2000) API recommended practice 2A-WSD—recommended practice for planning, designing and constructing fixed offshore platforms—Working Stress Design, American Petroleum Institute
20. Arneson LAM, Zeyenberge LW, Lagasse PF, Clopper PE (2012) Evaluating scour at bridges. Hydraulic Engineering Circular (HEC) No: 18, Publication number: FHWA-HIF-12-003. Federal Highway Administration, Department of Transportation, Washington, D.C.
21. CSI (2013) SAP 2000 Nonlinear analysis manual. Computers and Structures, Inc., California, Berkley
22. Priestley MJN, Seible F, Calvi GM (1996) Seismic design and retrofit of bridges. Wiley, New York

# Effect of Fire Hazard on Seismic Capacity of RC Frame Building

**Praveen Oggu, Swati Raman, and K. Gopikrishna**

## 1 Introduction

Fire hazard is characterized to be one of the most disastrous events, frequently encountered in urban cities as a primary or as a secondary event. This, if not controlled at early stages, can result in a catastrophe involving huge property damages and loss of life. This disastrous nature is more pronounced in areas of high incidence, coupled with densely populated urban environments. The occurrence of fire inside a building is an unexpected phenomenon and can break out at any time due to various reasons. Conflagrations following the 1906 San Francisco and 1923 Tokyo earthquakes led to serious discussions among engineers about simultaneous safety of buildings against earthquake and fire. The CITF world statistics reported that India loses about 2500 lives annually to fire with highest rate of deaths. Even though the occurrence of building fire is more frequent in comparison with other extreme events like earthquake and hurricanes, etc., it has not been given importance during structural design and detailing. The effect of fire on material properties and on structural members are the two important parameters for proper evaluation of damage of a building due to fire. Experimental studies in literature have reported material strength degradation at higher temperature as a major cause for damages in a fire event. The study of fire and earthquake can be done in two ways—pre-fire earthquake (structure subjected to fire after earthquake) and post-fire earthquake (fire affected structure subjected to an earthquake). The present study focuses on seismic behaviour of building subjected to post-fire earthquake. In reality, the fire leads to material strength deterioration, however, the extent of reduction in strength depends on exposure time and temperature. Moreover, the reduction in overall strength of building is uncertain, and therefore, the assessment of seismic performance of such fire affected structure becomes important. Hence, post-fire behaviour of structure is necessary for determining the

P. Oggu (✉) · S. Raman · K. Gopikrishna
National Institute of Technology Warangal, Warangal, Telangana, India

 
S. Kolathayar and S. C. Chian (eds.), *Recent Advances in Earthquake Engineering*, Lecture Notes in Civil Engineering 175,
https://doi.org/10.1007/978-981-16-4617-1_18

extent of structural damage post a fire event, necessary for repair and also for ensuring the safety and integrity of the structure.

Arioz (2007) studied the effect of high temperature on compressive strength of concrete with different types of aggregates and found significant reduction in compressive strength of concrete above 800 °C [1]. Topcu (2008) investigated performance of S220 and S420 reinforcing steel rebars under fire and found significant reduction in tensile strength of rebars above 950 °C [2]. Youssef and Moftah (2007) proposed stress strain model of concrete incorporating effect of temperature [3]. Jau and Hang (2007) studied the effect of non-uniform fire on strength of reinforced concrete column subjected to 2 and 4 h of fire exposure along with axial loading and biaxial bending. Kodur and Dwaikat (2008) developed the mathematical model for predicting the behaviour of reinforced concrete beams subjected to fire with the effect of spalling [4]. Tan and Yao (2004) presented simplified approach for predicting fire resistance of column subjected to different thermal boundary conditions and developed strength reduction factors for steel and concrete subjected to 1-face, 2-face, 3-face and 4-face heating by using SAFIR software tool [5–7]. Therefore, assessment of impact of fire hazard on the structural capacity of RC frame building is essential in mitigating this hazard. Hence, four-storey RC frame building perceived to be located in the city of Warangal, Telangana State has been selected to analyse thermal behaviour and post-fire seismic behaviour in this study.

## 2 Structural Modelling

The structural model considered consists of a regular RC building frame with G + 3, i.e. four storeys of uniform height 3.2 m having two bays with a bay width of 5 m (see Fig. 1). This represents the typical building configuration prevalent in the said location. The building frame has been designed confirming to code provisions of IS 456: 2000 and IS 1893 (Part 1) as located in seismic zone III with medium type of soil [8, 9]. The description of models and the details of loads considered for seismic analysis are given in Table 1 and Table 2, respectively. The structural design details of the model are given in Table 3. The structural modelling is performed using SAP2000, a commercial structural analysis software [10]. The structural elements (beams and columns) were modeled with concentrated plastic hinges, where the beams have only moment (M3) hinges, and the columns have an axial load and a biaxial moment (PMM) hinges as per FEMA 356 (2000) [11]. Additionally, rigid diaphragms were assigned for every storey level throughout the structure ignoring the flexibility of the floor. Mander et al. and Park et al. models were used in characterizing the stress–strain behaviour of concrete and steel rebars, as shown in Fig. 2 [10, 12]. Moreover, as per the recommendations of IS 1893 (2016), moments of inertia of beams and columns were reduced to 35% and 70% for beams and columns, respectively, while performing nonlinear structural analysis.

Fig. 1 3D representation of structural configuration investigated in this study

**Table 1** Design details of the selected four-storey building [13]

| Member | Storey | Width (mm) | Depth (mm) | Longitudinal Reinforcement | | Transverse reinforcement |
|---|---|---|---|---|---|---|
| | | | | Top | Bottom | |
| Beam | 1 | 250 | 450 | 5–20ϕ | 4–20ϕ | 8ϕ @ 150c/c |
| | 2 and 3 | 250 | 400 | 4–16ϕ | 3–16ϕ | 8ϕ @ 150c/c |
| | 4 | 250 | 300 | 4–12ϕ | 3–12ϕ | 8ϕ @ 150c/c |
| Column | 1–4 | 420 | 420 | 8–16ϕ | | 8ϕ @ 175c/c |

**Table 2** Description of the models considered

| S. No | Parameter | Data |
|---|---|---|
| 1 | Height of each storey | 3.2 m |
| 2 | Width of each bay | 5 m |
| 3 | Grade of concrete | M 25 |
| 4 | Poisson's ratio of concrete | 0.2 |
| 5 | Unit weight of concrete | 25 kN/m$^3$ |
| 6 | Modulus of elasticity of concrete | 25 GPa |
| 7 | Grade of rebar | Fe 415 |
| 8 | Modulus of elasticity of rebar | 200 GPa |

Table 3 Details of loads considered for seismic analysis

| S. No | Particulars | Description |
|---|---|---|
| 1 | Dead load of beams and columns | Self-weight of the structure |
| 2 | Live load on slab | 3 kN/m$^2$ |
| 3 | Dead load on slab including floor finish | 3.75 kN/m$^2$ |
| 4 | Dead load of wall (230 mm thick) | 14.2 kN/m |
| 5 | Seismic load | IS 1893 (Part 1): 2016 |
| 6 | Type of structure | RC frame structure |
| 7 | Importance factor | 1 |
| 8 | Response reduction factor | 3 for OMRF |
| 9 | Soil type | Medium |
| 10 | Zone factor | 0.16 (Zone III) |

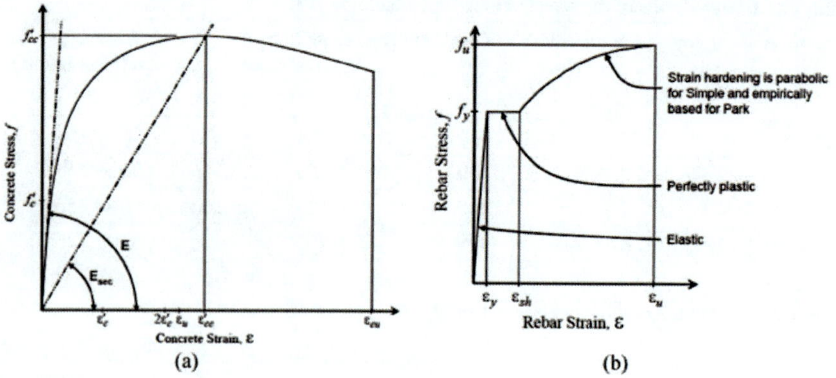


**Fig. 2** **a** Mander confined concrete stress–strain curve. **b** Park stress–strain model

## 3 Analysis Methodology

Thermal analysis of the structural model is performed using SAFIR, a thermal analysis software to obtain evolution of temperature within structural components. In addition, modified material properties for the structural components at various elevated temperatures are considered from EUROCODE [14]. Further, section analysis for various fire exposure times (30 and 60 min) with modified material properties is performed using XTRACT software. The outcome is moment rotation and interaction curve, necessary for defining the nonlinear hinge properties of the structural components (i.e. beam and column).

Finally, the pre-fire seismic inelastic capacity of the structure is assessed using nonlinear static analysis (NLS) for various fire scenarios described in terms of exposure times (30 and 60 min) on the structural models considered. The moment rotation and P-M interaction curves obtained from XTRACT software for various fire scenarios are considered as input for seismic behaviour assessment. Nonlinear static (or pushover) analysis is then performed, using the response spectrum specified in IS code for the specified site conditions (Zone III, Medium soil profile) at Warangal city, Telangana state. Pushover analysis is a static, nonlinear procedure in which the magnitude of the structural loading or displacement is incrementally increased in accordance with a certain predefined pattern. The loading is monotonic. With increase in magnitude of loading, weak links of the structures can be obtained. Pushover analysis estimates force and displacement capacity of structure along with sequential formation of hinges in the structure under analysis. The outcome of pushover analysis is usually represented in the form of base force (or base shear) versus roof displacement. This is popularly referred as capacity curve of structure. The outcome of the pre-fire nonlinear seismic analysis is a capacity curve defined in terms of base shear with roof. The overall methodology of the study is depicted as a flowchart in Fig. 3. In general, the seismic behaviour of RC building in terms of typical capacity curves representing original (or stronger) capacity and reduced (or weaker) capacity are shown in Fig. 4.

# 4 Results and Discussion

## 4.1 Thermal Analysis

The structural components, i.e. beams and column are first modelled, and their thermal capacities were estimated using SAFIR software. SAFIR performs the transient thermal analysis to determine the temperature distribution in the structure. The thermal profiles of the cross-sections of the structural components are shown in Fig. 5.

The sectional analysis results for the beam and column structural components in terms of moment curvature and interaction diagrams for fire scenarios of 30 min and 60 min duration were performed in XTRACT program. The reduced sectional capacities of the structural components are shown in Figs. 6 and 7.

## 4.2 Seismic Analysis

The reduced sectional properties of beams and columns obtained from thermal analysis are considered for defining hinge properties at elevated temperatures in the NLS of the structural model as discussed. The outcome of NLS analysis, the capacity curve developed in accordance with FEMA regulations for no fire, 30 min fire exposure

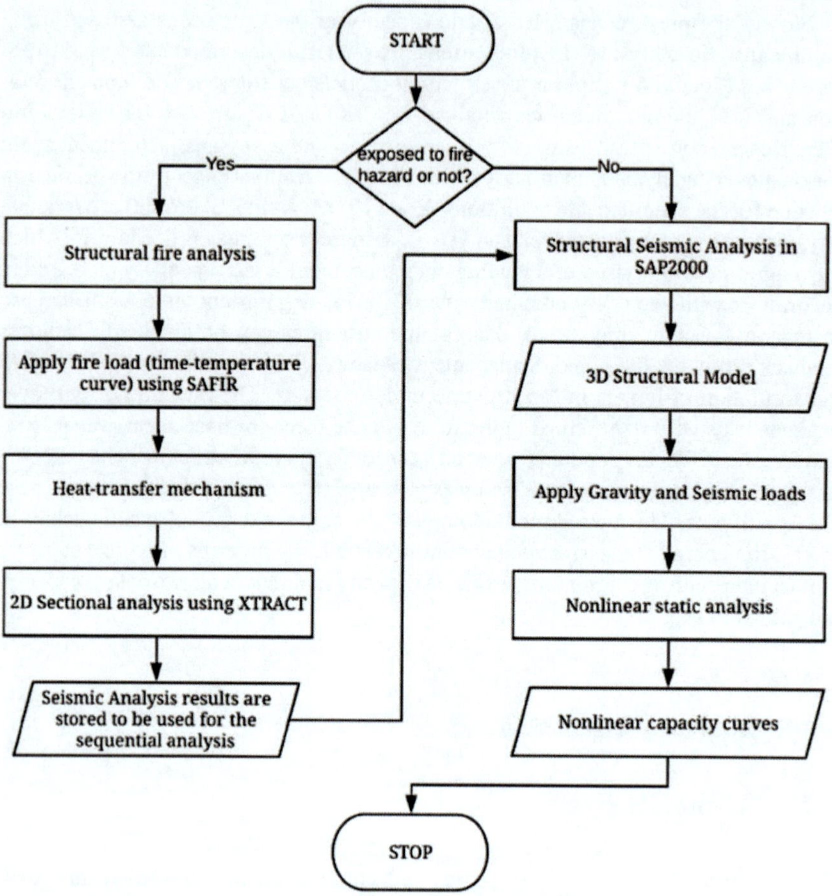


**Fig. 3** Flowchart of the methodology used in this study

and 60 min fire exposure durations is shown in Fig. 8. It can be observed that there is substantial reduction in base shear for the structural model due to fire hazard. This suggests reduction in capacity of the structure due to degradation in exposed material due to fire. Hence, there is an imminent need to assess the structural behaviour considering the effects of fire as it proves detrimental for the survival and functionality of the structure. Therefore, assessment of impact of fire hazard on the structural capacity of RC frame building is essential in mitigating this hazard.

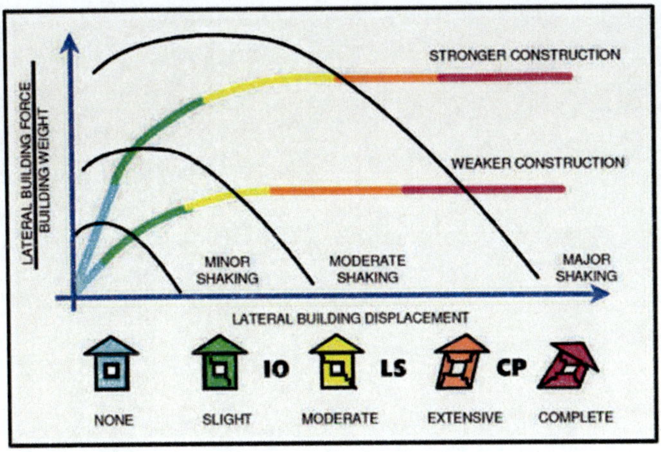


**Fig. 4** Seismic behaviour of RC building for different performance levels

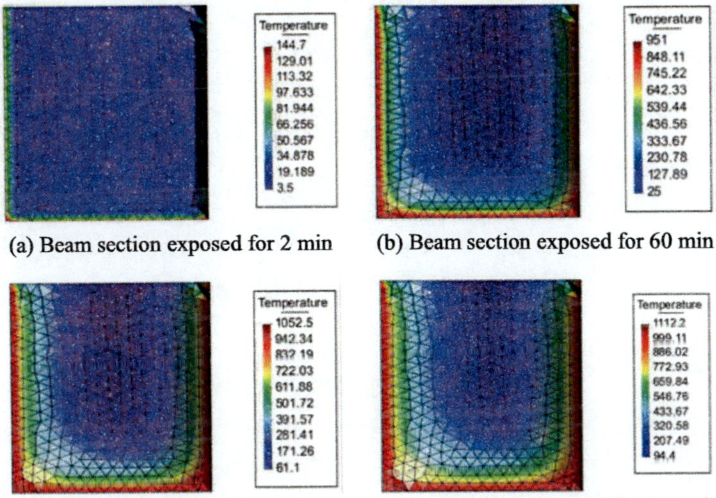


(a) Beam section exposed for 2 min (b) Beam section exposed for 60 min

(c) Beam section exposed for 120 min (d) Beam section exposed for 180 min

**Fig. 5** Beam section exposed to ISO 834 fire for various durations

## 5 Conclusions

Four-storeyed RC building frame is modelled using SAP2000 commercial structural software. The modified material and cross-sectional properties necessary for defining the hinge properties of the structural components due to various fire exposure times (i.e. 30 and 60 min duration) are developed using XTRACT software. Further, the pre-fire seismic behaviour is assessed by performing NLS analysis with modified

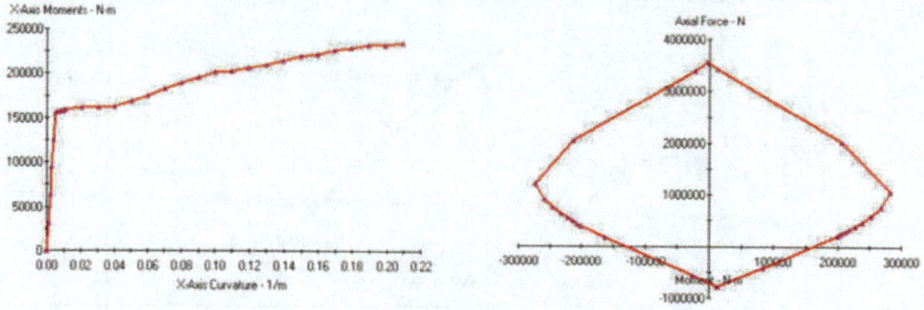


Beam 1

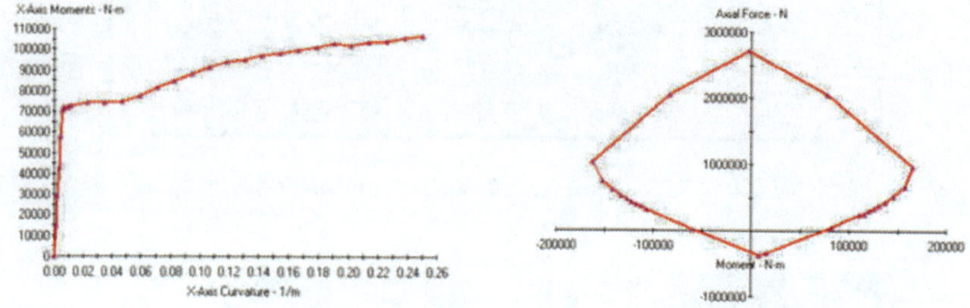


Beam 2 & 3

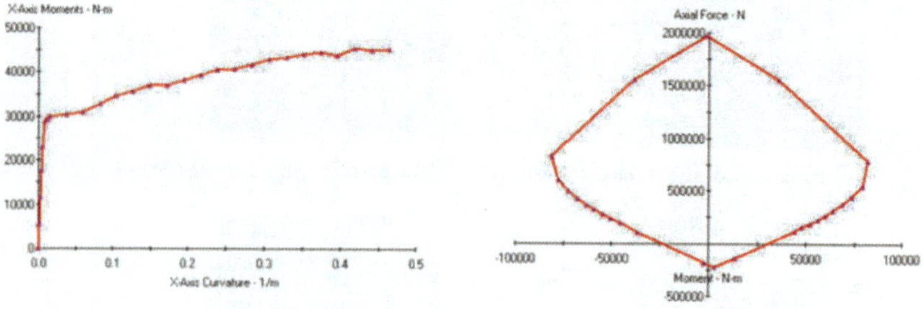


Beam 4

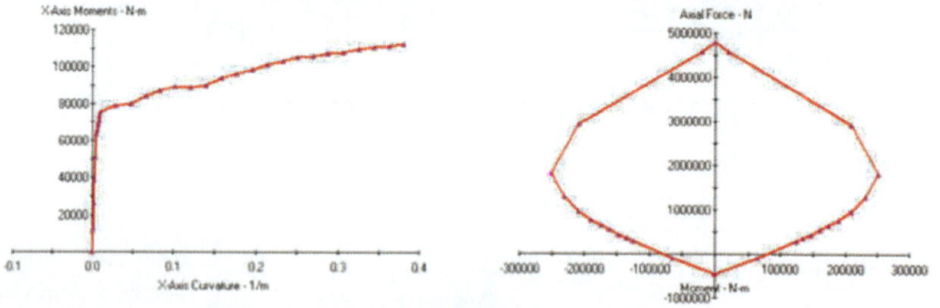


Column

**Fig. 6** Moment curvature and P-M interaction diagrams of beams and column, exposed for a duration of 30 min

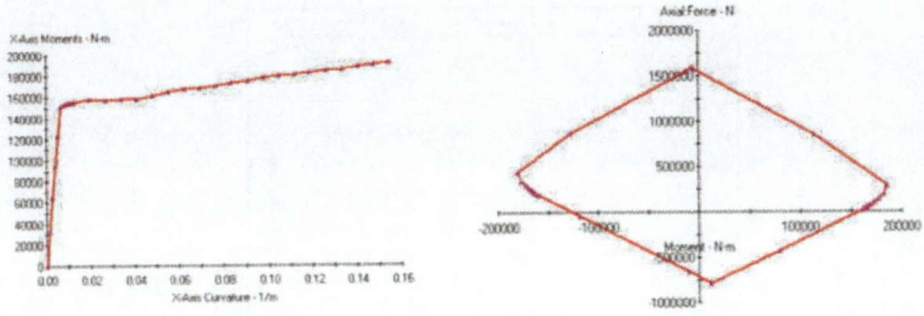


Beam 1

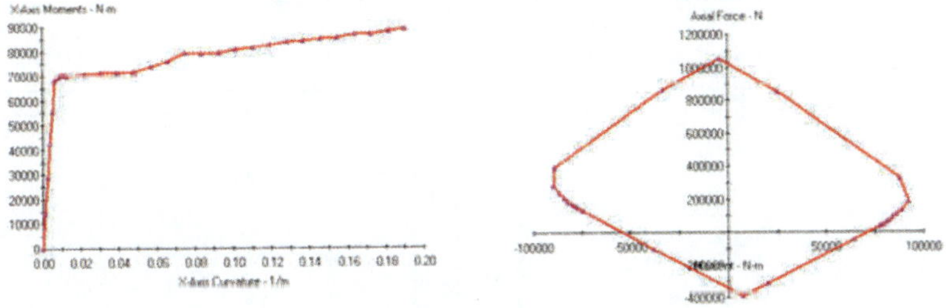


Beam 2 & 3

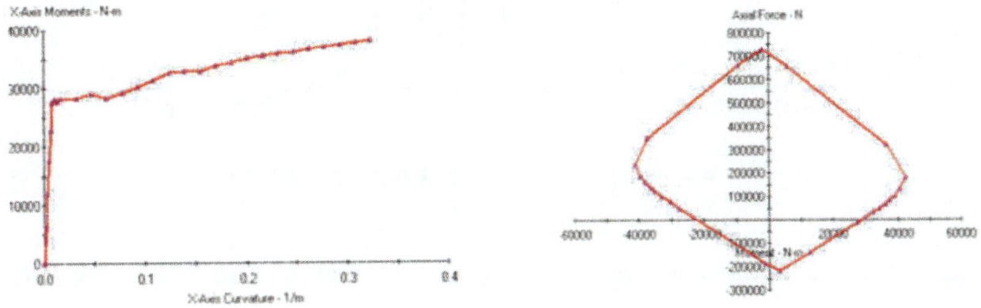


Beam 4

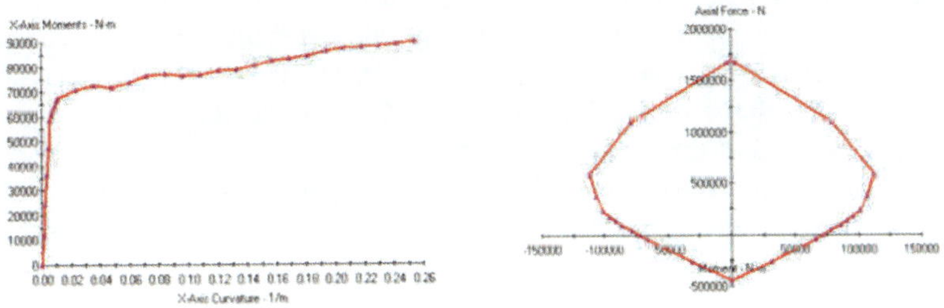


Column

**Fig. 7** Moment curvature and P-M interaction diagrams of beams and column, exposed for a duration of 60 min

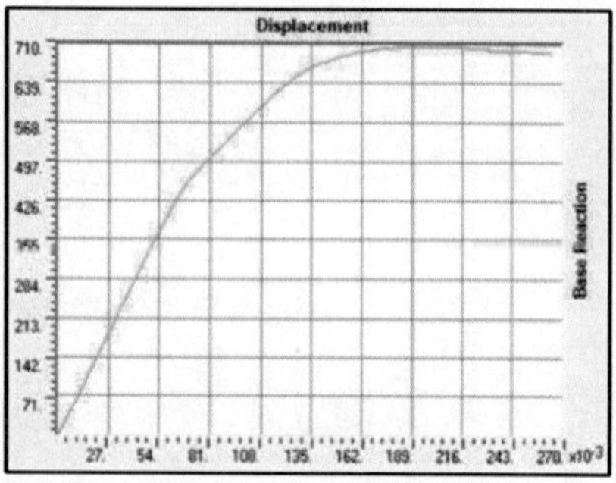


(a) 0 min or no exposure

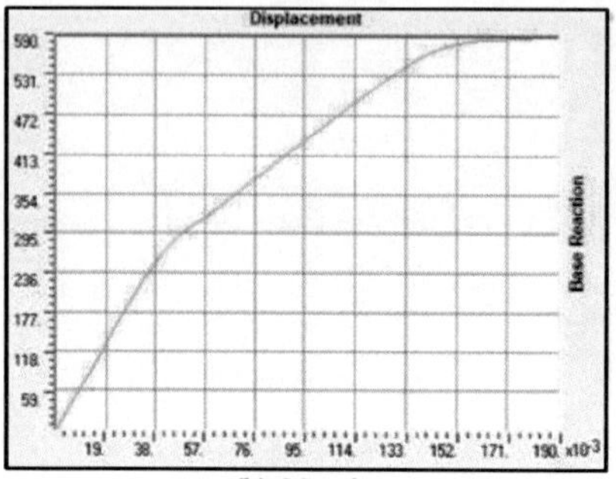


(b) 30 min

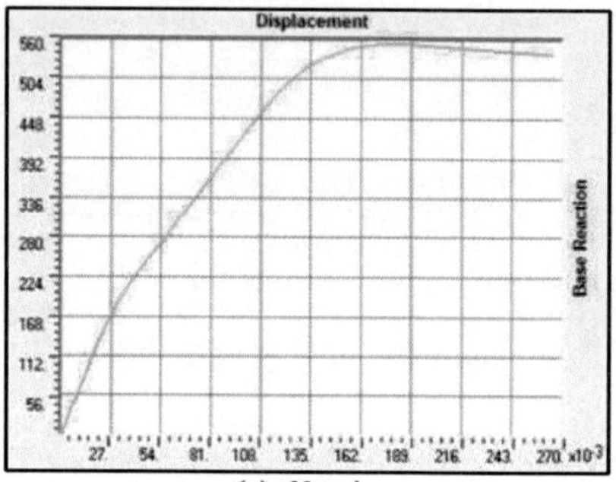


(c) 60 min

**Fig. 8** Capacity curves of the model exposed to different durations

hinge properties at elevated temperatures. The RC frame buildings are reported to be vulnerable to fire hazard in literature, this observation can be clearly envisaged in the study presented. From the capacity curves developed during nonlinear static analysis for the hypothetical structural model, it can be observed that certain structural components might have incurred damages contributing to reduced seismic capacity.

From this preliminary analysis, it can be concluded that fire hazard has significant effect on reducing the capacity of the structure and proves detrimental for the survival in case of existence of fire for prolonged duration. Hence, structural analysis, post a fire hazard is imminent for appropriate design of repair/rehabilitation measures to restore the functionality of the structure. This aids in building a safe and functional built environment.

**Acknowledgements** The authors express their gratitude to the Department of Civil Engineering, National Institute of Technology, Warangal for providing laboratory facilities to conduct the work, and the Ministry of Human Resource Development, India for providing fellowship to first and second authors. Also, the authors express their acknowledgment to CSi, New Delhi for extending a standalone license of SAP2000 software for this work.

## References

1. Arioz O (2007) Effects of elevated temperatures on properties of concrete. Fire Saf J 42(8):516–522
2. Topcu IB, Karakurt C (2008) Properties of reinforced concrete steel rebars exposed to high temperatures. Mater Sci 2008(814137):1–4
3. Youssef MA, Moftah M (2007) General stress–strain relationship for concrete at elevated temperatures. Eng Struct 29:2618–2634
4. Kodur VKR, Dwaikat M (2008) A numerical model for predicting the fire resistance of reinforced concrete beams. Cement Concr Compos 30(5):431–443
5. Tan KH, Yao Y (2004) Fire resistance of reinforced concrete columns subjected to 1-, 2-, and 3-face heating. J Struct Eng 130(11):1820–1828
6. SAFIR manual for fire analysis
7. GID manual for fire analysis
8. IS 456. Plain and reinforced concrete-code of practice. BIS, New Delhi, India (2000)
9. IS 1893. Criteria for earthquake resistant design of structures. Part 1 General provision and buildings. BIS, New Delhi, India (2016)
10. SAP2000, CSi. Analysis reference manual. Computer and Structures Inc., Berkeley (2016)
11. FEMA 356: pre-standard and commentary for the seismic rehabilitation of buildings. Federal Emergency Management Agency (2000)
12. Mander JB, Priestley MJ, Park R (1988) Theoretical stress-strain model for confined concrete. J Struct Eng 114(8):1804–1826
13. Dhir PK, Robin D, Pradip S (2018) Safety assessment of gravity load–designed reinforced concrete–framed buildings. ASCE-ASME J Risk Uncertain Eng Syst Part A Civil Eng 4(2):04018004
14. European committee for standardisation. Design of concrete structures: general rules: structural fire design. EN 1992-1-2, EUROCODE 2, part 1–2: Brussels (2004)

# Geotechnical Vulnerability Assessment

# Effect of Soil-Structure Interaction on the Seismic Response of Elevated Water Tank

**Pranitha Jogi and B. R. Jayalekshmi**

## 1 Introduction

An elevated water tank is a structure constructed for the purpose of storing water and to hold the water supply level at a definite height to provide sufficient pressure in the water distribution system. Elevated water tanks with circular container are commonly built. However, in the flat bottom, the thickness and reinforcement are found to be heavy. But in the case of intze tanks, in its domed bottom, the thickness and reinforcement in dome is normal. The main advantages of intze tanks are that the outward thrust from top of conical part is resisted by the ring beam and economical.

It is reported that in circular water tanks, the base shear and base moment increase in higher seismicity zones and decrease with an increase in soil stiffness [10]. Most of the analytical studies showed that the seismic behavior of elevated tanks suggested by Eurocode-8 gives the more accurate results than the ACI-371R-98 codal provisions [6, 8]. American codes provide a detailed classification of tanks, but Eurocode-8 and New Zealand do not have such classification. Soil–structure interaction provisions are provided in New Zealand and Eurocode-8 [9]. The frame structure supporting the elevated water tank must have sufficient strength to resist the axial loads, moment, and shear force due to lateral loads. The lateral forces due to earthquakes depend on the location of the structure, the type of soil in that location, and total weight of the structure. The total weight varies with the water level in the tank container. The interaction between soil and structure can affect the dynamic response of structures, especially for structures which are supported on relatively flexible soil strata. The consideration of the soil–structure interaction (SSI) effect is mainly important in the seismic analyses of structures which are located in severe seismic zones. Therefore, accurate representation and analysis of the soil–structure interaction effects are a

P. Jogi (✉) · B. R. Jayalekshmi
Department of Civil Engineering, National Institute of Technology Karnataka, Surathkal, India
e-mail: jogipranitha.197cv019@nitk.edu.in

 
S. Kolathayar and S. C. Chian (eds.), *Recent Advances in Earthquake Engineering*, Lecture Notes in Civil Engineering 175,
https://doi.org/10.1007/978-981-16-4617-1_19

crucial part. The soil–structure interaction may be more important for elevated tanks due to most of the liquid masses being lumped above the ground level and the foundation being supported on a relatively small area. The displacement and base shear of the structure increases with the increased flexibility of the supporting soil strata [7]. The consideration of foundation embedment is also necessary to review the dynamic responses. Tank roof displacements are significantly affected by the embedment in soft soil than the stiff soils. Except for soft soil strata, embedment did not affect the other response parameters, such as sloshing displacement of the tank [3]. Without proper site investigation, if the subsoil is assumed as rigid or flexible, the earthquake design of the RC elevated tanks may lead to a wrong assessment of the seismic base shear and over-turning moment. Base shears in the order of three or more times larger may be obtained, especially for soft subsoil [4]. The displacement response of the structure is generally altered, when the soil gets softer. The SSI effects are more effected for the flexible soil strata than the rigid one. Therefore, SSI effect has to be considered while designing the tank which is supported on the soft soil [5].

The present study focus on identifying the critical condition in intze tanks by comparing the seismic response of elevated intze water tanks considering different water levels and different supporting soil conditions.

## 2 Methodology

The finite element software ANSYS [1] was used to evaluate the SSI effects on the dynamic response of an elevated intze water tank of frame type staging. Figure 1 shows the vertical elevation of an elevated intze tank, which was analyzed and reported in [3].

The cross-section details of various components and the material properties are shown in Tables 1 and 2.

During the seismic activity, seismic waves are transmitted through the soil from the fault rupture to the structure of interest. The response of structure depends on the material properties of the soil medium, the source of excitation, and the type of foundation. The structural seismic response due to SSI differs significantly from the response of structure on a rigid support. The effect of interaction is generally considered to be favorable in the seismic design because the supporting soil medium provides more damping which can reduce the response. Even though SSI induces damping, it can also cause increased displacement of the overall structure. The SSI increases lateral displacements because of lengthened period and secondary forces associated with P-Δ effects. Hence, it is essential to take SSI into consideration during the analysis to represent the real field conditions.

Four types of soils, namely, rock, dense soil, stiff soil, and soft soil [2], were considered as supporting medium for this study. According to FEMA 273, various soil properties and corresponding damping values of these soils are noted and are shown in Tables 3 and 4. To incorporate the SSI effects, $40 \times 40 \times 20$ m of soil stratum supporting the tank was considered, which is 3–5 times the width (radius) of the raft.

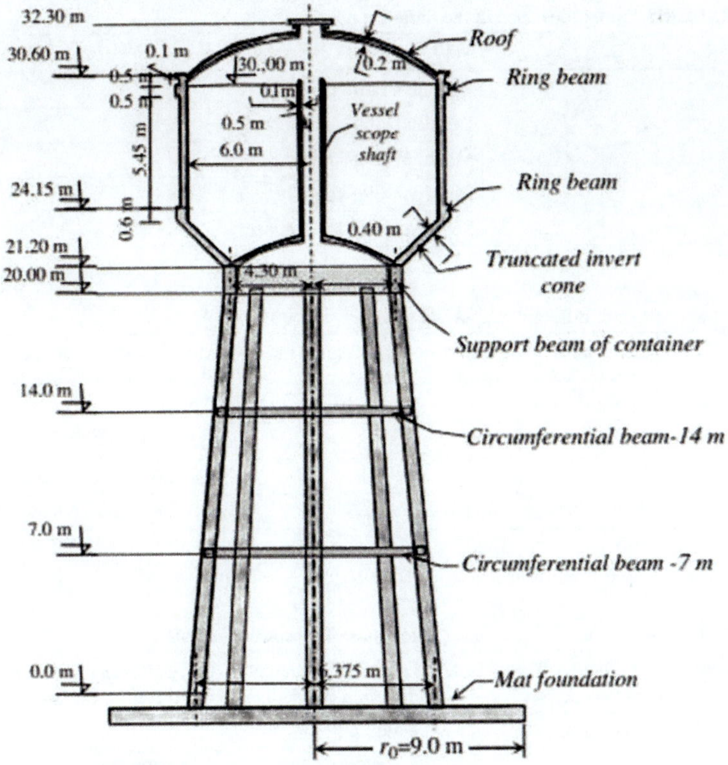


**Fig. 1** Elevation of the reinforced concrete elevated intze tank (*courtesy* R. Livaoglu [3])

**Table 1** Cross sectional details of the structural components of the intze water tank

| Component | Dimension (mm) |
|---|---|
| Top dome | 200 thick |
| Top ring beam | 300 × 500 |
| Cylindrical wall | 400 thick |
| Bottom ring beam | 450 × 600 |
| Circular ring beam | 500 × 1200 |
| Bottom dome | 200 thick |
| Conical dome | 400 thick |
| Braces | 350 × 1200 |
| Columns | 300 (diameter) |
| Raft | 300 Thick |

**Table 2** Material properties used in the finite element analysis

| Concrete | M30 |
|---|---|
| Water | Poisson's ratio: 0.15 |
| | Density: 2500 kg/m$^3$ |
| | Modulus of elasticity: 3.2E10 N/m$^2$ |
| | Density: 1000 kg/m$^3$ |
| | Sonic velocity 1500 m/sec$^2$ |

**Table 3** Properties of different supporting soils

| Soil class | Shear wave velocity VS (m/sec) | Density $\rho$ (kg/m$^3$) | Poisson's ratio | Modulus of elasticity E (kN/m$^2$) | Dilatational velocity Vp (m/sec) |
|---|---|---|---|---|---|
| Rock | 1200 | 2200 | 0.3 | 8,236,800 | 2245 |
| Dense soil | 600 | 2000 | 0.3 | 1,872,000 | 1122.5 |
| Stiff soil | 300 | 1800 | 0.35 | 437,400 | 624.5 |
| Soft soil | 100 | 1600 | 0.4 | 44,800 | 245 |

**Table 4** Damping values in X, Y, and Z directions of viscous boundary

| Soil class | Damping coefficient in X direction (kg/sec) | Dampingcoefficient in Y direction (kg/sec) | Damping coefficient in Z direction (kg/sec) |
|---|---|---|---|
| Rock | 10,560,000 | 5,280,000 | 9,877,976 |
| Dense soil | 4,800,000 | 2,400,000 | 4,489,989 |
| Stiff soil | 2,160,000 | 1,080,000 | 2,248,199 |
| Soft soil | 640,000 | 320,000 | 783,836.7 |

Viscous boundaries were provided at lateral boundaries to represent infinite soil stratum and its seismic behavior, i.e., the nodes are restrained in all three directions X, Y, and Z with springs having damping equal to [3]

$$C_{t1} = At_1 \times \rho \times \text{V}s$$

$$C_{\text{n}} = \text{An} \times \rho \times \text{V}p$$

$$C_{t2} = At_2 \times \rho \times \text{V}s$$

where

$\rho$ = Density of soil, V$p$ = Dilatational velocity of soil, V$s$ = Shear wave velocity of soil [2], and An, A$t$ = Area controlling viscous dampers.

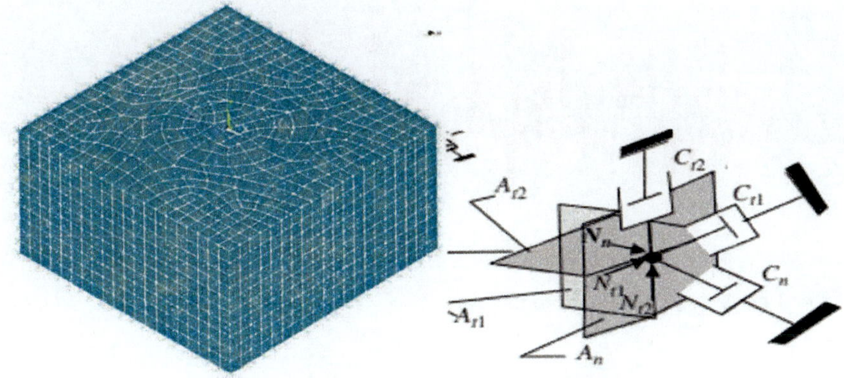


**Fig. 2** Viscous boundary considered in the 3-D finite element model

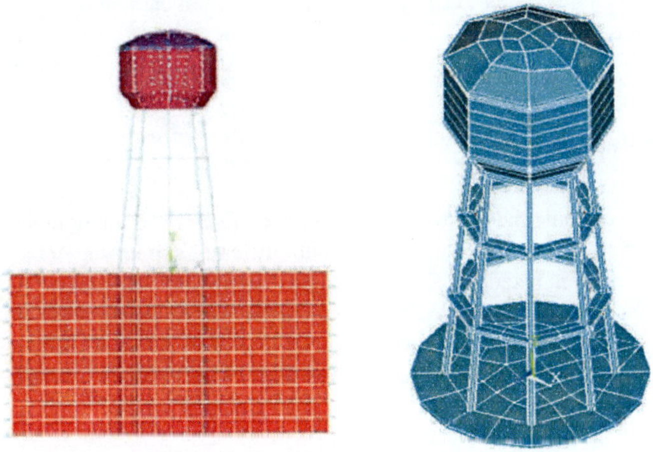

**Fig. 3** Finite element model of the elevated water tank

The subscripts n and t are the normal and tangential directions in the boundary. Figure 2 shows the viscous boundary representation of nodes, and Table 4 gives the damping values of springs.

## 3 Finite Element Model of Elevated Intze Tank

The finite element modeling of the elevated intze water tank was carried out in ANSYS mechanical APDL software. Beams and columns were modeled by structural 3-D two noded BEAM 188 elements having six DOF at each node, the container

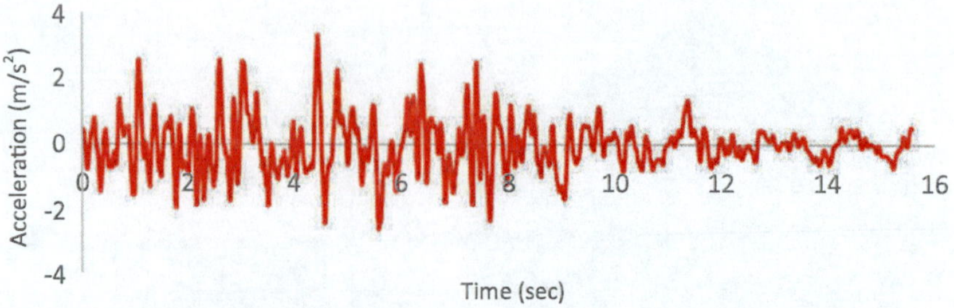


**Fig. 4** Time–history plot of Kobe earthquake

walls, raft, and top and bottom domes were modeled by 3D four noded structural SHELL 181 elements having 6 DOF at each node, and the soil medium was modeled by 3D eight noded structural solid BRICK 185 elements having three DOF at each node. Fluid is modeled by 3D eight noded acoustic fluid element FLUID 30 having five DOF at each node.

## 4 Transient Analysis

A series of transient analysis was carried out for 3D elevated water tank by varying the depth of water level in the tank and with different supporting soils. A total of 16 models were analyzed for Kobe earthquake data. Kobe earthquake happened in Japan in 1995 with a magnitude of seven and peak ground acceleration of 3.36 m/s$^2$. The time–history plot of this ground motion is shown in Fig. 4. This time history of acceleration was applied along the X direction of all the models in ANSYS, and the structural response was studied.

## 5 Results and Discussions

The seismic response parameters such as base shear, base moment at base of the tank, and structural displacements at the top of the tank shell are found with the variation of depth of fluid inside the tank for all the models. Figs. 17, 18 and 19 shows the maximum responses for the elevated tank having different water levels and supporting soil profiles. The base shear ratio representing base shear of each model with respect to its total weight is used for comparison. The time history plots are drawn and represented in following graphs.

## 5.1 Maximum Displacement

Figures 5, 6, 7, and 8 show the variation of top shell displacement for tank full, two third full, one third full and empty conditions for the tank supporting on rock, dense soil, stiff soil, and soft soil. The maximum shell roof displacement is observed for the one third full condition of the tank supporting on soft soil. Because in one third full condition, there are additional hydrodynamic pressures acting on the tank wall and the tank base which can increase the displacement of the top shell. From Figs. 17, 18, and 19, it is reported that for the tank supporting on rock the displacement is more in tank full condition because of its heavy water weight, but this is not following the same pattern for the tank supporting on dense, stiff, and soft soil. This observation

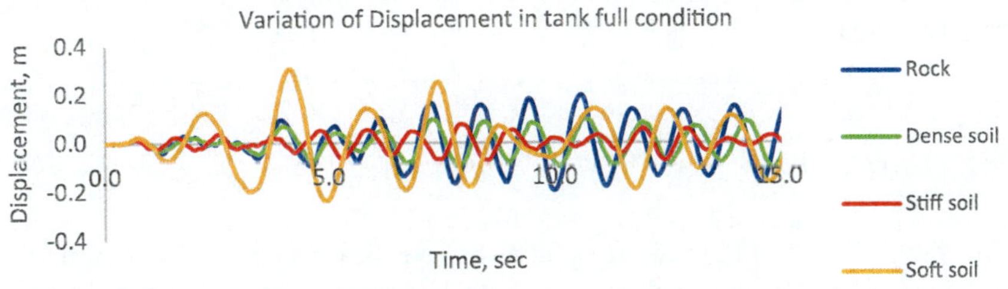


Fig. 5 Variation of top shell displacement in tank full condition

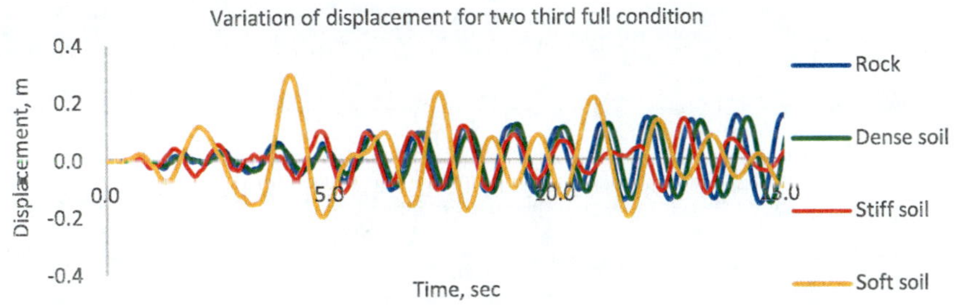


Fig. 6 Variation of top shell displacement in two third full condition

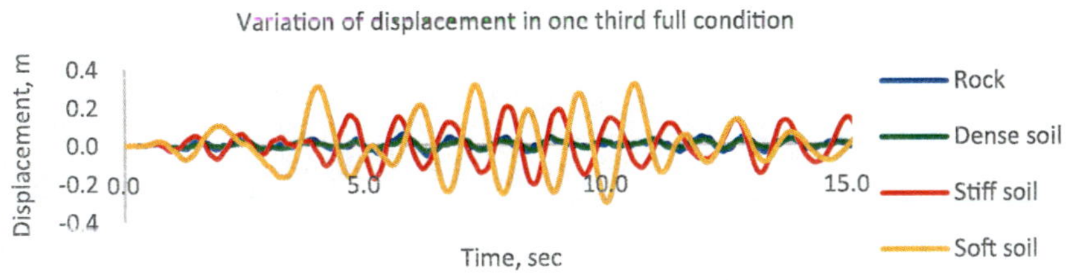


Fig. 7 Variation of top shell displacement in one third full condition

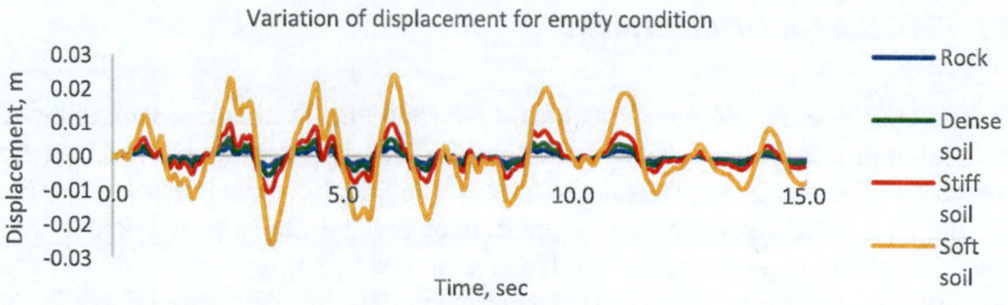


**Fig. 8** Variation of top displacement in empty condition

will lead to understand the hydrodynamic pressure effect is dependent not only on water level but also on the supporting soil too. In soft soil, these hydrodynamic pressures are more and it leads to the maximum displacements at the top.

## 5.2 *Base Shear*

Figures 9, 10, 11, and 12 show the variation of base shear for tank full, two third full, one third full, and empty conditions for the tank supporting on rock, dense soil, stiff soil, and soft soil. The base shear is maximum for the tank supporting on soft soil

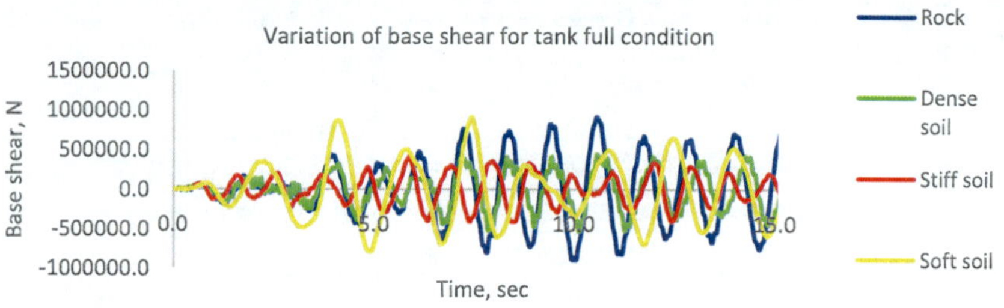


**Fig. 9** Variation of base shear in tank full condition

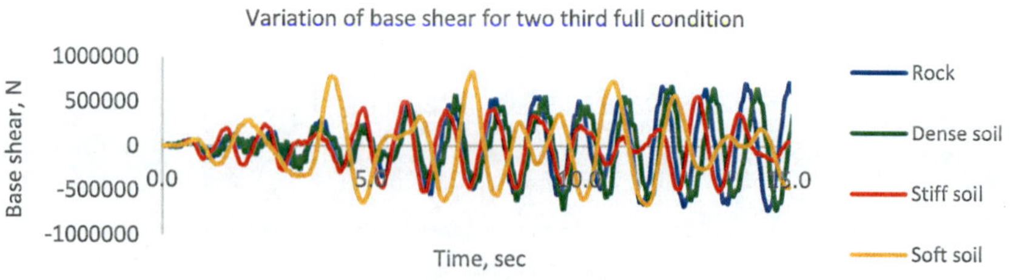


**Fig. 10** Variation of base shear in two third full condition

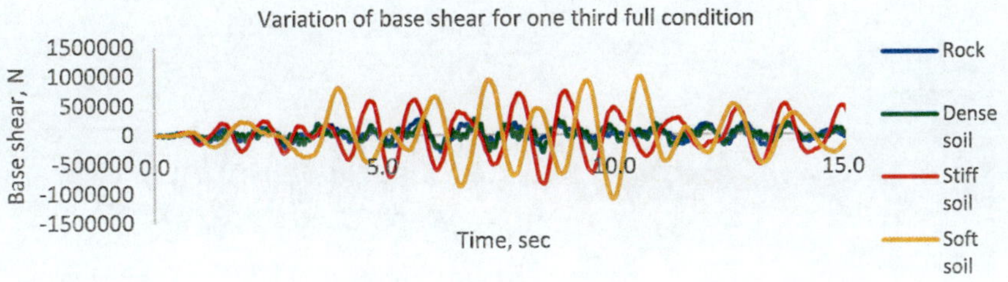


**Fig. 11** Variation of base shear in one third full condition

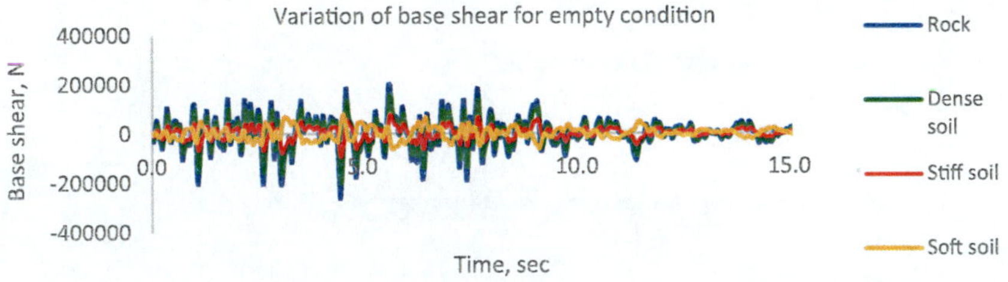


**Fig. 12** Variation of base shear in empty condition

for one third full condition, but there is not much difference in the base shear of the tank supporting on rock for tank full condition. The base shear ratio is represented by taking the base shear of the tank to its seismic weight, and it is very less for all the empty tanks supporting on different soils due to its lower seismic weight.

## 5.3 *Base Moment*

Figures 13, 14, 15, and 16 show the variation of base moment for tank full, two third full, one third full, and empty conditions for the tank supporting on rock, dense soil, stiff soil, and soft soil. The base moment is more in tank resting on soft soil for the one third fill condition. Base moment is observed generally more for tank full

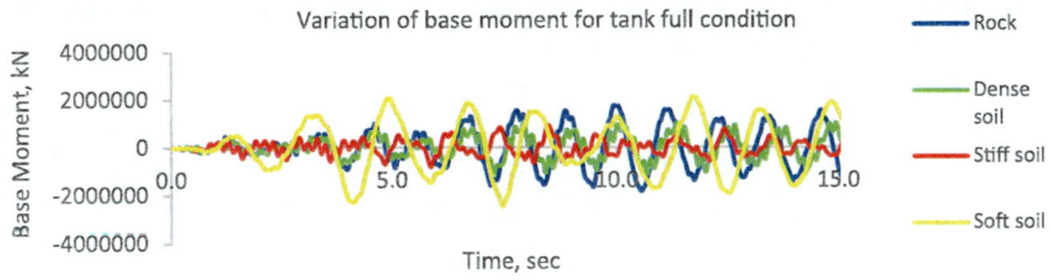


**Fig. 13** Variation of base moment in tank full condition

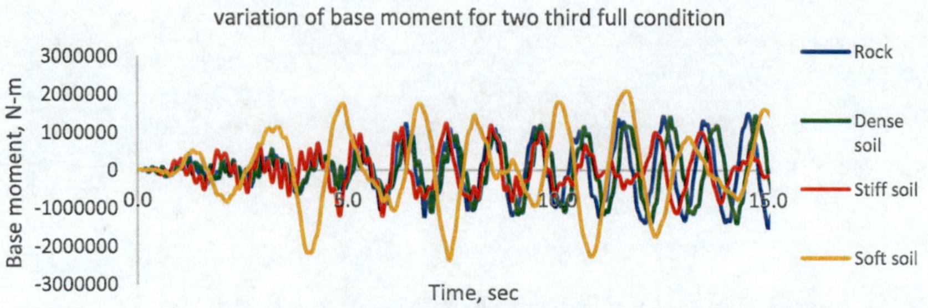


**Fig. 14** Variation of base moment in two third full condition

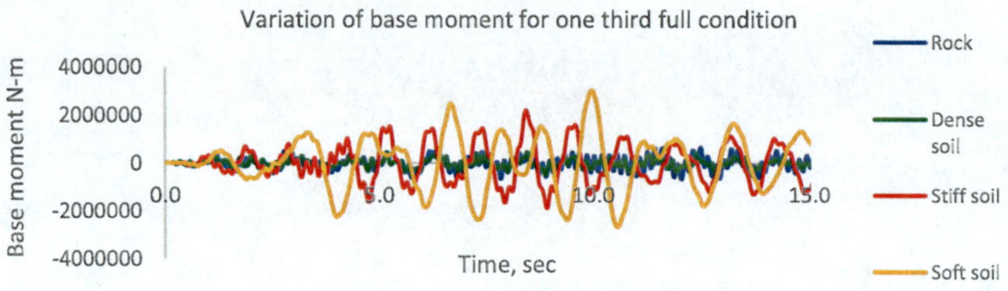


**Fig. 15** Variation of base moment in one third full condition

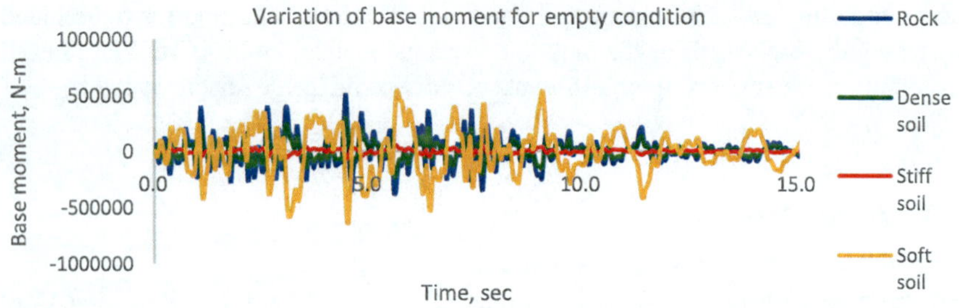


**Fig. 16** Variation of base moment in empty condition

condition, and it is reducing with the decrease in water level for the tanks resting on rock and dense soil. But for the tanks resting on stiff and soft soil, the base moments were increasing with the reduction in water level

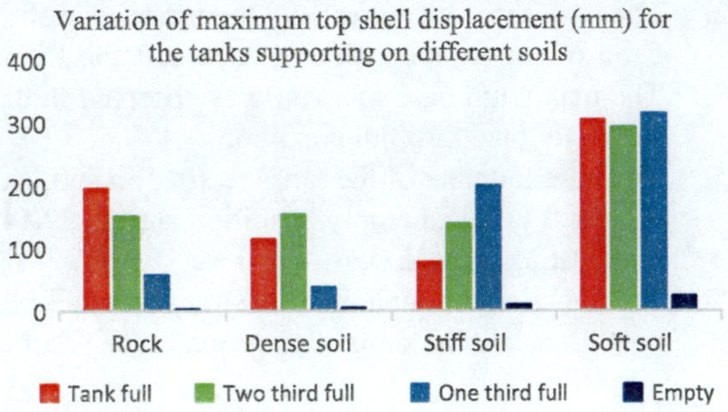


**Fig. 17** Variation of maximum top shell displacement (mm) for the tanks supporting on different soils

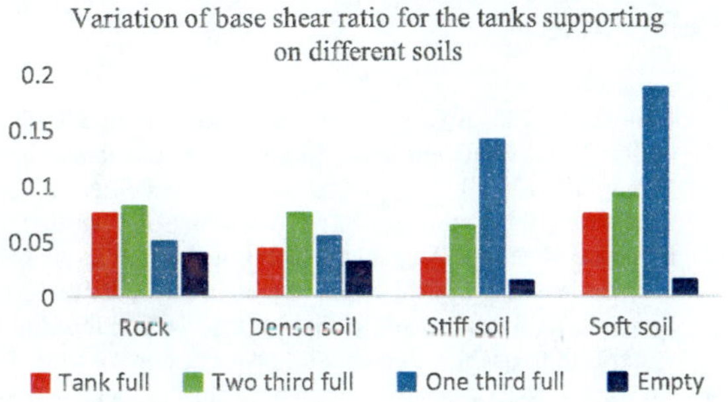


**Fig. 18** Variation of base shear ratio for the tanks supporting on different soils

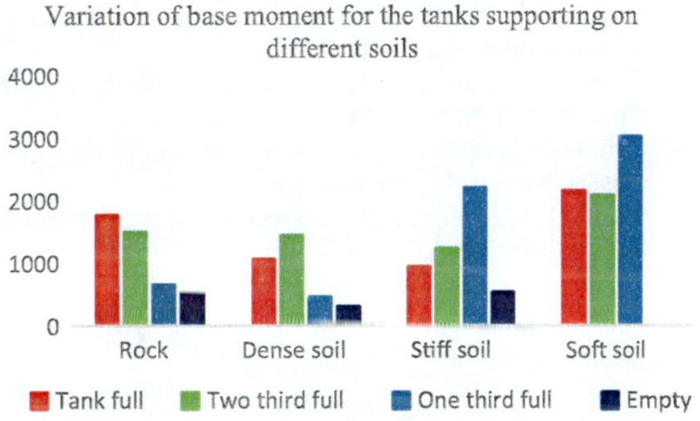


**Fig. 19** Variation of base moment for the tanks supporting on different soils

## 6 Conclusions

The dynamic response parameters such as the displacement of the top shell, base shear, and base moments at the base of the staging were calculated and the following conclusions are drawn:

1. The maximum shell roof displacement increases with the decrease in supporting soil stiffness. Soft soil condition produces the largest displacement for one third full condition.

2. The base shear increases with the increase in stiffness of soil. Base shear ratio is maximum for the tank resting on soft soil with the one third water level.
3. The maximum base moment was observed in the tank supported on soft soil strata, for one third full condition.
4. The base moment of the tanks resting on soft soil for tank full, two third full, one third full, and empty condition are 22, 39.14, 358, and 5% more than the tank resting on rock.
5. The most critical condition occurs when the liquid level is at one third full condition, and the supporting ground type is soft soil.

## References


1. Kohnke P (1994) ANSYS theory reference release 5.6. ISO 9001
2. (1997) NEHRP guidelines for the seismic rehabilitation of buildings, FEMA 273 October
3. Livaoglu R (2007) Effect of foundation embedment on seismic behavior of elevated tanks considering fluid–structure-soil interaction. Soil Dyn Earthquake Eng 27:855–863
4. Livaoglu R (2006) Dogangun, simplified seismic analysis procedures for elevated tanks considering fluid–structure–soil interaction. ELSEVIER J Fluids Struct 22:421–439
5. Livaoglu R (2008) Investigation of seismic behavior of fluid–rectangular tank–soil foundation systems in frequency domain. Soil Dyn Earthquake Eng 28:132–146
6. Livaoglu R et al (2006) Evaluation of seismic models for fluid elevated tanks systems suggested in codes. In: 7th international congress on advances in civil engineering, Yıldız Technical University, Istanbul, Turkey
7. Hassan A et al (2018) Effect of soil condition on seismic response of isolated base buildings. Int J Adv Struct Eng 10:249–261
8. Praveen K, Malhotra et al. (2000) Simple procedure for seismic analysis of liquid-storage tanks. Struct Eng Int 10(3):197–201
9. Chaduvula U et al (2013) Fluid-structure-soil interaction effects on seismic behaviour of elevated water tanks. Procedia Eng 51:84–91
10. Swamy K (2016) Seismic analysis and design of existing elevated RCs intze water tank at pedana. IJTRE 4

# A Study on Settlement Variation in Piled Raft Foundation Under Seismic Loads

S. Vinoda Krishna and B. R. Jayalekshmi

## 1 Introduction

A piled raft foundation is a type of sub structure which consist of three components namely piles, raft and subsoil. Such foundations are getting more and more popularity for high rise structures especially those present over soft soil. The soil-structure interaction phenomenon between the soil and piled raft foundation is very complex, which includes pile-soil interaction, pile-soil-pile interaction, raft-soil interaction, and finally the piles-raft interaction. Figure 1 provides a schematic representation of the combined piled raft system.

Even though piled raft foundation and pile group foundation are constructed using similar way, top portion of the foundation is called raft in case of piled raft foundation, and it is called as cap in case of pile group foundation. The difference in the piled raft and pile group is in the design stage of these foundations. The design of pile group usually does not consider the bearing pressure developed due to the contact between soil and pile cap, so the pile group is usually designed in conservative way by ignoring the bearing capacity of the pile cap. In case of piled raft foundation bearing pressure due to the raft is considered in the design. Generally, required bearing pressure is achieved by the raft alone; however, there is a chance of excessive settlement of the raft. Hence to counter this settlement Burland et al. (1977) mooted the idea of incorporating piles as settlement reducers.

Numerical, analytical and physical modelling methods have been used to evaluate the behaviour of piled raft foundations. Numerical analysis is usually done using three-dimensional finite element modelling [4]. A numerical method was developed by Reul to study the bearing behaviour of piled raft foundation [10]. Few studies combine the finite element analysis for the raft and the boundary element analysis for the piles [11]. The simplified Poulos-Davis-Randolph (PDR) method [6] is a

S. V. Krishna (✉) · B. R. Jayalekshmi
National Institute of Technology Karnataka, Surathkal, India



S. Kolathayar and S. C. Chian (eds.), *Recent Advances in Earthquake Engineering*, Lecture Notes in Civil Engineering 175,
https://doi.org/10.1007/978-981-16-4617-1_20

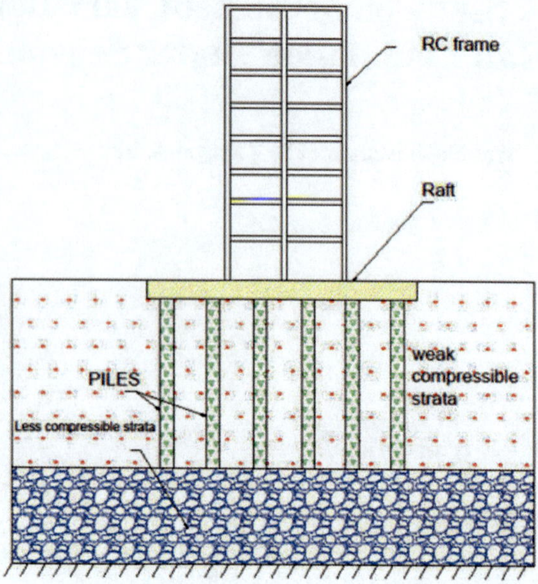


**Fig. 1** Schematic illustration of piled raft foundation

combination of analytical methods proposed by Poulos and Davis [7] and Randolph [4]. Clancy and Randolph suggested a 'plate on spring' method in which the raft is considered as a plate and the piles as springs [2]. A study on two dimensional numerical model using plate-on-spring theory was conducted by Poulos [8] to investigate the influence of geometrical parameters such as the raft thickness, pile length, and the number piles in a piled raft group on the bending moment in the raft and the load shared by the piles and the raft. Piled raft behaviour was also studied using physical modelling such as in centrifuge testing [3, 5]. Plane-strain analysis to study the effect of foundation geometry on total and differential settlement of the system is performed by Prakoso and Kulhawy [9]. This paper investigates the seismic performance of piled raft foundations with different stiffness in terms of raft settlement. The soil-structure system was modelled with the finite-element-based structural software ANSYS and analysed under dynamic loading conditions.

## 2 Numerical Modelling

Ten storey RC building with four bays in either direction resting on medium sand stratum of 32 m depth with piled raft foundation of different configuration were subjected to two ground motions. The bay length of the frames and the storey heights were taken as 3 m, and they were modelled with two-node linear beam element, BEAM 188. The cross-sectional dimensions of the beams and columns were 300mmx450mm and 500mmx500mm, respectively.

A square raft of 14 m width and 1 m thickness was modelled using plate elements. Twenty five numbers of fixed headed floating piles of 1 m in diameter, and a centre-to-centre spacing of 3 m, were modelled with BEAM 188 element in ANSYS. The properties of soil, pile and raft are shown in Table 1. A node to node connection is established between the raft elements and soil, thus preventing the relative sliding motion of the raft. Figure 2 shows 3-D finite element model of 10 storey building with and without soil stratum. The viscous boundary was provided at the lateral soil boundaries. The artificial boundary plays an important role in seismic analysis of the soil-pile-raft-structure system.

The soil model, which has dimensions of 110 m × 110 m × 32 m, was generated using eight node linear brick reduced integration element, i.e., SOLID 185 in ANSYS. Figure 2a, b shows the model discretized by finite-element mesh. A linear elastic model which is available in the ANSYS finite element program is used to model the wave propagation. To simulate the semi-infinite boundary condition of the soil, viscous boundaries were assigned on lateral sides of the soil. Viscous boundary condition prevents the reflection of the waves from the boundaries of the soil.

**Table 1** Material properties

| Specification | Dimension | Young's modulus of elasticity (MPa) | Poisson's ratio | Density (kg/m$^3$) ($\gamma_d$) |
|---|---|---|---|---|
| Medium sand | 110 m × 110 m × 30 m (B × L × H) | 35 | 0.4 | 1800 |
| Raft | 14 m × 14 m × 1 m (B × L × H) | 25,000 | 0.15 | 2500 |
| Pile | 1 m × 10 m (D × H) | 25,000 | 0.15 | 2500 |

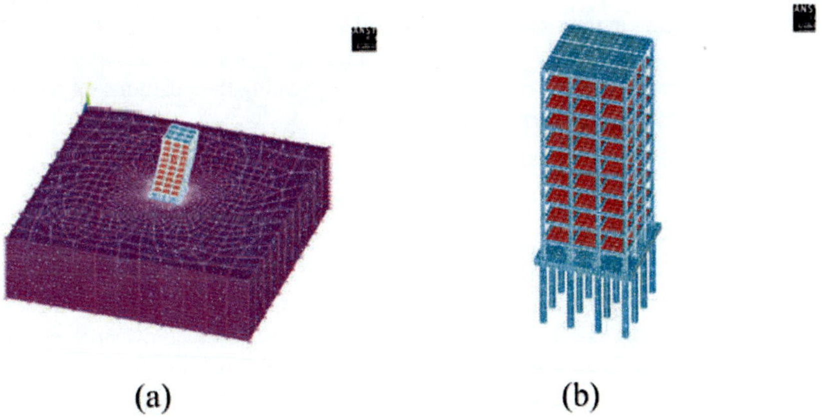


**Fig. 2** The 3-D finite element model of 10 storey building **a** Soil-pile-raft-structure system **b** building-pile-raft system

**Table 2** Pile configuration

| Model no | Pile arrangement | Length of the pile (m) | | |
|---|---|---|---|---|
| | | P1 | P2 | P3 |
| 1 | M1 | 12 | 9 | 6 |
| 2 | M2 | 12 | 10 | 8 |
| 3 | M3 | 8 | 8 | 8 |
| 4 | M4 | 8 | 10 | 12 |
| 5 | M5 | 6 | 9 | 12 |

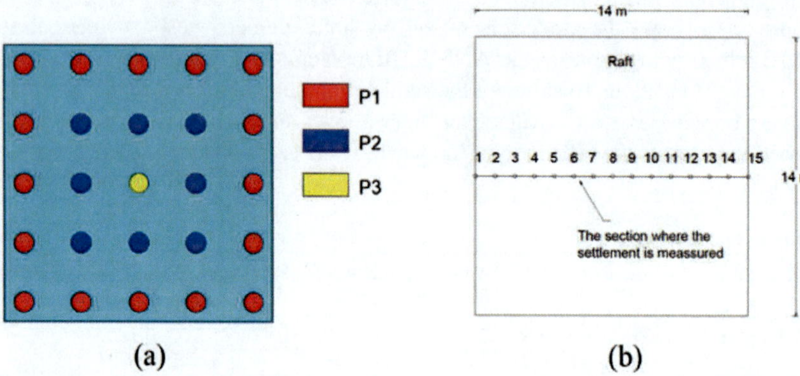


**Fig. 3** Plan view of raft foundation indicating **a** positions of the piles **b** locations of the settlements are measurement

Five different combinations of pile arrangements designated as M1, M2, M3, M4 and M5 as detailed in Table 2 were considered. These models have different stiffness due to the variation in the lengths of peripheral and interior piles P1, P2 and P3. M3 model represents the required pile length for a safe design.

Figure 3a, b shows the plan view of the raft indicating the positions of pile and locations where the raft settlement was measured.

## 3 Seismic Analysis of Soil-Pile-Raft-Building System

Seismic analysis of the multi-storey building was carried out to identify the best configuration among pile types M1 to M5 which resist the loading with less raft settlement. For this time history of acceleration of two ground motions were considered. The recorded accelerogram corresponding to the El Centro earthquake with 0.295 g PGA at the location and time history corresponding to response spectrum given in IS 1983 which was scaled down to 0.295 g to match the of El Centro PGA

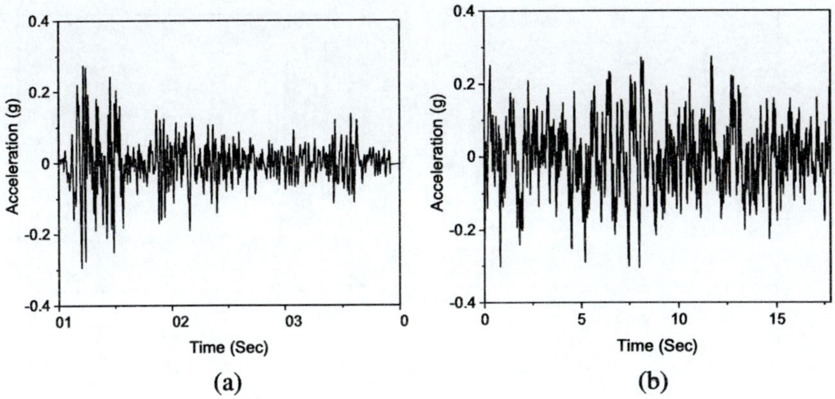


**Fig. 4** Time history of input ground acceleration record of **a** El Centro input motion **b** IS input motion

were applied to the models. These acceleration records are shown which is shown in Fig. 4a, b.

## 4 Results and Discussions

Results of time history analysis of 10 storey RC framed structures subjected to El Centro and IS input motion are focused in this section. Particularly time history of settlement behaviour of the raft at a section passing through mid-point of the raft along the direction of applied earthquake load is reported in this section.

### *4.1 Time History of Settlement*

Time history of the settlement is plotted for three points in the raft along the line passing through the centre of the raft in the direction of applied input motion. These three points are two edges of the raft and the centre point. Figure 5a shows the time history of displacement of the raft at the left edge (Point 1), Fig. 5b shows the deflection at the centre (Point 8), and Fig. 5c shows the deflection in the right edge corresponding to point no 15, when subjected to El Centro input motion. The minimum settlement at the edges is seen for model M5, and the minimum settlement at the centre happens in case of the M1 model. Total variation in the settlement between M1 and M5 model is 38.4% at both the edges of the raft and 36.1% variation in the settlement at the centre of the raft.

Similarly, the time history of the settlement is plotted at three points in the raft along the line passing through the centre of the raft in the direction of applied input

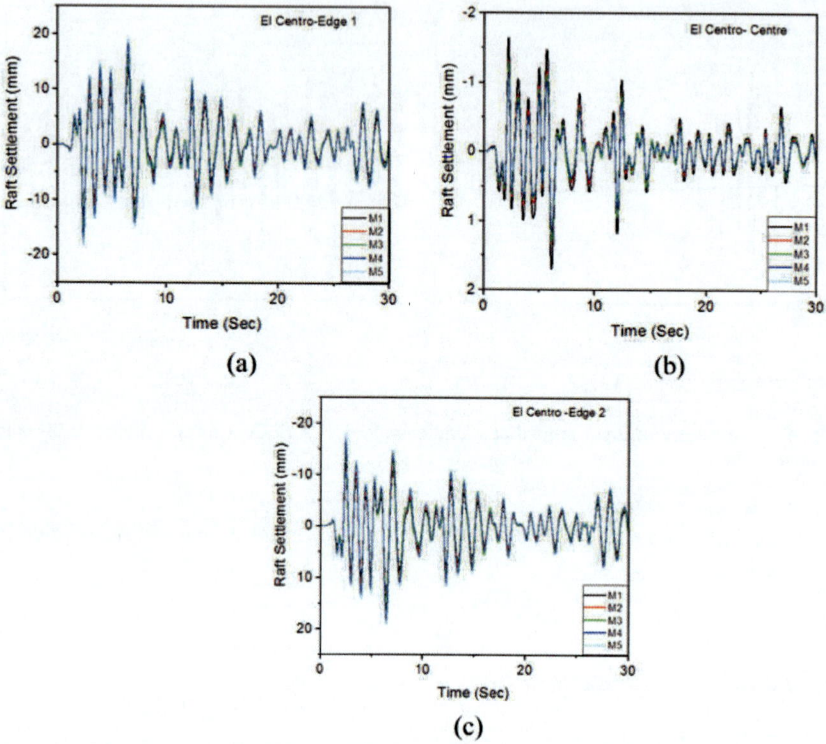


**Fig. 5** Time history of the settlement of the raft at **a** at the left edge, **b** at the centre and **c** at right edge for El Centro earthquake motion

motion for the same structure subjected to IS input motion, i.e., two edges of the raft and centre point of the raft. Figure 6a shows the time history of displacement of the raft at point 1, Fig. 6b shows the deflection at point 8, and Fig. 6c shows the deflection at point 15.

## 4.2 *Maximum Settlement*

Maximum values of the deflection of the raft at different points along the section passing through the centre of the raft are shown here. Time interval where the maximum deflection occurs is different for different points in the raft. Figure 7 shows the maximum raft settlement when the system is subjected to El Centro input motion, and Fig. 8 shows the settlement of the raft when subjected to IS input motion.

Results show that settlement is always maximum at point 2 and 14 in the raft for all the models when subjected to both seismic excitations; this is due to the tilting

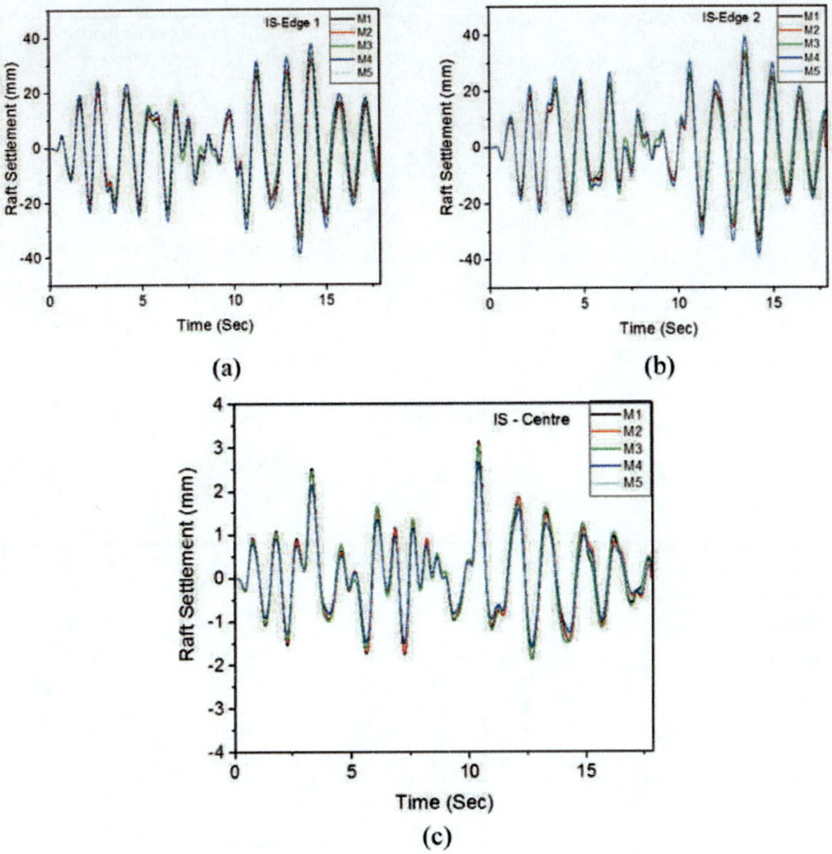


**Fig. 6** Time history of the settlement of the raft at **a** at the left edge, **b** at the centre and **c** at right edge for IS earthquake motion

action of the raft during seismic excitation. Columns transfer the loads to those piles which are present directly over the piles at points 2 and 14. Deflection is minimum at point 8 which is the centre point of the raft because this point lies in the central axis about which tilting of raft takes place. At points 2 and 14 maximum settlement is observed in model M5 and minimum in model M1, and the percentage variation in the settlement of raft between M1 and M5 model is 27.3% at those points for El Centro earthquake input and percentage variation in settlement between M1 and M5 at same locations for IS input motion is 17%. Whereas settlement at the centre of the raft is minimum for model M5 and maximum settlement at centre is seen in the case of raft with pile configuration M1. Percentage variation in the settlement at the centre is 36% in case of El Centro and 19% in case of IS input motion.

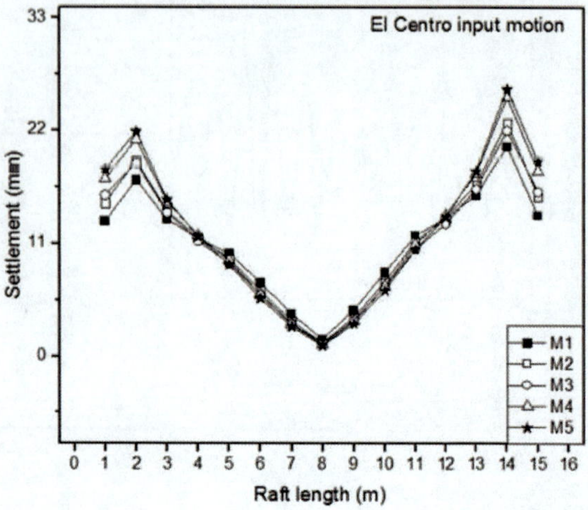


**Fig. 7** Maximum settlement of the raft at along mid-section for El Centro earthquake motion

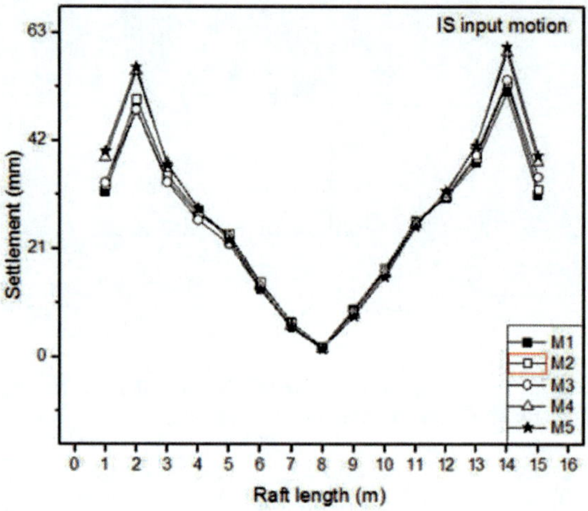


**Fig. 8** Maximum settlement of the raft at along mid-section for IS input motion

## 5 Conclusions

In the present study, 3D analysis was carried out using ANSYS software to investigate the behaviour of a multistorey building having different configuration of combined piled raft foundation under dynamic loading conditions. The settlements of the raft for at different locations for all those models are noted. Results show maximum settlement near the edges of the raft at points where the peripheral piles connect the raft, and minimum settlement is observed at the centre. Amongst the different pile configurations considered in the study, the models with longer piles provided at the edges of the raft helps in reducing the settlements of the raft to a considerable value.

## References


1. Borland JB, Broms BB, De Mello VFB (1977) Behaviour of foundations and structures. In: Proceedings of the 9th International Conference on Soil Mechanics and Foundation Engineering, vol 2. Tokio, Japan, pp 495–546
2. Clancy P, Randolph MF (1993) An approximate analysis procedure for piled raft foundations. Int J Numer Anal Meth Geomech 17:849–869
3. Horikoshi K, Watanabe T, Fukuyama H, Matsumoto T (2002) Behavior of piled raft foundations subjected to horizontal loads. In: Physical modelling in geotechnics: ICPMG'02, pp 715–721
4. Katzenbach R, Arslan U, Moorman C, Reul O (1998) Piled raft foundation: interaction between piles and raft. Darmstadt Geotech Darmstadt Univ Technol 4:279–296
5. Matsumoto T, Fukumura K, Pastsakorn K, Horikoshi K, Oki A (2004) Experimental and analytical study on behaviour of model piled raft in sand subjected to horizontal and moment loading. Int J Phys Model Geotech 4(3):1–19
6. Poulos HG (2001) Piled raft foundations: design and applications. Geotechnique 51(2):95–113. https://doi.org/10.1680/geot.2001.51.2.95
7. Poulos HG, Davis EH (1980) Pile foundation analysis and design (No. Monograph)
8. Poulos HG (2001) Methods of analysis of piled raft foundations. A Report Prepared on Behalf of Technical Committee TC18 of Piled Foundations
9. Prakoso WA, Kulhawy FH (2001) Contribution to piled raft foundation design. J Geotech Geoenviron Eng. https://doi.org/10.1061/(ASCE)1090-0241(2001)127:1(17),17-24
10. Reul O (2004) Numerical study of the bearing behavior of piled rafts. Int J Geomech 4(2):59–68
11. Ta LD, Small JC (1996) Analysis of piled raft system in layered soils. Int J Numer Anal Meth Geomech 20(1):57–72

# Seismic Behavior of Tunnel Structure Under Varying Surrounding Soil, Water Table, and Overburden Condition

**Chiranjib Sarkar and Sibapriya Mukherjee**

## 1 Introduction

Underground structures have now become the need of time with gaining popularity. Rapid expansion of transportation network and restrictions in surface movement are the main key reasons of increasing importance of tunnel structures. From past experiences and very recent incidents of earthquakes, it is proved that tunnel structures are quite vulnerable to earthquake events. Hence, the performance of this type of structures against natural hazards, like earthquakes, is required to be examined more in-depth. It has been noticed that tunnels structure in relatively soft soil at lower depth generally suffer more damages than tunnels structure at higher depth in intact rock (Sharma and Judd 1991; Power et al. 1998). The most important part of seismic design of tunnel structures is to estimate the actual deformation caused by the surrounding soil under seismic loading conditions. Seismic behavior of tunnel structures is mainly controlled by the response from surrounding soil under seismic loading (Wang 1993, Hashash et al. 2001; Arango 2008). The basic parameters which influence the damage to tunnel structures earthquake are magnitude of earthquake, depth of the tunnel, Peak Ground Acceleration (PGA), distance to epicenter, surrounding ground condition, etc.

The variation of seismic behavior in tunnel structure has been significantly reported for varying characteristics of surrounding soil, water table, and overburden depth with different seismic zones. The present paper covers a parametric study on a typical box structure of size 10.0 m wide × 6.0 m high, having different tunnel axis depths of 8 m, 10 m, and 12 m, encountering varying soil layers of loose to medium, medium to dense and very dense non-cohesive soil at different seismic

C. Sarkar (✉)
AECOM India Pvt Ltd, New Town, Kolkata 700156, India

S. Mukherjee
Civil Engineering Department, Jadavpur University, Kolkata 700032, India

 
S. Kolathayar and S. C. Chian (eds.), *Recent Advances in Earthquake Engineering*, Lecture Notes in Civil Engineering 175,
https://doi.org/10.1007/978-981-16-4617-1_21

zones. The variation of deformation, force, and moment for different tunnel depth with varying characteristics of surrounding soil and water table has been assessed for different seismic zones. A set of different cases of numerical models in Finite Element Method (FEM) have been further developed to study the non-linear behavior of tunnel structure under seismic condition.

## 2 Methodology

Seismic effect on the tunnel structure can be obtained from the required racking force to generate the deformation calculated with respect to the surrounding ground strain [1]. Racking deformation of a cut and cover rectangular box tunnels is produced due to shear distortions at time of earthquake events. The Bulletin of The New Zealand National Society for Earthquake Engineering (Vol. 13, No. 3, 1980) recommends that for underground structures with depths of fill less than or equal to the depth of the tunnel that the seismic load increment be estimated based on the pressure distributions [2]. For underground box structures it is unlikely that such an active wedge will procedure, and it is therefore solutions based on rigid retaining walls as developed by Wood are used. Present paper estimates the seismic behavior of tunnel structure adopting Racking analysis/Free-Field deformation and elastic dynamic solutions.

## 3 Model of the Study

A typical box metro tunnel structure of size 10.0 m wide × 6.0 m high has been considered for present study and analysis. As part of construction, cut and cover method with diaphragm wall has been considered where diaphragm wall is used as main body structure as well as earth retaining structure. A parametric study has been conducted on this typical cut and cover box tunnel structure having different depths of 8 m, 10 m, and 12 m, encountering varying soil layers of loose to medium, medium to dense, and very dense non-cohesive soil at different seismic zones. The variation of seismic behavior (deformation, force, moment, etc.) in tunnel structure have been noted for varying characteristics of surrounding soil, water table, and overburden depth with different seismic zones. A two-dimensional space frame model has been developed in STAAD.Pro [6] for different soil and different seismic conditions. As a part of soil-structure interaction, linear springs have been assigned with equivalent stiffnesses based on subgrade modulus of reaction as per IS: 2950 (Part I) [7]. A typical two-dimensional space frame model along with typical bending moment diagram developed in STAAD.Pro is shown in Fig. 1.

Apart from above stiffness-based models, a set of numerical models with 0.5 m × 0.5 m mesh size in Finite Element Method (FEM) have been further developed in MIDAS GTX NX software to study the non-linear behavior of tunnel structure under seismic condition. A Model view of cut and cover box tunnel cross section

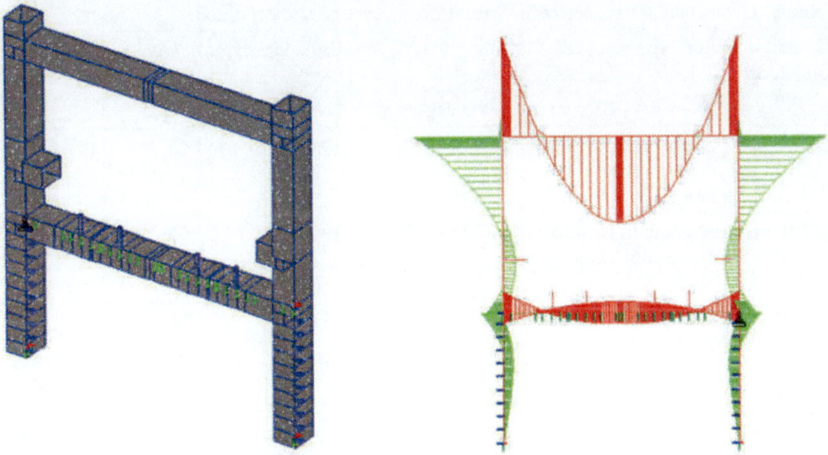

**Fig. 1** A typical 2-D space frame model along with typical bending moment diagram

along with geotechnical profile model developed in MIDAS GTX NX is shown in Fig. 2.

Three different ground soil conditions have been considered for present parametric study. Geotechnical parameters for three type of ground conditions are tabulated in Table 1.

Three different tunnel axis depths of 8 m, 10 m, and 12 m have been adopted with respect to backfill depths of 4 m, 6.4 m, and 8.2 m, respectively.

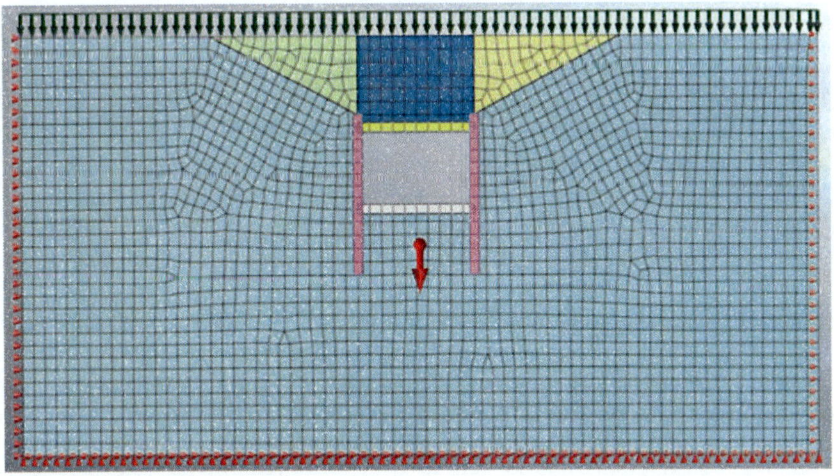

**Fig. 2** Model view of box tunnel cross section with geotechnical profile

**Table 1** Geotechnical parameters for three different ground soil condition

| Ground Condition | Soil type | SPT value (Avg.) | Φ' (Avg.) | Bulk density (Avg.) | E' (Avg.) | ν' (Avg.) |
|---|---|---|---|---|---|---|
| | | | deg | kN/m$^3$ | MPa | |
| GC-I | Loose to medium silty fine sand | 15–25 | 28–30 | 19.0 | 13.5 | 0.3 |
| GC-II | Medium to dense silty sand | 30–40 | 31–33 | 19.5 | 15.5 | 0.3 |
| GC-III | Very dense sand | 50–60 | 34–35 | 20.0 | 17.0 | 0.3 |

Three different seismic zones (zones II, zone III, and zone IV as per IS-1893) have been considered for present study. Considering important factor of 1.5, equivalent Peak Ground Acceleration (PGA) value of 0.18, 0.24, and 0.36 in Maximum Design Earthquake (MDE)/Maximum Credible Earthquake (MCE) have been adopted which are close to zone II, zone III, and zone IV, respectively.

Three different water table depths have been considered in present analysis which are water table at tunnel center level, water table at tunnel top level, and water table at 3 m below ground level.

For different tunnel depths, soil condition, and water table depth, the lateral earth pressure and vertical surcharge have been calculated and the same have been assigned on the STAAD and MIDAS GTX NX models with the appropriate load combinations.

## 4 Presentation of Results

Racking deformations of underground box structure at the same depth (10 m) with different seismic condition (PGA of 0.18, 0.24, and 0.36), and various ground conditions (GC-I, GC-II, and GC-III) have been plotted and shown in Fig. 3.

Further, racking deformation of underground cut and cover box structure with different depths of 10 m, 12 m, and 14 m for three different ground conditions (GC-I, GC-II, and GC-III) having two different seismic zones (zone-III and zone-IV) are shown in Table 2.

Bending moment comparison of underground cut and cover box structure with three different depths (8 m, 10 m, and 12 m) for three different ground conditions (GC-I, GC-II, and GC-III), and three different seismic condition (PGA of 0.18, 0.24, and 0.36) are shown in Table 3 (for wall and roof slab junction) and Table 4 (for wall and base slab junction).

Further, bending moment comparison of underground box structure with three different depths (8 m, 10 m, and 12 m) and different seismic conditions (non-seismic case and seismic zone-II, zone-III, and zone-IV) within the same ground condition

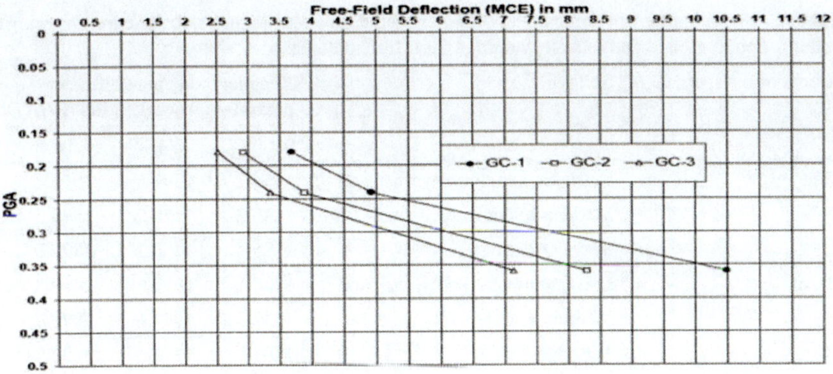


**Fig. 3** Racking deformations of underground box structure at same depth (10 m) with different seismic condition and various ground conditions

**Table 2** Racking deformation of underground box structure with different depths under different ground and seismic conditions

| Ground condition | Soil type | Depth of overburden (m) | Free–field Deflection (MCE condition), mm | |
|---|---|---|---|---|
| | | | Seismic Zone-III | Seismic Zone-IV |
| GC-I | Loose to medium silty fine sand | 10 | 5.105 | 10.888 |
| | | 12 | 4.914 | 10.481 |
| | | 14 | 4.723 | 10.074 |
| GC-II | Medium to dense silty sand | 10 | 4.035 | 8.608 |
| | | 12 | 3.884 | 8.286 |
| | | 14 | 3.733 | 7.964 |
| GC-III | Very dense sand | 14 | 3.474 | 7.410 |
| | | 12 | 3.344 | 7.133 |
| | | 14 | 3.214 | 6.856 |

(GC-I) are shown in Table 5 (for wall and roof slab junction) and Table 6 (for wall and base slab junction).

Comparison of bending moment for underground cut and cover box structure at different water table depths (water table at tunnel center level, water table at tunnel top level, and water table at 3 m) for three different seismic condition (PGA of 0.18, 0.24, and 0.36) within the same tunnel depth and ground condition are shown in Table 7.

**Table 3** Bending moment comparison of underground box wall at roof slab Junction (hogging) under different ground and seismic condition for different depths

| Depth of overburden | Soil type | Bending moment at different peak ground acceleration (PGA) | | |
|---|---|---|---|---|
| | | 0.18 | 0.24 | 0.36 |
| 8 m | Loose to medium silty fine sand | 1404.2 | 1592.5 | 1956.7 |
| | Medium to dense silty sand | 1339.2 | 1520.2 | 1889.5 |
| | Very dense sand | 1285.4 | 1460.2 | 1827.0 |
| 10 m | Loose to medium silty fine sand | 1728.1 | 1959.8 | 2418.8 |
| | Medium to dense silty sand | 1632.7 | 1854.6 | 2273.9 |
| | Very dense sand | 1567.9 | 1763.4 | 2227.3 |
| 12 m | Loose to medium silty fine sand | 1997.6 | 2251.8 | 2716.8 |
| | Medium to dense silty sand | 1896.9 | 2123.6 | 2600.0 |
| | Very dense sand | 1813.6 | 2035.2 | 2499.9 |

**Table 4** Bending moment comparison of underground box wall at base slab Junction (hogging) under different ground and seismic condition for different depths

| Depth of overburden | Soil type | Bending moment at different peak ground acceleration (PGA) | | |
|---|---|---|---|---|
| | | 0.18 | 0.24 | 0.36 |
| 8 m | Loose to medium silty fine sand | 558.7 | 673.5 | 842.4 |
| | Medium to dense silty sand | 478.3 | 585.4 | 768.0 |
| | Very dense sand | 421.6 | 521.2 | 705.3 |
| 10 m | Loose to medium silty fine sand | 576.0 | 625.1 | 674.5 |
| | Medium to dense silty sand | 468.9 | 511.9 | 576.9 |
| | Very dense sand | 393.1 | 430.7 | 486.9 |
| 12 m | Loose to medium silty fine sand | 753.7 | 840.1 | 995.9 |
| | Medium to dense silty sand | 626.9 | 706.1 | 853.7 |
| | Very dense sand | 535.9 | 608.9 | 746.1 |

**Table 5** Bending moment comparison of underground box wall at roof slab junction for different depths and seismic condition within the same ground

| Depth of overburden | Bending moment at non-seismic condition, kN-m | Bending moment of wall—roof slab junction location for different peak ground acceleration (PGA), kN-m | | |
|---|---|---|---|---|
| | | 0.18 | 0.24 | 0.36 |
| 8 m | 945.3 | 1404.2 | 1592.5 | 1956.7 |
| 10 m | 1265.2 | 1728.1 | 1959.8 | 2418.8 |
| 12 m | 1442.9 | 1997.6 | 2251.8 | 2716.8 |

**Table 6** Bending moment comparison of underground box wall at base slab junction for different depths and seismic condition within the same ground

| Depth of overburden | Bending moment at non-seismic condition, kN-m | Bending moment of wall—base slab junction location for different peak ground acceleration (PGA), kN-m | | |
|---|---|---|---|---|
| | | 0.18 | 0.24 | 0.36 |
| 8 m | 343.4 | 558.7 | 673.5 | 842.4 |
| 10 m | 444.2 | 576.0 | 625.1 | 674.5 |
| 12 m | 491.8 | 753.7 | 840.1 | 995.9 |

**Table 7** Bending moment comparison of box structure at different water table depths for three different seismic condition within the same tunnel depth and ground condition

| Depth of overburden | Water table depth | Bending moment of wall—roof slab junction location for different peak ground acceleration (PGA), kN-m | | |
|---|---|---|---|---|
| | | 0.18 | 0.24 | 0.36 |
| 10 m (6.4 m backfill height) | At tunnel center level | 2050 | 2200 | 2260 |
| | At tunnel top level | 2010 | 2060 | 2160 |
| | At 3 m below ground level | 1950 | 2000 | 2090 |

## 5 Discussions on Results

Based on different series of analyses done in the present study to understand the non-linear behavior of box tunnel structure under seismic condition, an attempt has been made to address the effect of variation in ground condition, tunnel depth, water table depth, and seismic condition. Also results obtained from different of set analysis have been further compared in the present study.

### *5.1 Racking Deformation of Box Tunnel*

- Based on the results from Table 2 and Fig. 3, it is clear that raking deformation decreases gradually with improvement of surrounding soil conditions. It is observed that deformation value reduced by 14–22% with the improvement of ground condition. Raking deformation value is reduced by 22% when ground soil condition improved from GC-I (loose to medium silty fine sand) to GC-II (medium to dense silty sand). Raking deformation value is further reduced by

14% when ground soil condition improved from GC-II (medium to dense silty sand) to GC-III (very dense sand).

- With an increase in tunnel axis depth, raking deformation decreases. Raking deformation value decreases by 4–7% with an increase in tunnel depth from 10 to 12 m and then to 14 m within the same ground soil condition.
- Once the seismic condition changes from lower to higher PGA, racking deformation values increase scientifically even for the same ground soil condition (GC-I, GC-II, or GC-III). With the changes in seismic level from zone-II to zone-III, racking deformation is increased by 33% in the same soil condition. Racking deformation is further increased by 113% when seismic level changes from zone-III to zone-IV.

## 5.2 Bending Moment Comparison of Box Tunnel Wall Under Different Surrounding Ground for Same Tunnel Depth and Same Seismic Conditions

- With improvement of surrounding soil conditions within the same seismic zone and the same tunnel depth, bending moment of wall at roof slab junction location is reduced by 4–6%. Bending moment value is reduced by 4% when ground soil condition improved from GC-I (loose to medium silty fine sand) to GC-II (medium to dense silty sand). Bending moment value is further reduced by 6% when ground soil condition improved from GC-II (medium to dense silty sand) to GC-III (very dense sand).
- With improvement of surrounding soil conditions within same seismic zone and same tunnel depth, bending moment of wall at base slab junction location is reduced by 15–22%. Bending moment value is reduced by 22% when ground soil condition improved from GC-I (loose to medium silty fine sand) to GC-II (medium to dense silty sand). Bending moment value is further reduced by 15% when ground soil condition improved from GC-II (medium to dense silty sand) to GC-III (very dense sand).

## 5.3 Bending Moment Comparison of Box Tunnel Wall Under Different Seismic Conditions for Same Tunnel Depth and Same Surrounding Ground Soil

- With the changes in seismic condition from lower to higher PGA, bending moment of wall at roof slab junction location is increased by 8–22%. Bending moment value is increased by 8% when PGA changed from 0.18 to 0.24. Bending moment value is further increased by 22% when PGA changed from 0.24 to 0.36. Also bending moment value of wall at roof slab junction is increased by 35–40% when seismic condition changes from non-seismic to seismic zone-II even for the same tunnel depth and the same surrounding ground soil.

- With the changes in seismic condition from lower to higher PGA, bending moment of wall at base slab junction location is increased by 15–22%. Bending moment value is increased by 15% when PGA changed from 0.18 to 0.24. Bending moment value is further increased by 22% when PGA changed from 0.24 to 0.36. Also bending moment value of wall at base slab junction is increased by 40–60% when seismic condition changes from non-seismic to seismic zone-II even for the same tunnel depth and the same surrounding ground soil.

Bending moment diagram from MIDAS GTX NX for cut and cover box tunnel of 10m tunnel axis depth at the same ground soil condition (GC-I) under three different seismic condition are shown in Fig. 4 (PGA of 0.18), Fig. 5 (PGA of 0.24), and Fig. 6 (PGA of 0.36).

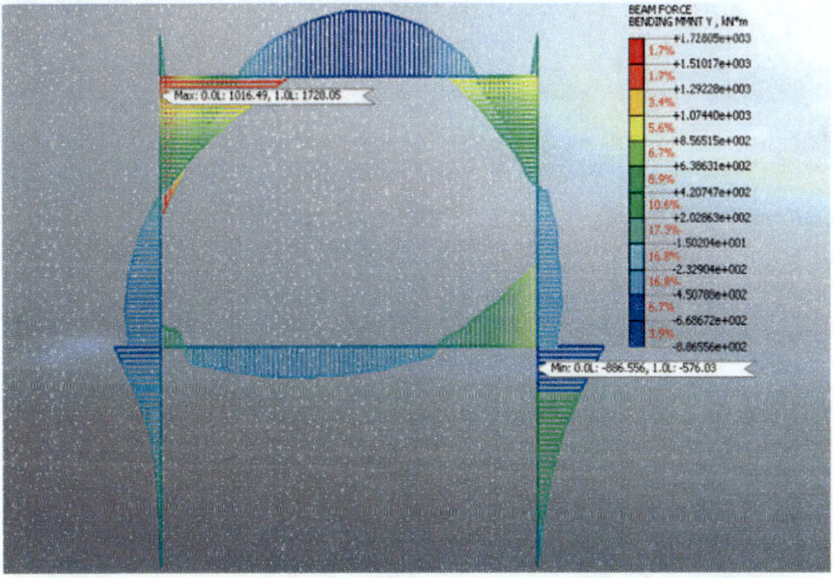


**Fig. 4** Bending moment diagram of box tunnel for 10 m tunnel axis depth at the same ground soil condition (GC-I) under seismic condition PGA of 0.18

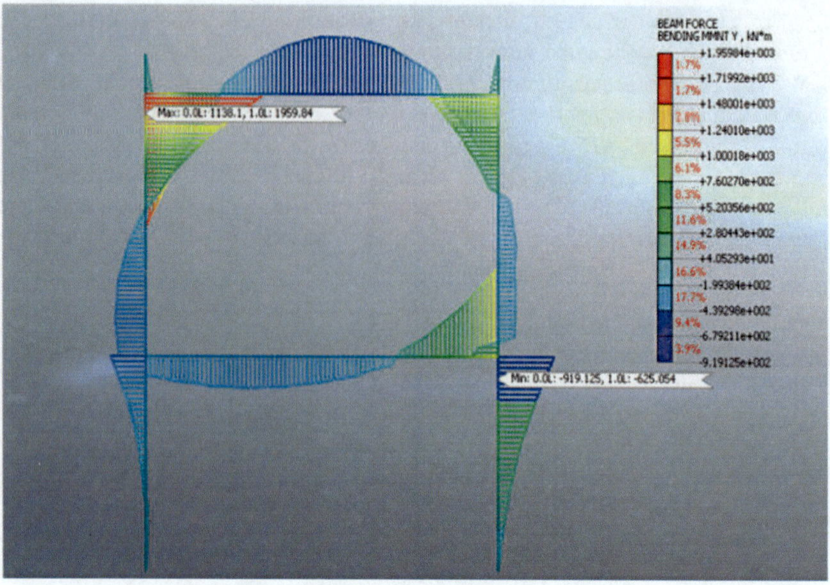


**Fig. 5** Bending moment diagram of box tunnel for 10 m tunnel axis depth at the same ground soil condition (GC-I) under seismic condition PGA of 0.24

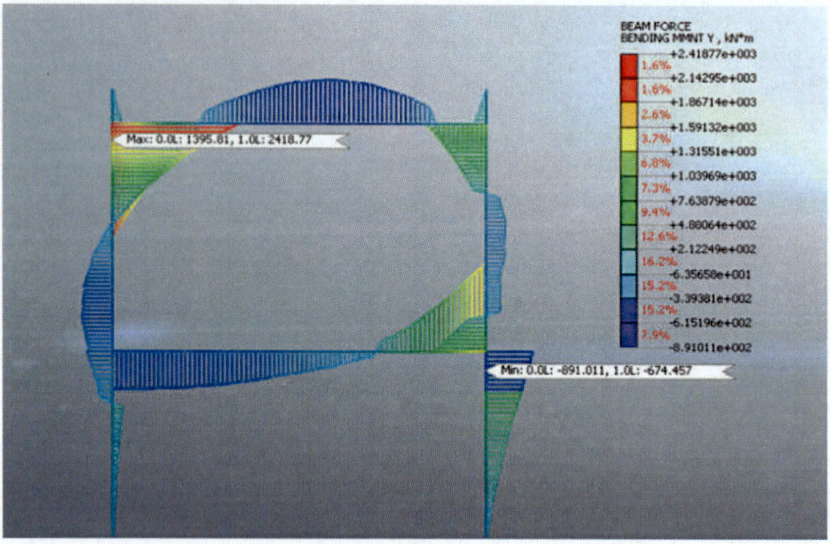


**Fig. 6** Bending moment diagram of box tunnel for 10 m tunnel axis depth at the same ground soil condition (GC-I) under seismic condition PGA of 0.36

### 5.4 *Bending Moment Comparison of Box Structure at Different Water Table Depths for Three Different Seismic Condition Within Same Tunnel Depth and Same Ground Condition*

- With the decreasing of water table depths, bending moment of wall is reduced by 3–5% within the same ground condition. Bending moment value is reduced by 5% when water table depth changes from tunnel center level to tunnel top. Bending moment value is further reduced by 3% when water table depth changes from tunnel top to 3 m below ground level.
- With the changes in seismic condition from lower to higher PGA, bending moment of wall by 3–6% due to combined effect of water table with seismic conditions.

## 6 Conclusions

In the present study, seismic behavior of box tunnel structure situated on varying soil parameters with different tunnel depths under different seismic loading conditions have been studied. A series of parametric studies have been carried out to observe the tunnel and soil interface under the same and different tunnel depth, ground soil conditions with different seismic conditions. From the current study the following conclusions may be drawn:

- Raking deformation decreases gradually with improvement of surrounding soil conditions. Deformation value reduced by 14–22% with the improvement of ground condition. Raking deformation value is reduced by 22% when ground soil condition improved from GC-I (loose to medium silty fine sand) to GC-II (medium to dense silty sand). Raking deformation value is further reduced by 14% when ground soil condition improved from GC-II (medium to dense silty sand) to GC-III (very dense sand).
- Raking deformation value decrease by 4–7% with an increase in tunnel depth from 10 to 12 m and then to 14 m within the same ground soil condition.
- With the changes in seismic condition from lower to higher PGA, racking defor mation values increase scientifically even for the same ground soil condition (GC-I, GC-II, or GC-III). With the changes in seismic level from zone-II to zone-III, racking deformation is increased by 33% in the same soil condition. Racking deformation is further increased by 113% when seismic level changes from zone-III to zone-IV.
- With improvement of surrounding soil conditions within the same seismic zone and the same tunnel depth, bending moment of wall at roof slab junction location is reduced by 4–6%. With improvement of surrounding soil conditions within the same seismic zone and the same tunnel depth, bending moment of wall at base slab junction location is reduced by 15–22%.

- With the changes in seismic condition from lower to higher PGA, bending moment of wall at roof slab junction location is increased by 8–22%. With the changes in seismic condition from lower to higher PGA, bending moment of wall at base slab junction location is increased by 15–22%.
- With the decreasing of water table depths, bending moment of wall is reduced by 3–5% within the same ground condition. With the changes in seismic condition from lower to higher PGA, bending moment of wall by 3–6% due to combined effect of water table with seismic conditions.

## References


1. Hashash YMA, Hook JJ, Schmidt B, Chiang Yao JI (2001) Seismic design and analysis of underground structures. Tunnelling Underground Space Technol 16:247–293
2. Matthewson MB, Wood JH, Berrill JB (1980) Seismic design of bridges section 9 earth retaining structures. Bull New Zealand Natl Soc Earthquake Eng 13(3)
3. Patil M, Choudhury D, Ranjith PG, Zhao J (2018) Behavior of shallow tunnel in soft soil under seismic conditions. Tunn Undergr Space Technol 82:30–38
4. Delhi Metro Railway Corporation, Technical specification: DMRC out line design specification for underground structure
5. Wang J-N (1993) Seismic design of tunnels. Parsons Brinckerhoff Quade & Douglas, Inc.
6. STAAD.Pro V8i (SELECT Series 6) 20.07.11.33.
7. IS: 2950 (Part I) (1981) Code of practice for design and construction of raft foundations, part 1: design
8. IS: 1893 (Part I) (2016) Criteria for earthquake resistant design of structures

# Seismic Response Study on Kaswati Dam

**Barnali Ghosh and S. K. Prasad**

## 1 Introduction

A Mw 7.7 earthquake struck the Kutch region, in the state of Gujarat, India, at 8:46 a.m., on January 26th, 2001. This was among the most harmful earthquakes in the history of India. More than 16,000 human beings were killed and more than 167,000 were injured [1]. There are as many as 170 earth dams in the Kutch region. Among them, seven dams medium in size and 14 dams small in size were severely damaged [3]. Longitudinal cracks, Subsidence of the crest, lateral spreading are a few typical failures suffered. Most of these dams were built in 1950s and 1960s [6] and were not designed as earthquake resistant structures due to the lack of knowledge in seismic design during those days. The seismic behaviour of earth dams under an earthquake requires proper analysis in order to understand the exact cause of failure so that necessary remedial measures and proper rehabilitation can be carried out in future.

Here, stability analysis of Kaswati dam is presented using PLAXIS software (version 8), a finite element package, considering two-dimensional plane-strain idealization to evolve dynamic response of earth dams [8]. PLAXIS was introduced for two-dimensional plane-strain analysis to find out the dynamic behaviour of earthen dams and embankments [9]. Griffith et al. (1988) showed that advanced soil models can be used to showcase the correct predictions for different loading conditions and have successfully used numerical approach of dams and embankments [4]. The analysis using the finite element method allows studying the behaviour of the dam considering changes in stresses, strains, accelerations and displacements at different points in the body of the dam using a real accelerogram.

B. Ghosh (✉)
East Point College of Engineering and Technology, Bangalore, India

S. K. Prasad
Vidyavardhaka College of Engineering, Mysuru, India



S. Kolathayar and S. C. Chian (eds.), *Recent Advances in Earthquake Engineering*, Lecture Notes in Civil Engineering 175,
https://doi.org/10.1007/978-981-16-4617-1_22

## 2 Project Description

Kaswati Dam, an earth dam, was constructed in 1973. The maximum height of the dam is 8.8 m and the crest length of 1455 m [7]. Similar to other dams in Gujarat, the water level in Kaswati Reservoir was almost near the dead storage level during the Bhuj Earthquake [5] and the soil below that level was completely saturated. The devastating earthquake activated shallow sliding mainly near the toe of the upstream side slope of the dam [2] and liquefaction near the upstream side of the toe of the dam was considered the main culprit for this failure. The EERI reconnaissance team reported that there are developments of cracks in the upstream slope side of the dam and cracks are as deep as 1.5 to 1.7 m [7]. It was also reported that there was a loss of stability near the upper part of the downstream side slope after the earthquake. Even the appearance of the longitudinal cracks on the crest of the dam may also be related to the liquefaction of the foundation soil. However, loss of stability in the upper part of the downstream slope is unlikely because of the liquefaction of foundation soils [5]. The deformed shape of the Kaswati Dam is shown in Fig. 1 with its pre-failure layout and Fig. 2 represents the idealization using PLAXIS [7].

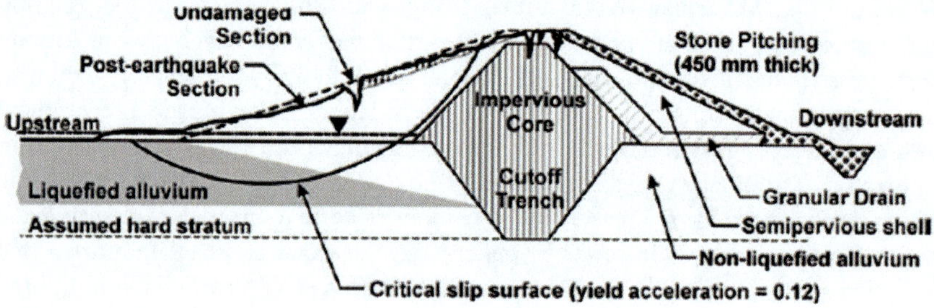


**Fig. 1** Cross section of Kaswati Dam (modified from EERI 2001, Singh et al. (2005))

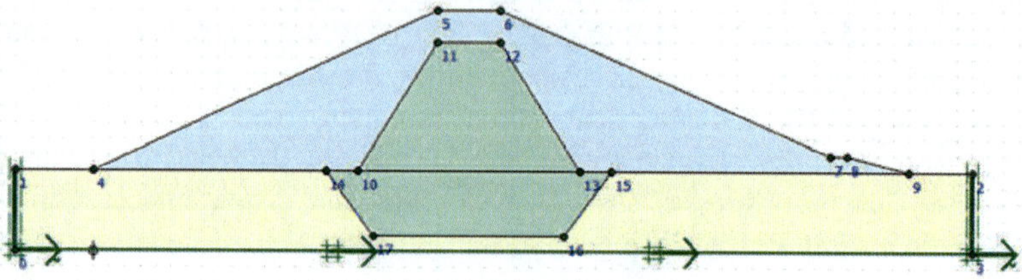

**Fig. 2** Pre failure Cross section of Kaswati dam by PLAXIS 2D

**Table 1** Soil properties of Kaswati dam

| Property | Semi pervious shell | Impervious core | Foundation |
|---|---|---|---|
| $\gamma$ (kN/m$^3$) | 18 | 20 | 18 |
| $\gamma_{sat}$ (kN/m$^3$) | 20 | 22 | 20 |
| $K_x$ (m/day) | 1.0 E$^{-7}$ | 1.0 E$^{-9}$ | 1.0 E$^{-6}$ |
| $K_y$ (m/day) | 1.0 E$^{-7}$ | 1.0 E$^{-9}$ | 1.0 E$^{-6}$ |
| $\upsilon$ | 0.33 | 0.33 | 0.33 |
| E (kN/m$^2$) | 50,000 | 70,000 | 60,000 |
| C (kN/m$^2$) | 0 | 0.2 | 0 |
| Ø (°) | 34 | 20 | 20 |

## 3 Material Properties of Kaswati Dam

As shown in Fig. 2, Kaswati dam is split into 4 separate layers, each representing individual soil material stretch. The soil stretches of the dam portion are grouped by using Eq. (1) suggested by Sawada and Takahashi. The equation shows the effect of height differential on shear modulus and the shear wave velocity of the zone materials situated at several parts of the dam cross section.

$$V_s = 140Z^{0.34} \tag{1}$$

Here, $V_S$ indicates shear wave velocity and Z is the dam height from the crest. The relationship between shear wave velocity, shear modulus and Modulus of Elasticity (*G* and *E*) is shown in Eq. (2).

$$G = \rho \times V_s^{-2}$$

$$G = E/2(1 + \mu) \tag{2}$$

Here, μ is the Poisson's ratio. The soil parameters related to the four zones of soil found using the above two equations and the following results given by [12] are listed in Table 1. These results are used as input parameters in this study.

## 4 Dam Simulation

The simulation of the dam starts with identifying the clusters (Kaswati Dam consists of 4 clusters) and finding the properties associated to each cluster. Figure 3 shows the section of a generated mesh of the Kaswati dam.

The impact of the input motion on the seismic response of the Kaswati dam is presented below. The nearest station was the IITR station in Ahmedabad with

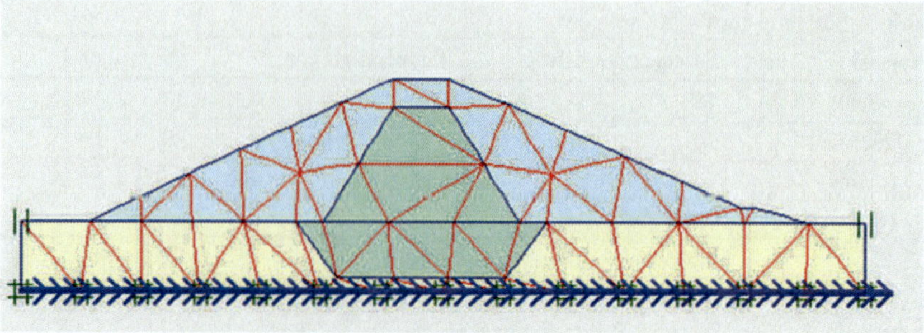

**Fig. 3** Section of generated Mesh of Kaswati dam by PLAXIS 2D

a hypocentral distance of 239 km. This accelerogram cannot provide an accurate result as the distance is too large. Hence, the dam was excited with the accelerogram obtained from Strong-Motion Virtual Data Centre (VDC), which was having modified shaking and almost similar effect on dam (Chilie earthquake, 27/02/2010). It was applied for a period of 20 s as shown in Fig. 4.

The numerical analysis with PLAXIS generally involves 3 parts [11]. In the initial phase, plastic analysis is performed by adding different dam materials and applying suitable material properties. The next phase consists of plastic analysis under the self-weight of the dam and the third step includes dynamic analysis under seismic loading. The last and final phase loading is put in the form of an accelerogram file input. The total deflections, displacements in the horizontal direction and vertical

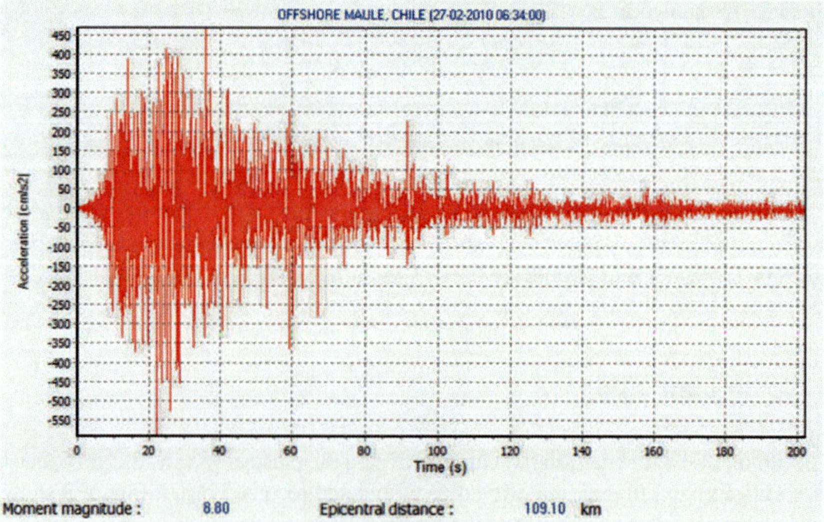


**Fig. 4** Acceleration time history used for the analysis

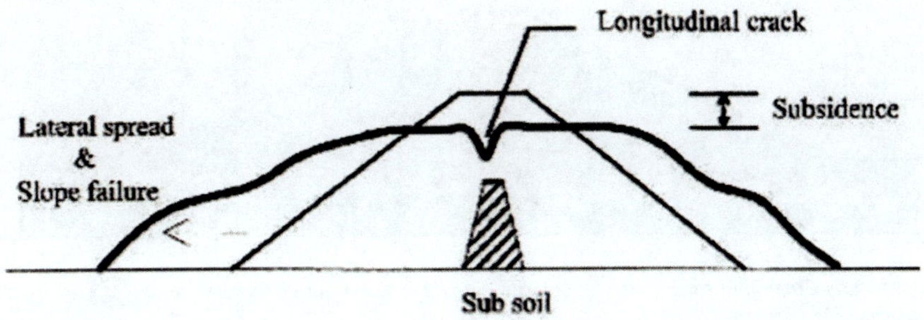


**Fig. 5** Typical failures observed in earth dams during Bhuj earthquake (modified from EERI 2001)

direction of the dam are found as the result of the analysis, and diagrams are shown in Figs. 6 and 7.

The total displacements are the complete collected displacements combined from horizontal and vertical displacement components at all the nodes displayed on the geometrical figure. Similarly, the horizontal (x) and vertical (y) displacement components at all the nodes are found at the end of calculation. Figure 6 shows the large lateral displacements (Extreme total displacement is 1.94 m) on both upstream and downstream sides. Sliding of slopes associated with a major drop down in the crest elevation also can be seen. Slope failure also can be observed (Fig. 8).

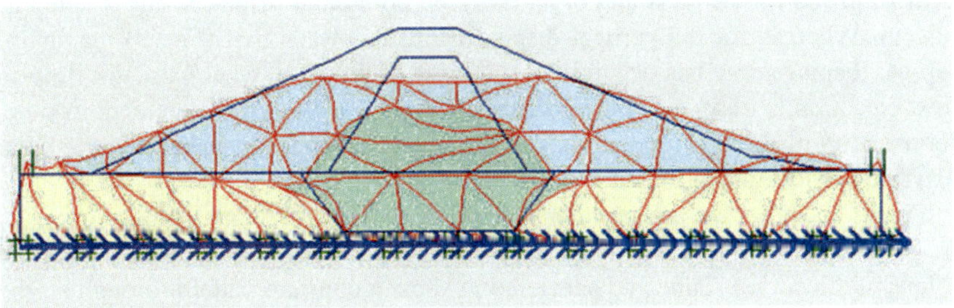

**Fig. 6** Deformed mesh of Kaswati Dam (extreme total displacement was 1.94 m, displacement scaled up 2.00 times)

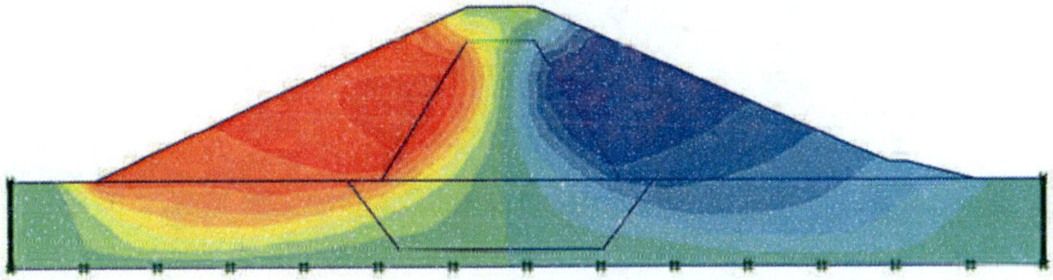

**Fig. 7** Distribution of total horizontal displacements (Ux), maximum 1.32 m

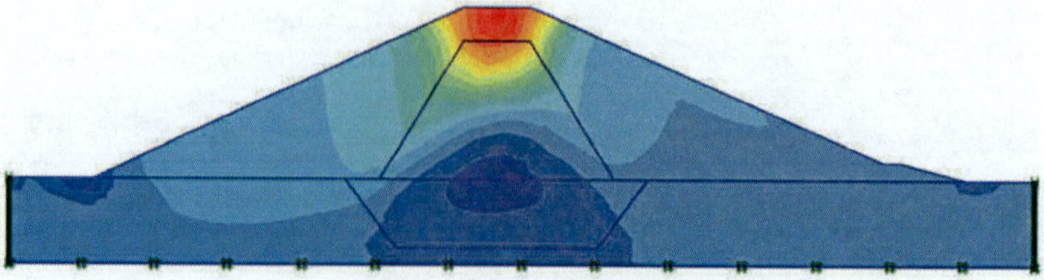

**Fig. 8** Distribution of total vertical displacements (Uy), maximum 1.94 mm

## 5 Time Displacement Analysis

Before starting the calculation in PLAXIS, reference points are chosen on cross section of the dam. These points are chosen at different levels of the dam. Point D is chosen at the centre of the dam, Point A is at the top, points B and C are chosen on upstream and downstream sides of the dam and points E and F are at the base. Horizontal displacement–time curves related to the above mentioned points at different levels of the cross sections are shown for Kaswati Dam. Figure 9 shows the location of nodes for time-displacement curves.

Figure 10 shows time-displacement curve for various points of the dam. From Fig. 10, it is seen that point B and C (red and yellow colour curve) which is at the upstream side and downstream side of the dam shows maximum horizontal displacement followed by the crest and base. Extreme horizontal displacement is noted as 1.22 m. Whereas for the vertical displacement, it is seen that (Fig. 9) maximum vertical displacement has occurred at the crest of the dam, where a major drop in crest elevation is observed (blue colour in the curve). With the increment in time, deformation of the crest increases and ultimately leads to the failure of the dam. Extreme vertical displacement is noted as 1.8 m (Fig. 11).

Figure 12 shows the time acceleration curve at different points of the dam. From the graph, it is seen that maximum acceleration is the inside core of the dam (pink colour in the curve). Time and acceleration show a constant pattern throughout the duration of shaking.

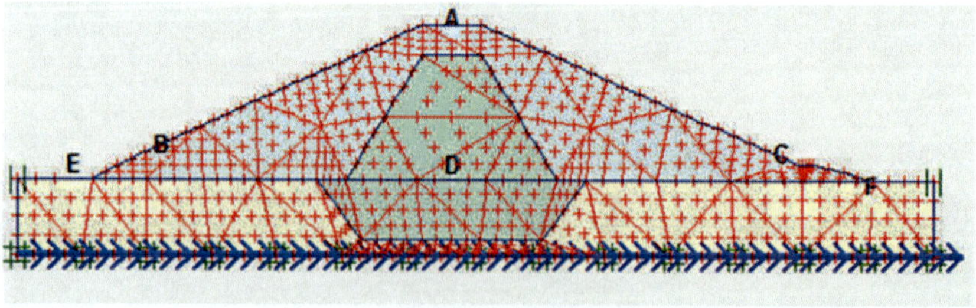


**Fig. 9** Location of nodes for time-displacement curves

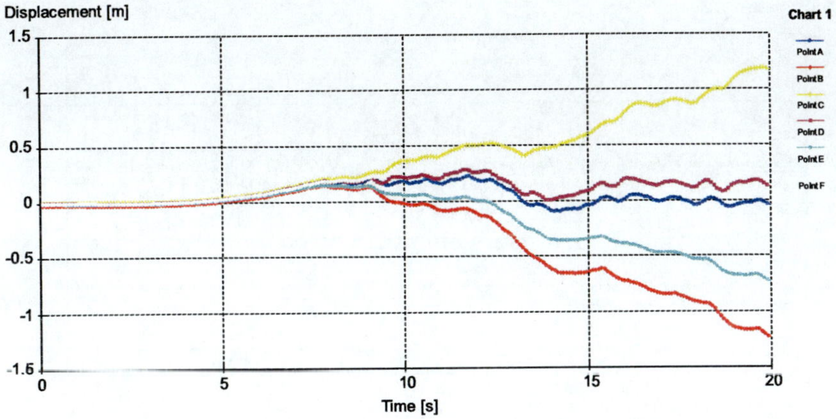


**Fig. 10** Time displacement curves for different points (horizontal displacement)

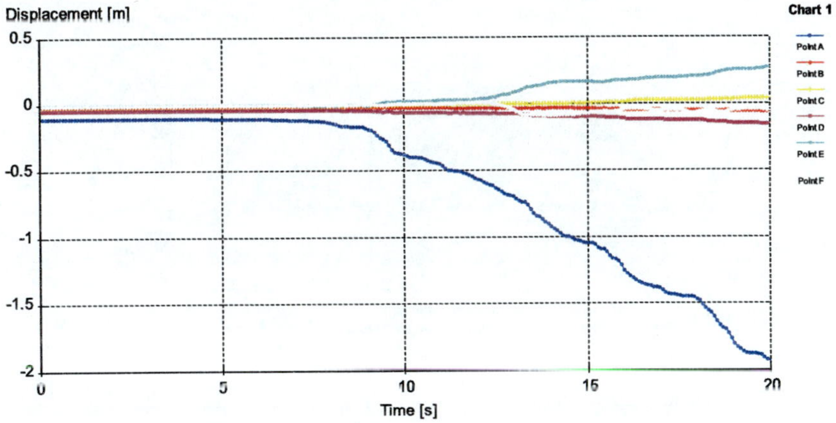


**Fig. 11** Time displacement curves for different points (vertical displacement)

## 6 Stress Analysis

The extreme effective principal stress is $-292.42$ kN/m$^2$. Pore water pressure is additionally developed due to seepage of water through the dam. Hence, active pore water pressure includes steady state water pressure, seepage pressure and excess pore water pressure developed during shaking. For this analysis, active pore water pressure was found to be $-50$ kN/m$^2$.

During an earthquake, the development of excess pore pressure causes liquefaction resulting in collapse [10]. The downstream slope of the dam did not show any evidence of sand boils, lateral spread, or liquefaction related effects [1], but possible liquefaction is found in the foundation near the upstream toe. In Fig. 13, it is seen

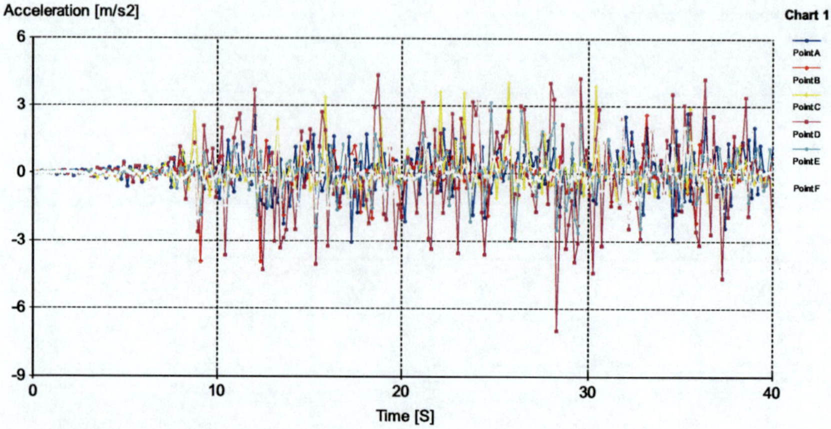


**Fig. 12** Time acceleration curves for different points

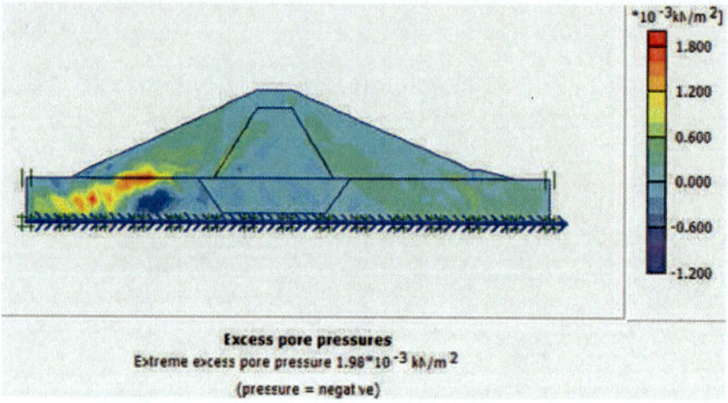


**Fig. 13** Excess pore water pressure

that extreme excess pore water pressure was generated near the upstream toe of the dam.

## 7 Concluding Remarks

The present study focussed on comparing the findings from numerical analysis with those of actual behaviour during the Bhuj earthquake. In the numerical analysis, it was found that the vertical displacement was the maximum at the crest indicating the subsidence. The lateral displacements were maximum at toe and heel, but of opposite natures. The core was subjected to maximum acceleration during shaking indicating the effect of shaking on dis-similar materials. The excess pore water pressure was the maximum near the upstream in foundation soil indicating the possible liquefaction. All these observations matched the actual conditions during the earthquake. Hence, the present study infers that seismic safety analysis of all existing dams in India may result in useful findings and the analysis using finite element approach with 2D plane-strain idealization may be sufficient to gather enough information about the safety of dams. This can be an excellent tool for preparedness for future big earthquakes.

## References


1. EERI Special Earthquake Report (2001) Preliminary observations on the origin and effects of the January 26, 2001 Bhuj (Gujarat, India) Earthquake
2. Earthquake Engineering Research Institute (EERI) (2001) Bhuj, India Republic Day January 26, 2001. Earthquake ReconnaissanceReport
3. Chakraborty D, Choudhury D (2009) Investigation of the behavior of tailings earthen dam under seismic conditions. Am J Eng Appl Sci 2(3):559–564
4. Griffiths DV, Prevost JH (1988) Two and three dimensional dynamic finite element analyses of the long valley dam. Geotechnique 38:367–388
5. Rampello S, Cascone E, Grosso N (2009) Evaluation of the seismic response of a homogenous earth dam. Soil Dyn Earthquake Eng 29:782–798
6. Singh R, Roy D, Jain SK (2005) Analysis of earth dams affected by the 2001 Bhuj Earthquake. Eng Geol
7. Earthquake Spectra (2002) 2001 Bhuj, India Earthquake Reconnaissance Report, EERI Publication No. 2002–01
8. Singh SK, Bansal BK, Bhattacharya SN, Pacheco JF, Dattatrayam RS, Ordaz M, Suresh G, Hough SE (2003) Estimation of ground motion for Bhuj (26January 2001; Mw 7.6) and for future earthquakes in India. Bull Seismol Soc Am 93:353–370
9. Clough RW, Chopra AK (1966) Earthquake stress analysis in earth dams. J Eng Mech ASCE 92:197–211. http://nisee.berkeley.edu/elibrary/Text/300451
10. Towhata, Prasad SK, Honda T, Chandradhara GP (2002) Geotechnical reconnaissance study on damage caused by 2001Gujarat earthquake. India Soils Found 42(4):77–88
11. Plaxis, 2D (2010) Tutorial manual. Delft University of Technology & PLAXIS bv, The Netherlands
12. Huang TK (1996) Stability analysis of an earth dam under steady state seepage. Comput Struct 58(6):1075–1082

# Influence of Reservoir Level on the Dynamic Behaviour of Concrete Gravity Dam

**P. N. Biju and Glory Joseph**

## 1 Introduction

The evaluation of dynamic behaviour and seismic safety has been an area of study for many researchers, as the failure of these structures may cause dangerous consequences. The Dynamic behaviour of dam depends on the interaction among dam reservoir and dam-reservoir-foundation systems. In gravity, dam construction monoliths are constructed with joints for facilitating movements in upstream and downstream as well as upward and downward directions. During an Earthquake, inertia and hydrodynamic force acts from upstream to downstream as well as from downstream to upstream. Similarly, the vertical inertia force acts from downward to upward as well as from upward to downward direction. Hence, the structure is to be analyzed based on the worst combination of horizontal and vertical inertia force.

## 2 Dam Reservoir Foundation Interaction

Reservoir considerably affects the dynamic response of dams during earthquakes. Three approaches are generally used to consider reservoir effects in the analyses. The Westergaard added mass approach, Eulerian approach and LagrangianApproach. Since the behaviour of fluid structure is expressed in terms of displacement, the equations of motions of the fluid system can easily be established through the Lagrangian approach and are commonly used for the analysis [1–3].

P. N. Biju (✉)
Cochin University of Science and Technology, Cochin 682022, India

G. Joseph
School of Engineering, Cochin University of Science and Technology, Cochin 682022, India
e-mail: glorybaby@cusat.ac.in



S. Kolathayar and S. C. Chian (eds.), *Recent Advances in Earthquake Engineering*, Lecture Notes in Civil Engineering 175,
https://doi.org/10.1007/978-981-16-4617-1_23

The complete finite element discretized equation for the fluid structure interaction (FSI) problem for the dam-reservoir-foundation system can be obtained by coupling the acoustic fluid and the structural matrices and can be represented [4] as

$$\begin{bmatrix} [M_S] & [0] \\ \rho_0[R]^T & [M_F] \end{bmatrix} \begin{Bmatrix} \{\ddot{u}\} \\ \{\ddot{p}\} \end{Bmatrix} + \begin{bmatrix} [C_s] & [0] \\ [0] & [C_F] \end{bmatrix} \begin{Bmatrix} \{\dot{u}\} \\ \{\dot{p}\} \end{Bmatrix} + \begin{bmatrix} [K_S] & -[R] \\ [0] & [K_F] \end{bmatrix} \begin{Bmatrix} \{u\} \\ \{p\} \end{Bmatrix} = \begin{Bmatrix} f_S \\ f_F \end{Bmatrix}$$

where $[M_S]$, $[C_S]$ and $[K_S]$ are the mass, damping and stiffness matrices of the structure (including dam and foundation), respectively, and $[M_F]$, $[C_F]$ and $[K_F]$ are the mass, damping and stiffness matrices of the acoustic fluid. The fluid density is denoted by $\rho_O$ and $[R]$ gives the coupling matrix which represents the coupling conditions on the interface between acoustic fluid and structure. $\{f_s\}$ and $\{f_F\}$ are structural and fluid load quantities produced at fluid structure interface in terms of unknown nodal displacement $\{u\}$ and pressure $\{p\}$.

## 3 Cross Section Details and Finite Element Implementation

The Dam structure selected for the study is Koyna Dam which has been affected by the Koyna Nagar earthquake of 1967. The Koyna dam has 103mheight and 70 m wide base non overflow section. The dam is assumed to be rest over 350 × 140 m foundation as per Federal Energy Commission Guidelines, [5] for incorporating the effect of stress, based on the structural dimensions. The bottom of the foundation is assumed to be fixed. The extent of the reservoir considered in the analysis is 140 m in length, i.e. two times the width of the dam at the bottom. The cross section of the Koyna dam along with the reservoir and dimensions of the foundation for the analysis is given in Fig. 1.

Modelling and analysis of the dam is done using ANSYS16. Plane 182 element is used to model foundation and Dam and Fluid 29 acoustic elements for the reservoir. The non overflow section is considered in the plane stress condition and the foundation is considered as plane strain condition.

The material properties of concrete considered for the dam structure [6] are modulus of elasticity 31,027 MPa, tensile strength 2.9 N/mm$^2$, compressive strength 29 N/mm$^2$, mass density 2643 kg/m$^3$ and Poisson's ratio 0.2. Parameters of the foundation are modulus of elasticity 62,054 MPa, mass density 3300 kg/m$^3$ and poison's ratio 0.33. Reservoir parameters include bulk modules 2250 MPa, mass density 1000 kg/m$^3$ and sonic velocity 1440 m/s. Self-weight of the structure, water pressure from the reservoir and uplift pressure are the loads considered other than seismic load.

Uplift pressure distribution along the base of the dam is based on Indian Code [7], without the tail water condition. The variation of uplift pressure with respect to reservoir level is depicted in Fig. 2.

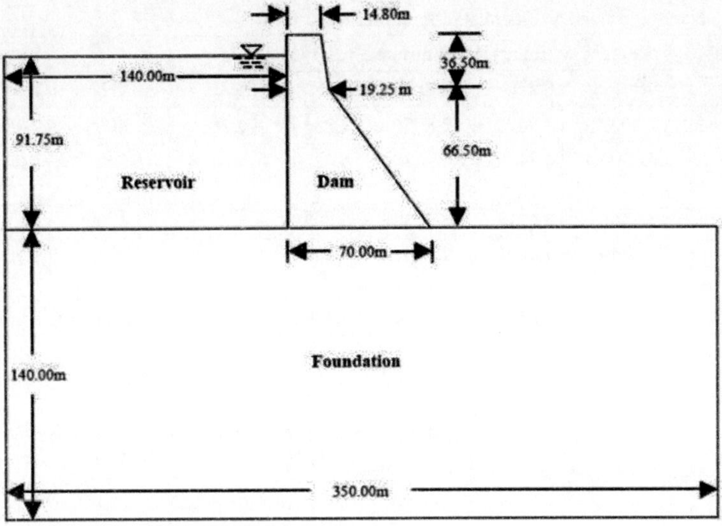


**Fig. 1** Cross section of Koyna Dam

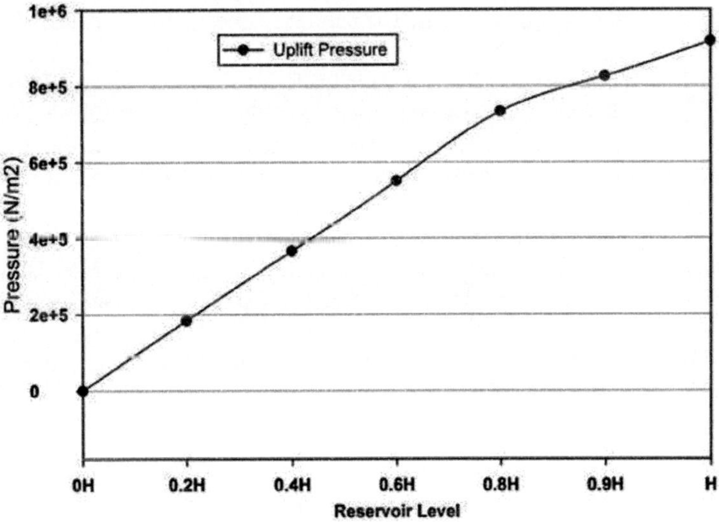


**Fig. 2** Uplift pressure variation at the upstream side

**Table 1** Natural frequency of dam at different reservoir level

| Mode of vibration | Natural frequency (Hz) at different reservoir levels | | | | | | |
|---|---|---|---|---|---|---|---|
| | 0H | 0.2H | 0.4H | 0.6H | 0.8H | 0.9H | H |
| First mode | 2.8032 | 2.8049 | 2.8056 | 2.8044 | 2.805 | 2.8065 | 2.804 |
| Second mode | 6.6003 | 4.8158 | 4.9670 | 5.0219 | 5.0504 | 5.0601 | 5.0682 |
| Third Mode | 7.3868 | 6.6214 | 6.6137 | 6.6088 | 6.6073 | 6.6096 | 6.6129 |
| Fourth mode | 8.1549 | 7.4608 | 7.4137 | 7.3836 | 7.3413 | 7.2874 | 7.1045 |
| Fifth Mode | 9.4431 | 8.2343 | 8.1804 | 8.1295 | 8.0141 | 7.8321 | 7.5654 |

# 4 Results and Discussion

## 4.1 *Modal Analysis*

Finite Element cross sections of the dam fluid foundation system are modelled varying from empty conditions to full reservoir level (H) with reservoir storage as 0H, 0.2H, 0.4H, 0.6H, 0.8H, 0.9H and H. Modal analysis was performed to find the natural frequency of the systems. The natural frequency for the dam at different reservoir level is presented in Table 1.

Even though there is not much change in the frequency of the first mode in all the cases of reservoir level, for the second mode onwards, a pattern of decreasing the frequency of the vibration from empty condition to full reservoir condition is seen and from the third mode onwards it is more evident. The percentage of reduction in natural frequency at the third mode of vibration is 10.5% as the reservoir level increases from empty to full level. This is because of the interaction between water and dam body which causes an increase in effective mass. Also, the reservoir moves along with the structure which leads to a change in the dynamic characteristic of the connecting system.

The first mode shape of the dam at various reservoir heights are depicted in Fig. 3a–g

## 4.2 *Seismic Behaviour of Gravity Dam*

Time history analysis was carried out on the dam sections with time history records of the Koyna Earthquake, 1967 (moment magnitude 6.6, peak horizontal acceleration 0.4896 g and peak vertical acceleration 0.245 g). The time history recorded in both Horizontal and Vertical directions has been applied simultaneously. The natural frequency obtained in the modal analysis is used for finding the Rayleigh Stiffness damping factor for the analysis. Dam Structure and fluid structure interactions

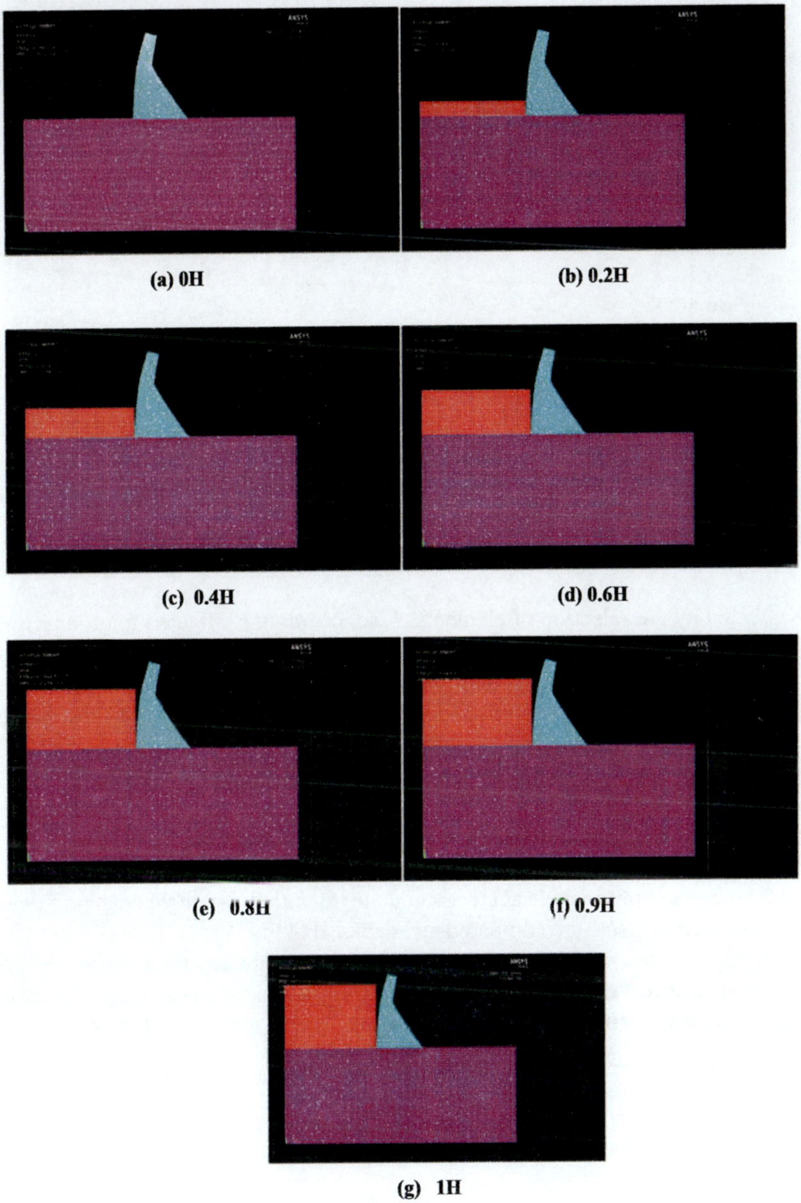


**Fig. 3** First mode shape of dam at different reservoir level

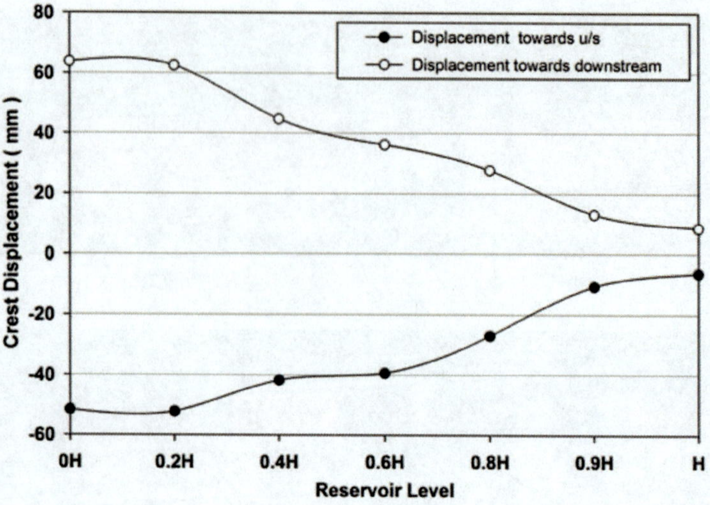


**Fig. 4** Crest displacement

are affected by the selection of elements. Also, coupling is affected for connecting common nodes. Time increment in the analysis is taken as one by twenty of frequency. The maximum response like displacement, normal vertical stress, principal tensile stress, principal compressive stress, stress intensity at salient sections of the dam is studied.

### 4.2.1 Crest Displacement and Node Below Slope Changing Point

The displacement of the node at crest and node just below the slope changing point on the downstream side are compared and depicted in Fig. 4.

The response towards the displacement decreases as the reservoir storage height increases because of the effective mass increase in the system as expected. The same pattern is obtained at the downstream node of the slope changing point and is depicted in Fig. 5.

### 4.2.2 Vertical Stress Variation at Heel

Variation of maximum vertical stress at heel at different reservoir heights are plotted in Fig. 6. As the reservoir height increases, only marginal variation in stress is noticed. The reason may also be attributed to the addition of effective mass increment. It may also be noted that the stress in the downstream movement is higher than that of upstream movement as the inertia effect is higher towards the downstream side.

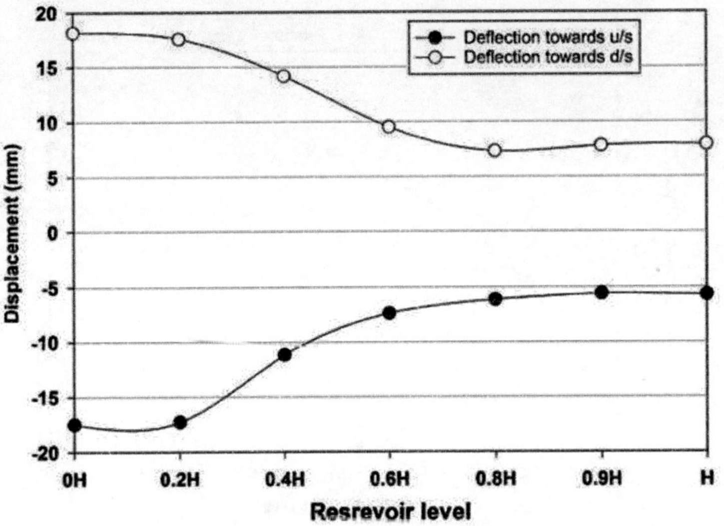


**Fig. 5** Displacement of the node adjacent to the slope changing point

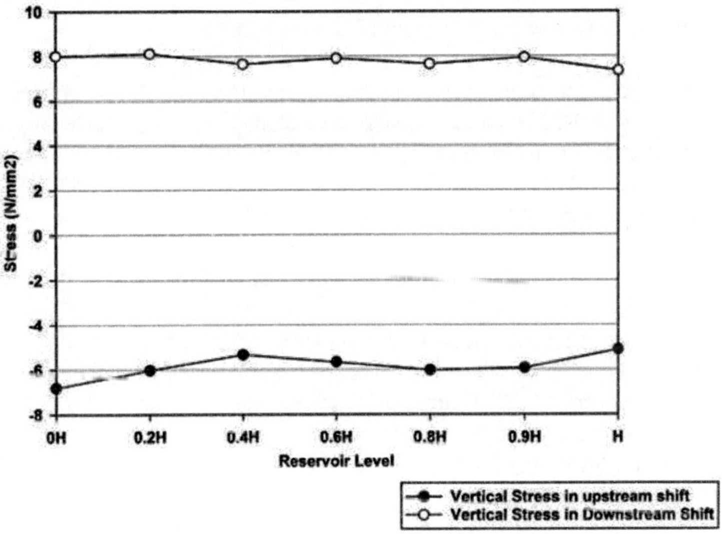


**Fig. 6** Vertical Stress variation at Heel

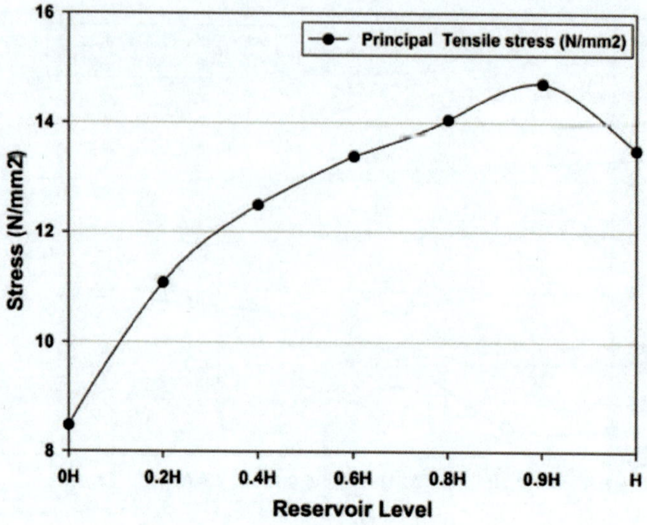


**Fig. 7** Variation of maximum principal tensile stress at the heel

### 4.2.3 Principal Tensile and Compressive Stress at Heel

Maximum principal tensile stress and compressive stress developed at heel due to seismic loading of the Koyna earthquake is noticed. Figure 7 indicates that as the reservoir storage height increases the principal tensile stress also increases gradually up to a reservoir level of 0.9H and then reduces at the full reservoir level. The highest tensile stress is noticed at the reservoir level of 0.9H is 9.6% more than that at the full reservoir level. The tensile stress developed at the heel indicates the necessity of increased base width and modification of dam cross section along with provisions for steel reinforcement.

The maximum principal compressive stress at heel due to transient loading of the Koyna earthquake at various reservoir levels is also plotted in Fig. 8. It can be noticed that the maximum compressive stress occurs when the reservoir level is at 0.8H and the stress is 16% more than that occurs at the full reservoir level. This infers that the full reservoir level may not be the critical case when earthquake loading is also encountered in the structure.

## 5 Conclusions

The following conclusions are drawn from the modal analysis and seismic loading due to the Koyna earthquake on the dam-reservoir-foundation system for various water levels in the reservoir.

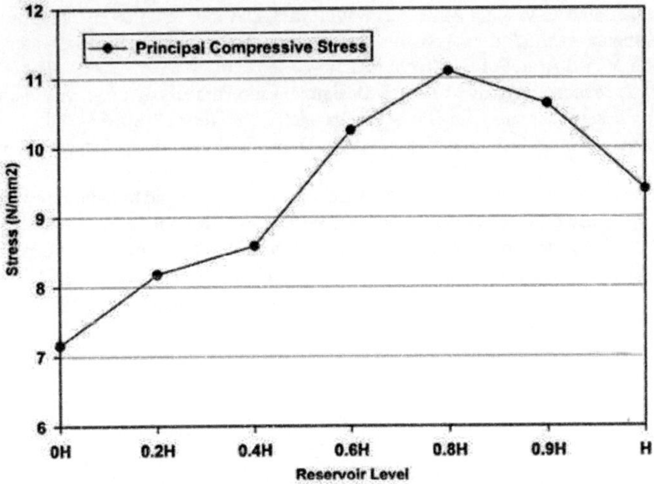


**Fig. 8** Variation of Maximum principal compressive stress at the heel

1. Variation of the natural frequency with variation in reservoir level at the first mode of vibration is marginal. However, at higher modes, the frequency of vibration gradually decreases as the reservoir level increases. The percentage of reduction in natural frequency at the third mode of vibration is 10.5% as the reservoir level increases from empty to full level.
2. The deflection at crest due to seismic loading towards downstream side is more compared to upstream side due to increase in effective mass.
3. Vertical stress at heel is observed to be reduced as the reservoir level increases attributable to the effective mass through fluid structure interaction.
4. The maximum principal tensile stress and compressive stress at the heel increases as water in the reservoir increases from empty condition to a certain level of water and thereafter decreases. The maximum response of the dam due to seismic loading may not always occur at full reservoir conditions.
5. The maximum principal tensile stress at heel due to seismic loading of Koyna earthquake occurs at reservoir level of 0.9H and compressive stress at 0.8H which are, respectively, 9.6–16% more than that occurs when the reservoir is at its full level.

## References

1. Caliyar Y, Dumanoglu AA, Bayraktar A (1996) Earthquake analysis of gravity dam reservoir using Eulerian and Lagrangian approaches. Comput Struct 58(5):877–890
2. Bayraktar A, Dumanoglu AA, Caliyar Y (1996) Asynchronous dynamic analysis of Dam reservoir foundation system by the Lagrangian approach. Comput Struct 58(5):925–935

3. Khan Mohammad K, Vaseghi Amiri J, Navayineya B, Davoodi M (2009) Evaluation of Eulerian and Lagrangian method in analysis of concrete gravity dam including dam water foundation interaction. World Acad Sci Eng Technol 3:10–20
4. Khosravi S, Mehate Heydari M (2015) Design and modal analysis of gravity dam by Ansys parametric design language. Eng Phys Sci Walailak J S & Tech 12(2):167–180
5. Federal Energy Commission Guideline (1999) Division of dam safety inspection, Washington DC
6. Sarkar R, Paul DK, Stempniewski L (2007) Influence of reservoir and foundation on the nonlinear dynamic response of concrete gravity dams. J Earthquake Technol 44:377–389
7. IS 6512–2019. Criteria for design of solid gravity dam, Bureau of Indian Standards
8. IS 1893–1984. Criteria for earthquake resistant design of structures, Bureau of Indian Standards

# Assessment of Seismic Displacement of Quay Walls

K. Pushpa, S. K. Prasad, and P. Nanjundaswamy

## 1 Introduction

Quay walls are the basic elements of any marine structure and form basic elements of infrastructure. Despite the advances in earthquake geotechnical engineering, there have been many failures of port structures during strong earthquakes. Damage to these structures results in serious economic and physical consequences. The typical seismic failures usually are lateral sliding, overturning, and settlement of quay walls leading to port facilities being non-functional for a long time. Hence, a lot of insight is necessary for the performance of these structures. Port structures that were subjected to strong earthquakes at Niigata (1964), Tokachi-Oki (1968), Nemuro-Hanto-Oki (1973), Kushiro-Oki (1993), and Hyogoken Nanbu (1995) had been designed employing seismic coefficients whose values were about half the values of the peak accelerations actually experienced during the earthquakes (Pianc 2001). Even though the design was so similar to the actual conditions, the extent of damage was very severe. The damage was observed to be dependent on the liquefaction of the soil in the surrounding region which was highest during the Hyogo-ken-Nambu and Niigata earthquakes. It clearly indicates that the effect of liquefaction at site and pore pressure build-up have a high impact on the performance of port structures.

K. Pushpa (✉)
Department of Construction Technology & Management, J.S.S. Science & Technology University, Mysuru 570006, India
e-mail: pushpa_k@sjce.ac.in

S. K. Prasad
Vidyavardhaka College of Engineering, Mysuru 570006, India
e-mail: skprasad@vvce.ac.in

P. Nanjundaswamy
Department of Civil Engineering, J.S.S. Science & Technology University, Mysuru 570006, India
e-mail: pnswamy@sjce.ac.in



S. Kolathayar and S. C. Chian (eds.), *Recent Advances in Earthquake Engineering*, Lecture Notes in Civil Engineering 175,
https://doi.org/10.1007/978-981-16-4617-1_24

The very existence of water alongside the retaining walls influences the magnitude of seismic pressure on it. The liquefaction of backfill soil of port structures is one of the major damages caused due to the occurrence of earthquakes, may it be moderate or strong.

Various researchers Al-Homoud and Whitman [1], and various others have conducted numerical analysis on retaining walls subjected to dynamic earth pressures. Choudhury and Chatterjee [5] made use of the experimental results to verify that the seismic active earth pressure distribution relies on the type and amount of wall movement. Nanjundaswamy et al. [12] have employed the use of finite element software FLAC to model and analyze the seismic performance of quay walls. Richard-Elms gave the design of retaining walls on the basis of allowable permanent displacement of the retaining walls. All these studies on retaining walls show that the design of retaining walls based on allowable displacement reduces the extent of the damage.

## 2 Analysis of Displacement of Quay Walls

The factors responsible for the performance of water retaining structures are mainly backfill and foundation soil properties, geometry of the wall, boundary conditions, groundwater level, characteristics of earthquake motions, and more. A typical Quay wall with properties as mentioned in Table 1 is modeled using the software. These properties are used as input parameters in the software.

A typical quay wall modeled using Quake/w feature of GeoStudio software is as shown in Fig. 1. It consists of a retaining wall constructed on a foundation of dense sand and a backfill composed of loose sand on one side and seawater on the other side. The boundary conditions are taken as fixed X along with the depth of foundation and backfill and fixed XY along the length of the foundation are specified. The presence of water outside the retaining wall will impose a dynamic pressure on the face of the retaining wall, while the presence of water in the backfill soil adds on the dynamic pressure on the back of the retaining wall. There is a cyclic loading due to the water pressure. Westergaard [15] gives a method to estimate the hydrodynamic pressure.

In addition to the lateral earth pressure and the hydrodynamic pressure on the quay wall, a ground motion due to the earthquake is applied. Considering the ground motion waves as sinusoidal with a frequency of 1 Hz and peak ground acceleration varying from 0.05 to 1 g the analysis is conducted. The various parameters attributed to the deformation or permanent displacement of the quay wall are studied.

## 3 Parametric Study

The various parameters which influence the permanent displacement of the wall studied here are amplitude of acceleration of ground motion and liquefaction of

**Table 1** Properties of backfill and foundation soil

| Property | Units |
|---|---|
| Backfill soil properties | |
| Unit weight | 16 kN/m$^3$ |
| Angle of internal friction | 30$^0$ |
| Cohesion | 0 |
| Damping ratio | 0.6 |
| Poisson's ratio | 0.25 |
| Coefficient of lateral earth pressure, $K_o$ | 0.5 |
| Foundation soil properties | |
| Unit weight | 20 kN/m$^3$ |
| Angle of internal friction | 40$^0$ |
| Cohesion | 0 |
| Damping ratio | 0.6 |
| Poisson's ratio | 0.25 |
| Coefficient of lateral earth pressure, $K_o$ | 0.5 |
| Quay wall properties | |
| Unit weight | 24 kN/m$^3$ |
| Shear modulus, G | 21GPa |
| Height of the wall | 8 m |
| Width of the wall | 6 m |

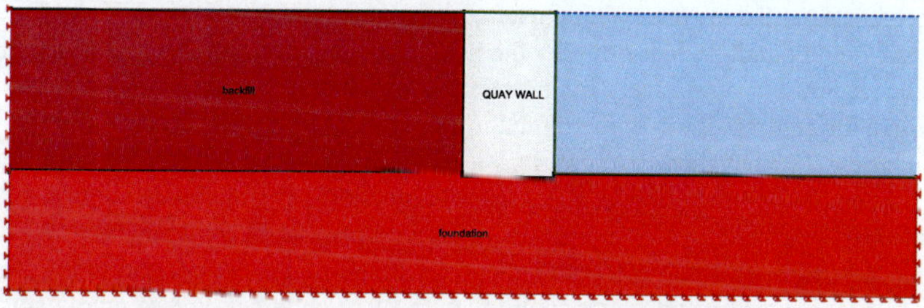


**Fig. 1** A typical quay wall model with backfill soil, foundation, and seawater

backfill soil. Mononobe–Okabe's equation [11] and Westergaard's equation [15] are used to calculate seismic earth pressure and dynamic water pressure on a retaining wall. The backfill soil and water system shows movement along the base of the retaining wall when the ground acceleration is greater than yield acceleration.

Richard-Elms gave a method of calculating the allowable or permanent displacements in the same way as that of the Newmark sliding block method as

$$d_{perm} = 0.087\frac{v_{max}^2 a_{max}^3}{a_y^4} \quad (1)$$

Towhata and Islam [13] gave a method to estimate the displacements of retaining walls by employing Newmark's sliding block theory. Towhata (1993) proposed another type of simplified method to predict the permanent displacement of gravity type quay walls.

In order to analyze the allowable permanent displacement, the present study employs the use of a model of quay wall developed using quake/w feature of finite element software GeoStudio. The methodology behind the calculation of displacement in Quake/w is double integration of the acceleration versus time increment record. The strain components are related to x and y displacements, u and ν, as follows:

$$\varepsilon_x = \frac{\partial u}{\partial x} \tag{2}$$

$$\varepsilon_y = \frac{\partial v}{\partial y} \tag{3}$$

$$\gamma_{xy} = \frac{\partial u}{\partial y} + \frac{\partial v}{\partial x} \tag{4}$$

### 3.1 Boundary Effect

It becomes essential to ensure that the length of the analytical model is sufficient such that the fixed boundary does not alter the response of an infinitely long system in the middle region. For a relatively long model, it can be assumed that the deformation of soil in the central region of the model is close to that of the prototype with infinite boundaries. The physical or numerical model has restricted length, but represent infinite length or field. Hence, the focus of preparing the model is to plan such infinite physical length which does not alter the behavior of the quay wall region. Hence, arbitrarily in Table 2, length beyond quay wall till the boundary ranging from 6 to 41 m were tested and it was found that 256.5 m on either side of quay wall as

**Table 2** Variation of horizontal displacement with an increase in length of foundation

| Length of foundation (m) | Displacement (m) (before liquefaction) | Displacement (m) (after liquefaction) |
|---|---|---|
| 15 | 0.930 | 2.12 |
| 31 | 0.879 | 2.10 |
| 48 | 0.870 | 2.17 |
| 56 | 0.874 | 2.20 |
| 64 | 0.873 | 2.20 |
| 85 | 0.871 | 2.01 |

**Table 3** Variation of displacement of quay wall with an increase in mesh size

| Mesh size (m) | Displacement (m) |
|---|---|
| 0.2 | 0.69 |
| 0.5 | 0.69 |
| 1.0 | 0.70 |
| 1.5 | 0.70 |
| 2.0 | 0.71 |
| 2.5 | 0.70 |

shown in Fig. 1 was sufficient. Hence, the foundation length is varied in order to study the boundary effect on the analytical model as shown in Table 2.

The above observations indicate that there is an increase in the displacement after liquefaction compared to that before liquefaction. But, the changes in displacement before and after liquefaction with an increase in the foundation length are insignificant, and hence the boundary effects are taken care.

## 3.2 Convergence Check

Discretization or meshing is one of the fundamental aspects of finite element modeling besides defining boundary conditions and material properties. Discretization involves defining the geometry, distance, area, and volume of the mesh. GeoStudio ensures mesh compatibility within a region and for the most part ensures mesh compatibility across adjacent regions, but it is still possible to create a situation whereby mesh incompatibility exists. Table 3 shows the variation of displacement of quay wall with an increase in mesh size.

It can be observed that the increase in mesh size does not have a measurable impact on the values of displacement and hence the convergence effect is taken care of.

## 3.3 Effect of Acceleration of Ground Motion on Quay Walls

**Before liquefaction of backfill soil**

From the study of earthquake records, it is found that the predominant frequency of most of the earthquakes that cause severe damage were in the range of 1 to 3 Hz and peak average amplitude of acceleration was around 0.5 g. [2]. Hence, in the present analytical study, the amplitude of ground acceleration is varied from 0.05 g to 1 g at a frequency of 1 Hz. The magnitude and frequency of acceleration have a direct impact on the lateral displacement of the quay walls. Quake/W computes the displacement by double integration of the acceleration versus time increment record.

**Fig. 2** Variation of horizontal displacement of quay wall with change in acceleration

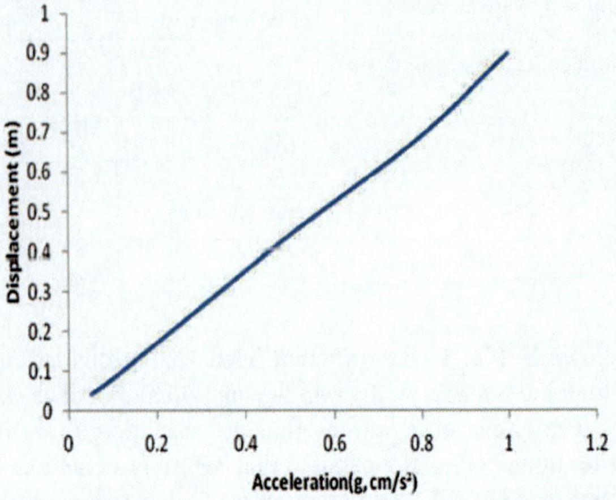


Figure 2 shows a linear variation of displacement with an increase in the magnitude of the acceleration of ground motion when the backfill is not liquefied.

**After liquefaction of backfill and foundation soil**

The most widespread source of damage due to an earthquake on the port facility has been the liquefaction of loose, saturated, and sandy soils. It has been observed over the years that significant liquefaction and ground movement and damages have not only occurred under very strong ground movements, but under moderate shaking also. Since the backfill soils are usually cohesionless, there are higher chances of liquefaction even under moderate earthquake motion. As a result, the modes of failure of the quay walls are usually lateral sliding, settlement, and rotation.

The sliding displacement of the quay wall when the backfill liquefies is analyzed by the analytical approach. The observations are presented in graphical form in Fig. 3.

It can be observed that the sliding displacement increases exponentially with the increase in the horizontal acceleration of ground motion. The magnitude of displacement is higher than the case without backfill liquefaction by several folds.

## 4 Comparison of Analytical Test with Model Test Results

In an earthquake, geotechnical engineering model testing is one of the important methods of recreating the field scenario in the laboratory. Model studies give a path to procure special data like excess pore pressure changes, flow of liquefied ground, and amplification in ground motion. Both practical and theoretical investigations are necessary to understand the problems linked with seismic failure of quay walls considering the complexities and uncertainties involved. It is essential to evaluate the true mode of the wall movement and impact of liquefaction both in backfill and foundation soil regions.

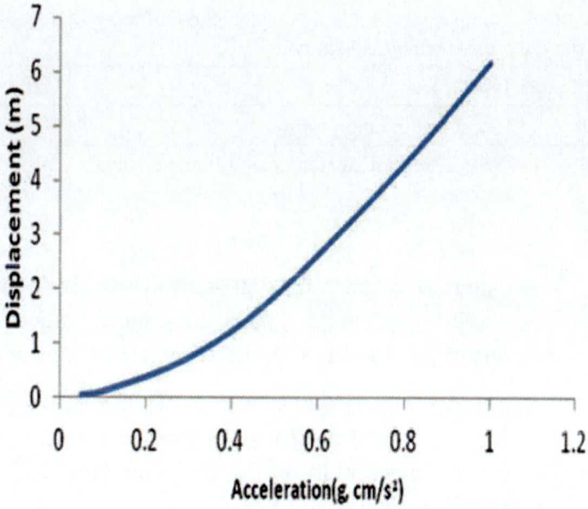


**Fig. 3** Variation of sliding displacement of quay wall with the acceleration of ground motion

The details and results obtained from model tests on the shaking table conducted by Nanjundaswamy [12] as a part of his Doctoral thesis are compared with those obtained from the finite element software. Model studies were conducted at normal gravitational environment (1-g test). The transparent model container and manual shaking table developed at Earthquake Engineering Laboratory of S. J. College of Engineering, Mysore, is shown in Fig. 4.

**Fig. 4** Assembly of manual shaking table with a transparent model container

**Table 4** Comparison of sliding displacement of quay wall from the physical model test with that from the present analytical study

| Sl no | Type of test | Horizontal sliding displacement (mm) |
|---|---|---|
| 1 | Model test in laboratory | 80 |
| 2 | Present Numerical study using Finite element analysis | 66 |

The materials that are used as well as the size of the shaking table used in the shaking table play a major role in the ground motion study. The configuration and dimensions of the manual shaking table used are as detailed below [12].

- Two wooden panels 600 mm wide, 1800 mm long, and 25 mm thick were used such that one of them formed the base and the other acted as a platform.
- Four steel plates 350 mm long, 550 mm wide, and 2 mm thick were provided to act as springs.
- The connections between plates and wooden panels were provided through steel bolts and angle sections.
- A handle was provided at the end to apply harmonic sinusoidal input force along the longitudinal direction.
- Rubber membranes of 3 mm thick were provided between the floor and table and model container and table in order to prevent relative slip between the components.
- Shaking table was designed to vibrate at around 2 Hz. with 0.5 g level of acceleration at a payload of around 7 kN.
- The overall cost for the assembly of the entire shaking table did not exceed Rs. 10,000/- at the time of fabrication.

Table 4 presents the comparison of sliding displacements obtained from the physical model test with that of the numerical model using the Finite Element approach in the present study. In both the cases, similar conditions such as dense foundation soil, loose backfill and medium-heavy quay wall were considered.

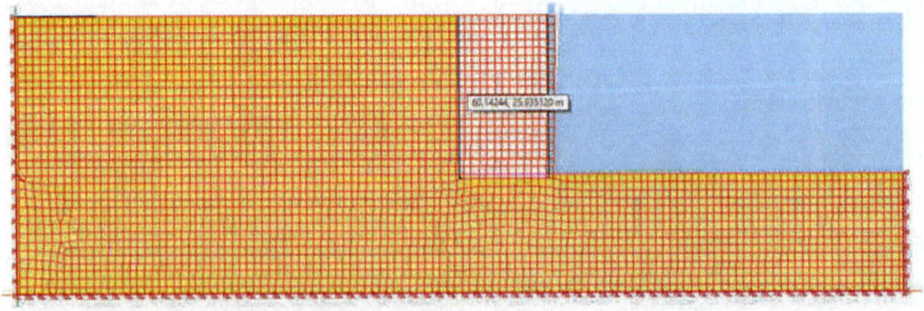

The results from the analytical model obtained from quake/w model as shown above indicates that the quay wall is displaced by 66 mm while the model test results show that there is a sliding displacement of 80 mm of the quay wall. There is a considerable closeness in both the approaches.

## 5 Conclusions

1. The study conducted from Quake/w module of finite element software Geostudio shows that the effect of liquefaction of backfill has a pronounced effect on the permanent displacement of the quay wall.
2. The finite element software GeoStudio serves as an effective tool in the analysis of the performance of quay walls during an earthquake.
3. The amplitude of acceleration of ground motion has a considerable impact on the performance of quay wall, especially when the backfill liquefies.
4. The results obtained by the analysis of the model developed using the finite element software are comparable with those of model tests conducted using the shaking table with similar properties of backfill and foundation soil, and quay wall.

## References


1. Al-Homoud AS, Whitman RV (1999) Seismic analysis and design of rigid bridge abutments considering rotation and sliding incorporating non-linear soil behavior. Soil Dyn Earthquake Eng 247–277
2. Chandradhara GP, Prasad SK (2000) An approach to assess damage during earthquakes. In: Proceedings of Indian geotechnical conference, IGC-2000, Quest Publications, Mumbai, pp 359–360
3. Chandradhara GP, Prasad SK, Nanjundaswamy P (2002) Seismic response of model embankments from shaking table tests. 12SEE-2002
4. Choudhury D, Nimbalkar SS (2006) Pseudo-dynamic approach of seismic active earth pressure behind retaining wall. Geotech Geol Eng Springer, Netherlands 24(5):1103–1113
5. Choudhury D, Nimbalkar SS (2007) Sliding stability and seismic design of retaining wall by pseudo dynamic method for passive case. Soil Dyn Earthquake Eng 27:497–505
6. Ebeling RM, Morrison EE Jr (1992) The seismic design of waterfront retaining structures , US Army Technical Report ITL-92-11, Washington DC
7. Iai S (1998) Seismic analysis and performance of retaining structures. Geotechnical earthquake engineering and soil dynamics III, Geotechnical Special Publication No.75, ASCE, 1020–1044
8. Geo-Slope (1992) User's Manual. Geo Slope Office, Geo-Slope International, Calgary, Alberta, Canada
9. Inagaki H, Iai S, Sugano T, Yamazaki II and Inatomi T, (1996) Geotechnical aspects of the January 17, 1995 Hyogoken-Nambu earthquake: performance of Caisson type Quay Walls at Kobe port, Soils and Foundations (Special Issue), pp 119–36
10. Kramer SL (1996) Geotechnical earthquake engineering. Pearson Education Inc., New Jersey
11. Mononobe N, Matsuo H (1929) On the determination of earth pressures during earthquakes. In: Proceedings of the world engineering congress, Tokyo, Japan, vol. 9, paper no. 388
12. Nanjundaswamy P (2008) A study on seismic response of quay walls,Doctoral thesis submitted to Kuvempu University, India
13. Towhata I, Islam S (1987) Prediction of lateral movement of anchored bulkheads induced by seismic liquefaction. Soils Found 27(4):137–147
14. Toyoto H, Towhata I, Imamura S, Kudo K (2004) Shaking table tests on flow dynamics in liquefied slope. Soils Found 44(5):67–84
15. Westergaard HM (1933) Water pressures on dams during earthquakes. Trans ASCE 98:418–433

# Retrofitting and Rehabilitation

# Seismic Response Elimination of Structure by Using Passive Devices: An Overview

Rushikesh Jadhav, Digambar Patil, and Prachi Sohoni

## 1 Introduction

A planet like Earth has seen several catastrophic earthquakes over the last few decades, contributing to a growing loss of human life due to building collapse and severe structural damage [18]. The occurrence of this destruction throughout disasters indicates the high potential hazard, and structures such as high rises, shelter buildings, necessary parts, and commercial buildings should be designed very attentively to protect them from earthquakes [19]. Civil engineering has now widely accepted and implemented the structural modelling approach using seismic response control [23]. In recent years, substantial attention was paid to the research and development of dynamic control methods, such as passive control systems, active control systems, and semi-active control systems to enhance the wind and seismic response of structures, bridges, and piping systems [20]. Passive control devices do not require a power source. External power and function based on structurally linked sensors are needed for active systems. Semi-Active systems having external power supply and operations using structurally assembled sensors are used to corporate passive and active control systems [1]. However, passive control mechanisms control the vibration of objects while there is no power source. For heavy wind motion and earthquakes, all control systems may be used. Various piping materials provide minimal damping to the system when the stress is within standard code limits [2].

This review paper focuses mainly on passive dampers to apply to the piping system. It only studies application, advantage, and disadvantage, i.e., merits and demerits of passive dampers to 2-D and 3-D piping system and building structures. This review paper suggests the best applicable vibrational control device by comparing various criteria, as discussed in Tables 1, 2 and 3. Also, the paper gives

R. Jadhav (✉) · D. Patil · P. Sohoni
Department of Civil Engineering, Walchand College of Engineering, Sangli, India



S. Kolathayar and S. C. Chian (eds.), *Recent Advances in Earthquake Engineering*, Lecture Notes in Civil Engineering 175,
https://doi.org/10.1007/978-981-16-4617-1_25

**Table 1** Technical and economic comparison of dampers

| Sr. No | Parameter | Damper's | | | | |
|---|---|---|---|---|---|---|
| | | X-Plate Damper (XPD) | Fluid Viscous Damper (FVD) | Visco-Elastic Damper (VED) | Tuned Mass Damper (TMD [17]) | Multiple-Tuned Mass Damper (MTMD) |
| 1 | Complexity | Simple mechanical devices | Complex | Complex | Simple mechanical devices | Complex |
| 2 | Number of damped resonances | Damps all | Damps all | Damps only one resonance | Damps only one resonance | Damps all |
| 3 | Sensitivity to resonance frequency changes | Sensitive | Not-Sensitive | Sensitive | Sensitive | Sensitive |
| 4 | Power or Electricity [1] | Do not require. (It works even when power gets off) | Do not require power | Do not require power | Do not require power | Do not require |
| | | However, for sensors, an external power supply is required | | | | |
| 5 | Manufacturing cost | Low | Medium | High | Low | Medium |
| 6 | Implementation cost | Medium | High | Low | High | High |
| 7 | Mass | Proportional to vibration energy | Proportional to stiffness | Proportional to stiffness | Proportional to target mass | Proportional to vibration energy |
| 8 | Implementation | Simple | Simple | Complex (tuning)/ engineering | Complex (tuning)/ engineering | Complex (tuning)/ engineering |

**Table 2** Comparative study of different types of passive dampers

| Sr. No | Types of damper | Capabilities | Limitations |
|---|---|---|---|
| 1 | X-Plate Damper (XPD) | (1) Adds extra stiffness to the system<br>(2) X shape of the device results in constant strain variation<br>(3) Capable of a high level of energy dissipation<br>(4) No frequent inspection is required [14]<br>(5) Iterative response spectrum method is the best method for seismic response prediction of a piping system [6] | (1) Natural device frequency is dependent on the original XPD stiffness of XPD<br>(2) The Energy dispersed through a pipe structure is dependent on XPD's thickness [14]<br>(3) Dissipated Energy in a piping device depends on the rotation of the land<br>(4) The efficacy of XPD even reduces, dissipated by the energy of the XPD [14]<br>(5) Energy dissipated by damper is reliant on outside seismic energy impact to the structure that is i/p ground motion [15] |
| 2 | Fluid Viscous Damper (FVD) [14] | (1) Added additional dampers increase the overall input energy entirely dissipated by the rise in damping Energy (Ed)<br>(2) The damping coefficient and the number of structural dampers used in the building shall reduce, compared to the non-linear one that stays stable for an $\alpha = 0.2$, to more effective use of this instrument and the economy<br>(3) Non-linear damper specifications for the FVD damping constant, 125 times less than that needed for a linear system, minimizing the overall damping force required to maintain the highest response rate | (1) Filled types of fluid affects workability of FVDs<br>(2) Fluid speed varies with change in flow characteristics, i.e., depends on the piston's shape<br>(3) When $\acute{\alpha} \neq 1$, then FVD is not behaving as a liner<br>(4) Linear FVDs are disadvantageous because they do not achieve a further reduction in the system response essentially |

(continued)

**Table 2** (continued)

| Sr. No | Types of damper | Capabilities | Limitations |
|---|---|---|---|
| 3 | Visco-Elastic Damper (VED) [16] | (1) Passive control systems are viscoelastic dampers (VEDs), which can be integrated with relative ease in construction systems compared with other passive monitoring devices<br>(2) VEDs are healthy substitutes in the renovation of old or demolished buildings to base insulation<br>(3) It was appealing for steel frame structures<br>(4) VEDs are used as suitable damping equipment for building buildings against dynamic charges such as earthquakes and wind charges due to their cost efficiency, high reliability, and sufficient energy discharges<br>(5) VED is widely used for aerospace and military operations also [12] | (1) VED's results vary with changes in viscous material such as co-polymer and glassy substances<br>(2) VED's should be investigated under 2D and 3D excitation for natural earthquakes, as they may vary results for displacement, acceleration, and support reactions of the piping system<br>(3) VEDs are sensitive to effectiveness error in the natural period of real EQ and time history data |
| 4 | Tuned Mass Damper (TMD) [7] | (1) Easy computational implementation<br>(2) Involves no complicated mechanism [17]<br>(3) Energy dissipation occurs by the structural motion [25]<br>(4) Successfully implemented all over the world for buildings, bridges, other Civil, and Mechanical systems | (1) Absorber mass takes up vibratory energy, leaving the central mass (building) almost static<br>(2) Not very useful for earthquake excitations, which occur over the wide frequency range |

(continued)

an idea on technical vs economic study with consideration of the piping system's stability, strength, and structural stability.

**Table 2** (continued)

| Sr. No | Types of damper | Capabilities | Limitations |
|---|---|---|---|
| 5 | Multiple-Tuned Mass Damper (MTMD) [7] | (1) MTMDs improve the response of structures during earthquakes [22]<br>(2) These dampers with minimizing mass would offer better control during the wide range of frequency and eliminate the waste space during the installation of MTMDs<br>(3) MTMD has created the uniform distribution of masses for economic reason<br>(4) MTMDs are simple, more economical, and more authentic tactics for structural vibration control | (1) It is prone to error efficiency in the system's standard frequency and damping ratio<br>(2) MTMDs damps response of the system over a particular bandwidth of frequencies |

# 2 Literature Review

Structural management systems improve structures' power losses during shaking by transforming motorized energy into thermal energy. The following are various types of mechanisms for energy dissipation:

## *2.1 Passive Response Control Systems*

Kumar et al. [16] commented that due to the increasing number of destructive earthquakes, research and seismic response control systems' production has taken on primary importance [21]. Passive control systems are now available worldwide, and thus accurate, useful, and economical devices and component modelling continue to be established in this domain for research. This document starts with the definition of static, active, and semi-active control systems and their contrast. In comparison, passive control mechanisms have benefits over others. A literature review of passive devices covers the devices' recorded history, complex nature, research, and study of these instruments integrated into the structural models. These systems' benefits and drawbacks are often addressed in several structures when retrofitting structures and first and recent implementations. The passive reaction control systems mentioned contain viscoelastic dampers, damping, viscous damping, friction dampers, modified mass dampers, tuned liquid dampers, tuned column dampers, super elastic

**Table 3** Comparative performance of dynamic vibrational control passive devices

| Damper | Criteria | | | |
|---|---|---|---|---|
| | Seismic and vibrational response elimination effect | Displacement reaction of structure | Energy acceleration displacement by the structure | Support Reactions |
| Tuned Mass Damper (TMD) [17] | (1) Impact sensitivity to a structure error or one in the TMD damping ratio<br>(2) XPD is made up of Aluminum or steel, effectively minimizing the building's seismic response [15]<br>(3) XPD is also more effective against blast load and seismic load. It decreases structural response, i.e., maximum displacement and story displacement [12] | (1) TMDs are less effective than MTMDs for reducing displacement and accelerating the complex piping system | (1) As compared to MTMDs, having less capability to energy acceleration displacement by structure | (1) TMDs are less effective than MTMDs for reducing the support reaction of the complex piping system |
| Multiple-Tuned Mass Damper (MTMD) [14] | (1) The use of MTMD for different characteristics has been proposed to improve the seismic response elimination of piping systems | (1) Displacement of a node of the piping system up to 29% in X and Z direction<br>(2) MTMDs are more helpful in reducing displacement, acceleration of the complex piping system | (1) Significantly more Energy acceleration displacement by the structure as compared to a single TMD | (1) MTMDs are more advantageous than TMDs for reducing the support reaction of the complex piping system |

(continued)

**Table 3** (continued)

| Damper | Criteria | | | |
|---|---|---|---|---|
| | Seismic and vibrational response elimination effect | Displacement reaction of structure | Energy acceleration displacement by the structure | Support Reactions |
| X-Plate Dampers (XPD) [14] | (1) Acceleration response reduction up to 44%<br>(2) XPD decreases the seismic response of the pipe structure more efficiently<br>(3) Vertical damper is more effective than a horizontal damper | (1) XPD effectively reduces displacement of inelastic behavior of the chosen structure and dissipates the i/p seismic energy of the structure [13]<br>(2) More adaptive to a and b with a lower answer observed at a higher value of b<br>(3) The pipe system's response has decreased to = 40 mm and t = 3.5 mm<br>(4) The reaction reduces as the XPD's thickness increases | (1) In the operated piping device, the maximum energy is dissipated by XPDs<br>(2) In contrast to Horizontal XPDs, higher Energy is consumed by vertical XPDs<br>(3) The Energy of XPD distributed by the piping structures will depend on the XPD's thickness<br>(4) The Energy of XPD distributed through the piping device is dependent on the feedback of the ground motion data<br>(5) For XPDs with a lower value of one (i.e., half the height of XPDs and a higher value of b), a proportion of Energy dissipated by XPD is higher in a regulated piping device | (1) Significant reduction of support reaction up to 50% of without damper |

(continued)

**Table 3** (continued)

| Damper | Criteria | | | |
|---|---|---|---|---|
| | Seismic and vibrational response elimination effect | Displacement reaction of structure | Energy acceleration displacement by the structure | Support Reactions |
| Elasto-Plastic Damper (EPD) [10] | (1) Response reduction is about 40%<br>(2) EPD used to decrease the seismic response of the piping system [9]<br>(3) It is an improved and cost-effective substitute for costly snubbers [9] | (1) Displacement at the horizontal damper's damping location is about $\frac{1}{3}$ rd—the displacement at the damper location of the vertical damper | (1) Significant amounts of Energy consumed by longitudinal damping deformation on the X-Plate | (1) Multiple Elasto-plastic dampers can be effectively applied for reducing the response of support reaction of the complex piping system subjected to 3-D EQ excitation |
| Fluid Viscous Damper (FVD) [14] | (1) Initial value of the damping coefficient (cd/cp) up to 5 rates of response reduction significantly. While the lower rate of reduction for further increases in the value of (cd/cp) | (1) Reduced from 0 to 20 in various time histories to improve the damping coefficient ratio (cd/cp) | (1) Important decrease in acceleration, moving, and reaction help for an FVD piping structure<br>(2) Hysteresis loop shows Energy absorbed by the damper was good [24] | |

*Note* Where a = breadth of dampers and b = width of dampers, (cd/cp) = damping coefficient dampers, such as memory metal dampers, and insulators. Active variable stiffness (AVS) systems using stand-by bracings, locking, and unlocking devices effectively minimize inter-storey drifts [4].

## 2.2 Passive Devices

Kumar et al. [14] suggest X-plate damper, viscous damper, viscoelastic damper, tuned mass damper, and a variety of tuned mass dampers among many passive control devices, and they are used for minimizing the 3-D pipe seismic reaction. Detailed experiments in this paper investigate the efficacy of dampers in 3-D piping structures exposed to the high intensity of artificial earthquakes. Compared with the similar scientific findings available on the Wen model, the scientific results indicated that this was in line with the X-plate's planned damping study. There is a significant reduction in the seismic interest response, such as relative propagation, acceleration, and assisting passive instrument pipes' responses. In general, passive devices are wildly successful and can be used for seismic response elimination, vibratory control, and the seismic reclassification of piping structures under some optimum criteria, including steepness damping.

## 2.3 Passive and Semi-active Devices

Kumar et al. [16] in their article aim to minimize seismic response and vibration regulation of piping systems in operation, fossil fuel industry, and passive and semi-active supplementary instruments. This article discusses an analysis of passive or semi-active additional systems' efficiency because of improvements in equipment parameters, various damper control algorithms, and the resulting optimal unit parameters. These instruments' efficacy in minimizing the piping system's reactions is tested by contrasting unregulated reactions with growing amplitudes under four distinct artificial earthquake motions. The findings revealed that the seismic response reduction, vibrational regulation, and seismic requalification of the piping structures with specific optimal parameters are highly successful and are feasible.

## 2.4 Tuned Mass Dampers (TMDs)

Rahimi et al. [17] studied various vibration control devices. Because of the growing demand for buildings and high-rise buildings, it seems more necessary than ever to manage the structural vibrations under earthquake and other external dynamic forces. The vibration control instruments can be divided into passive control schemes, active, and hybrid. Popular technology for managing vibrations, minimizing damage, and

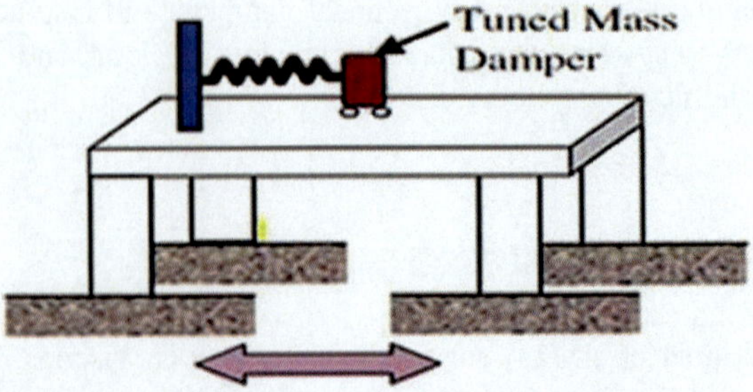


**Fig. 1** Tuned mass damper and direction of displacement [11]

enhancing structural stability usually involves damping, vibration insulation, and management of excitation forces and vibration absorbers, but not limited to them. Due to their very basic concepts and relatively short performance optimization, tuned mass dampers (TMDs) have become a common means for protecting structures from unexpected vibrations, as demonstrated in many successful recent applications. This study provides fundamental analysis and compares the performance and comparative advantages and drawbacks of active, passive, foundational, and hybrid TMD control systems to protect buildings against earthquake or wind strength.

Notwithstanding the significance and recent development in this area, previous reports concentrated mainly on passive or active TMDs. Thus, this analysis examines the historical context of all forms of TMDs and the structural, methodological, functional, and economic variations in their systemic management implementation. Furthermore, a variety of knowledge limitations in the current studies within the field of study are identified and stressed. Between these research boundaries, we find that existing practices need to be strengthened to assess TMDs principal natural occurrence. Moreover, more sophisticated analytical methods for TMD and structures that take their non-linear behavior into account are increasingly needed, as this may significantly increase the forecasting of structural reaction and, in turn, maximize TMD. Figure 1 shows an actual configuration of TMD.

## 2.5 X-Plate Damper (XPDs)

Reddy et al. [8] studied the application of various dampers and discussed the elastic–plastic pressure of additional metal components with plastic deformation properties as the most efficient method for dissipating energy in a vibrant environment. An X-plate damper (XPD) is a system able to tolerate several periods of steady outcomes in high energy dissipation or damping. The paper focuses on numerical studies investigating the seismic effectiveness of an XPD on tubing, especially for industrial plants or facilities such as atomic and nuclear power plants, such as the chemical and petrochemical industry. In subjective floor motions, the pipes' seismic activity is measured by significant parametric changes of the damper properties (e.g., height, width, and thread of XPD). Research is reported on an XPD-equipped industrial piping system,

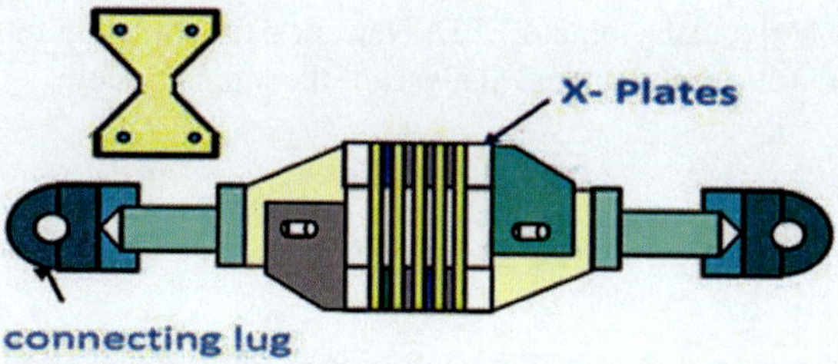


**Fig. 2** Schematic representation of XPD [9]

and the respondent interest quantities include relative displacements, absolute accelerations, and piping system supporting responses. To calculate the XPD's earthquake performance, the managed device's sensitive volumes (XPD) are compared with the corresponding unregulated (XPD) pipe networks. The earthquake discharge in the piping system represented by the hysterical energy of the XPD is also measured and equal. The seismographic responses of piping systems can be studied very effectively by XPDs. Also, parametrically changing XPD characteristics and manipulating pipe system responses are challenging to achieve the desired XPD properties for a given piping system and ground movements. Hysteretic energy dissipation is then recommended to be used by an XPD to achieve the optimal XPD characteristics. The impact of the XPD features on the piping system's free vibration characteristics, which are critical for the design of piping systems using XPD, is also presented.

As shown in Fig. 2, XPD combines single or multiple X-Plate combinations [3]. A review of the XPD-type research revealed that XPDs are mainly used in vibration management science and typically found on the structures' top floor. XPDs are especially strong for lightweight systems under wind excitement. Regarding XPDs' performance during the earthquake powers, it is weaker because not all modes of fundamental structural frequencies have been taken into account, and it is only tuned to the dominant frequency and is prone to uncertainty.

## 2.6 *Elasto-Plastic Damper (EPD) and Form Memory Alloy Damper (SMAD)*

Parulekar et al. [10] observed snubbers in piping systems of the nuclear power plants as they are costly, complicated, and need regular maintenance. Efficient, simple, economical, and trustworthy passive dampers like Elasto-Plastic Damper (EPD) and Form Memory Alloy Damper (SMAD) are inexpensive. They can be used to replace snubbers in piping systems because they preserve pipe stability and improve damping. A complicated piping model having internal pressure with and without two EPDs on the shake table is evaluated in this paper to measure the efficacy of multiple elastoplastic dampers in practical piping structures with 3-D earthquake excitation. Tests are often carried out with pipings only with a vertically attached EPD and only a horizontally bound EPD. Newark's time integration method performs linear and non-linear dynamical analysis of the piping system.

## 2.7 Viscoelastic Damper

Umachagi et al. [1] studied friction dampers as a robust friction system in energy dissipation. VED consists of a viscoelastic layer bonded with a steel plate with viscous material, i.e., Co-Polymers or Glassy Substances [1]. The simple steel structure behaves partially as a friction damper under seismic pressure. Experimental and computer findings indicate that the friction damper will improve the vibrational reaction to conventional architecture in innovative designs and the effectiveness of existing buildings when subjected to ground movement and friction to investigate the piping process. Dampers are effective based on their degree of excitation and environmental temperature. With the elevated temperature, energy dissipation reduces proportionally. Also, the authors correctly estimated the seismic response of structures using the stress-energy approach at different temperatures. Figures 3 and 4 show an assembly model of VED.

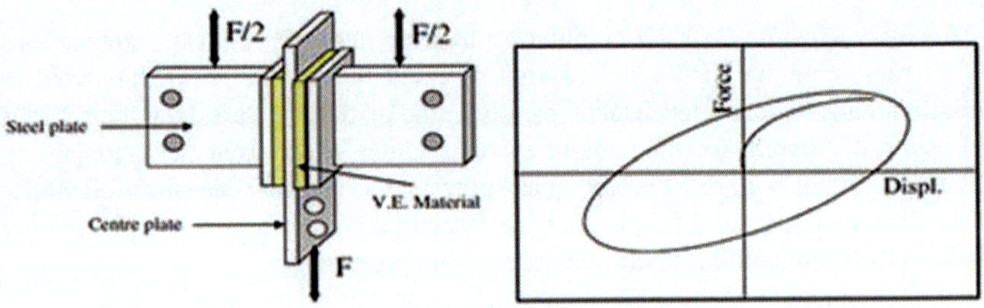


**Fig. 3** Assembly model of VED and hysteric loop graph [11]

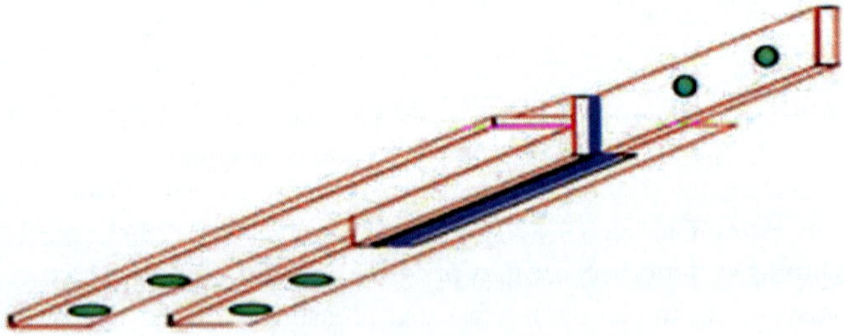

**Fig. 4** Assembly model of viscoelastic dampers [13]

# 3 Practical Applications of Passive, Active, and Semi-active Dampers

## 3.1 Practical Applications of Passive Control Devices

(1) The lead-rubber bearing is the choice in Japan, the USA, and New Zealand, while vicious and steel hysteretic dampers are used in Italy.
(2) In the USA, the application has been mostly retrofitted to current structures with a seismic exclusion system, including Oakland's City Halls (LRB), Los Angeles (HDRB), San Francisco (LRB), San Francisco US Court of Appeals (Friction Pendulum).
(3) Two major earthquake isolation projects have been completed in New Zealand. This consisted of the restoration of the NZ Parliament building's separation, the corresponding Assembly Library, and the current NZ Museum. In this system, LRB is used.
(4) The US Club of Teaching Hospital in Los Angeles (LRB) in 1994, the Ministry of Post and Teleggy's data center (LRB) in 1995, and the Matsumura Science Institute building in Kobe in 1995 were the extraordinary success of three isolated buildings during earthquakes. These three buildings demonstrate the tremendous benefits of seismic isolation in actual earthquakes that these systems will continue to function without any disruption during and directly after an earthquake.

## 3.2 Practical Applications of Active Control Devices

(1) AMD at Chuo-Ku, Tokyo's 4 m frontages, with a total height of 33 m, is located inside Kyobashi Seiwa house. The 11-story structure consists of rigid steel panels. The upper floor is spaced apart with two AMDs. The first AMD is 4-tonne weight, and the second has 1-tonne weight. Two AMD's were also intended to monitor the structure's torsional reaction. A hydraulic pump of 22 kW powers the masses. On the 6th floor and the 11th floor, the controls are located in the basement. On the first level, the machine is supplied. It is the first AMD on the house in the world.
(2) Deux is situated in Chiyoda-Ku, Tokyo, on ANDO Nighikicho. It is constructed of solid steel frameworks and has 14 floors and two floors of the basement. The weight of the construction is 2600 t-f above the ground. Two-way AMD is positioned at the top of a TMD on the top floor of the building. A petroleum damping system damps time. The TMD is 18 t-f, and the AMD's have two t-f each. The Deux method uses the idea that the TMD has the minimum power, at least if the active control system fails.
(3) Shinjuku Tower's Trigon is situated in the 264.100 $m^2$ floor area of Tokyo's Shinjuku-Ku Tower. The construction is made of steel and concrete frameworks

partly strengthened. There is a building of 52 floors plus five basements with an overground weight of 130,000 t-f. The roller pendula weight on the 36$^{th}$ floor is mounted on three control systems. Every control weight is 110 t. The pendulum has an overall length of 110 cm. The modified time is 3,7258 s, with a pendulum engine power of 75 kW.

### *3.3 Practical Applications of Semi-active Control Devices*

(1) The first semi-active controlled 5-story Kajima Shizuoka Building was constructed in Shizuoka, Japan, in 1998. Each damper can develop a maximum damping force of 1000 KN.
(2) In Osaka, Japan, a 27-story Luxa Osaka building was constructed in 1999 and employed semi-active TMD.
(3) An 11-story building, CEPCO Gifu constructed in Gifu, Japan in 2000, used semi-active dampers.

## 4 Conclusions

Every passive device has been used for different piping system geometry, support condition, and EQ forces. The permanent consideration for analysis depends on the minimum optimal response in support, displacement, and accelerations. The minimum optimal parameter will be, to some extent, varying with types of industrial problems. Hence, for minimum optimal solutions, every type of analysis is required to be done. It has been observed that:

(i) Passive dampers effectively decrease the bending moment, shear force, and effectively increase the structure's axial force [13]
(ii) Viscous, Viscoelastic, and X-plate dampers can reduce the seismic responses of the piping structure more effectively [26]
(iii) MTMD and TMD frequency changes with the fundamental model frequency of the piping system [7]
(iv) MTMD and TMD are less powerful to minimize the seismic reaction of the piping system than other passive instruments [7]
(v) Hence, the piping system is made per industrial demand to meet seismic and vibrational control of the complex piping system. For that purpose, X-Plate damper, VDs, and VEDs are more preferable.
(vi) VDs are not affected by natural frequency, but on the other hand, VEDs are affected by the same.

This article has observed that active TMD performance is significantly better than passive TMDs in reducing the structural response, particularly in storage displacement. Even when the dominant excitation frequency is similar to the framework's fundamental frequency, the active TMDs showed good quality. Active TMDs for human relaxation modes such as hear and torsional modes should also be considered. All these recommendations endorse active TMDs in contrast to passive TMDs as the optimal alternative. However, it should be noted that while AMDs are comparatively cheaper, they have higher running costs, mainly as a result of the energy source needed for their activation [25].

The appropriate design and implementation of MTMDs for adequate damping, and the required mass ratio for torsional and shear modes in structures are vital topics frequently overlooked in the literature. MTMDs value is that the optimum parameters can be better obtained due to their spatial distribution, but it was evident that experimental evidence on their optimal architecture and operation is minimal [7].

## References


1. Umachagi V, Venkataramana K, Reddy GR, Verma R (2013) Application of dampers for vibrational control of structures: an overview. Int J Res Eng Technol (IJRET), pp 6–11
2. Praveen Kumar RS, Jangid, Reddy GR (2013) Response of piping system with semi-active variable stiffness damper under tri-directional seismic excitation. Nuclear Eng Design, pp 130–143
3. Nirmala G, Manchalwar A, Machalwar S (2019) Vibrational control of water tank staging equipped X-Plate damper. IJRTE, pp 5551–5554
4. Yang JN, Wu JC, Li Z (2019) Control of seismic-excited buildings using active variable stiffness systems. Eng Struct 18(8):589–596
5. Omprakash Reddy N, Machalwar A (2019) Performance of moment resisting RC building equipped with X-plate damper under seismic and blast loading. Int J Innov Technol Explor Eng (IJITEE), pp 2758–2762
6. Ravikan A, Dubey PN, Agrawal MK, Reddy GR, Vaze KK (2013) Evaluation of inelastic seismic response of piping system using a modified iterative response spectrum method. J Pressure Vessel Technol ASME, pp 1–8
7. Rahimi F, Aghayari R, Samali B (2020) Application of tuned mass dampers for structural vibration control: a state-of-the-art review. Civil Eng J 6(8):1622–1651
8. Ravi Kiran A, Reddy GR, Agrawal MK, Raj M, Sajish SD Ratcheting based seismic performance assessment of a pressurized piping system: experiments & analysis. Int J Pressure Vessel Piping, pp 1–9
9. Dubey PN, Gupta C, Reddy GR, Vaze KK, Ghosh AK (2009) Performance of X-plate elasto—plastic dampers, a passive seismic support for nuclear piping under cyclic loading. In: 20th international conference on structural mechanics in reactor technology (SMiRT 20), pp 1–10
10. Parulekar YM, Reddy GR, Vaze KK, Ghosh AK, Kushwaha HS, Babu R (2017) Seismic response control of complex piping systems using elasto-plastic dampers- experiments and analysis. Conference paper, pp 1–8
11. Parulekar YM, Reddy GR (2009) Passive response control systems for seismic response reduction: a state-of-the-art review. Int J Struct Stab Dyn pp 151–177
12. Nishant Kishore R, Reddy GR, Ramanujam S, Venkatraj V, Agrawal P (2009) Seismic response control system for structures. Defen Sci J 59(3):239–251. DESIDOC
13. Nirmala G, Manchalwar A (2020) Behavior of the water tank staging with aluminium & X-Plate damper. E3S Web of conference 184, 01078 (2020) ICMED 2020, pp 1–8
14. Kumar P, Jangid RS, Reddy GR (2015) Comparative performance of passive devices for piping system under seismic excitation. Nuclear Eng Design 121–134

15. Manchalwar A, Bakare SV (2016) Performance of RC structure equipped with steel & aluminum X-Pate damper. J Inst Eng Ser A, pp 1–10
16. Kumar P, Jangid RS, Reddy GR (2013) Seismic response of piping system with passive & semi-active supplemental device under 3-D seismic excitation. BARC Newsletter, Res Article 24–31
17. Rahimi F, Reza Aghayari B, Samali B (2020) Application of tuned mass dampers for structural vibration control: a state-of-the-art review, pp 1622–1651
18. Kobori T, Takahashi M, Nasu T, Ogasawara K (1993) Seismic response controlled structure with active variable stiffness system. Earthq Eng Struct Dyn 925–941
19. Nagarajaiah S, Sahasrabudhe S (2006) Seismic response control of smart sliding isolated buildings using variable stiffness systems: an experimental and numerical study. Earthq Eng Struct Dyn 35(2):177–197
20. Kumar P, Jangid RS, Reddy GR (2012) Response of piping system with semi-active magnetorheological damper under tri-directional seismic excitation. Int J Appl Sci Eng 99–111
21. Patil SJ, Reddy GR (2012) State of art review -base isolation systems for structures. Int J Emerg Technol Adv Eng 2(7):438–453
22. Bakre SV, Jangid RS (2004) Optimum multiple tuned mass dampers for base-excited damped main system. J Struct Stab Dyn 4:527–542
23. Bakare SV, Jangid RS, Reddy GR (2007) Response of piping system on frictional support to bi-directional excitation. J Nuclear Eng Design 124–136
24. Symans MD, Constantinou MC (1998) Passive fluid viscous damping systems for seismic energy dissipation. ISET J Earthq Technol 35(4):185–206
25. Symans MD, Constantinou MC (1999) Semi-active control system for seismic protection of structures a state-of-the art review. Eng Struct 21:469–487
26. Parulekar YM, Bakre GR, Jangid RS, Reddy GR (2006) Optimum X-plate dampers for seismic response control of piping systems. Int J Pressure Vessel Piping 672–685

# Response Reduction of Secondary Piping Systems in Base-Isolated Buildings

**Vishal Kamble, Shiv Dayal Bharti, Mahendra Kumar Shrimali, and Kedar Kumbhojkar**

## 1 Introduction

The members used in load-bearing and frame structures for functioning the precarious operations, but are not a member of the primary structure are generally known as secondary systems (SS). There are many secondary systems in several specific facilities, such as nuclear power plants, hospitals, data centers, and telecommunication buildings. Secondary systems play a crucial role in the performance of these special buildings. The safety of the SS is therefore as essential as the structure themselves during an earthquake. The SS is affected by the floor motion when fitted to the primary structure (PS) and interacts with the primary structure. It has been noted that whenever the PS is subjected to ground vibration, the SS could fail due to the tuning or almost tuning of the natural frequencies of the PS and SS, even in small-scale earthquakes. There are many instances where the SS has suffered major damage, while the PS itself has suffered an earthquake attack. For the seismic protection of the SS, one of the approaches that are now commonly used is base isolation. Buildings with multiple SS are generally base isolated.

Base isolation decreases the acceleration and other time histories of the SS due to the reduction in the interstorey drift and floor acceleration of the PS. In this technique, the planning, using the base isolator, is used to decouple the superstructure from the substructure just so the horizontal component of the earthquake ground motion is less transferred to the structure. The isolation technique is co-ordinated with both energy absorption capacity and flexibility. The fundamental frequency of the structure is decreased due to its flexibility in the horizontal direction, which decreases the loading on the structure due to the earthquake. The use of the building isolation technique

V. Kamble (✉) · S. D. Bharti · M. K. Shrimali · K. Kumbhojkar
National Centre for Disaster Mitigation and Management, Malaviya National Institute of Technology, Jaipur 302017, India



S. Kolathayar and S. C. Chian (eds.), *Recent Advances in Earthquake Engineering*, Lecture Notes in Civil Engineering 175,
https://doi.org/10.1007/978-981-16-4617-1_26

relies on shifting the building's fundamental frequency out of a prevailing region of ground movement frequency and reducing the responses of both SS and PS.

In modern industrial structures, piping systems are the most common non-structural arrangement amongst the major secondary systems. For the smooth running of the industrial structure, its operation is of the greatest priority. Strict criteria are implemented on the seismic efficiency of the piping system at the nuclear power plant. Normally, a piping system is connected to various locations of the primary structure. Throughout a seismic event, the primary structure causes the attachment points to move non uniformly developing high stresses in the pipeline.

One of the earlier research on securing the SS against the earthquake was that by Kelly and Tsai. [10] evaluated the response of the light equipment mounted on the base-isolated structure. They concluded that the base isolation technique with a lead plug significantly reduces the seismic response of the light internal equipment. Tsai and Kelly [15] did a dynamic analysis of internal equipment attached inside a base-isolated building. According to the study, time history analysis using the classical mode method gives less response compared to the analysis using the complex mode method. [9] calculated the stochastic response of auxiliary systems attached to the base-isolated primary structure. The analysis found that the auxiliary system by the Teflon-coated sliding bearings was adequate in minimizing the response. [4] presented the findings evaluated by involving several isolation systems for the seismic response analysis of secondary systems attached in base-isolated structures considering the equipment-structure interaction. Among the various base isolation systems, it has been observed that, in most practical situations, the linear LRB method provides the lowest peak response for secondary systems.

Juhn and Manolis [5] studied the intensity of stochastic root mean square (r. m. s.) spectra of secondary systems to calculate the response of a SS connected to a six-storied building. The results show that uncertain system parameters affect the time history of the secondary system. [13] analyzed the impact of the isolation technique attached at the attachment point of the secondary structure. Results show that the time history of the response of support-isolated secondary system with fixed base primary structure is less. [8] carried out an experiment on the shaking table to determine the response of the secondary system connected to the sliding base-isolated primary structure. It was observed from experimental results that the base isolation technique minimized the time history of the secondary system's response. Agrawal [1] studied the seismic response of equipment fitted over a sliding base torsionally coupled primary structure for the bidirectional earthquake. From the analysis, it was found that the response of the equipment fixed over the sliding primary structure was less than the equipment fixed over the primary system with no slip. Khechfe et al. [11] did an experimental analysis of the base-isolated secondary system. For the isolation, laminated rubber isolation system was used. The results of the study conclude that the time history of the response of the secondary system was minimized by the appropriate design of the laminated rubber isolator. [3] explain the various types of base isolation techniques that are used for minimizing the response of the light secondary system. It was shown from experimental results that different types of isolation techniques reduced the time history of the light secondary system response in a broad frequency region.

Contento and Di Egidio [2] examined a multi-storeyed building on which a rigid body was mounted for various PGA levels. The efficiency of the base isolation in avoiding failure during an earthquake was studied. The study findings showed that base isolation minimized the response of the rigid body. Jeon et al. [7] studied the seismic fragility analysis of a piping system attached at the base-isolated nuclear power plant. NRC-BNL model was adopted for the seismic analysis, which involved the isolation device. For the piping system, a quasi-static analysis was conducted.

The literature indicated that the seismic behavior of a piping system installed on a 3D base-isolated building subjected to a bidirectional earthquake has not been reported in much work. In the case of a bidirectional earthquake, connection points of the piping system to the primary structure are more disapprovingly stressed than the unidirectional earthquake. Besides, the type and nature of the earthquake also greatly affect the response of the secondary system considerably. In the present work, a piping system attached along one side of a six-storey building which is base isolated is taken for the non-linear time history analysis under bidirectional earthquake. The different parameters of the type of earthquake (near-field and far-field) and PGA level (0.2, 0.3, 0.4, 0.5, 0.6 g) of the earthquake are considered for analysis. The ratio between the bidirectional ground motion ($\lambda$) is taken as 1:3/4. The main aim of the study is to examine the seismic response of the secondary piping system and the amount of response control obtained due to the base isolation of the structure. The response quantities of pipe involve absolute accelerations, relative displacements, stresses developed at the attachment points of the piping system, drifts, and floor response spectrums of the building.

## 2 Modelling of the Isolators

A circular cross-section of laminated rubber bearing (LRB) is used as isolators. With the help of the widely used Wen's model, the hysteretic behavior of the isolator is modeled where a bilinear curve defines the force–deformation hysteretic backbone curve as shown in Fig. 1. Three significant parameters are defined in the model,

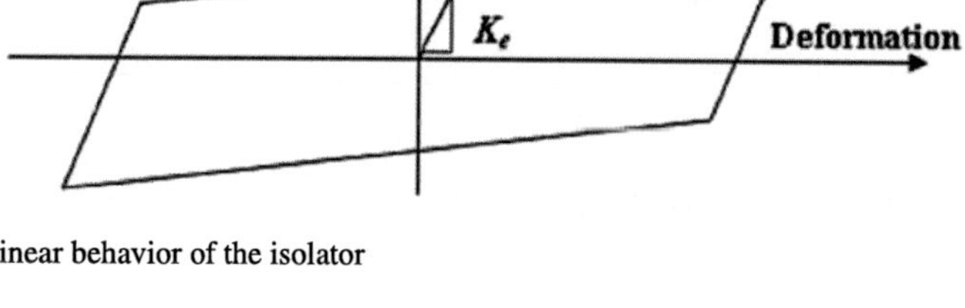


**Fig. 1** Bilinear behavior of the isolator

namely, post-yield stiffness ratio ($k_2$ or $k_d$), the characteristic strength (Qd), yield strength (Fy). The yield strength of the lead in shear ($f_{py}$) provides a characteristic strength of the LRB. The sectional area of the lead plug Ap is calculated by the following equation

$$Q_d = f_{py} \times A_p \tag{1}$$

For a particular design displacement (D) and post-yield stiffness (k2), the effective stiffness (Keff) of the LRB is calculated by.

For the design of the isolator, the $K_{eff}$ given by Eq. 2 is used. Note that the hysteretic property of circular isolators is the same in all directions. As a result, the impact of bidirectional interaction on the yielding of the isolators occurring because of the two-component earthquake does not alter the hysteretic property of the isolator.

## 3 Modelling of the (P-S) System

For non-linear time history analysis, a 3D modeling of the P-S system is carried out. To examine the efficiency of the base isolator in minimizing the responses of the secondary system, both fixed base and base-isolated buildings are taken into account in the study. Beams and columns are modeled as linear elements that contain six degrees of freedom at both ends. The slab is modeled as a shell area element. The pipe is also modeled as the shell area element. Seismic design of building structures is done as per IS 1893(2016) in a high seismic zone, having a response reduction factor (R) as 5, importance factor (I) as 1, zone factor (Z) as 0.36. In choosing the response spectrum, medium soil condition is taken into account. The non-linearity at the yield member is presented by assigning plastic hinges at a distance of 0.1L from the two ends of the column and for beams. The hysteretic property of the plastic hinge is allocated according to ASCE 41-13, which considers bending moment (M3) hinges for beams and provides P-M2-M3 hinges for columns that take into account the interaction of the axial force and bending moments into consideration. In the material property section, the yield of the pipe is considered by assigning a yield stress value of 450 N/mm$^2$.

## 4 Analysis Using SAP 2000

Building analysis for bidirectional earthquake ground motion shall be carried out using a 3D primary and secondary system model as described above. By considering the full interaction effect between the primary and secondary structures, the analysis is carried out. With SAP 2000, a non-linear time history analysis (NTHA) is performed. It implements the Hibler Hughes Taylor integration scheme by considering the P delta effect, taking the value of $\beta = 0.25$ and $\gamma = 0.5$. Rayleigh damping is evaluated for 5% damping, corresponding to the 1st and 2nd modes of the structural vibrations.

## 5 Numerical Example

Two six-storeyed buildings, one is a fixed base and the other is a base-isolated, are taken as the primary structures for comparison to study the use of the base isolation method to reduce the seismic demand of the piping systems carrying fluid. IS 456 is used for designing both buildings. The plan of the building is rectangular with two bays in both principle directions, having a center-to-center distance of 6 m; height between two floors is 3.00 m as shown in Fig. 5. Beam is 0.5 m × 0.25 m in cross-section, while the plinth beam size is 0.6 m × 0.3 m, column sizes are taken as 0.7 m × 0.7 m. The depth taken for the slab is 0.18 m. Also, 9kN/m self-weight is imposed as a dead load (due to a wall, etc.), 2 kN/m is imposed on the beam as the live load. M20 concrete and HYSD 500 steel are used. The fluid-carrying pipe is attached vertically to the center column on one of the outer faces of the building, connected from the 1st floor to the 6th floor as a secondary system as shown in Fig. 5. The pipe's external diameter is 166 mm with a 5 mm thickness. For connecting the pipe with the primary structure, a zero link element is used. As per I.S.1239 (Part–I)-2004, the material of the pipe is Grade A carbon steel with a unit weight of 7.85 gm/cm$^3$. The fluid density is taken as 1.2 gm/cc and encompassed in the mass analysis of the pipe. By using the SAP 2000, an analysis is done with the 3D model of the primary and secondary system for the bidirectional earthquake ground motions.

Based on time period separation, the isolators are designed. For the design of isolators, the time period shall be considered three times of the fixed base time period of the building. Three kinds of isolators are designed for vertical load coming from the columns. Thus, the isolators (LRB-1, LRB-2, LRB-3) as shown in Fig. 2 are designed for vertical loads of 1150, 1820, 2850 KN. Isolator design details are shown in Table 1. For the study, three types of the earthquake are used, namely, the near-field earthquakes with directivity (NFD) and fling step effects (NFF), far-field earthquake (FF). Details of ground motion are given in Table 2. The record of ground motion is collected from the PEER Strong Motion Database of Barkley University [https://ngawest2.berkeley.edu]. Based on Joyner Boore epicenter distance, classification of the near-field and far-field earthquakes is made. Near field earthquakes are typically believed to be classified on the basis of less than 15 km of epicenter distance. In the present analysis, the earthquake classification as provided in Table 2 is based on the above assumption. The plot of the time histories of the ground motion of superstition hill (Near-field with forwarding directivity) in both directions are shown in Figs. 3 and 4. Each earthquake is scaled to 0.2, 0.3, 0.4, 0.5, and 0.6 g.

Mass participation factors, corresponding frequencies, and the first two mode shapes of the fixed base and base-isolated buildings are shown in Fig. 5. The fundamental frequency in the U1 direction is shown to be relatively minimum compared to the U2 direction. Further, mode shapes in both U1 and U2 directions are pure

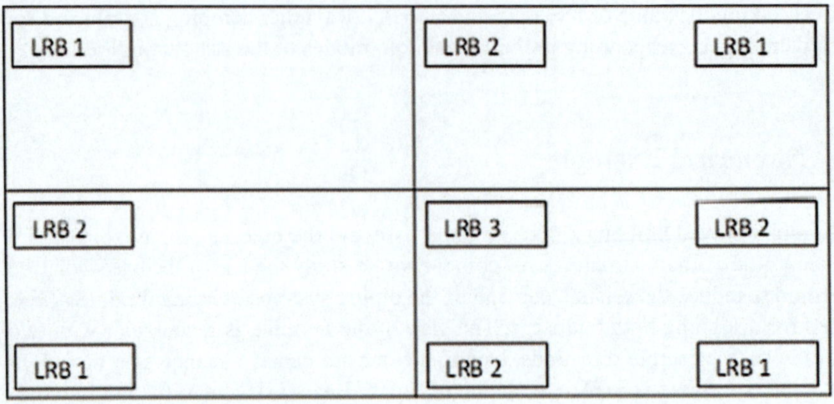


**Fig. 2** (Plan of base-isolated building)

**Table 1** Properties of LRB

| LRB | $K_{eff}$ (kN/m) | $K_1$ (kN/m) | $F_y$ (kN) | Post yield stiffness ratio | Design Disp (m) |
|---|---|---|---|---|---|
| LRB-1 | 740 | 5866 | 41 | 0.1 | 0.213 |
| LRB-2 | 1170 | 9284 | 65 | 0.1 | 0.213 |
| LRB-3 | 1835 | 14,540 | 102 | 0.1 | 0.213 |

translational modes in the respective directions. The fundamental natural frequencies of the pipe on the inflexible supports are maximum as associated with the natural frequencies of the building. As a result, the secondary method is generally dictated by the dynamics of the primary system.

## 6 Results and Discussions

Figures 6 and 7 indicate the percentage reduction of the relative displacements, absolute accelerations, stresses at the connection points, percentage reductions in storey drifts, and their variations along with the height of the building.

Figure 7 indicates the comparisons between the floor response spectrums of the isolated and fixed base buildings. Subsequently, bidirectional earthquake is used, the percentage reductions of time histories are shown for both U1 and U2 directions. Two significant parameters, namely, the type of earthquakes and PGA of earthquakes are varied. The percentage reduction of time histories is shown with respect to the variation of these parameters.

Figure 6a–c shows the variations of percentage reduction in peak relative displacements (U1 and U2), peak absolute accelerations (U1 and U2), and peak stresses at the

**Table 2** Detail records of ground motion

| S.No | Year | Earthquake | $M_w$ | Station | Component | PGA (g) | PGV (cm/s) | PGD (cm) | $R_{jb}$ Km |
|---|---|---|---|---|---|---|---|---|---|
| Near Field Record (forward directivity effect) (NFD) | | | | | | | | | |
| 1 | 1987 | Superstition Hill | 6.5 | parachute test site_ | 315 | 0.31 | 58.86 | 44.06 | 0.95 |
| | | | | | 270 | 0.36 | 55.66 | 24.95 | |
| Far-Field Record (FF) | | | | | | | | | |
| 2 | 1994 | Northridge | 6.7 | Beverly hills | MULH009 | 0.42 | 58.91 | 13.18 | 37.47 |
| | | | | | MULH279 | 0.48 | 66.71 | 12.17 | |
| Near Field Record (Fling step effect) (NFF) | | | | | | | | | |
| 3 | 1999 | Chi-Chi | 7.6 | TCU 072 | E | 0.46 | 83.6 | 209.7 | 0 |
| | | | | | N | 0.37 | 52.52 | 43.31 | |

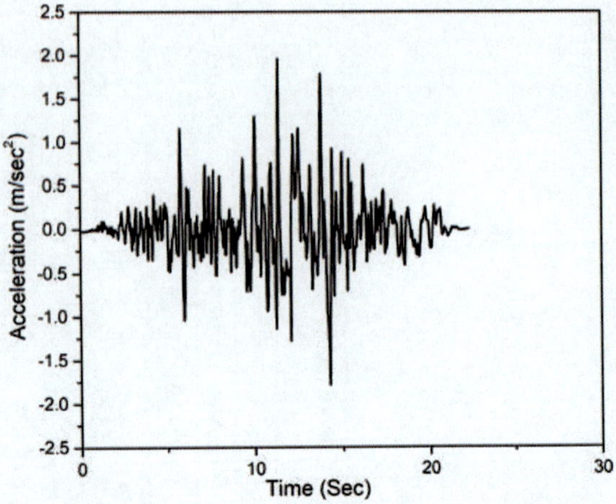


**Fig. 3** Time history of the ground acceleration (Superstition Hill 315)

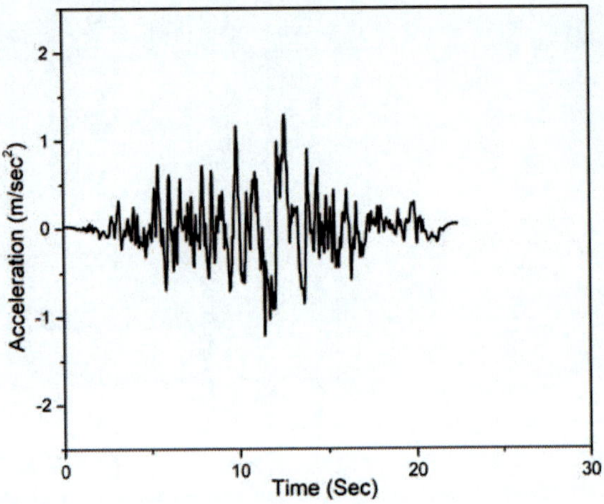


**Fig. 4** Time history of the ground acceleration (Superstition Hill 225)

top attachment point with the peak ground acceleration. It is shown from Fig. 6a that perhaps the percentage reduction in relative displacement for FF is not that sensitive to the PGA variation (especially beyond 0.3 g). The highest percentage reduction is between 75–80%, but the pattern of percentage reduction in relative displacement with the PGA remains the same. Notice that for both U1 and U2, the percentage reduction and its variability with the PGA are almost the same.

As for the FF, the percentage reduction is minimum as associated with the near-field earthquakes. Besides, unlike the near-field earthquake, the variation in percentage reduction is known to be vulnerable to the PGA variation. It is commonly found that the percentage of reduction rises with the rise in the PGA. Also, the percentage reduction is not the same for U1 and U2 directions; the percentage reduction in the U1 direction is greater.

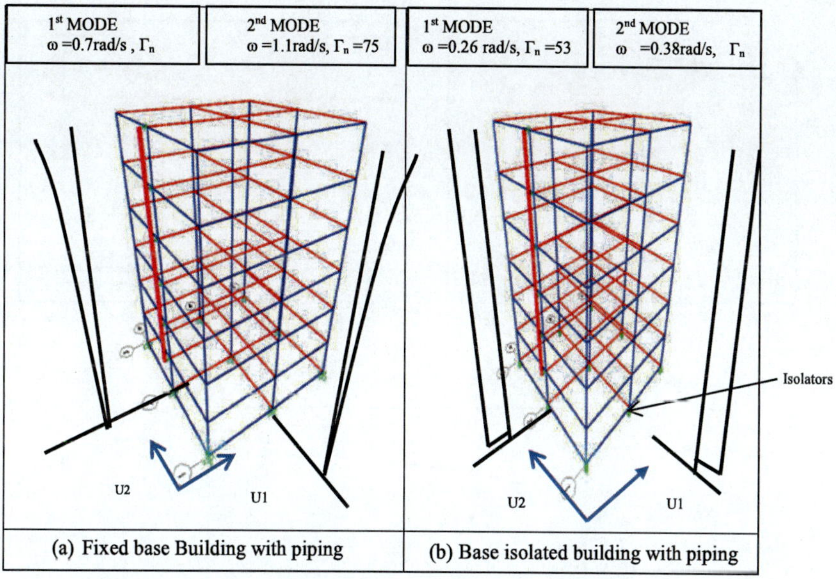


**Fig. 5** 3D Models of fixed and base-isolated buildings

Figure 6b shows the pattern of the percentage reduction at the top attachment point in absolute acceleration with the PGA. No clear trend of variation of the percentage reduction with the PGA is observed for the near-field earthquake, unlike the relative displacement. In general, the percentage reduction rises to 0.6 g with the value of the PGA. For near-field earthquakes, the maximum percentage reduction is 75%.

As for the far-field earthquake, the trend of variation is comparable to that of the relative displacement; with the rise in the PGA, it rises continuously. The highest percentage reduction shall be of the order of 62%.

Figure 6c shows the variation of the reduction of peak stresses at the top attachment point with the PGA. The peak stresses are the resultant stresses generated at the connection point of the anchor rod with the pipe due to the movements of the point in the U1 and U2 directions. It is observed from the figure that, except for the NFD and the stress in the U1 direction, other stresses for the near-field earthquake remain nearly insensitive to the variation of the PGA. Maximum stress reduction occurs in the U1 direction of the NFF earthquake, which is around 65% to 70% depending on the PGA value.

As for the FF, for the PGA up to 0.4 g, the percentage reduction increases dramatically, and then stays almost stationary for the PGA greater than 0.4 g. Besides, the stress reduction in the U1 and U2 directions remains virtually the same.

Therefore, from Fig. 6a–c, it is clear that with the two earthquake parameters, the variations in response may not show any regular trend. This is the situation because the responses created in the pipe are a complicated subject arising from the earthquake's frequency content. The PGA level also defines whether or not the

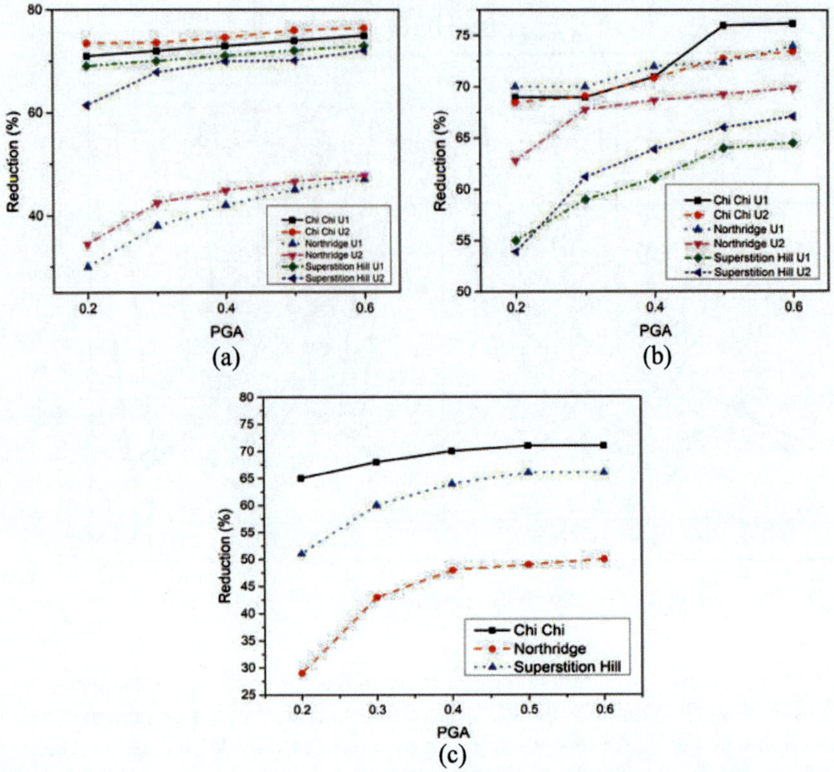


**Fig. 6** Variations of the % reduction in U1 and U2 direction at the top attachment point **a** Relative displacement **b** Absolute acceleration **c** Stress

primary structure stays in an elastic state. Underneath the inelastic excursion of the primary system, the behaviors of the pipe become significantly different from those when the building stays in the elastic system. Notice that at the PGA levels considered, the pipe does not experience an inelastic excursion.

Figure 7a shows the discrepancy of the percentage reduction of the relative displacement of the attachment point along with the height of the building for a case, namely, PGA = 0.4 g (1:3/4). The figure shows that the variations between the percentage reductions (both U1 and U2) obtained for different earthquakes increase more with the floor height with the higher value of PGA = 0.4 g and the lower value of λ. For the NFD for PGA = 0.4 g, the maximum reduction in relative displacement (U1) at floor level occurs, although the maximum reduction at floor level is almost the same for both forms of near-field earthquakes. Besides, the discrepancy between the percentage reductions in relative displacement (U1 and U2) is shown to vary more at higher floor levels for a higher PGA value than for a lower PGA value. This is because more plastification of the building takes place at the higher PGA level, leading to more discrepancies between the U1 and U2 displacements. The percentage

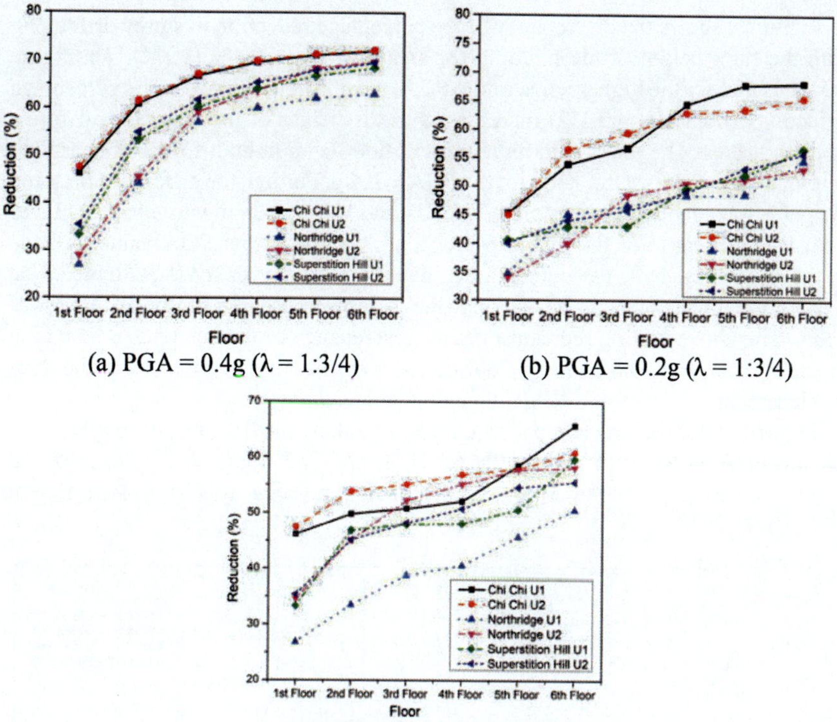


(c) PGA = 0.4g (λ = 1:3/4)

**Fig. 7** Variations of the % reduction in U1 and U2 direction at the attach point along with the floor height of the building **a** Relative displacement **b** Absolute acceleration **c** Storey drift

reduction in displacement U1 is found to be more at the highest floor level for the higher value of PGA relative to the U2 displacement. An inverse trend is observed for the lower PGA value.

Figure 7b shows the discrepancy of the percentage reduction in absolute acceleration along with the floor height of the building for cases, PGA = 0.2 g (1:3/4). The figure indicates that there is again a significant difference in the percentage reductions in absolute acceleration for all floor levels between various forms of the earthquake. The percentage reduction discrepancy does not follow any definite trend. However, it is found that, in general, with the rise in floor height, the percentage reduction in absolute acceleration increases. For the earthquake with the NFF, the highest percentage reduction is found for all floors. Furthermore, it is noted that the difference is decreased for the lower PGA value, but the pattern of the percentage reduction variation stays the same as that seen for the higher PGA value. The difference between the absolute accelerations in the U1 and U2 directions also decreases the lower PGA value at all floor heights.

Figure 7c shows the discrepancy of the percentage reduction in storey drift along with the floor height of the building for a case, PGA = 0.4 g (1:3/4). The figure indicates that for the higher value of the PGA, the difference between the percentage reductions (both U1 and U2) increases with the height of the floor for all forms of earthquakes. The maximum reductions at floor level in both forms of near-field earthquakes are almost the same for the PGA = 0.4 g. Further, the difference between the percentage reductions in storey drift (U1 and U2) is seen to vary more at higher floor levels because of the higher PGA value. For the higher PGA value, relative to the U2 storey drift, the percentage reduction in storey drift U1 is found to be more at the highest floor level. An opposite pattern is observed for the PGA's lower value. The above finding indicates that the discrepancy of the percentage reduction in storey drift along the building floor height is similar to that observed in the floor displacement.

Figure 8a–c shows the comparative analysis, among the floor response spectrums for the PGA = 0.4 g (1:3/4) at the 6th, 4th, and 2nd floors of the isolated and fixed base buildings. The figure shows that the floor response spectrums for different

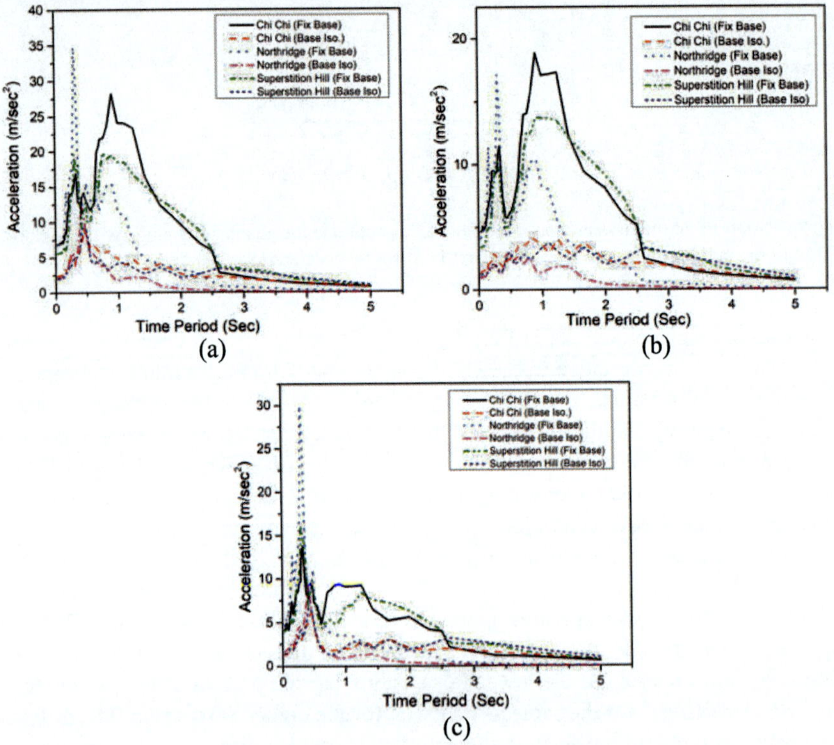


**Fig. 8** Comparison between floor response spectrums of isolated and fixed base buildings for PGA = 0.4 g (λ = 1:3/4) for U1 and U2 direction **a)** 6th floor **b)** 4th floor **c)** 2nd floor

earthquake types vary significantly in both magnitude and shape. For fixed and base-isolated buildings, the trend of variation in the floor response spectrum is different for different floors. It can be noted that there are ordinates in the floor response spectrum for the base-isolated building, which is much lower compared to the fixed base building.

## 7 Conclusion

To minimize the responses of a secondary system, taken as a running pipeline and the efficacy of base isolation of a 3D building model is investigated for bidirectional ground motions. The beams and columns of the building are modeled as linear elements, the slab is modeled as the shell area element, and the pipe is also modeled as the shell area element. At each floor level, the anchor connected to the building frame for connecting the pipe is modeled as a zero link element. Using SAP 2000, responses are obtained for both base-isolated and fixed base buildings. The percentage reduction in relative displacement, absolute acceleration, storey drift, and stresses in the connection points of the secondary system with the primary system are determined from the responses thus obtained. Besides, the variation of the percentage reduction in these quantities along the building height is studied. The floor response spectrum of the base-isolated building and fixed base building is evaluated. The variable parameters include the PGA and the type of earthquake. The numerical study findings lead to the following conclusions:

(1) Percentage reductions in the absolute accelerations, relative displacements, storey drift, and stresses at the attachment point of the SS system, in general, increase with the increase in the PGA; in the case of the near-field earthquake, the percentage reductions in displacement and acceleration remain fairly insensitive to the variation of the PGA.
(2) Percentage reductions of the response quantities in both U1 and U2 directions remain nearly the same for most of the cases.
(3) Compared to the far-field earthquake, the percentage reduction of responses is higher for the near-field earthquake; the maximum percentage reduction of all response amounts usually ranges between 60 and 75% for the near-field earthquake.
(4) Variation of the percentage reduction along the floor height does not show a consistent pattern. The percentage reduction generally increases along with the floor height.
(5) The floor response spectrums of the base-isolated building vary greatly in both shape and magnitudes of the ordinates from those of the fixed base building; the base-isolated building has less value of the ordinates of the response spectrum.
(6) Comparison of the peak ordinates of the fixed base and base-isolated building spectrums reveal that depending on the type of earthquake and the PGA value, there may be a cumulative decrease of up to 82%.

## 8 Future Scope

The present study considers pipe as a secondary system. This is one type of secondary system, along with that some vibrating machines and static equipments are generally installed on the building, which can be considered as a secondary system. The response of the machine which is in motion along with a bidirectional earthquake can be analyzed in SAP2000. The behavior of the machine in motion subjected to bidirectional earthquake can be considered as the future scope of the study.

## References

1. Agrawal AK (2000) Behaviour of equipment mounted over a torsionally coupled structure with sliding support. Eng Struct 22:72–84
2. Contento A, Di Egidio A (2014) On the use of base isolation for the protection of rigid bodies placed on a multi-storey frame under seismic excitation. engineering structures. Elsevier Ltd 62–63:1–10. https://doi.org/10.1016/j.engstruct.2014.01.019
3. Dolce M, Cardone D (2003) Seismic protection of light secondary systems through different base isolation systems. J Earthq Eng ISSN 2469. https://doi.org/10.1080/13632460309350447
4. Fan FG, Ahmadi G (1992) Seismic responses of secondary systems in base-isolated structures. Eng Struct 14(1):35–48. https://doi.org/10.1016/0141-0296(92)90006-C
6. Huang Y-NASWMCCSM (2007) Seismic demands on secondary systems in base-isolated nuclear power plants. Earthq Eng Struct Dynam 41(11):1549–1568. https://doi.org/10.1002/eqe
7. Jeon B, Choi H, Hahm D (2014) Seismic fragility evalution of base isolated nuclear power plant', ICTWS2014-0901; SEISMIC (May 2015)
5. Juhn G, Manolis GD (1992) Stochastic sensitivity and uncertainty of secondary system in base isolated structures. J Sound Vib 159:207–222
8. Juhn BG, Manolis GD (1993) Experimental study of secondary system in base-isolated structure. J Struct Eng 118(8):2204–2221
9. Juhn G, Manolis GD, Constantinou MC (1992) Stochastic response of secondary systems in base-isolated structures. Probab Eng Mech 7:91–102
10. Kelly JM, Tsai H-C (1985) Seismic response of light subsystems on inelasitc structures. Earthq Eng Struct Dyn 13:711–732
11. Khechfe H et al (2002) An experimental study on the seismic response of base-isolated secondary systems. J Press Vessel Technol, Trans ASME 124(1):81–88. https://doi.org/10.1115/1.1445795
12. Kim S et al (2018) Seismic fragility evaluation of the base-isolated nuclear power plant piping system using the failure criterion based on stress-strain. Elsevier Ltd., Nuclear Engineering and Technology. https://doi.org/10.1016/j.net.2018.10.006
13. Kim YS, Lee DG (1993) Seismic response of support-isolated secondary structures in a multistorey structure. Eng Struct 15(5):335–347. https://doi.org/10.1016/0141-0296(93)90037-5
14. Pardalopoulos SI, an. S. J. P. (2014) Seismic response of nonstructural components attached on multistorey buildings. Earthq Eng Struct Dyn 41(11):1549–1568. https://doi.org/10.1002/eqe
15. Tsai H-C and Kelly JM (1988) Non-classical damping in dynamic analysis of base-isolated structures with internal equipment. Earthq Eng Struct Dyn 16(1):29–43. https://doi.org/10.1002/eqe.4290160104

# Seismic Analysis of Isolated and Non-isolated Railway Bridges with Slab Track—A Comparative Study

A. Shivaraj, G. Sudarshan, G. Sridevi, and B. Umesh

## 1 Introduction

Highway and railway bridges are vital components in the infrastructure of transportation networks. These structures experience significant structural and non-structural damage during the earth quake. Earthquake can cause damage not only on account of the shaking of the super structure but also due to permanent lateral displacement because of either soil settlements or ground lateral displacements. It is, therefore, important to take necessary precautions in the planning and design of structures so that they are safe against such secondary effects also. The damage caused to these structures during earthquakes engenders the functionality of the railways, thus leading to significant losses. The behavior of a railway bridge system depends on the interaction of the three subsystems: the rolling stock, the rail track, and the subsoil. The nonlinear behavior of these subsystems makes the assessment of the system, as a whole, very complex. The base isolation attempts to decouple the structure from the damage caused due to ground movement at the time of earthquake.

Xiang and Li [1] conducted a practical test to identify the sliding behavior of LRB with specific configurations in China. Shen et al. [2] introduced a new seismic isolation system.

A. Shivaraj (✉) · G. Sudarshan · G. Sridevi · B. Umesh
B. V. Raju Institute of Technology, Narsapur, India
e-mail: Shivaraj.a@bvrit.ac.in

G. Sudarshan
e-mail: Sudarshan.g@bvrit.ac.in

G. Sridevi
e-mail: Sridevi.g@bvrit.ac.in

B. Umesh
e-mail: Umesh.b@bvrit.ac.in

 
S. Kolathayar and S. C. Chian (eds.), *Recent Advances in Earthquake Engineering*, Lecture Notes in Civil Engineering 175,
https://doi.org/10.1007/978-981-16-4617-1_27

Yuan et al. [3] conducted real-time hybrid simulation tests on a multi-span continuous highway bridge under different seismic ground motions.

Research has shown that traditional isolation methods are efficient in reducing the harmful effects of earthquakes on bridges [4], among various low-loss flexible design options.

Haque et al. [5] analyzed the seismic response of a base-isolated highway bridge with seismic responses of a base-isolated highway bridge with different isolators. Non-linear dynamic analysis of the multi-span bridge was carried out with three types of LRB, namely, natural, high dumping, and lead rubber bearing. The impact of laminated rubber bearings on significantly reducing the reaction of rigid bridge systems is significant.

Di Sarno [6] carried out the comprehensive linear and nonlinear dynamic analyses of a typical railway bridge system. The introduction of lead rubber bearings greatly improved the earthquake bridge response; however, the isolation equipment does not meet the serviceability requirements because of the rails. Steel dampers as isolators are recognized as an advantageous construction system in terms of service efficiency and final boundary states. Dimitriadou [7] studied the seismic performance of a bridge with LRB provided at the top of the pier. The dynamic response was studied before and after the introduction of lead rubber bearings. As peak ground acceleration (PGA) levels increase, the bridge response in the piers increases with the corresponding values in the rest of the deck at the same level. Couture Zella et al. (2014) studied the effect seismic isolation system of a bridge for Kobe earthquake of 1995 in both lateral and longitudinal directions.

Alberto et al. [8] analyzed the effects of the isolation devices on the Post-Yielding Stiffness. Four different yielding forces of the hysteric damper (Fy) were considered, keeping the stiffness properties constant. The behavior of the bridge is analyzed for maximum and residual displacements by changing the device's yielding force.

## *1.1 Need of Seismic Isolation in Bridges*

Isolation systems do not absorb seismic energy, but instead dissipate it through the dynamics of the system. In this way, a structure is separated from its foundation, and superstructure of the bridge is separated from the piers.

In bridges, the concept of decoupling is very different than in buildings. In bridges, most of the load is concentrated in a parallel plane, superstructure.

## *1.2 High Damping Rubber Bearings*

HDRB reinforces rubber sheets and steel plates in the isolator. These components allow HDRB to separate the bridge superstructure from the substructure due to its high horizontal orientation and allow up to 16% of the energy dissipation. Effective

HDRB isolators have the ability to transmit vertical loads due to live loads, dead loads and accidental impacts, the capacity to support horizontal loads by allowing very little displacement relative to the ground. HDRB does not damage the remaining structure due to its elastic response to high magnitude earthquakes without interruption in structural performance.

### 1.3 Slab Track (Ballastless Track)

A slab track means that the ballast is recoupled by a reinforced concrete slab track that transmits load and enhances stability. It is, also known as ballastless track, a modern form of track used for high-speed lines, heavy rail, and tram systems. Slab track technology offers higher performance and longer service life than traditional ballasted track.

### 1.4 Aim of the Work

- To find the response time of decoupled bridges and Non-Isolated bridges for same material, structural properties, and study the performance of isolator in prolonging the response time subjected to seismic loads.
- To find the displacement of bridge deck of Isolated and non- isolated bridges in both directions.

## 2 Methodology

In the present study, two railway bridges, one with base isolation and one without base isolation having 39 m in length and 12.16 m width are designed by LIMIT STATE OF DESIGN method. SAP2000 V14 is a standout amongst the most refined and easy to understand programming. Both models are analyzed by RESPONSE SPECTRUM METHOD in SAP2000. High damping rubber isolator for isolated bridge is designed as per AASTHO code. Dynamic responses of models are compared and studied.

Slab to support ballastless broad gauge of track width 1.676 m is analyzed and designed by considering live load 80.9 kN/m on both the rails as per IRS Standards. SAFE is the ultimate tool for designing RCC floor and foundation systems. In this present study, structural and geometrical design of slab on which rails are resting has been done by SAFE.

## 2.1 Response Spectrum Method

Earthquake analysis and the design of a structure to be built at a specific location require real-time history records of earthquakes. However, it is not possible to have such records everywhere. Furthermore, seismic analysis of structures cannot be done only on the basis of the maximum value of the earth acceleration because the response of the structure depends on the frequency of the earth motion and its dynamic properties. To overcome the above difficulties, the most popular tool in seismic analysis of structures used is seismic response spectrum. This method calculates only the maximum values of displacement and member forces in each vibration mode.

## 2.2 Governing Factors

- Energy dissipation
- Epi-central distance
- Soil condition
- Richter magnitude
- Damping
- Time period.

## 2.3 Description of Model

Dynamic live load on rail tracks is converted into static load and applied on both the rails. As per IRS standards, live load on each track of broad gauge of width 1.6765 m is 80.9 kN/m. Along with live load on track, self-weight of track is also considered as per IRS standards, i.e., 0.58 kN/m. In this study, ballast less slab of width 6.08 m is analyzed and designed for above mentioned load configuration in SAFE software. As a result, intensity and nature of various stresses that have been generated in slab model is studied and designed for the maximum stresses.

In this study, 2-span elevated RC-Bridge of each span length 19.5 m is considered. Typical circular piers of height 10 and 1 m in cross section are supporting pier cap of size 0.4 m × 0.9 m, over which longitudinal girders of size 1.3 m × 0.8 m are resting. And also, to enhance lateral stability of super structure, cross girder of size 0.65 m × 0.4 m is introduced at center of each span of bridge. Center-to-center distance between piers in lateral direction is 4.955 m. And total thickness of slab designed and considered is 0.35 m.

High damping rubber bearing is introduced at the top of pier cap, i.e., between pier cap and longitudinal girders. Load on each HDRB is calculated and designed as per AASTHO shown in Table 4. Here, HDRB is used to decouple the superstructure and substructure to avoid the transfer of hazardous seismic ground motions to superstructure by means of energy dissipation. When structure gets excited at certain frequency during earthquake, kinetic energy is transferred to the super structure through sub structure. When isolators are introduced in between superstructure and substructure, kinetic energy will get dissipated by converting into heat energy. And also, these isolators have got good restoring capacity because of their own material properties to regain its original position.

### 2.4 *Description of Bridge*

The Super structure model of bridge structure chosen for analysis is modelled in SAFE is shown in Fig. 1. The bridge was modelled in SAP 2000 and the model is shown in Fig. 2. The geometrical properties of the slab, unit weight of the material are presented in Table 1, bridge components data in Table 2, and selected criteria for the analysis in Table 3 and the properties of High damp rubber bearing are given in Table 4.

Model has been analyzed for severe earthquake conditions.

**Designed Properties of High Damping Rubber Bearing (As per AASTHO).**

## 3 Results and Discussions

The results of non-isolated bridge model, i.e., time period and frequency, static–dynamic percentage, and displacement are shown in Tables 5, 6, and 7, respectively, and Fig. 3 represents the response spectrum curve of non-isolated bridge.

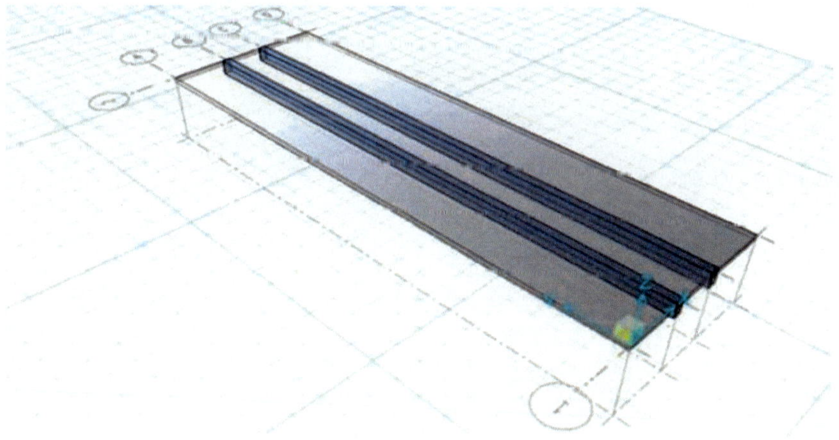

**Fig. 1** Slab model in SAFE

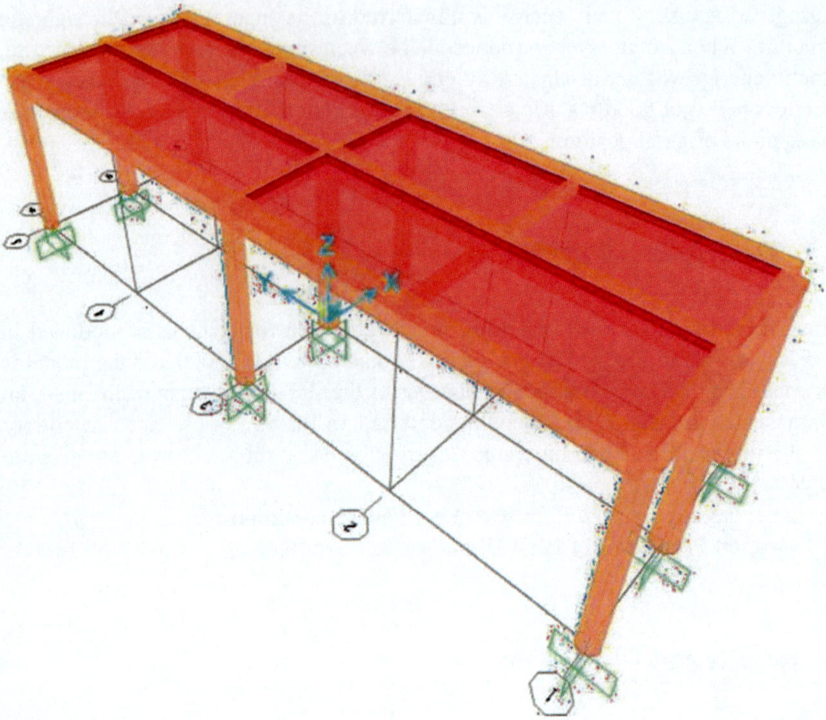


**Fig. 2** Model in SAP

**Table 1** Geometrical properties of slab

| | |
|---|---|
| Length | 19.5 m |
| Width | 6.08 m |
| Thickness | 0.20 m |
| Load on each track | 80.9 kN/m |
| Weight of rail | 0.58 kN/m |

**Table 2** Bridge data

| | |
|---|---|
| Total length | 39 m |
| Width | 12.16 m |
| Pier height | 10.0 m |
| Dimensions of pier | 1.0 m in diameter |
| Dimensions of pier cap | 0.4 m * 0.9 m |
| Clear span length | 19.5 m |
| Cross section of longitudinal girder | 1.3 m * 0.8 m |
| Cross section of cross girder | 0.65 m * 0.4 m |
| Thickness of slab | 0.350 m |
| Distance between piers | 4.955 m |

**Table 3** Criteria's

| | |
|---|---|
| Zone | V |
| Importance factor-I | 1.5 |
| Response reduction factor-R | 3 |
| Soil type | II (Medium) |
| Damping of structure | 5% |

### *3.1 Non-isolated Bridge model*

The results of isolated bridge model, i.e., time period and frequency, static–dynamic percentage, and displacement are shown in Tables 8, 9, and 10, respectively, and Fig. 4 represents the response of Isolated Bridge.

## 4 Observations and Conclusions

By the observation of the RESPONSE SPECTRUM curves of both isolated and non-isolated bridges, the time periods or response time of isolated bridges is observed to get postponed from 0.40288 to 0.4742 s because of the use of HDRB rubber bearing for isolation for same seismic, material, and structural parameters.

Model participation ratios for both non-isolated and isolated bridges are 99.99%. Displacement of bridge deck or superstructure for non-isolated bridge is 26.2 mm in X-direction and 73.5 mm in Y-direction. Whereas for the isolated bridge, it is 26.3 mm in the X-direction and 0.8 mm in Y-direction. The reduction in the displacement is not observed in X-direction because of higher stiffness of the bridge in that particular direction.

The reduction in displacement in Y-direction is observed significantly and a reduction of 91.87% is observed.

The study of 12 mode shapes of dynamic response show the lateral displacement of bridge in both the directions, i.e., X and Y for the first two mode shapes, respectively. Most importantly, the torsion is observed in the bridge model at 3 mode shape of the response.

**Table 4** Properties of HDRB

| | HDBR for (6089.91 kN) | HDBRfor (11593.96 kN) | HDBR for (12119.04 kN) | HDBR for (23066.36 kN) |
|---|---|---|---|---|
| Diameter of Rubber (d) (m) | 0.9944947 | 1.372 | 1.4029 | 1.93547 |
| No. of layer (N) | 8.0 | 6.0 | 6.0 | 4.0 |
| Steel Plate thickness (ts) (m) | 0.0023678 | 0.0032671 | 0.0033403 | 0.0046083 |
| Height of bearing (m) | 0.1486072 | 0.148607 | 0.148607 | 0.1486072 |
| Base shear Vb (kN) | 4384.7352 | 8347.6512 | 8725.7088 | 16607.779 |

**Table 5** Period and frequency

| Period (s) | Frequency (Cyc/s) |
|---|---|
| 0.40288 | 2.4821106 |
| 0.40233 | 2.4855317 |
| 0.33778 | 2.9604959 |
| 0.19966 | 5.0085891 |
| 0.1649 | 6.0643451 |
| 0.16261 | 6.1495877 |
| 0.15978 | 6.258487 |
| 0.12185 | 8.2066327 |
| 0.10948 | 9.1343455 |
| 0.06248 | 16.006078 |
| 0.03127 | 25.980709 |
| 0.03002 | 27.020723 |

**Table 6** Static and dynamic percentage for non-isolated bridge

| | Item | Static percent | Dynamic percent |
|---|---|---|---|
| Acceleration | X-direction | 100 | 99.998 |
| | Y Direction | 100 | 99.999 |

**Table 7** Displacements for non-isolated bridge

| X-direction | Y-direction |
|---|---|
| 6.2 mm | 73.5 mm |

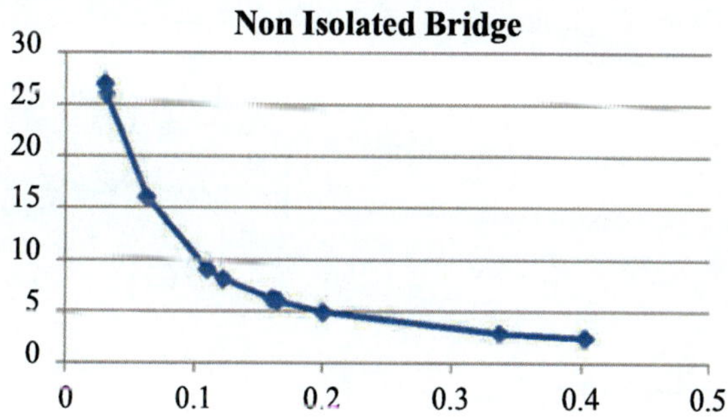


**Fig. 3** Response curve for non-isolated bridge (time period v/s frequency)

**Table 8** Time period and frequencies

| Period (s) | Frequency (Cyc/s) |
|---|---|
| 0.4742 | 2.4821106 |
| 0.47131 | 2.4855317 |
| 0.37473 | 2.9604959 |
| 0.20067 | 5.0085891 |
| 0.17528 | 6.0643451 |
| 0.16517 | 6.1495877 |
| 0.16016 | 6.258487 |
| 0.13399 | 8.2066327 |
| 0.12227 | 9.1343455 |
| 0.09272 | 16.006078 |
| 0.06989 | 25.980709 |
| 0.05947 | 27.020723 |

**Table 9** Static and dynamic percentage for isolated bridge

| | Item | Static percent | Dynamic percent |
|---|---|---|---|
| Acceleration | X-direction | 100 | 99.9999 |
| | Y-direction | 100 | 99.9999 |

**Table 10** Displacements for isolated bridge

| X-direction | Y-direction |
|---|---|
| 26.3 mm | 0.8 mm |

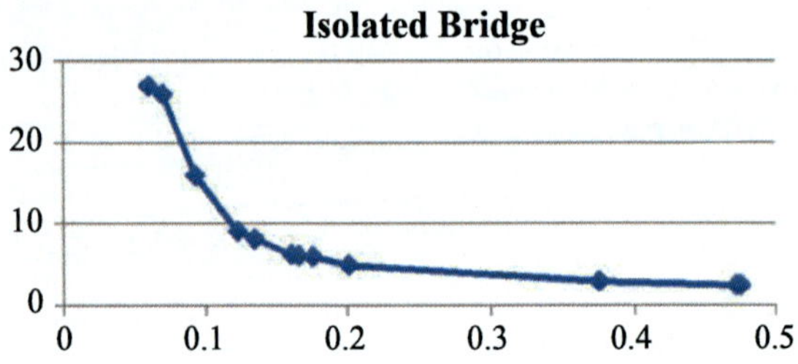


**Fig. 4** Response curve for isolated bridge (time period v/s frequency)

## References

1. Xiang N, Li J (2017) Experimental and numerical study on seismic sliding mechanism of laminated-rubber bearings. Eng Struct 141:159–174
2. Shen X, Wang X, Ye Q, Ye A (2017) Seismic performance of transverse steel damper seismic system for long span bridges. Eng Struct 141:14–28
3. Yuan Y, Wei W, Igarashi A, Tan P, Iemura H, Zhu H (2017) Experimental and analytical studies of seismic response of highway bridges isolated by rate-dependent rubber bearings. Eng Struct 150:288–299. https://doi.org/10.1016/j.engstruct.2017.06.020

4. Li Y, Astroza R, Conte JP, Soto P (2017) Nonlinear fe model updating and reconstruction of the response of an instrumented seismic isolated bridge to the 2010 Maule Chile earthquake. Earthq Eng Struct Dyn 46(15):1–18
5. Haque MN, Bhuiyan AR, Alam MN (2010) Seismic response analysis of base isolated highway bridge: effectiveness of using laminated rubber bearings. IABSE-JSCE Joint Conf Adv Bridge Eng-II. ISBN: 978-984-33-1893-0, 336–343
6. Di Sarno L (2013) Base isolation of railway bridges. Int J Mech 3(7)
7. Dimitriadou O (2007) Effect of isolation on bridge seismic design and response, Pavia, Italy European School for Advanced Studies in Reduction of Seismic Risk
8. Alberto L (2008) Institute superior technico. file:///Resumo_Alargado_Inglês_FINAL.pdf

# Seismic Performance Analysis of Regular and Irregular RCC Framed Building with Dampers

**Laxmi M. Ramdas and M. Helen Santhi**

## 1 Introduction

Dampers are energy dissipating systems. Dampers dissipate the wave energy inside a building developed due to seismic waves and thereby diminish the oscillation of the building. Different types of dampers are available in the market such as friction damper, fluid viscous damper, viscoelastic damper, yielding damper, magnetic damper, tuned mass damper, etc. In fluid viscous damper, a kind of silicon oil passes through piston cylinder arrangement which absorbs seismic energy. In this paper, a G + 19 storey regular and irregular building with and without damper is analysed using the response spectrum method. Fluid viscous damper of clevis-base plate type manufactured by Taylor device inc. is used as damper in both the types of buildings.

Many research studies have been carried out to investigate the seismic behaviour of buildings with different types of dampers. Lee et al. [14] had studied the conventional analysis methods for building structures with viscoelastic dampers using Metrics condensation method and found that the more accurate method was mode superposition method. Aydin et al. [1] had worked on the optimal damper arrangement using different objective functions. Optimal storey displacement and drift were reduced when damper placement was optimal. George et al. [9] had studied the response of structures with fluid viscous damper and concluded that it is practically possible to assess the maximum velocity using elastic equivalent. Landi et al. [13] had studied about efficacious straight up distributions of the damping coefficients of nonlinear viscous dampers and suggested that energy method was good in case of practical design. Hessabi and Mercan [10] had investigated the implementation of GMD and

L. M. Ramdas (✉) · M. Helen Santhi
School of Civil Engineering, Vellore Institute of Technology, Chennai, India
e-mail: laxmiramdas.m2020@vitstudent.ac.in

M. Helen Santhi
e-mail: helensanthi.m@vit.ac.in

 
S. Kolathayar and S. C. Chian (eds.), *Recent Advances in Earthquake Engineering*, Lecture Notes in Civil Engineering 175,
https://doi.org/10.1007/978-981-16-4617-1_28

some other types of dampers for effectiveness in reducing the vibration of building structures. Energy balance equation and time history analysis were used, and the conclusion was that GMD can increase the time period and improve performance. Liu et al. [15] had analysed the effectiveness of FVDs along with inter-storey isolation using time history method. By the optimal placement of FVD, the seismic response was improved. Ying et al. [22] had studied about analytical model for viscous wall damper (VWD) and found that it is effective in reducing seismic damages to structures. Purayil et al. [18] had designed the linear fluid viscous damper using 1st order multi objective optimization method. The method was good for optimal distribution of damper. Domenico et al. [6] had studied about design methods of viscous dampers to protect buildings from earthquake and found that the energy-based method was most efficient and economical. Aydin et al. [2] had analysed the efficiency of passive dampers in multistorey buildings. Optimum design was investigated and maximum damping coefficient value was found. Walsh et al. [20] had isolated the base of the building using a stiffness damper. Resetting passive stiffness damper (RPSD) was effective to bring down the maximum base drift along with maintaining the response of the building within its elastic stage. It reduced peak base drift but was not efficacious in decreasing the floor drift. Pérez et al. [19] had studied the optimum design of friction dampers in buildings based on the reliability in seismic regions. Building was modelled as finite element 3D structure and linear frames were provided with nonlinear friction damper.

Wang et al. [21] had studied the seismic response diminution of building structures with a vibro impact dual mass damper (VIDM). VIDM showed good results similar to that of Öexisting ones. Eldin et al. [7] had worked on seismic retrofit of reinforced concrete buildings with friction damper using egoistical PC frames. Non linear time history, incremental dynamic analysis and fragility analysis were conducted. There was decrease in drift. Nabid et al. [17] had worked on energy optimization-based seismic design of friction dampers in reinforced concrete structures. Pushover method and nonlinear time history method were used for analysis. The drift and damage got reduced using dampers. Milanchian and Hosseini [16] had studied about torsional control in plan-mass asymmetric vertical seismic isolation system. Response history analysis was conducted. It was found that the large links reduced the damping effect and plan-wise distribution was not much effective. Kumar and Chakraborty [12] had studied the reduction of seismic vibration in multistorey structures retrofitted with nonlinear viscous damper using mode summation method. In each floor, there was NVD, and the results showed reduction in the induced vibration energy. Hassan and Reyes [8] had analysed mid-rise concrete building with and without damper using pushover method. In the case of the building without damper, MPA moderately overestimated the peak floor displacement over the frame height. In the case of building with damper, MPA predicted roof displacement history accurately. The objective of the current study is to investigate the influence of damper in RCC framed regular and irregular building using response spectrum analysis.

## 2 Structural Modelling and Analysis

### 2.1 *Modelling and Analysis of Building Without Damper*

For the analysis, a G + 19 storey RCC framed building, regular in plan and irregular in plan are modelled in ETABS 2018 software. Plan of regular RCC framed building consists of six spans of 4.0 m each along the X-axis and four spans of 4.0 m each along the Y-axis (Fig. 1a). Plan of irregular building is shown in Fig. 1b. The 3D view of regular and irregular building is shown in Fig. 1c and 1d, respectively.

Each floor has a height of 3.0 m. Size of column is 0.75 m × 0.75 m, size of beam is 0.3 m × 0.45 m, thickness of slab is 0.150 m and wall thickness is 0.230 m. The material used for the frame is concrete of grade M 25. Ordinary burnt clay brick masonry is used for the wall. The soil type chosen is type II soil, and the seismic zone is III with an importance factor of 1.2 and response reduction factor of 5. Response spectrum method of analysis is carried out to understand the seismic performance of regular and irregular building.

### 2.2 *Modelling of Building with Damper*

For the modelling of building with damper, fluid viscous damper manufactured by Taylor device is used. The properties of damper are taken as provided by the manufacturing company. The damper consists of a force of 500 kN, and a mass of 98 kg is used in the study. The position of dampers in the regular and irregular RCC framed building is shown in Fig. 2a and 2b, respectively. The building models are analysed with the earthquake loading as specified in the IS code 1893:2016.

## 3 Results and Discussion

### 3.1 *Response of G + 19 Storey Regular Building*

From the modal analysis, the time period is observed as 1.539 s, and the maximum storey displacement of 189.42 mm is obtained along the Y-direction from the response spectrum analysis. The storey displacement plot of the building along X and Y directions is shown in Fig. 3a and 3b, respectively. It is found that the displacement is reached around 80% of the maximum value of displacement allowed, i.e. 240 mm for a G + 19 storeyed building as per codal provision.

Figure 3c shows the storey drift along x direction, and Fig. 3d shows the drift along y direction due to seismic load obtained from response spectrum analysis. Here, the

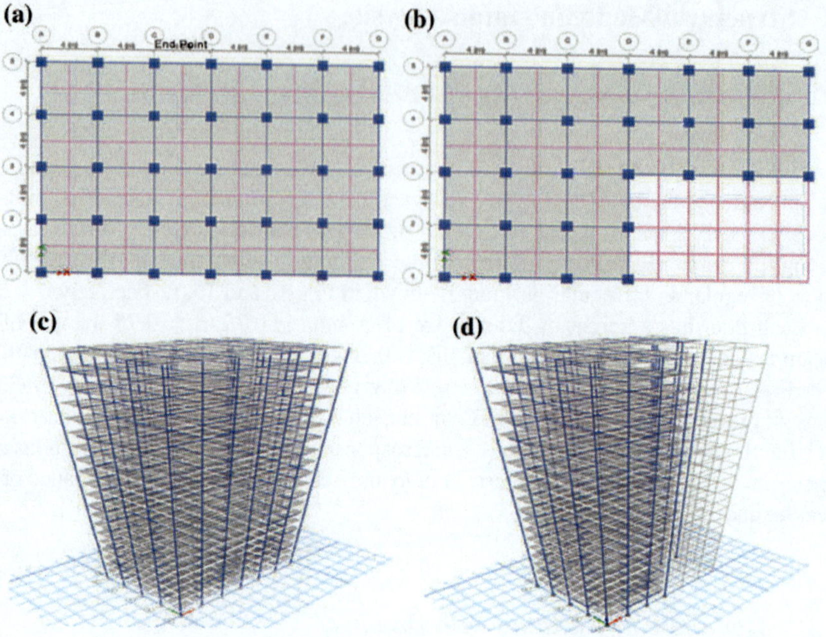


**Fig. 1** **a** Plan of regular RCC building, **b** Plan of irregular RCC building, **c** 3D view of regular RCC building, **d** 3D view of irregular RCC building

max drift is found as 0.005367 which is not within the allowable limit (0.004) as per IS 1893.

Figure 3e and 3f represent the storey shear plot for regular building along X and Y direction, respectively, and the overturning moment along X and Y direction is displayed in Fig. 3g and 3h, respectively.

### 3.2 *Response of G + 19 Storey Irregular Building*

The fundamental time period of the building is obtained as 1.588 s. The maximum value of storey displacement is observed as 205.77 mm. The displacement plot obtained from the analysis along X-direction is shown in Fig. 4a and along Y-direction is shown in Fig. 4b. When compared to the regular building, the displacement of the irregular building is nearly 10% more.

The storey drift along X and Y axes due to seismic load obtained from response spectrum analysis is shown in Fig. 4c and 4d, respectively. The maximum value of drift is 0.00484 which is along Y-axis. Here also, the drift exceeded the maximum allowable drift as given in IS Code.

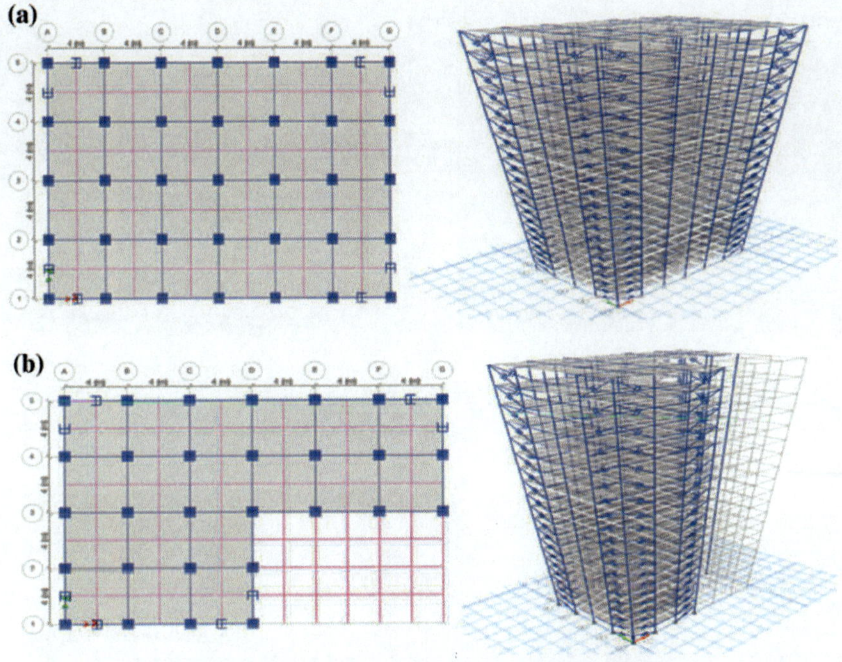

Fig. 2 **a** Regular building with damper, **b** Irregular building with damper

The storey shear along X and Y axes is shown in Fig. 4e, f, and the overturning moment obtained after analysis is shown in Fig. 4g, h.

### 3.3 Response of G + 19 Storeyed Regular Building with Damper

The fundamental modal time period is found as 1.17 s and the maximum storey displacement is 121.44 mm. The displacement plot along X and Y directions obtained from the analysis is shown in Fig. 5a and 5b, respectively. The displacement is 36% lesser than the displacement of regular building without damper and it is found as 50% of the allowable limit. Similar kind of response was reported by Kumar and Chakraborty [12] and Hua et al. [11].

The drift plot along X and Y directions is presented in Fig. 5c and 5d, respectively. It is observed that the value of storey drift came down to 0.0027 from 0.005367. The drift value is within the allowable limit after the placement of damper.

The storey shear along X and Y directions obtained from the analysis is shown in Fig. 5e, 5f, and the overturning moment plot is shown in Fig. 5g and 5h, respectively.

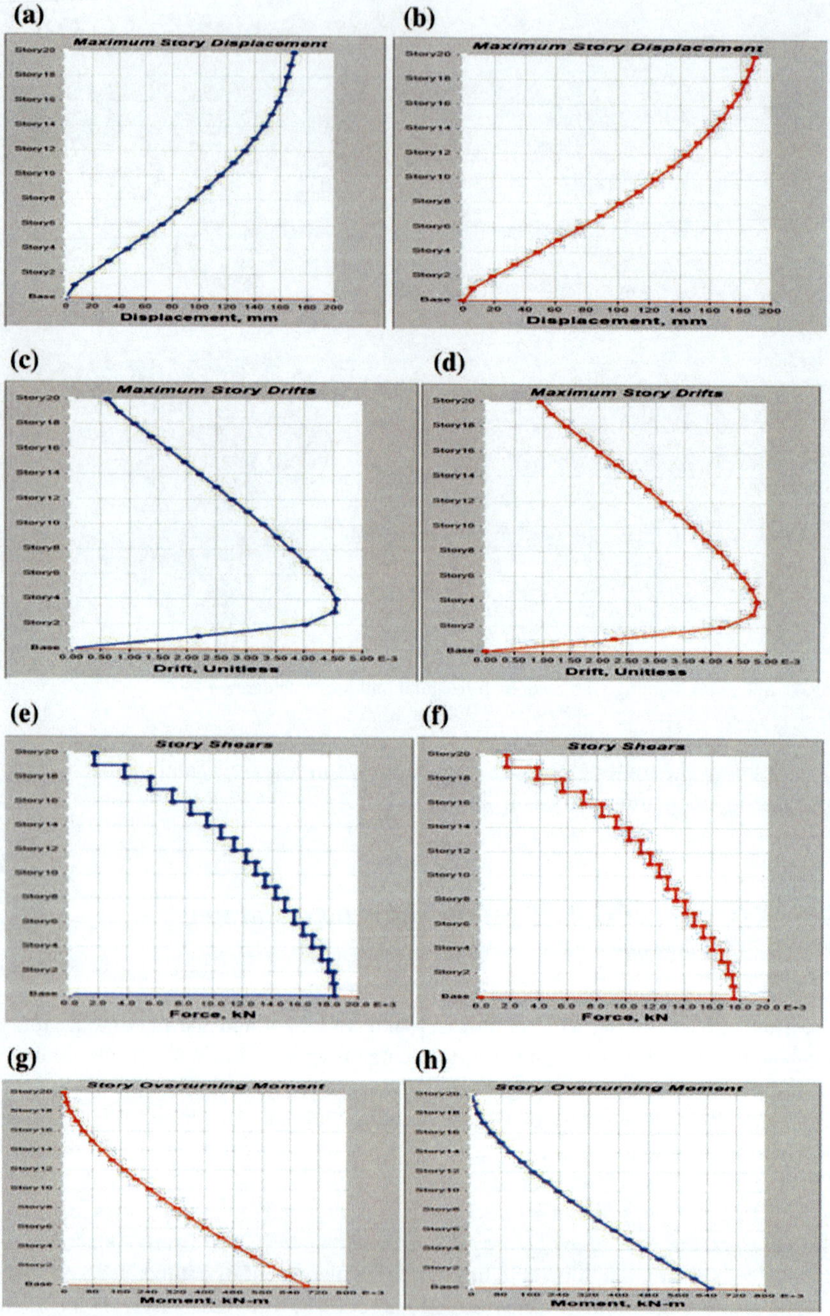


**Fig. 3** **a** Displacement along X-direction, **b** Displacement along Y-direction, **c** Drift plot along X-direction, **d** Drift plot along Y-direction, **e** Storey shear along X-direction, **f** Storey shear along Y-direction, **g** Storey moment along X-direction, **h** Storey moment along Y-direction

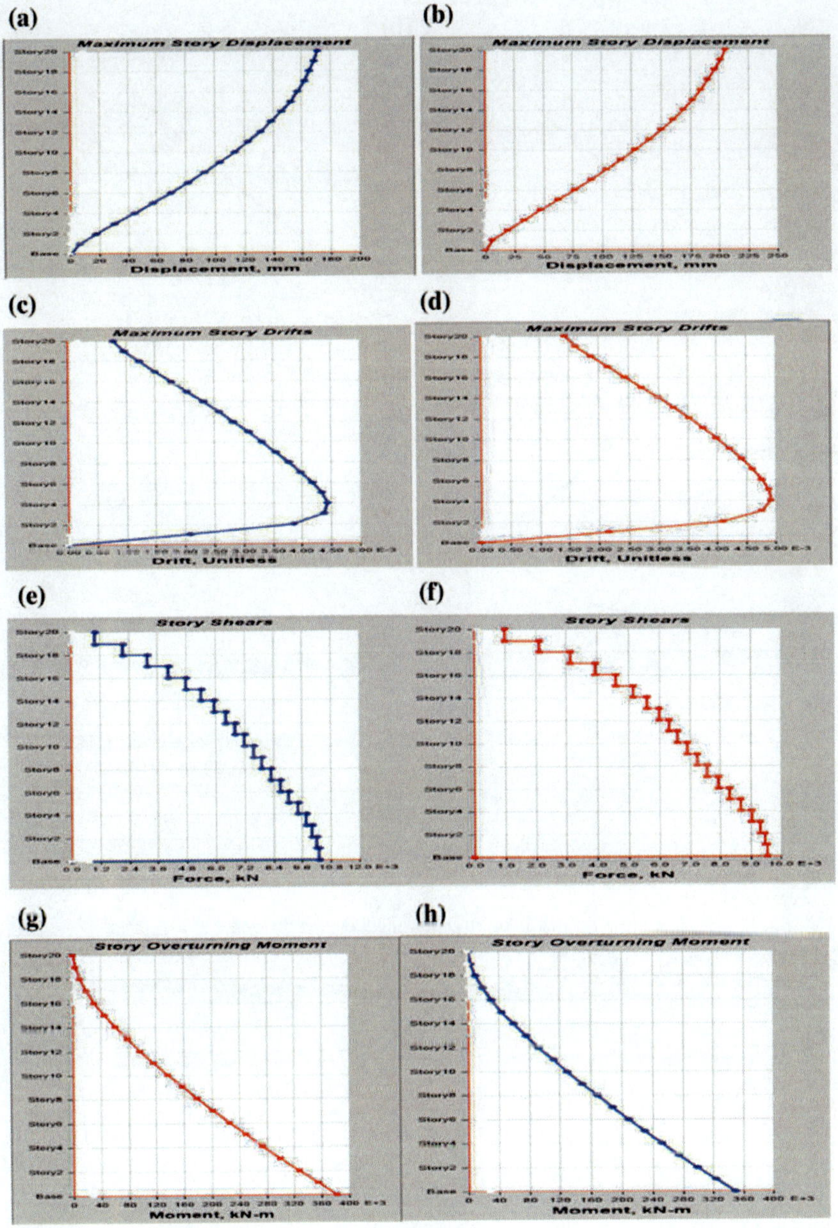


**Fig. 4** **a** Displacement along X-direction, **b** Displacement along Y-direction, **c** Storey drift along X-direction, **d** Storey drift along Y-direction, **e** Storey shear along X-direction, **f** Storey shear along Y-direction, **g** Storey moment along X-direction, **h** moment along Y-direction

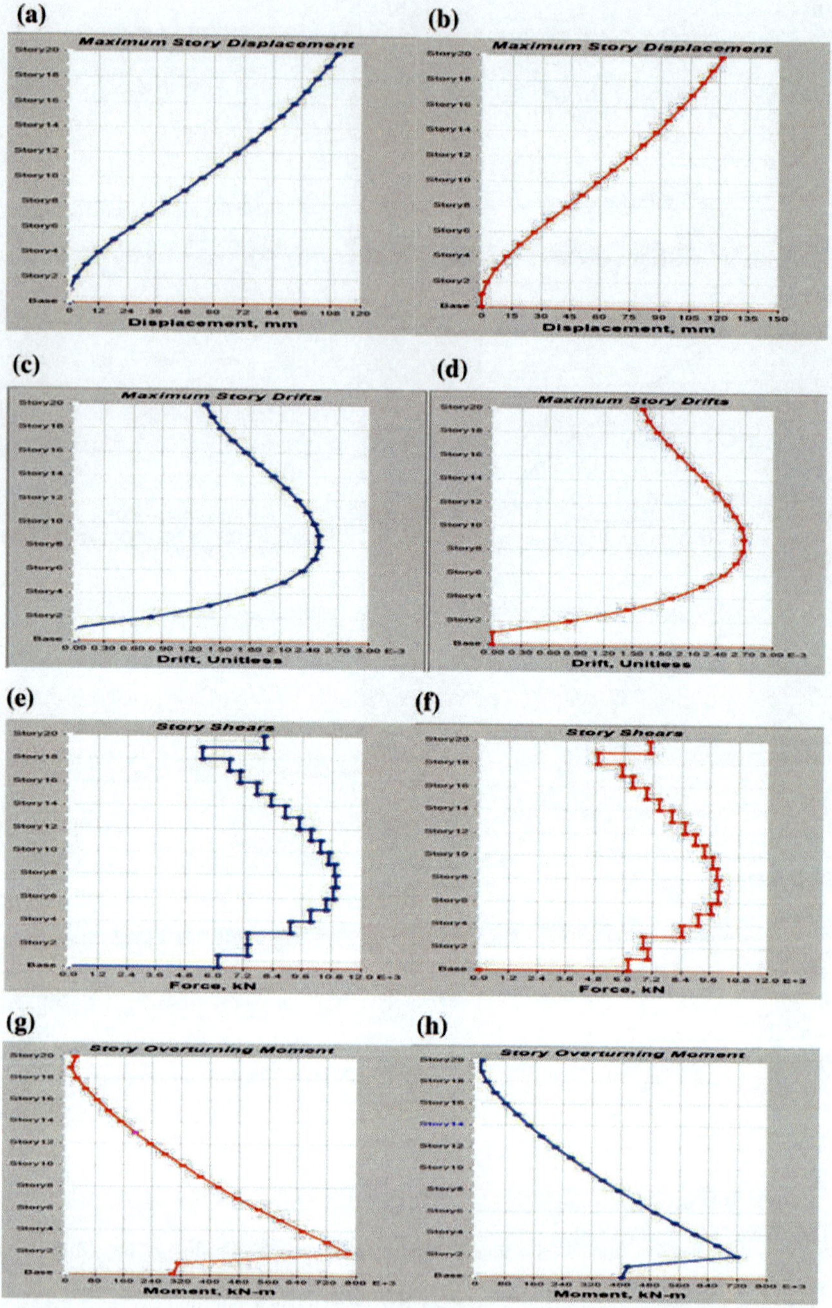


**Fig. 5** **a** Displacement along X-direction, **b** Displacement along Y-direction, **c** Drift along X-direction, **d** Drift along Y-direction, **e** Storey shear along X-direction, **f** Storey shear along Y-direction, **g** Storey moment along X-direction, **h** Storey moment along Y-direction

### 3.4 *Response of G + 19 Storeyed Irregular Building with Damper*

The modal time period obtained from the analysis is 1.171 s and the displacement value is 126.57 mm. The displacement plot along X and Y directions obtained from the response spectrum analysis is shown in Fig. 6a and 6b, respectively. The displacement value is nearly 38% lesser than that of the irregular building without damper.

The storey drift plot along X and Y axes is shown in Fig. 6c and 6d, respectively. The value of drift obtained is 0.0027 which got reduced as compared to that of irregular building without damper. Provision of damper brought down the drift to the allowable limit, 0.004.

The storey shear with respect to X and Y directions is presented in Fig. 6e and 6f, and the overturning moment along X and Y directions due to seismic load is presented in Fig. 6g and 6h, respectively.

Table 1 gives the comparison of maximum displacement and maximum drift between the buildings under study with and without damper. In both the types of buildings, without damper, the maximum drift value exceeded the allowable limit according to the code, i.e. 0.004. After providing the fluid viscous damper in the buildings, the drift value is reduced considerably and found within the allowable limit. This shows the efficiency of the dampers in controlling the vibration in the buildings.

## 4 Conclusion

In this paper, G + 19 storeyed regular and irregular buildings are modelled and analysed for earthquake loading using ETABS software version 2018. Response spectrum method as specified in the seismic code IS 1893:2016 is adopted to obtain the responses of the buildings with and without damper. The results reveal that there is a considerable reduction in the maximum storey displacement in the range of 35–45% after incorporating dampers in the buildings. The maximum drift in the buildings without damper is more than the allowable drift, 0.004 as given in the code. But after providing damper in the buildings, the drift is reduced significantly and it is found well below the drift limit. From the study, it is concluded that the fluid viscous damper of 500 kN is effective in reducing the displacement and drift, and bringing it within the allowable limit, and making the buildings safer under seismic conditions.

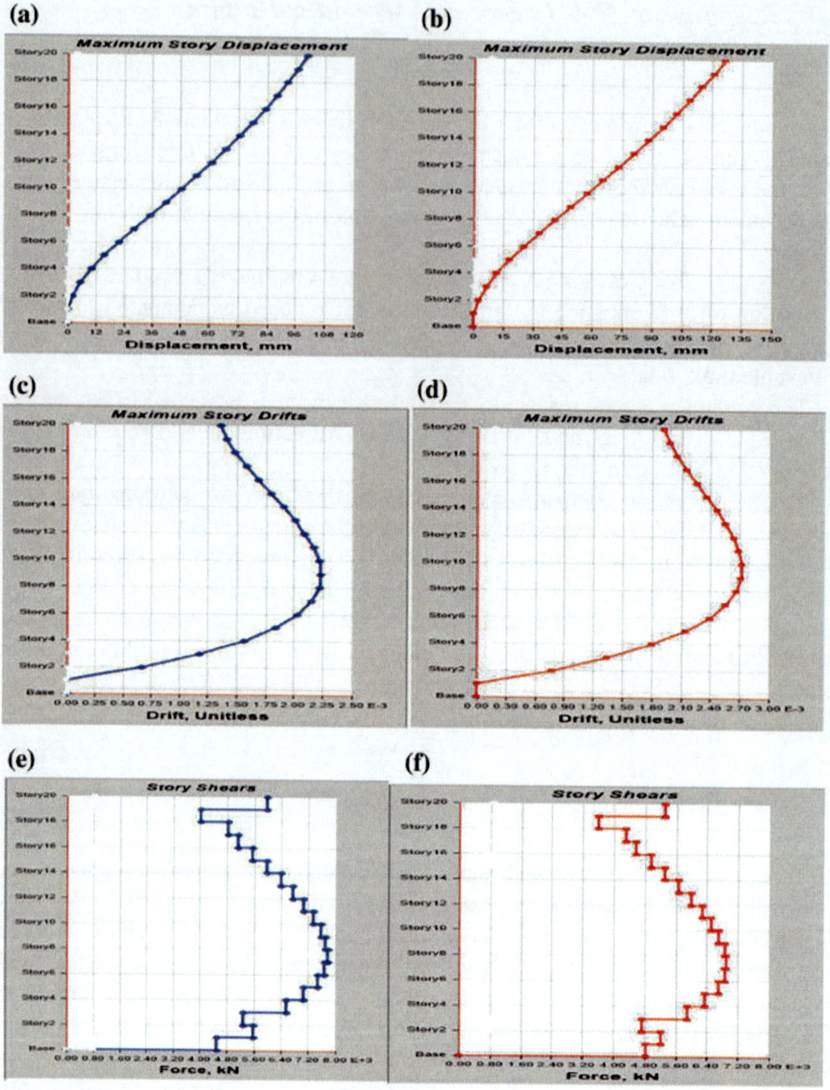


**Fig. 6** **a** Displacement along X-axis, **b** Displacement along Y-axis, **c** Drift along X-direction **d** Drift along Y-direction, **e** Shear along X-direction, **f** Shear along Y-direction, **g** Moment along X-direction, **h** Moment along Y-direction

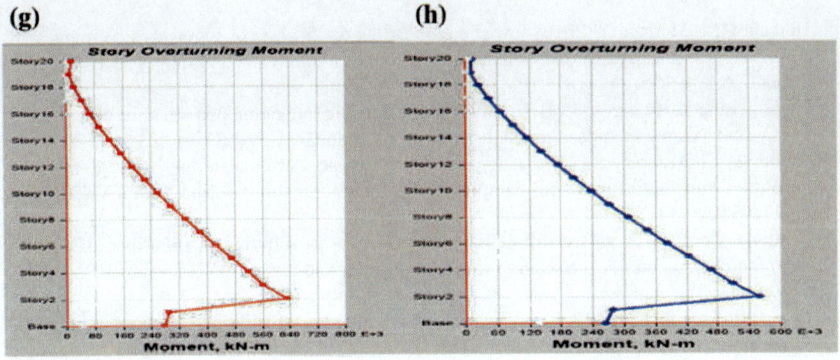


**Fig. 6** (continued)

**Table 1** Comparison of displacement and drift between buildings with and without damper

| Building type | Displacement (mm) | | Drift | |
|---|---|---|---|---|
| | Along X-direction | Along Y-direction | Along X-direction | Along Y-direction |
| Regular building without damper | 169.88 | 189.42 | 0.004553 | 0.004844 |
| Regular building with damper | 110.38 | 121.44 | 0.0025 | 0.0027 |
| Irregular building without damper | 170.84 | 205.77 | 0.004439 | 0.004915 |
| Irregular building with damper | 100.63 | 126.57 | 0.0022 | 0.0027 |

## References


1. Aydin E, Boduroglu MH, Guney D (2007) Optimal damper distribution for seismic rehabilitation of planar building structures. J Eng Struct 29:176–185
2. Aydin E, Oztürk B, Dutkiewicz M (2019) Analysis of efficiency of passive dampers in multistory buildings. J Sound Vib 439:17–28
3. Criteria for earthquake resistant design of structures (IS 1893 (part 1))
4. Code of practice for design loads (other than earthquake) for buildings and structures (IS 875 (part II) 1987
5. Ductile detailing of reinforced concrete structures subjected to seismic forces-code of practice (IS 13920 (2016))
6. Domenico De, Ricciardi G, Takewaki I (2019) Design strategies of viscous dampers for seismic protection of building structures: a review. J Soil Dyn Earthq Eng 118:144–165
7. Eldin MN, Dereje AJ, Kim J (2020) Seismic retrofit of RC buildings using self-centering PC frames with friction dampers. J Eng Struct 208
8. Hassan WH, Reyes JC (2020) Assesment of modal pushover analysis for mid-rise concrete buildings with and without viscous dampers. J Build Eng 29
9. Hatzigeorgiou GD, Pnevmatikos NG (2014) Maximum damping forces for structures with viscous dampers under near-source earthquakes. J Eng Struct 68:1–13

10. Hessabi RM, Mercan O (2016) Investigation of the application of gyro mass dampers with various types of supplementary dampers for vibration control of building structure. J Eng Struct 126:174–186
11. Hu G , Wang Y, Huang W, Li B, Luod B (2020) Seismic mitigation performance of structures with viscous dampers under near-fault pulse-type earthquakes. Eng Struct 203:109878
12. Kumar S, Chakraborty SK (2020) Reduction of seismic vibration in multi-storey structures retrofitted with nonlinear viscous dampers using mode summation method. J Appl Math Model 86:294–310
13. Landi L, Conti F, Diotallevi PP (2015) Effectiveness of different distributions of viscous damping coefficients for the seismic retrofit of regular and irregular RC frames. J Eng Struct 100:79–93
14. Lee DG, Hong S, Kim J (2002) Efficient seismic analysis of building structures with added viscoelastic dampers. J Eng Struct 24:1217–1227
15. Liu Y, Wu J, Donà M (2018) Effectiveness of fluid-viscous dampers for improved seismic performance of inter-storey isolated buildings. J Eng Struct 169:276–292
16. Milanchian R, Hosseini M (2020) Torsional response reduction of plan-asymmetric vertical seismic isolation by appropriate distribution of viscous and viscoelastic dampers. J Struct 27:962–974
17. Nabid N, Hajirasouliha I, Margarit DE, Petkovsk M (2020) Optimum energy based seismic design of friction dampers in RC structures. J Struct 27:2550–2562
18. Purayil AM, Lavan O, Dhakala RP (2019) Multi-objective loss-based optimization of viscous dampers for seismic retrofitting of irregular structures. J Soil Dyn Earthq Eng 129(2):1–12
19. Pérez SPO, Miguel LFF, Riera JD (2019) Reliability-based optimum design of passive friction dampers in buildings in seismic regions. J Eng Struct 190:276–284
20. Walsh KK, Sallar G, Steinberg EP (2019) Hybrid base-isolation of a nonlinear building using a passive resettable stiffness damper. J Eng Struct 178:206–211
21. Wang J, Wang B, Liud Z, Lie H, Zhange C (2020) Seismic response mitigation of building structures with a novel vibro-impact dual-mass damper. J Eng Struct 215
22. Zhou Y, Chen P, Zhang D, Gong S, Wensheng L (2018) A new analytical model for viscous wall dampers and its experimental validation. J Eng Struct 163:224–240

# Seismic Control of Structures with Mass Irregularities Through Optimally Placed Active Tendons Using Multi-objective Genetic Algorithm

F. Rather and M. Alam

## 1 Introduction

The performance of a structural building in a seismic event is affected by the distribution of its mass, stiffness, and strength both in lateral and vertical directions [1]. IS 1893 (Part 1): 2016 classifies irregularity into two types, viz., plan irregularity and vertical irregularity. Mass irregularity is considered to exist in structural buildings when the seismic weight of any floor is more than 150% than that of the floor below [2]. How does mass irregularity influence the performance of a framed structure has been studied in this research. Al-Ali and Krawinkler [3] investigated the effect of mass irregularity on inelastic response parameters of a shear building frame like storey drift, ductility demand, or energy dissipation. Mass irregularity was introduced through mass modification factors varying from 0.25 to 4. Small variation in mass has insignificant effect on these inelastic response parameters. Valmundson and Nau [4] considered 5, 10, and 20-storey shear buildings and varied mass of one of their floors and kept mass of remaining floors constant. For 1.5 mass ratio, they observed an increase of up to 20% in ductility demand. Irregularity in lower floors proved more critical. Michalis et al. [5] performed incremental dynamic analysis on buildings with mass of some floors twice the mass of the corresponding floors in reference regular frame and found that the effect of mass irregularity is more on inter storey drift. Vinod et al. [6] through their study on irregularity concluded that both its extent and location had an effect on storey drift as well as floor displacements of a structural building. The effect is more when the mass of the lower and upper storeys increases as compared with increase in mass of middle storeys. Poncet and Trembler [7] considered eight-storey concentrically braced steel frame with setbacks to study

F. Rather (✉) · M. Alam
Jamia Millia Islamia, New Delhi, India

M. Alam
e-mail: malam1@jmi.ac.in



S. Kolathayar and S. C. Chian (eds.), *Recent Advances in Earthquake Engineering*, Lecture Notes in Civil Engineering 175,
https://doi.org/10.1007/978-981-16-4617-1_29

its performance due to mass irregularity. Two mass ratios and three locations of mass irregularity were considered. These irregular structures showed lower performance than the regular reference frame. Magliulo et al. [8] studied the behaviour of a structure having mass, stiffness, and strength irregularities and subjected to earthquake loading. They found that the plastic demand is affected by strength irregularity and not by mass irregularity. The strength irregularity had no impact on seismic demand of columns, however; it increased the seismic demand in beams. Yael Daniel and Oen Lavan [9] used multi-tuned mass damper in irregular structures for seismic control. They presented a method to optimize the location and size of these dampers. Total mass of the damper was considered as an objective function and minimized while floor accelerations were used as constraints. The method proved effective in seismic response control of the irregular structures and was found applicable to all types of irregularities. Nazarimofrad and Mehdi [10] controlled the response of an irregular multi-storey building subjected to seismic load using an active tendon system. Soil structure interaction effect was also considered and control forces were generated through linear quadratic regulator control algorithm. The results indicated that the method is less efficient in reducing the responses of the buildings located in soft soils.

The optimal location of actuators had been found applying different optimization methods in seismic control of both regular and irregular buildings by researchers. Nazarimofrad et al. [11] used multi-objective genetic algorithm to obtain the optimum number and location of active tendons in 3D irregular buildings with plan irregularity. The method proved effective in reducing actuator requirement by 50%. Rao and Sivasubramanian [12] used multiple start-guided neighbourhood search algorithm and proposed a method to get the optimal actuator location in active seismic control. Rao et al. [13] did research on optimal placement of actuators in tall buildings applying the genetic algorithm. For the same purpose, Liu et al. [14] used a discrete nonlinear optimization method and genetic algorithm. Askari et al. [15] used multi-objective genetic algorithm in active control and magnetorheological dampers in semi-active control simultaneously.

In this research, the relative performance of a regular and an irregular deficient shear building frame having mass irregularities at different floors is first investigated, compared and then an optimized active tendon system operated through LQR control algorithm is used to control its response to an earthquake using multi-objective genetic algorithm. The two objectives minimized in Pareto optimization are the ratio of the controlled and uncontrolled base shear and the number of tendons/actuators (cost of control system).

## 2 Problem Formulation

A model of the deficient frame that is regular having active tendons in each storey with the bay width and height 4 m and 3 m, respectively, is shown in Fig. 1. The mass of each floor and stiffness of each storey are 17,100 kg and $2.68 \times 10^7$ N/m, respectively. The tendons are inclined at $36.87^0$ with the floors and have stiffness equal to $2.30 \times 10^6$ N/m. The size of both beams and columns is the same, i.e. 300 mm × 300 mm and 5% damping has been assumed. The free body diagrams have also been shown with the frame. In matrix form, the equation of motion is:

$$M\ddot{X} + C\dot{X} + KX = \delta \ddot{X}g + \gamma U \tag{1}$$

where

$$M = \begin{bmatrix} m_1 & \cdots & 0 \\ \vdots & \ddots & \vdots \\ 0 & \cdots & m_{10} \end{bmatrix}, C = \begin{bmatrix} c_1 + c_2 & -c_2 & \cdots & 0 \\ -c_2 & \ddots & \cdots & \vdots \\ \vdots & \vdots & \ddots & -c_{10} \\ 0 & \cdots & -c_{10} & c_{10} \end{bmatrix}$$

$$K = \begin{bmatrix} k_1 + k_2 & -k_2 & \cdots & 0 \\ -k_2 & \ddots & \cdots & \vdots \\ \vdots & \vdots & \ddots & -k_{10} \\ 0 & \cdots & -k_{10} & k_{10} \end{bmatrix}$$

In the above equation, $m_i$, $k_i$, and $c_i$ denote, respectively, the mass, stiffness, and damping of the *i*th storey where i = 1, 2, 3, …, 10. X(t), and U(t) are floor displacements and control forces, respectively. [γ] denotes the location of the tendons and {δ} is coefficient vector for ground acceleration. P, α, and $k_c$ are the prestressing force, inclination, and stiffness of tendons, respectively.

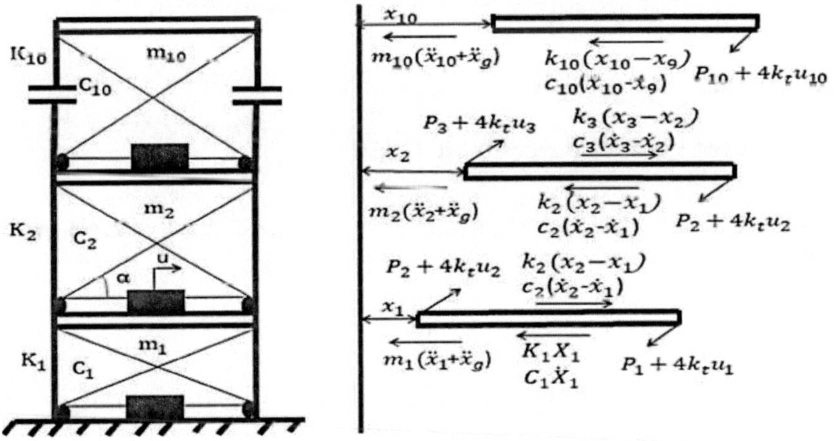


**Fig. 1** Ten-storey shear building frame and its free body diagram

**Table 1** Mass modification factors

| Case | 1 | 2 | | | 3 | | | 4 | | | 5 | | | 6 | | | 7 | | |
|---|---|---|---|---|---|---|---|---|---|---|---|---|---|---|---|---|---|---|---|
| MF | 0 | 1.5 | | | 2 | | | 2.5 | | | 3 | | | 3.5 | | | 4 | | |
| Sub-cases | 0 | a | b | c | a | b | c | a | b | c | a | b | c | a | b | c | a | b | c |
| Storey number | 0 | 1 | 2 | 3 | 1 | 2 | 3 | 1 | 2 | 3 | 1 | 2 | 3 | 1 | 2 | 3 | 1 | 2 | 3 |

$$\{\delta\} = -M * I = [-m_1, -m_2 \ldots m_{10}]^T$$

$$U(t) = [u_1, u_2, \ldots u_r]^T, X(t) = [x_1, x_2, \ldots, x_{10}]^T, \gamma = \overline{\gamma}\theta$$

$$\overline{\gamma} = 4k_c\cos\alpha \begin{bmatrix} -1 & 1 & \ldots & 0 \\ 0 & \ddots & \ldots & \vdots \\ \vdots & \ldots & -1 & 1 \\ 0 & \ldots & 0 & -1 \end{bmatrix}, \theta = \begin{bmatrix} 1 & \cdots & 0 \\ \vdots & \ddots & \vdots \\ 0 & \cdots & 1 \end{bmatrix}$$

I represents the influence coefficient vector.
In state-space Eq. 1 is written as:

$$\dot{Z} = AZ + B_r\ddot{X}_g + B_uU \tag{2}$$

where $A = \begin{bmatrix} [0]_{10\times10} & [1]_{10\times10} \\ \left[\frac{-K}{M}\right] & \left[\frac{-C}{M}\right] \end{bmatrix}_{20\times20}, B_r = \begin{Bmatrix} \{0\} \\ \left\{\frac{\delta}{M}\right\} \end{Bmatrix}, B_u = \begin{Bmatrix} \{0\} \\ \left\{\frac{\gamma}{M}\right\} \end{Bmatrix}_{20\times r}$

Mass irregularity has been introduced in the above frame using six modification factors 1.5, 2, 2.5, 3, 3.5, and 4 in the first, second, and third storeys separately under different cases and sub-cases as shown in Table 1.

## 3 Linear Quadratic Regulator (LQR)

In LQR, a quadratic performance index shown in Eq. (3) is minimized to obtain the control force 'U'.

$$J = \int_0^{t_f} \left(Z^T(t)QZ(t) + U^T(t)RU(t)\right)dt \tag{3}$$

In the above equation, $t_f$ is the duration of an earthquake. Z is given by Eq. (2). Q is a $2n \times 2n$ positive semi-definite matrix, and R is an $n \times r$ positive definite matrix, n is the degrees of freedom, and r is number of actuators employed. In the present optimization, Eq. (2) is used as a constraint and control force is obtained as given in Eq. (4).

$$U(t) = \left(\frac{-1}{2}\right)R^{-1}B^{T}PZ(t) = -GZ(t) \quad (4)$$

Equation (5) given below is called matrix Riccati equation provides the value of P

$$\left[PA - \left[\frac{1}{2}\right]PBR^{-1}B^{T}P + A^{T}P + 2Q\right] = 0 \quad (5)$$

## 4 Multi-objective Genetic Algorithm

Genetic algorithm is a method of optimization that works on the principles of genetics and natural selection. This method can handle large number of variables and is also suitable for non-differentiable objective functions. The three main operations in genetic algorithm are selection of chromosomes, crossover, and mutation. Multi objective genetic algorithm is a type of optimization where more than one conflicting objective functions are optimized simultaneously using genetic algorithm. It gives a set of optimal solutions termed as pareto front. In this research, multi-objective genetic algorithm is used to minimize the ratio of controlled to uncontrolled base shear of the frame ($J_1$) and number of tendons/actuators to be installed ($J_2$). Maximum storey drift and maximum floor displacement are used as constraints and the design variables (X) are the position of tendons.

$$\text{Minimize}: \quad J_1(X), J_2(X) \quad (6)$$

$$\text{Subjected to constraints:} \quad CSD(X) \le SD_P,\ CFD(X) \le FD_P \quad (7)$$

$SD_p$ and $FD_p$ are the permissible limits of maximum storey drift and maximum floor displacement.

## 5 Results and Discussion

The simulation of the deficient shear building frame with and without mass irregularities has been done in Mat Lab 2019 in this study. The comparisons of uncontrolled responses, viz., floor displacements and storey drifts with mass irregularities at different floors of the cases 2–7 with that of the regular frame of case 1 are presented in Fig. 2. Uncontrolled peak displacements, in most of the cases, of modification factors have decreased when the mass irregularity is present in the second storey and third storey. The maximum reduction is 7% and has occurred in case 4c. Mass irregularity in the first storey leads to increase peak floor displacement for all modification

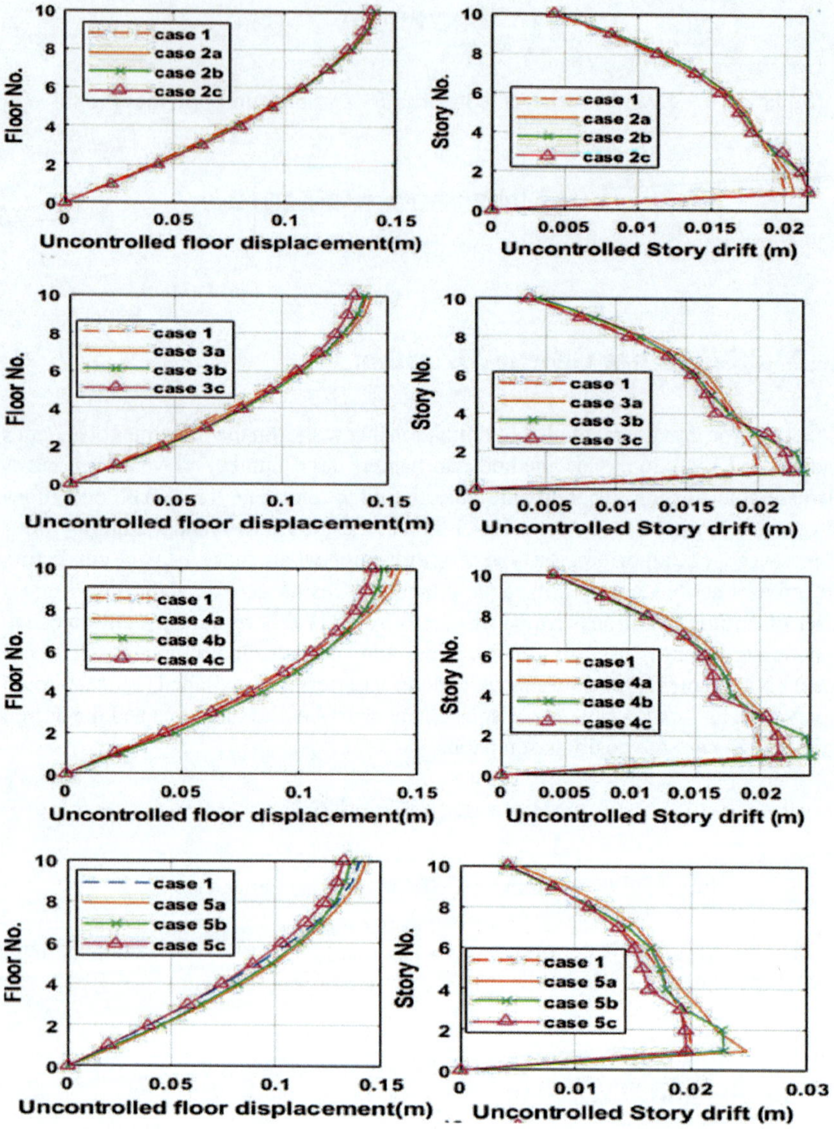


**Fig. 2** Comparison of the uncontrolled floor displacement and storey drift of irregular frames with regular one

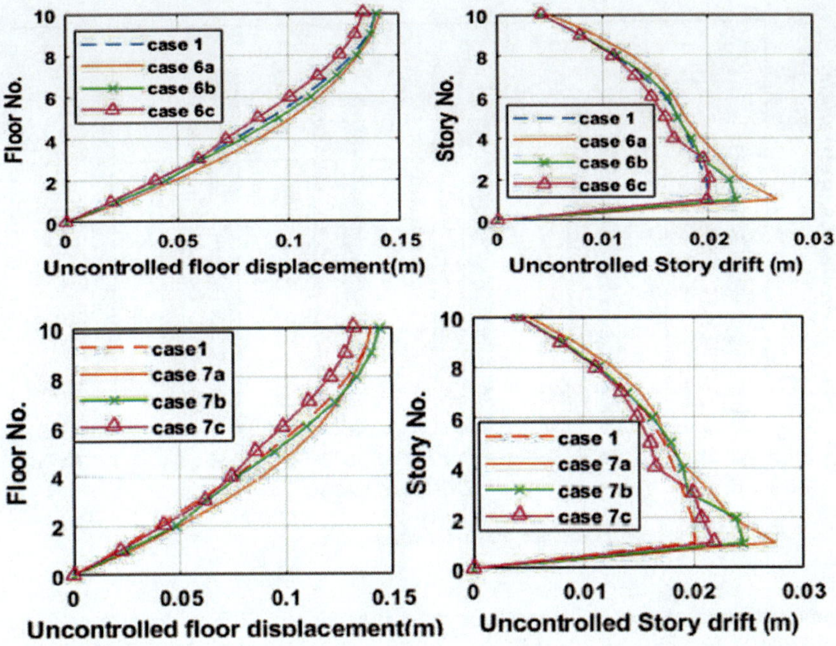


**Fig. 2** (continued)

factors. The maximum increase of peak floor displacement occurs in case 7b and is about 4% as shown in Fig. 2. Drift in first, second, and third storeys has increased in almost all cases of mass irregularity as compared with reference regular frame. The maximum increase in peak storey drift is 36.31% and has occurred in case 7a (mass irregularity in first storey with modification factor of 4) as shown in Fig. 2.

The base shear in almost all cases with mass irregularity has increased as compared with reference regular frame as shown in Fig. 3. The maximum increase of base shear has taken place in case 7a (mass irregularity in first storey with MF 4) and is equal to 37%.

In all the cases discussed, it is found that the maximum floor displacement and maximum storey drift of both, i.e. regular and irregular frames are beyond their prescribed permissible limits, i.e. 60 mm and 12 mm, respectively, and thus need to be controlled. For this purpose, linear quadratic regulator (LQR) control algorithm is used in this study to obtain control force required for active tendon system present in some storeys of the frame. This active tendon system is a combination of active tendons and actuators. The optimal location of these tendons is obtained through multi-objective genetic algorithm. The two conflicting objectives minimized for all cases are the controlled to uncontrolled base shear ratio of the frame and the number of tendons (cost of control system) against maximum floor displacement and storey drift as constraints. All the possible combinations of the two conflicting objectives known as Pareto curves are obtained and presented in Fig. 4. In this study, the selected

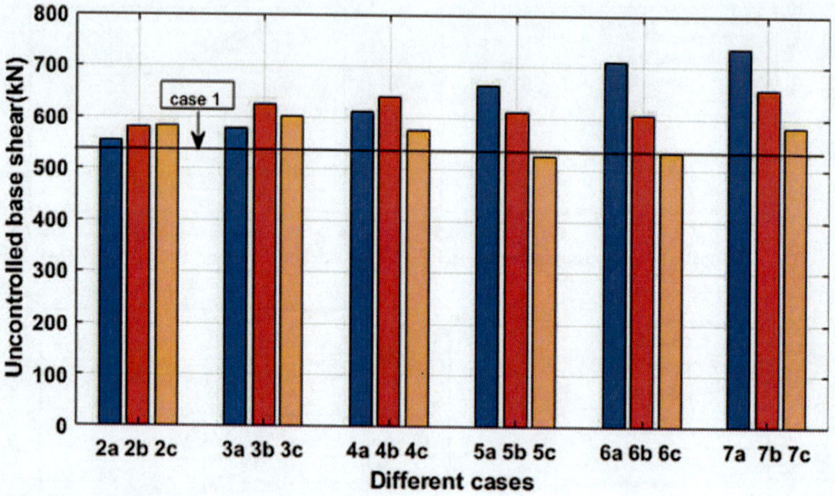


**Fig. 3** Uncontrolled base shear of the irregular frames

combination in each case having base shear reduction of 70% with corresponding optimal position of the tendons is given in Table 2. In the table, 'X' and '✓' denote the absence and presence of tendon in a storey, respectively. The table indicates that MF greater than or equal to 2 especially at the second floor requires an extra tendon for the same reduction of responses as that of the reference regular frame. There is no need to place tendons in the upper three storeys in almost all the cases of irregularities considered and placement of tendon in the first storey is indispensable.

The controlled and uncontrolled responses in all cases with tendons placed optimally as per Table 2 are compared in Fig. 5. In the figure, UC, C, and P stand for uncontrolled, controlled, and permissible, respectively. As can be seen in the figures, the optimal control system keeps the responses within the permissible limits in all the cases of mass irregularity. Also, the base shear in each and every case has reduced by at least 70% (Fig. 6).

The control forces in each of the optimally placed tendons for the considered seven cases of mass irregularity with modification factors up to 4 are shown in Table 3.

## 6 Conclusion

This study is done on active seismic control of structural buildings with mass irregularities. For this purpose, an active tendon system with tendons/actuators placed optimally using multi-objective genetic algorithm is employed through linear quadratic regulator. The numerical analysis is done on a 10-storey deficient shear frame with mass irregularity in the first, second, and third storeys introduced with modification

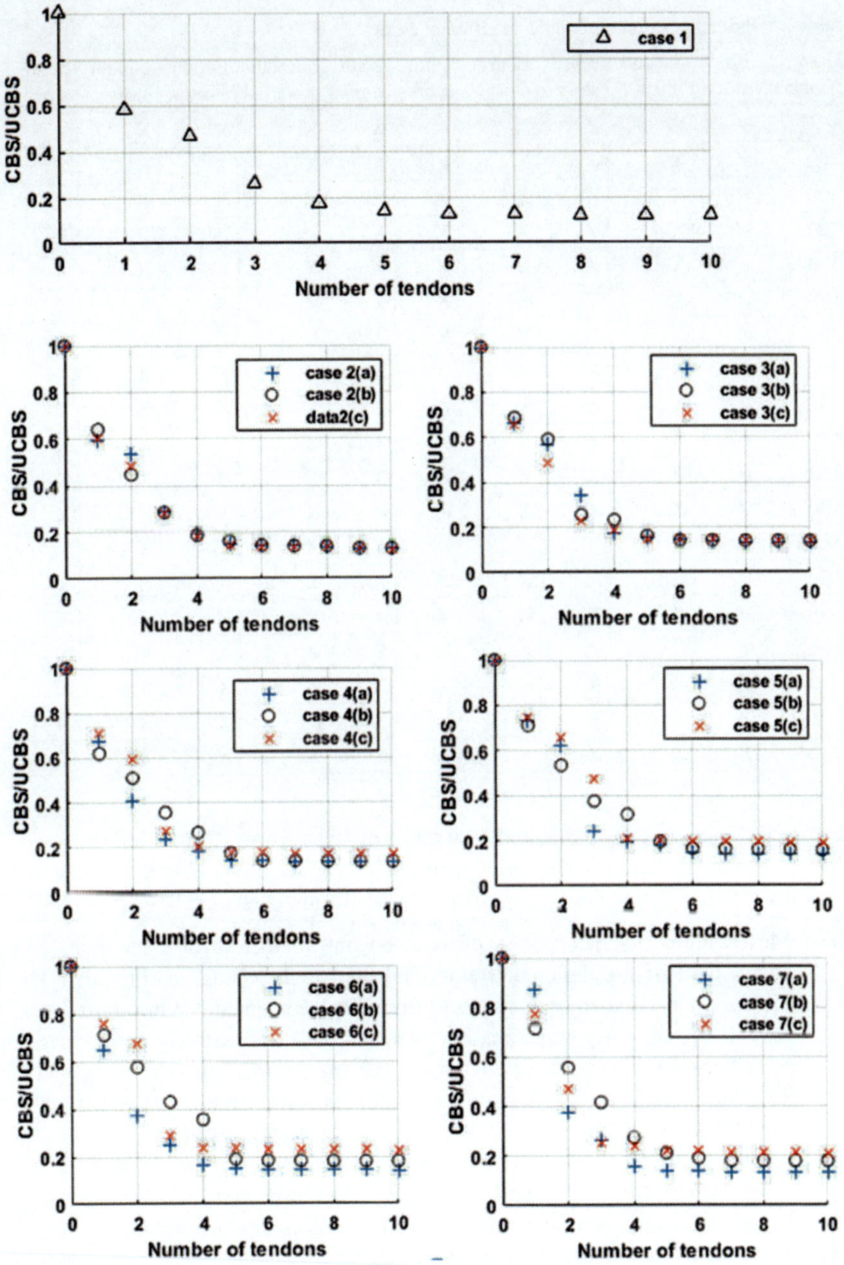


**Fig. 4** Pareto curves

**Table 2** Optimal position of tendons in different cases

| Case Number | First story | Second story | Third story | Fourth story | Fifth story | Sixth story | Seventh story | Eighth story | Nineth story | Tenth story |
|---|---|---|---|---|---|---|---|---|---|---|
| 1 | ✓ | X | ✓ | X | ✓ | X | X | X | X | X |
| 2(a) | ✓ | X | ✓ | X | ✓ | X | X | X | X | X |
| 2(b) | ✓ | X | ✓ | X | ✓ | X | X | X | X | X |
| 2(c) | ✓ | X | ✓ | X | X | X | ✓ | X | X | X |
| 3(a) | ✓ | X | ✓ | X | ✓ | ✓ | X | X | X | X |
| 3(b) | ✓ | X | ✓ | X | X | X | ✓ | X | X | X |
| 3(c) | ✓ | X | ✓ | X | X | ✓ | X | X | X | X |
| 4(a) | ✓ | X | X | ✓ | ✓ | X | X | X | X | X |
| 4(b) | ✓ | X | ✓ | X | ✓ | ✓ | X | X | X | X |
| 4(c) | ✓ | X | ✓ | X | ✓ | X | X | X | X | X |
| 5(a) | ✓ | X | X | ✓ | ✓ | X | X | X | X | X |
| 5(b) | ✓ | X | ✓ | X | ✓ | ✓ | X | X | X | X |
| 5(c) | ✓ | ✓ | X | ✓ | X | X | X | ✓ | X | X |
| 6(a) | ✓ | X | X | ✓ | ✓ | X | X | X | X | X |
| 6(b) | ✓ | ✓ | X | X | X | ✓ | X | X | ✓ | X |
| 6(c) | ✓ | X | ✓ | X | X | X | ✓ | X | X | X |
| 7(a) | ✓ | X | X | ✓ | ✓ | X | X | X | X | X |
| 7(b) | ✓ | ✓ | X | ✓ | X | X | ✓ | X | X | X |
| 7(c) | ✓ | X | ✓ | X | X | ✓ | X | X | X | X |

factors (MF) 1.5, 2, 2.5, 3, 3.5, and 4 as seven different cases. Controlled to uncontrolled base shear ratio and the number of tendons are minimized with constrained peak floor displacement and storey drift. The results are presented below:

(1) Mass irregularity has influenced both uncontrolled floor displacements and storey drift of the deficient frame subjected to El Centro earthquake. The maximum increase in peak values of floor displacement and storey drift due to the mass irregularity considered are 4% and 36%, respectively. In few cases, peak floor displacement has decreased with maximum reduction being 7%.
(2) Top three storeys do not require any tendon, however, placement of tendon in the first storey is indispensable as per the optimization results.
(3) An extra tendon is required in case of mass irregularity in the second and third storeys of the frame with modification factor greater than 2 for the same reduction of responses as compared with that of the regular frame.
(4) Even deficient frames with mass irregularity can be made safe without retrofitting using active control system.
(5) The maximum force in a tendon under any case is 49.03 kN. The maximum total control force is 140.79 kN and is generated when the mass irregularity is present in the third storey with modification factor equal to 3.

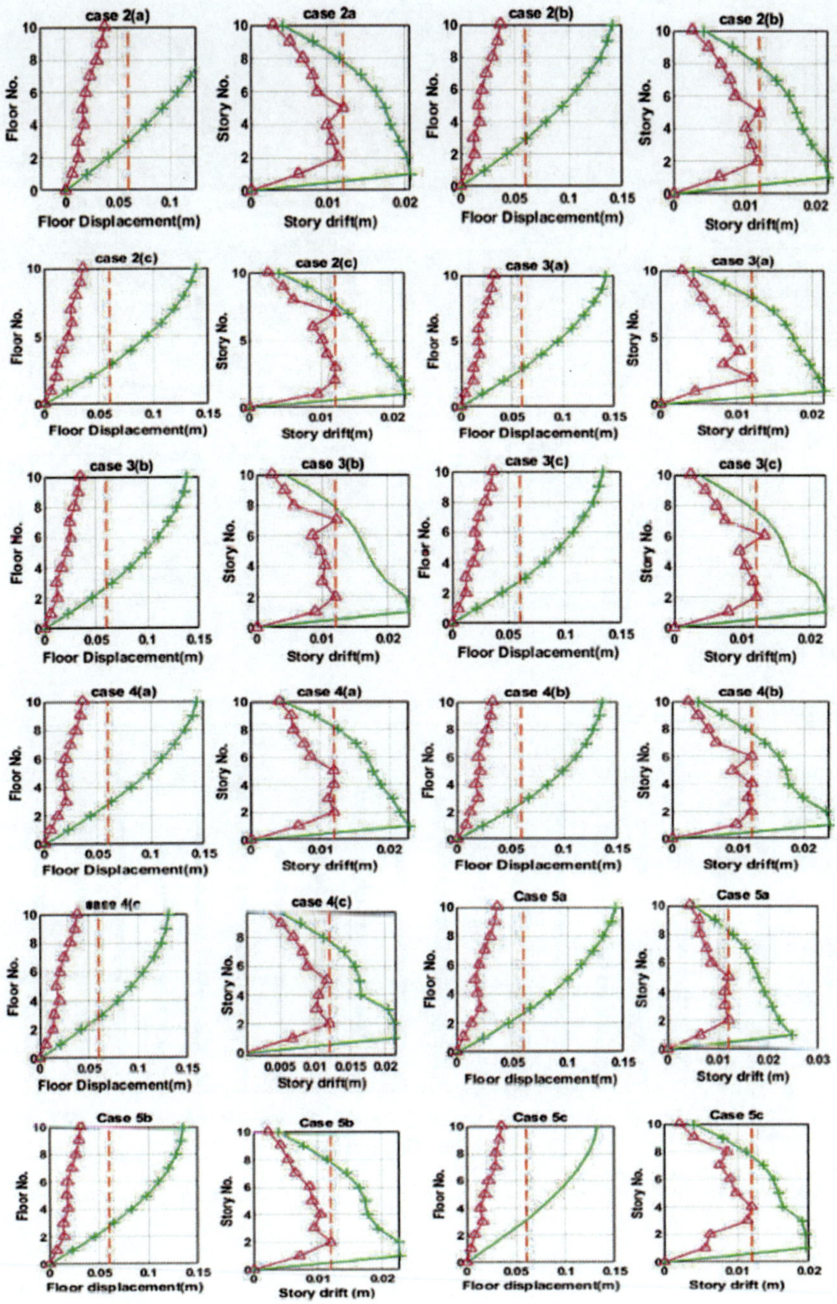


**Fig. 5** Comparison of controlled and uncontrolled responses

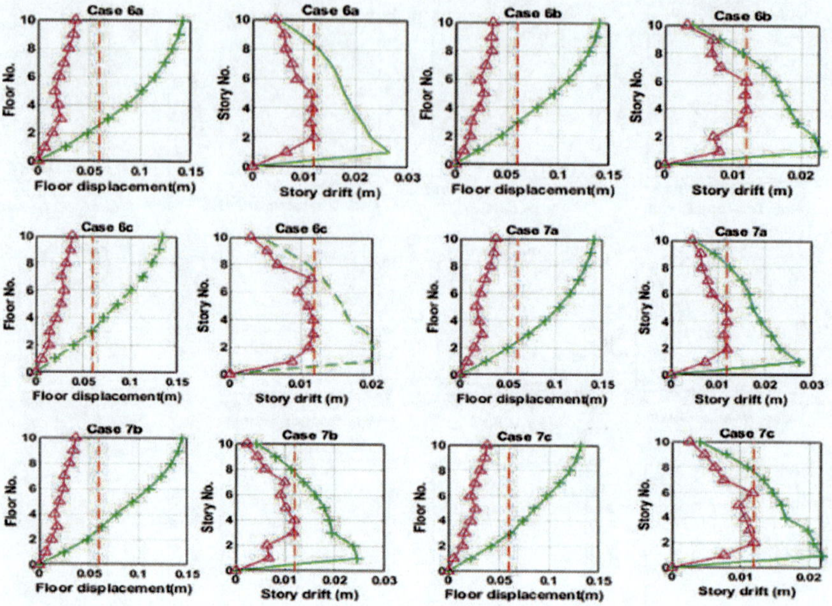


**Fig. 5** (continued)

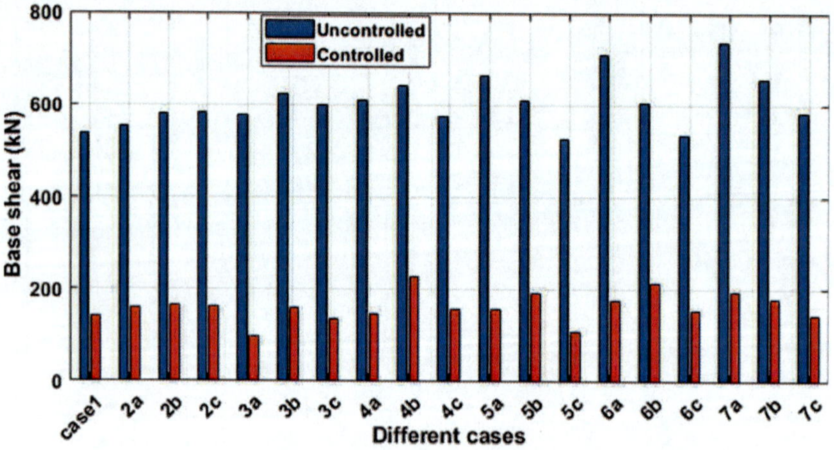


**Fig. 6** Comparison of controlled and uncontrolled base shear of the frame for different cases

**Table 3** Control force in tendons under different cases

| Cases | Control forces (kN) | | | | |
|---|---|---|---|---|---|
| | First tendon | Second tendon | Third tendon | Fourth tendon | Total |
| 1 | 40.05 | 39.51 | 37.63 | 0 | 117.19 |
| 2a | 27.19 | 40.44 | 33.11 | 0 | 105.74 |
| 2b | 40.99 | 40.52 | 33.1 | 0 | 119.61 |
| 2c | 45.43 | 30.15 | 31.63 | 0 | 107.21 |
| 3a | 13.66 | 35.69 | 30.32 | 27.4 | 112.07 |
| 3b | 45.59 | 45.39 | 31.247 | 0 | 122.727 |
| 4b | 40.93 | 36.72 | 30.62 | 27.53 | 135.9 |
| 4c | 42.03 | 16.49 | 40.6109 | 0 | 99.1309 |
| Sa | 15.51 | 41.45 | 33.06 | 0 | 95.02 |
| 5b | 42.53 | 37.23 | 30.37 | 27.75 | 133.33 |
| 5c | 33.41 | 33.42 | 40.33 | 23.63 | 140.79 |
| 6a | 13.4 | 42.16 | 33.52 | 0 | 94.03 |
| 6b | 14.2 | 14.26 | 39.63 | 16.43 | 34.57 |
| 6c | 49.03 | 13.33 | 35.39 | 0 | 93.3 |
| 7a | 11.9 | 42.94 | 39.02 | 0 | 93.36 |
| 7b | 9.77 | 9.3 | 40.94 | 29.07 | 39.53 |
| 7c | 46.55 | 45.51 | 40.09 | 0 | 132.15 |

**Acknowledgements** First, I owe God for allowing me to complete the present work. Second, I am grateful to my guide for his perpetual support and advice. Finally, I am thankful to Jamia Millia Islamia (A Central University) for allowing me to do my research.

## References

1. Bhosale AS, Davis R, Sarkar P (2016/2017) Vertical irregularity of buildings: regularity index versus seismic risk. ASCE-ASME J Risk uncertainty Eng Syst Part A: Civ Eng 3(3); 2(5):99–110
2. IS 1893 (Part 1):2016, Criteria for earthquake resistant design of buildings
3. Ali AAK, Krawinkler H (1998) Effects of vertical irregularities on seismic behavior of Building Structures. Rep. No. 130. Department of Civil and Environmental Engineering, Standford University San Francisco
4. Valmundson EV, Nau JM (1997) Seismic response of building frames with vertical structural irregularities. J Struct Eng ASCE 123(1):30–41
5. Michalis F, Vamvatsikos D, Monolis P (2006) Evaluation of the influence of vertical irregularities on the seismic performance of a nine-storey steel frame. Earthq Eng Struct Dyn 35:1489–1509
6. Vinod KS, Gregory AM, Brucel D (2009) Determination of structural irregularity limits-mass irregularity example. Bull NZ Soc Earthq Eng 42:288–301

7. Tremblay R, Poncet L (2005) Seismic performance of concentrically braced steel frames in multistorey buildings with mass irregularity. J Struct Eng 131:1363–1375
8. Magliulo G, Ramasco R, Realfonzo R (2002) A critical review of seismic code provisions for vertically irregular frames. In: Proceedings of the third European workshop on the seismic behavior of the irregular and complex structures, CD ROM, Florence
9. Yael D, Lavan O (2013) Allocation and Sizing of multiple tuned mass dampers for seismic control of irregular structures. In: Lavan O, De Stefano M (eds) Seismic behavior and design of irregular and complex civil structures. Geotech, Geol Earthq Eng 24. Springer
10. Nazarimfrad E, Mehdi S (2016) Seismic control of irregular multistorey buildings using active tendons considering soil-structure interaction effect. Soil Dyn Earthq Eng 89:100–115
11. Narzarimofrad E, Farahani S, Zahrai SM (2018) Multiobjective optimal placement of active tendons to control irregular multistorey buildings with soil-structure interaction. Struct Des Tall Spec Build e1581.10
12. Rao AR, Sivasubramanian K (2008) Comparative study on multi-objective genetic algorithms for seismic response controls of structures. J Sound Vib 311(1–2):133
13. Rao SS, Pan TS, Venkayya VB (1991) Optimal placement of actuators in actively controlled structures using genetic algorithms. AIAA J 29(6), 942.12 (1991)
14. Liu DK, Yang YL, Li QS (2003) Optimum positioning of actuators in tall buildings using genetic algorithm. Comput Struct 81(32), 2823(2003)
15. Askari M, Li J, Samali B (2017) Cost-effective multi-objective optimal positioning of magneto rheological dampers and active actuators in large nonlinear structures. J Intell Mater Syst Struct 28(2)

# Seismic Performance of UHPFRC-Strengthened RC Beam–Column Joints Using Damage Plasticity Model—A Numerical Study

**K. Sai Kubair and J. S. Kalyana Rama**

## 1 Introduction

Beam–column joints are the critical sections in any structure especially when they are subjected to lateral loads. Under severe shaking, beam–column joints become vulnerable to cracking and failure. Several studies have been carried out to strengthen the beam–column joints using materials like fiber-reinforced polymer (FRPs). Ultra High-Performance Fiber Reinforced Concrete (UHPFRC) is one of the recent findings in the last decade, which can be used as a potential strengthening material of reinforced concrete structures. Tayeh et al. [1] explored the possibility of using UHPFRC as a suitable rehabilitation material for a conventional structure subjected to seismic load. The results indicated that UHPFRC is proven to be an excellent material for repair and rehabilitation because of its enhanced durability over other concretes and its low porosity characteristics. Also, the working time of UHPFRC makes it advantageous to be used as a rehabilitation material. Lampropoulos et al. [2] assessed the efficiency of strengthening a reinforced concrete (RC) beam using UHPFRC. Full-scale experimental study on the strengthened beams had been performed using three different strengthening techniques. The results obtained from the study indicated that UHPFRC was efficient in improving the load-carrying capacity of the RC beam. Prem et al. [3] investigated the flexural performance of the pre-damaged RC beam with varying cross section using UHPFRC. UHPFRC strip was attached to the tension face of RC beam with epoxy, and the flexural testing was carried out. The results obtained after testing these beams indicated that UHPFRC was efficient in improving the strength properties of the damaged RC beam to a greater extent. Rahman et al. [4] documented several investigations on the use of UHPFRC

K. Sai Kubair
Delft University of Technology, Delft, Netherlands

J. S. Kalyana Rama (✉)
Ecole Centrale School of Engineering, Mahindra University, Hyderabad, India

 
S. Kolathayar and S. C. Chian (eds.), *Recent Advances in Earthquake Engineering*, Lecture Notes in Civil Engineering 175,
https://doi.org/10.1007/978-981-16-4617-1_30

for the construction of high-rise structures, retrofitting and rehabilitation of structures. Chen et al. [5] studied the structural performance of UHPFRC I-Girders using concrete damage plasticity model of a finite element software ABAQUS/CAE. From the results obtained, it was concluded that the CDP model was efficient in studying the linear and non-linear response of concrete structures. Sai Kubair et al. [6] assessed the performance of RC-framed structure strengthened using UHPFRC with varying thickness. Results indicated that UHPFRC strip with 20 mm thickness enhanced the load-carrying capacity of the structure.

From the existing studies, it can be noted that UHPFRC serves a good material in terms of both strength and durability and can be used for strengthening, retrofitting and rehabilitation of structures because of its enhanced ductility. Concrete damage plasticity model has also gained popularity in the last decade and the same is used in the present study. The influence of variation in the CDP parameters on the behavior of RC beam–column joint has been studied numerically using ABAQUS/CAE. An attempt has been made to assess the nonlinear performance of UHPFRC strip in strengthening RC beam–column joints using the CDP model. The effect of variation in the thickness of the UHPFRC strip on the load-carrying capacity of the RC beam–column joint is also addressed in the study.

## 2 Modeling Concrete Using the CDP Model

### 2.1 Compressive Behavior of Concrete

*Elastic modulus of concrete*: According to IS: 456:2000, the elastic modulus of a particular grade of concrete can be calculated using

$$E_{cm} = 5000(f_{ck})^{0.5} \tag{1}$$

where $f_{ck}$ is the characteristic compressive strength of concrete.

*HSU and HSU model*: This model is used to generate the stress–strain curve of concrete up to a point in the descending part where the stress is equal to 0.3 times the peak stress. The yield stress is equal to half the peak stress value. The maximum strength for this model is 62 MPa. The formulations involved in this model for normal strength concrete are as given in Eqs. 2 and 3

$$\sigma_c = E_{cm}\varepsilon_c \tag{2}$$

Equation (2) is valid up to yield point

$$\sigma_c = \left(\frac{\beta\frac{\varepsilon_c}{\varepsilon_o}}{\beta - 1 + (\frac{\varepsilon_c}{\varepsilon_o})^{\beta}}\right)\sigma_{cu} \tag{3}$$

Equation (3) is used post the yield point.

Where $\varepsilon_c$ and $\varepsilon_o$ are the strain at any point and strain at peak stress in concrete, respectively, $\sigma_{cu}$ is the peak stress in concrete in kip/in$^2$ and β is a parameter that decides the nature of the stress–strain curve. These parameters are obtained using Eqs. 4 and 5.

$$\varepsilon_0 = 8.9 \times 10^{-5}\sigma_{cu} + 2.114 \times 10^{-3} \tag{4}$$

$$\beta = \frac{1}{1 - \left(\frac{\sigma_{cu}}{E_{cm}\varepsilon_o}\right)} \tag{5}$$

## 2.2 *Tensile Behavior of Concrete*

The maximum tensile strength of concrete is calculated based on the formula mentioned in the EUROCODE 2 is given by

$$f_{ctm} = 0.3(f_{ck})^{2/3} \tag{6}$$

where $f_{ctm}$ is the maximum tensile strength of concrete.

This value is given as an input in the concrete damage plasticity (CDP) model in the tension part and the maximum cracking strain value of concrete is taken as a constant equal to 0.01. These two values are given as an input in the tension part of the CDP model.

## 2.3 *Plasticity Parameters in the CDP Model*

There are some parameters used for addressing the plasticity nature of concrete in the CDP model. The values of these parameters used in the present study in modeling concrete using the CDP model are shown (Table 1).

## 2.4 *Predicting the Damage Variables in the CDP Model*

The damage variable in compression (dc) is calculated based on the damage theory as the ratio of inelastic strain in compression (crushing strain) at a particular point to that of the maximum strain allowed in concrete

**Table 1** Plasticity parameters considered in the CDP model

| Parameter | Value |
|---|---|
| $\Psi$ | $34^0$ |
| $K_c$ | 2/3 |
| $f_{b0}/f_{c0}$ | 1.16 |
| $\varepsilon$ | 0.01 |
| $\mu$ | 0 |

$$d_c = \frac{\varepsilon_c^{in}}{\varepsilon_c^{max}} \tag{7}$$

where, $\varepsilon_{cu}$ is the maximum strain in compression that can be allowed in concrete calculated as per the HSU and HSU model.

The maximum value of damage variable in tension ($d_t$) is again taken as a constant, which is equal to 0.9 and the damage value at the yield stress will be equal to zero.

## 2.5 *Modeling Steel Using the Plasticity Theory*

Steel reinforcement inside the beam–column joint is modeled using the plastic theory of metals, which requires yield stress and plastic strain values of steel as an input. These values are taken based on direct tension test performed on steel. Fe 415 is considered for the entire analysis of the present study. The data considered for modeling the steel reinforcement are given in Table 2.

**Table 2** Yield stress and plastic strain data of Fe 415 steel

| Yield stress (MPa) | Plastic strain |
|---|---|
| 332 | 0 |
| 352 | 0.0001 |
| 373 | 0.0003 |
| 394 | 0.001 |
| 435 | 0.002 |
| 435 | 0.003 |
| 440 | 0.005 |
| 435 | 0.01 |
| 400 | 0.03 |
| 370 | 0.06 |

### 2.6 Modeling UHPFRC Using the CDP Model

UHPFRC generally has higher performance in terms of strength and workability when compared with that of normal concrete. UHPFRC is modeled with a varying thickness similar to that of plain concrete and the corresponding linear and non-linear material properties are used. The data required for modeling this concrete using CDP are taken from an experiment conducted by Prem et al. [3] and the details are given in Table 3. Density: 2400 kg/m$^3$, elastic modulus: 40000 MPa, Poisson's ratio: 0.18.

## 3 Designing the RC Beam–Column Joint

The beam–column joint used in the present study for analysis is designed as per IS: 13,920:2016, the Indian Standard Code of Practice for earthquake-resistant design of structures. The grade of concrete used for analysis is M20. The cross-sectional dimensions of beam and column considered are 300 mm × 450 mm and 300 mm × 530 mm, respectively. The beam–column joint is designed in STAAD PRO V8i according to IS: 1893 for earthquake-resistant design. The parameters considered for the design purpose are as follows:

i. Earthquake zone: V.
ii. Soil type: loose soil.
iii. Damping ratio: 5%.
iv. It is considered to be a highly important structural component.

**Table 3** CDP data for UHPFRC

| Compression behavior of UHPFRC | | | Tension behavior of UHPFRC | | | |
|---|---|---|---|---|---|---|
| Compressive stress (MPa) | Inelastic strain | Damage parameter | Tensile stress (MPa) | Crack opening (mm) | Damage parameter | Displacement, (mm) |
| 107.33 | 0 | 0.000 | 13.5 | 0 | 0 | 0 |
| 114.65 | 0.0032 | 0.032 | 5.5 | 1.235 | 0.564 | (1.983 |
| 124.43 | 0.0038 | 0.123 | 0 | 3.786 | 0.988 | 3.786 |
| 113.32 | 0.0044 | 0.221 | | | | |
| 93.54 | 0.0054 | 0.343 | | | | |
| 64.76 | 0.0063 | 0.564 | | | | |
| 41.32 | 0.0074 | 0.724 | | | | |
| 25.92 | 0.0083 | 0.872 | | | | |
| 14.41 | 0.0092 | 0.954 | | | | |
| 9.76 | 0.012 | 0.972 | | | | |

The occurrence of the earthquake is assumed to be in all four possible directions, i.e. + X, −X, +Z & −Z. The corresponding earthquake load definitions are assigned to the structure. The static load details used for this design are as follows:

i. *Dead loads (DL):*Self-weight = wall load = 12 KN/m.
- *Live loads (LL):*

i. Floor load = 3 KN/m$^2$.

Based on IS: 1893—Part-II, all the load combinations are generated depending on the nature of the general load cases that are applied to the structure.

### *3.1 RCC Design Details of the Beam–Column Joint*

The details of reinforcement that are provided inside beams and columns of the beam–column joint are given in Tables 4 and 5.

## 4 Finite Element Modeling of the Beam–Column Joint

This beam–column joint designed as per IS: 13,920:2016 is modeled in ABAQUS/CAE. Beams and columns are modeled using solid element and the reinforcing bars, stirrups and links are modeled using wire elements. The concrete part is meshed using the C3D8R (eight-noded linear brick) element and the reinforcement (steel bars and stirrups) are meshed using the T3D2 (two-noded truss) element. A

**Table 4** Beam reinforcement details in the beam–column joint

| S. No | Grade of concrete (MPa) | Beam reinforcement details | | |
|---|---|---|---|---|
| | | Top reinforcement (mm$^2$) | Bottom reinforcement (mm$^2$) | Stirrups |
| 1 | M20 | 4–16 ϕ | 4–22 ϕ | 8 mm ϕ@200 mm |

**Table 5** Column reinforcement details in the beam–column joint

| S. No | Grade of concrete (MPa) | Location of links | Column reinforcement details | |
|---|---|---|---|---|
| | | | Main Reinf (mm$^2$) | Links |
| 1 | M20 | Near the joint | 6–20 ϕ | 8 mm ϕ@150 mm |
| 2 | | Away from the joint | | 8 mm ϕ@80 mm |

*Reinf.* means Reinforcement

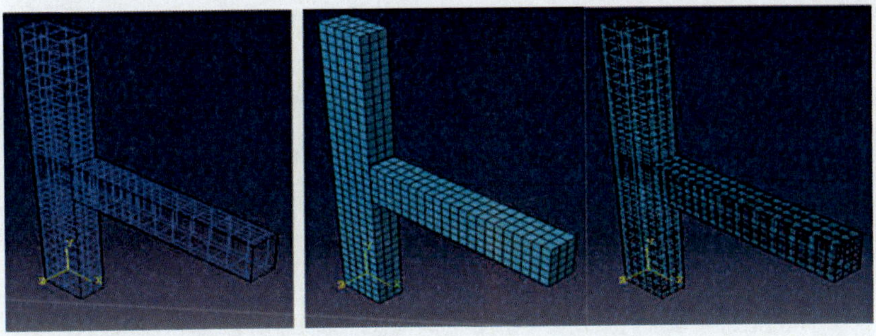

**Fig. 1** Reinforcement layout and meshing of concrete and steel reinforcement

constant element size of 100 is maintained throughout the analysis for all the components of the structure. The reinforcement assembly inside the beam column, the mesh details of the beam–column joint and the reinforcement are shown in Fig. 1.

## 4.1 Loading Details

The response of beam–column joint is studied by subjecting it to Elcentro, 1940 earthquake ground motion at the Imperial Valley in Southern California. The load is applied as a ground acceleration at the base of the beam–column joint. First 5 s of the data is used for simulating the joint as the peak ground acceleration is maximum in the first 5 s. The acceleration time history of the Elcentro earthquake considered in the present study is shown in Fig. 2.

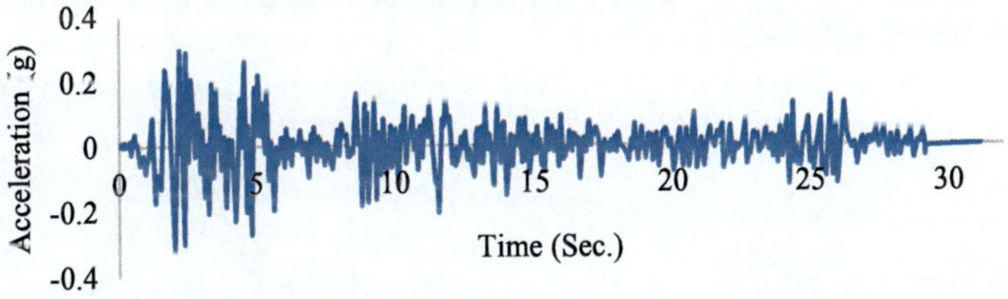


**Fig. 2** Elcentro earthquake time history data

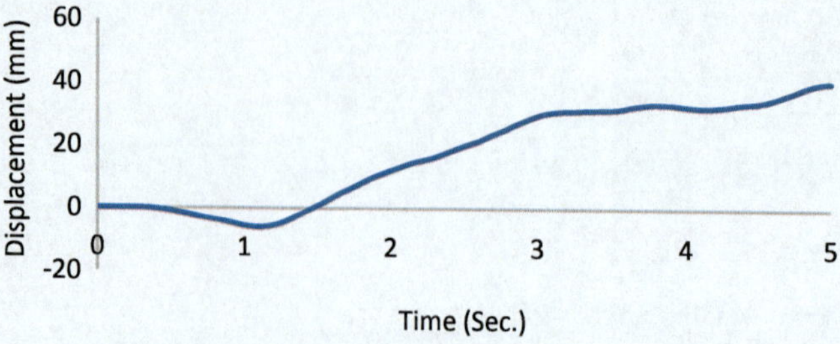


**Fig. 3** Displacement—time response of the beam–column joint for 5 s Elcentro earthquake

## *4.2 Response of the Beam–Column Joint*

The RC beam–column joint modeled based on a finite element approach using ABAQUS is assessed for its seismic response by subjecting it to Elcentro earthquake. The displacement–time response of the beam–column joint when excited with this earthquake of 5 s duration is shown in Fig. 3.

It can be observed that the beam–column joint has undergone a backward sway during the first one and half seconds of the earthquake approximately and a forward sway thereafter. The displacement of the beam–column joint is negative when the ground acceleration is negative and kept on increasing as the ground acceleration increases.

The displacement of the beam–column joint is almost constant when the ground acceleration is reduced. There are some cracks observed at the center and the base of the beam–column joint, which are the most critical parts. The reinforcement at the base of the beam–column joint also yielded under earthquake ground excitation. Cracks in the concrete of the beam–column joint appeared at the base first and later propagated to the center of the joint. This behavior of the beam–column joint obtained is as expected and it can be said that the CDP model is effective in predicting the behavior of beam–column joint subjected to seismic loads. The failure patterns in concrete and steel due to earthquake loads are as shown in Fig. 4.

# 5 Parametric Studies on the Behavior of the Beam–Column Joint

## *5.1 Dilation Angle*

The influence of varying dilation angle of concrete on the behavior of the beam–column joint is also investigated. Five different dilation angles 13, 20, 34, 36 and 40° are chosen for the study. All other parameters during this analysis are kept constant

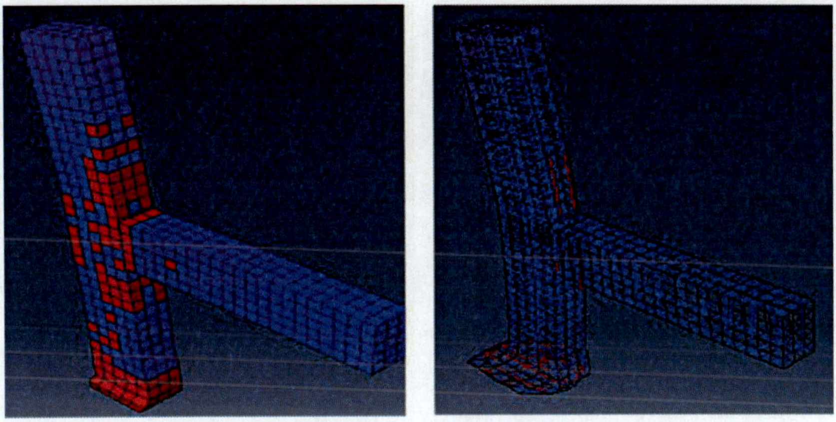

**Fig. 4** Cracks in concrete and yielding of steel due to the seismic load

as shown in Table 1. The displacement–time responses of the beam–column joint due to variation in the dilation angle of concrete are shown in Fig. 5.

It is observed that the increase in dilation angle increased the displacement of beam column. The difference in the displacement values is less when the dilation angles are 13 and 20°. It can also be noted that the nature of displacement–time responses of the beam–column joint with dilation angles of 34, 36 and 40° is similar. ABAQUS User's Manual [7] suggests the dilation angles to be any value in between 34 and 40°. So, any value among 34, 36 and 40° can be considered for analyzing the beam–column joint subjected to earthquake load as these values are almost same.

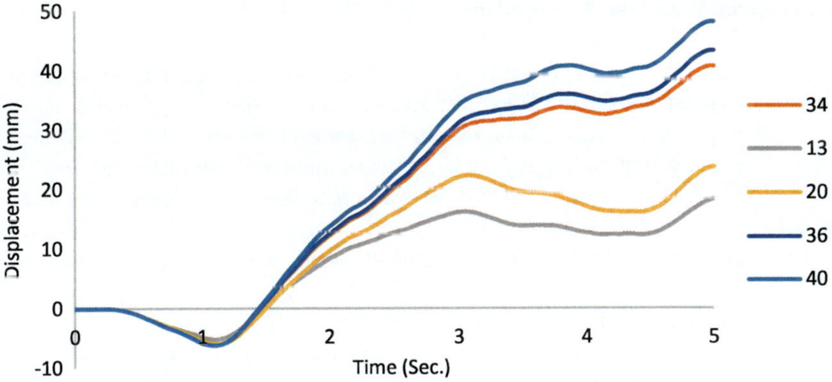


**Fig. 5** Displacement—time response of the beam–column joint for various dilation angles

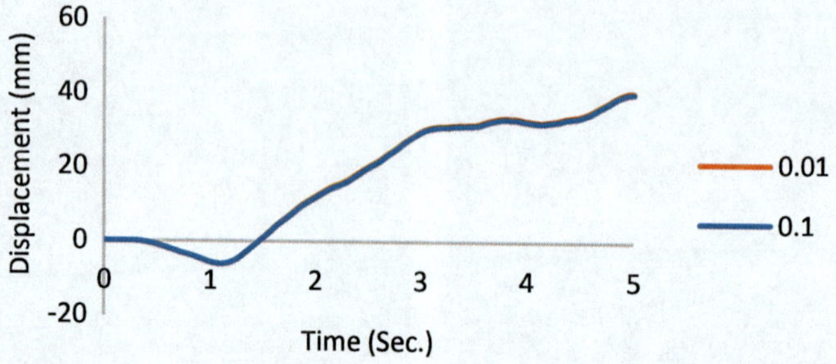


**Fig. 6** Displacement—time response of the beam–column joint for two different eccentricities

## 5.2 *Eccentricity*

The effect of the varying eccentricity of concrete on the behavior of the beam–column joint is analyzed. Two eccentricity values suggested for concrete 0.01 and 0.1 are considered for the study. The displacement–time responses of the beam–column joint due to variation in the eccentricity of concrete are shown in Fig. 6. From the analysis, it is observed that there is a negligible variation in the response using two values of eccentricity.

# 6 Using UHPFRC for Strengthening the RC Beam–Column Joint

Based on the CDP data mentioned above, UHPFRC in the form of strip is modeled in ABAQUS and wrapped around the beam–column joint. It is wrapped up to half length of the column and half length of the beam. Generally, these types of strips are attached using epoxy adhesives, etc. But, in that case, there might be a problem of delamination of the strip. But in this study, the strip is directly tied to the beam–column joint allowing no slippage, which means that the delamination effects are neglected. The plasticity parameters for normal concrete and the UHPFRC are taken from Table 1.

Three different thicknesses of the UHPFRC strips, i.e. 5, 10 and 20 mm are considered for strengthening. The wrapped UHPFRC strip and the assembly of the beam–column joint with the strip are shown in Fig. 7. The strip is meshed using the eight-noded cubic element (C3D8R). The element size used for meshing the strip is also 100 mm so as to maintain a node-to-node connectivity. The mesh details of the strip are shown in Fig. 7. The beam–column joint strengthened with UHPFRC is tested for its seismic performance by applying a first 5 s Elcentro earthquake ground acceleration at its base. The results obtained after the analysis are shown in Fig. 8.

From Fig. 8, it can be noted that as the thickness of the UHPFRC strip increases, the displacement of the beam–column joint decreases. This means that as the thickness

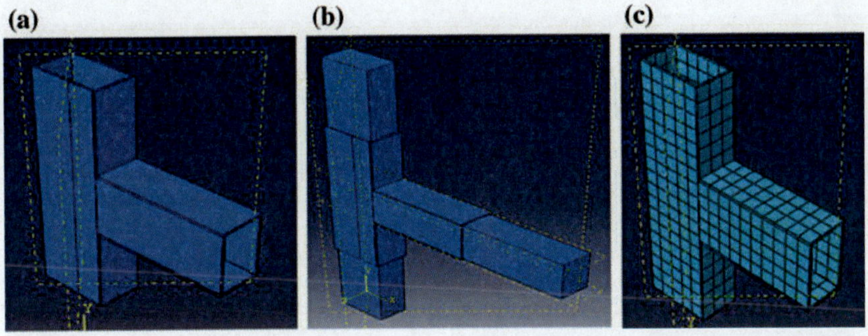


**Fig. 7** **a** Wrapped HPFRC strip. **b** Assembly of the beam column joint and **c** Meshing the UHPFRC strip

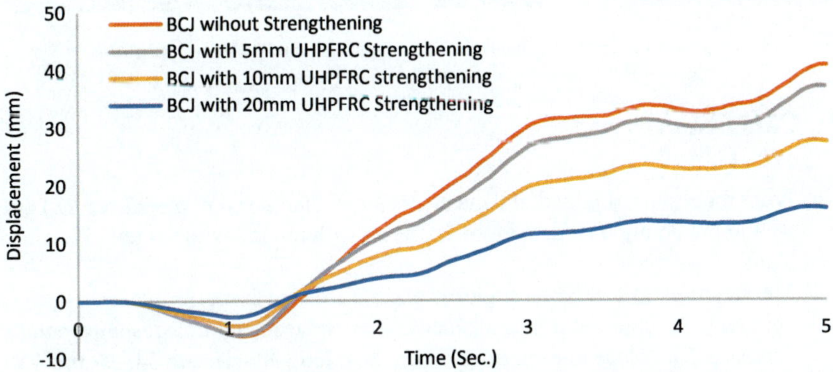


**Fig. 8** Responses of the beam–column joint for various thicknesses of strengthening with UHPFRC

of the strip increases there is an improvement in the performance of the beam–column joint against earthquake loads. There is also a sudden improvement in the performance of the beam–column joint when it is strengthened using 10 and 20 mm UHPFRC strips when compared with the beam–column joint without strengthening.

The nature of the response is observed to be the same, irrespective of the thickness of strip used for strengthening the beam–column joint. Similar crack patterns as in the case of beam–column joint without strengthening are observed, i.e. cracks appeared at the base of the beam–column joint first and then they propagated to the center. From the failure patterns, it can be said that the ductility of UHPFRC material is more than that of normal concrete as the first crack in UHPFRC strip after the cracking of normal concrete at the center of the strengthened beam–column joint. The failure patterns in the strengthened beam–column joint and the UHPFRC strip are shown in Fig. 9.

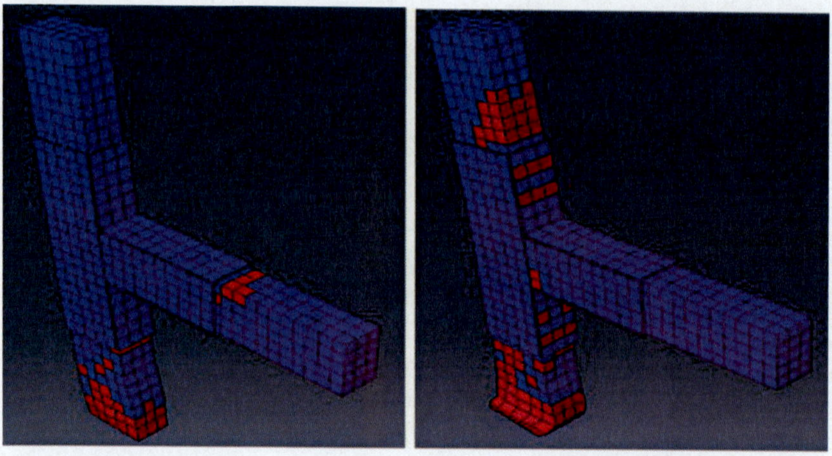

**Fig. 9** Failure patterns in the strengthened beam–column joint and UHPFRC strip

## 7 Conclusions

1. From the results obtained, it can be concluded that the CDP model worked out well in predicting the behavior of RC beam–column joints subjected to seismic loads.
2. The variation in the dilation angle of concrete affected the performance of beam–column joint subjected to seismic loads. The displacement–time response values and nature are close to each other when the dilation angles are 34, 36 and 40°. The difference in responses of the beam–column joint is also less in the cases when the dilation angles are 13 and 20°.
3. There is no effect of variation in the eccentricity values on the performance of the beam–column joint. The displacement–time responses are almost same for eccentricity equal to 0.1 and 0.01.
4. Ultra High-Performance Fiber Reinforced Concrete can be successfully used to strengthen RC beam–column joints subjected seismic loads.
5. As the thickness of the UHPFRC strip used for strengthening the RC beam–column joint increases, the performance of the beam–column joint against seismic loads improved.
6. From the failure patterns observed in the strengthened beam–column joint, it can be concluded that this UHPFRC has higher ductility when compared with normal concrete and hence can be used to strengthen structures effectively in earthquake-prone areas.

**Acknowledgements** The second author would like to acknowledge the computer center funded by DST-FIST at Vignana Bharathi Institute of Technology.

## References


1. Tayeh BA, Bakar BA, Johari MM, Voo YL (2013) Utilization of ultra-high performance fibre concrete (UHPFC) for rehabilitation–a review. Procedia Eng 54:525–538
2. Lampropoulos AP, Paschalis SA, Tsioulou OT, Dritsos SE (2016) Strengthening of reinforced concrete beams using ultra high performance fibre reinforced concrete (UHPFRC). Eng Struct 106:370–384
3. Prem PR, Murthy AR, Ramesh G, Bharatkumar BH, Iyer NR (2015) Flexural behaviour of damaged RC beams strengthened with ultra high performance concrete. Adv Struct Eng. Springer India 2057–2069
4. Rahman S, Molyneaux T, Patnaikuni I (2005) Ultra high performance concrete: recent applications and research. Aust J Civ Eng 2(1):13–20
5. Chen L, Graybeal BA (2011) Modeling structural performance of ultrahigh performance concrete I-girders. J Bridg Eng 17(5):754–764
6. Kota SK, Rama JS, Murthy AR (2019) Strengthening RC frames subjected to lateral load with Ultra High-Performance fiber reinforced concrete using damage plasticity model. Earthq Struct 17(2):221–232
7. Dassault (2014) SIMULIA Abaqus analysis user's manual version 6.14. Dassault Systems

# Assessment and Retrofitting Augmentation Methods of Seismic Performance on Existing RC Buildings Across India

Ankeeta Karmakar

## 1 Introduction

The super-continent Gondwana had broken into African, Antarctica, Australian, and Indian tectonic plates about 140 million years ago [3]. The Himalayan mountain belt is the seismically active part of this entire Indian tectonic plate. The plate has experienced major earthquakes over the last few decades—Shillong (1897), Kangra (1905), Bihar–Nepal (1934), Assam (1950), Gujarat (2001), Kashmir (2005) [4], Sikkim (2006) and (2011), Imphal (2016).

Gujarat earthquake (2001) was one of its kinds with a magnitude of 7.7 and duration of 22 s, which caused devastating effects in 21 districts, 7904 villages, and 182 administrative subdivisions of the district. The fatalities associated counted to 20,005, whereas 20,717 were seriously injured [5]. In the densely populated city, Ahmedabad, about 50 reinforced concrete buildings collapsed and a damage of $7.5 billion was estimated in the entire state [6]. An important facet of the ground motion was that it caused the collapse of RC building located at 250–300 km away from the epicenter [7].

Sikkim, a state that lies on the Himalayan foothills, has encountered two devastating forms of earthquakes—(a) February 2006 of 5.7 magnitude and (b) September 2011 of 6.9 magnitude. The earthquake of 2006 caused primary damages to RC buildings due to poor construction practices and design methodology. The second major earthquake that lasted for 40 s caused casualties of about 148. A damage of about $14 billion was estimated. The capital city—Gangtok experienced the complete collapse of two under construction five-storied buildings [8]. The region encountered aftershocks of 5.0, 4.5, and 4.2 magnitudes due to the severe earthquake [9].

A. Karmakar (✉)
University of Auckland, Auckland, New Zealand
e-mail: akar688@aucklanduni.ac.nz



S. Kolathayar and S. C. Chian (eds.), *Recent Advances in Earthquake Engineering*, Lecture Notes in Civil Engineering 175,
https://doi.org/10.1007/978-981-16-4617-1_31

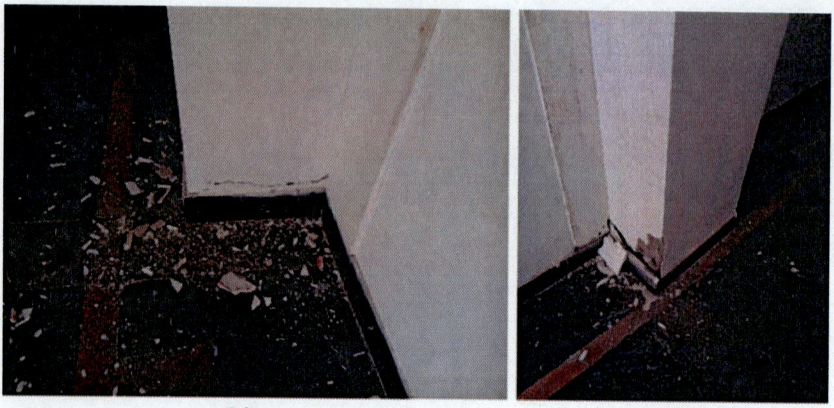

(a) (b)

**Fig. 1** Columns (fourth level) of Royal Group of Institution in Guwahati that faced damages due to the Imphal earthquake 2016

The Sumatra tsunami (2004) of magnitude 9.1 hits the Andaman and Nicobar Island of India causing fatalities of about 1500 and affecting more than 14 countries. The earthquake caused huge destruction on RC infrastructures, whereas the buildings whose major structural members were timber and masonry withstood the severe ground motions [10].

The land of Manipur was shaken by an earthquake of 6.7 magnitude in January 2016, which hits the capital city of Imphal, destroying more than 2000 three-story and four-story offices and residential buildings [11]. Due to the sparse population in the region, the reported fatalities were very minimal [12]. Newly built RC structures were damaged, a majority of publicly funded, and government buildings were affected badly due to failure to complying with the Indian standard code of Practice for seismic design, IS 1893 [13]. The earthquake was felt in few cities of Assam, which caused few damages to the RC buildings in the region of Silchar (260 km away) and Guwahati (490 km away). Figure 1 shows columns of the five-story Royal Group of Institution building located in the region of Guwahati, which experienced damage due to the severe ground motions of the Imphal earthquake.

## 2 Seismic Performance of Existing Buildings by Zones

According to BIS, India is divided into various seismic zones based on past seismic activity. According to the IS 1893 (Part 1): 2016, the country is divided into four seismic zones, II through V with increasing magnitude of intensity. The majority of India's northeastern and northwestern geographical area lies under seismic zones IV and V. The fifth revision of seismic zonation map of India resulted after the 1993 Latur earthquake, which shook the region of Maharashtra at an intensity of IX (magnitude 6.3) causing huge destructions [14]. In the previous edition, Latur laid in seismic zone I, but later after the revision of 2002, it lies between seismic zones II and III. The probability of exceedance of all the seismic zones of India is 10% in 50 years with Peak Ground Acceleration (PGA) of 0.1 g to 0.4 g.

Table 1 shows the four major earthquakes that the country experienced with their respective seismic zones along with the varied observed PGA. During these four major earthquakes, maximum reinforced concrete-framed structures did not perform well. The observed deformations were gravity load failure, open story failure, and component deformation modes being flexural, flexural shear, lap splice, sliding shear. The loss of gravity load capacity has been observed in all the major earthquakes in India. The deformation is witnessed when the structural element fails to bear its gravity load-carrying capacity, which often occurs due to shear failure in the member. In RC columns, a loss of axial load carrying capacity can give rise to a gravity failure, which can further extend to the total collapse of the structure.

The flexural failure in concrete members occurs in two modes as brittle and ductile. When the ultimate capacity of the compression zone is reached, the member deforms in ductile flexural failure. On the other hand, when the concrete member crushes before yielding the reinforcement, brittle flexural failure gets generated. When the member deforms due to the availability of no shear resistance provided by transverse reinforcement, shear failure occurs. Flexure shear is one of the major types of failure observed in RC-framed structure experiencing severe ground motions. The flexural-shear failure rises by the formation of cracks due to flexural tensile stress, which propagates diagonally in the member, giving rise to increased tensile stress from concrete tensile strength and leading to entire member crushing in shear. Few deformations in the RC buildings were observed due to splicing of reinforcement at critical locations, which gave rise to failure at beam–column junction.

According to IS 1893 (Part 1):2016, the westernmost state of India, Gujarat lies on three seismic zones—III, IV, and V. The epicenter of the Gujarat earthquake (2001) was in Bhuj, which lies on zone V. The reason for collapse of maximum of buildings during the Bhuj earthquake (Gujarat) was due to substandard construction practices and failure to comply with IS code of practices. Severe structural damage in RC-framed buildings was seen 300 km away from epicenter in the city of Ahmedabad [18]. The majority of the failures occurred in the buildings having no infill wall in the first story, which was adopted to provide facilities for parking. In Sikhara region of Ahmedabad, a newly constructed "H"-shaped 10-story building saw the collapse

**Table 1** Description of few major earthquakes across India

| Region | Date | Time | Seismic Zone | Magnitude | PGA (g) |
|---|---|---|---|---|---|
| Imphal | 4 January 2016 | 4:35 | V | 6.7 | 0.11—0.34 [13] |
| Sikkim | 18 September 2011 | 18:10 | IV | 6.9 | 0.2—0.45 [15] |
| Andaman and Nicobar Island | 26 December 2004 | 7:28 | V | 8.7 | 0.1 [16] |
| Gujarat | 26 January 2001 | 8:46 | V | 7.7 | 0.38 [17] |

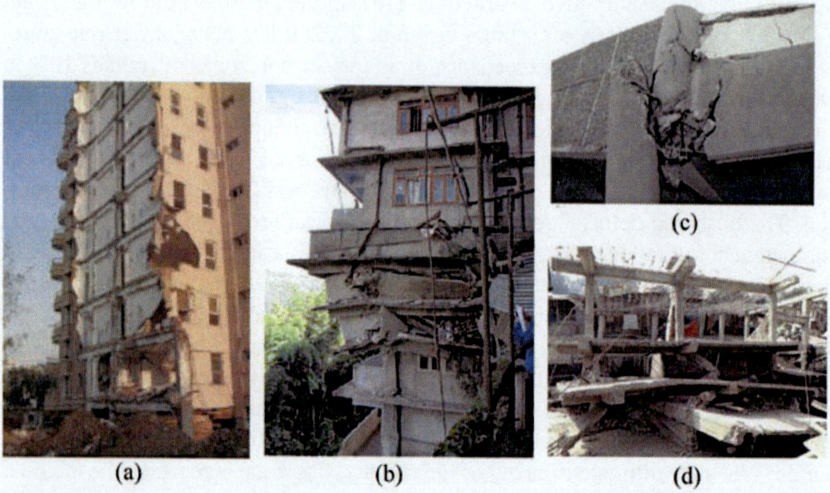

Fig. 2 **a** Collapse of an H-shaped 10-story newly constructed apartment building in the region of Sikhara, Ahmedabad [18], **b** Collapse of middle stories due to the 6.9 magnitude Sikkim earthquake (2011) in the region of Balwakhani, Gangtok [9], **c** Failure due to short column mechanism in few columns of New Passenger Terminal, Port Blair [19], **d** Complete collapse of a four-story building in the region of Dewlahland, Imphal [13]

of an entire wing of one of the open arms, which caused the lives of 89 persons, as shown in Fig. 2a. The open first story with no infill wall created a soft story mechanism where the lateral load resistance was significantly lesser than the upper stories. The presence of infill wall on the above-ground stories made the levels stiffer and resulted in attracting a huge amount of earthquake forces compared with the open ground story that caused large-scale destruction in the entire city of Ahmedabad.

The epicenter of the Andaman and Nicobar earthquake (2004) was in Sumatra, an Indonesian Island, which lies on seismic zone V. A study conducted showed that the majority of RC building collapsed due to improper reinforcement detailing, workmanship, and quality control during the construction of the infrastructure [19]. Old unreinforced building withstood the ground motions, but RC buildings whose ground story was an open story encountered severe damages and total collapse. Due to the presence of infill masonry wall on upper stories, little or no damage was observed. A study has shown that during the ground motion, the infilled frames acted as shear wall till the infill materials reached the brittle failure [20]. This might have also occurred during the earthquake, leading to short column mechanism. Few columns of the entire New Passenger terminal deformed due to the short column mechanism as shown in Fig. 2b. Due to the inappropriate detailing of ductile reinforcement, the columns failed in shear and flexural resistance on the beam–column joints.

Sikkim is in the northeastern part of India whose majority of the land is on hilly terrain and lies on seismic zone IV. Huge construction practices are carried out in the capital city Gangtok to accommodate the increasing population on sloping grounds.

This type of construction practice has given rise to irregular and unsymmetrical plans both vertically and horizontally [21]. Figure 2c shows the gravity load failure of a nine-story building in the region of Balwakhani, Gangtok during the Sikkim earthquake (2011). Various failures occurred due to lapping of longitudinal reinforcement at inappropriate locations, no adequate confining reinforcement, extensive extension of upper story over cantilever beam, inferior material quality, and poor workmanship.

One of the northeastern states, Manipur, lies on the seismic zone V. During the earthquake of 2016, huge destructions were seen. The three-story RC building of Sports Authority of India experienced severe damages, which led to its demolition at a few months of its life. Figure 2d shows a four-story building that collapsed completely during the earthquake. The Asia's largest all women's market "New Market building" (Ima Keithel) experienced major damages [13]. In the building, PVC pipes of 150 mm were concealed inside the circular columns, which jeopardized the confinement of exterior columns causing the reduction in strength and spalling of concrete. Failures in RC buildings were seen majorly due to no ductile detailing, inadequate spacing of transverse reinforcement, unequal clear cover, lack of confining reinforcement, inferior material quality, and failure to complying with IS code of practices. On the other hand, the old traditional wooden buildings showed better performance as compared with the RC buildings.

The observed drawbacks in the construction practices that led to the failure of RC buildings were poor construction materials, failure to comply with IS codes, no ductile detailing, inadequate longitudinal and transverse reinforcement, and widely spaced stirrups and ties, to name a few. Many schools and government buildings which were erected with strict supervision ended up being damaged during the earthquake. Newly constructed buildings like the H-shaped RC building in Gujarat and Passenger Terminal in Andaman and Nicobar Islands have also shown unexpected failures. All the failure mechanisms and deformations observed are similar across the various regions of the country. The difference in earthquake magnitudes and reported PGA did not have any significant impact on the failure mechanism shown by these RC buildings.

## 3 Analysis and Retrofitting Techniques

Various seismic retrofit techniques for strengthening a damaged-reinforced concrete member have been studied and adopted in practical applications across the country. Diverse methods include few conventional techniques like braces, jacketing, and infills and modern techniques like the use of advanced materials like Fiber-reinforced Polymers (FRP) and Shape Memory Alloys (SMA) [22]. Even though the greatest disadvantages of using retrofit techniques are cost, feasibility, and material availability [22], these techniques have proven to potentially increase the degraded strength. FRP is a system that constitutes fibers embedded in a polymeric matrix, which is used to increase the strength and ductility of the damaged member and provides effective confinement to the member [23]. Concrete jacketing is another

**Fig. 3** Different retrofitting measures adopted to strengthen—**a** Mani Mandir, Morbi [23], **b** H-shaped 10-story apartment building, Sikhara [18]

form of retrofitting method, which is used widely in the construction field due to the cost efficiency, ease of implementation, and availability of materials [24]. When a new layer of reinforced concrete is applied to the deformed member, the capacity of the member increases in terms of stiffness, ductility, axial, and shear strength in uniform and distributed manner [23, 25].

During the Bhuj earthquake (Gujarat) 2001, the Mani Mandir complex located at Morbi at 125 km away from the epicenter experienced severe damages. The building was constructed during 1930s. Dynamic analysis was carried out to observe the seismic performance, and retrofitting technique was implemented. The building was constructed using yellow sandstone, but after carrying out the dynamic analysis, few members like wall, bastion were retrofitted using reinforced concrete skin walls of 150 mm thickness, as shown in Fig. 3a [26]. As mentioned in Sect. 2, an entire wing of a 10-storied H-shaped building in Sikhara collapsed. The columns at the open ground story of the other wing of the building have been strengthened using concrete jacketing as shown in Fig. 3b.

During the Sikkim earthquake 2011, the Moonlight School of Chungthang region, constructed in the year 1985, encountered collapse of the second story, which propagated to the upper story, as shown in Fig. 4a. During the non-linear static pushover analysis, two forms of design analysis were applied—Design Basis Earthquake (DBE) bearing PGA of 0.18 g and Maximum Considered Earthquake (MCE) bearing PGA of 0.36 g. Using the guidelines of ATC-40, the performance point was found out and was compared with FEMA-365 guidelines. The analysis resulted in hinges at the region of immediate occupancy and life safety for the maximum considered earthquake at 22 columns of the corridor of the school building. Glass Fiber-Reinforced Polymer (GFRP) was used to strengthen the columns of the structure, which generated hinges during the non-linear static pushover analysis [25].

During the Imphal earthquake 2016, the New Market building (Keithel 3) constructed in the year 2006 encountered severe damages in various structural components due to poor workmanship, substandard quality of material, and failure

(a) (b)

**Fig. 4** **a** Total collapse at the second story leading to extensive damage at the third story of Moonlight School building at Chungthang, Sikkim [25], **b** Damages in the columns due to short column effect in the New Market building, Imphal [24]

to compliance with Indian Standard guidelines as shown in Fig. 4b. For the non-linear analysis with PGA of 0.36 g, the non-linear flexural hinges were assigned to the beams and columns using the definition generated by Paulay and Priestley [27]. The shear hinges were assigned to the model using the shear force—shear deformation (V-Δ) models explained in IS 456. As the region of Imphal is located at the remotely hilly terrain, to have an easy, convenient, and quicker solution with implementation of local resources, concrete jacketing was adopted to strengthen the column that failed in flexure and shear. The pushover curves generated before and after retrofitting showed that the stiffness and lateral strength increased remarkably as shown in Fig. 5. The load and moment curve for the most critical column (C6-GR) showed an increasing capacity by three times as shown in Fig. 6. The capacity of all the columns increased sufficiently and the demand pairs were laid inside the interaction curve [24].

The northeastern part of India, Assam, lies on a seismic belt of zone V with its capital city, Guwahati, being the most densely populated region in the entire state.

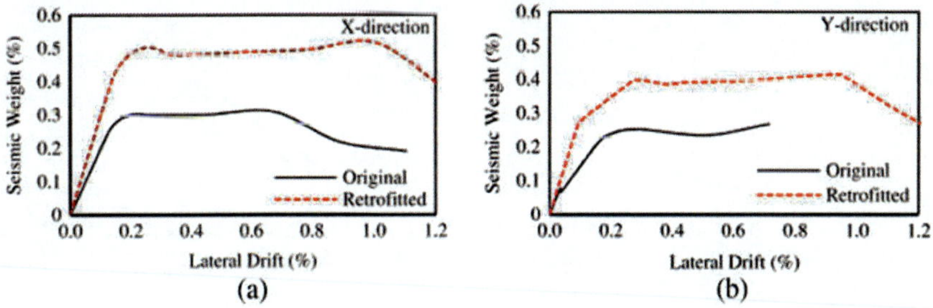


**Fig. 5** Pushover curve obtained for the original and retrofitted building after the analysis in the direction of **a** X axis, **b** Y axis [24]

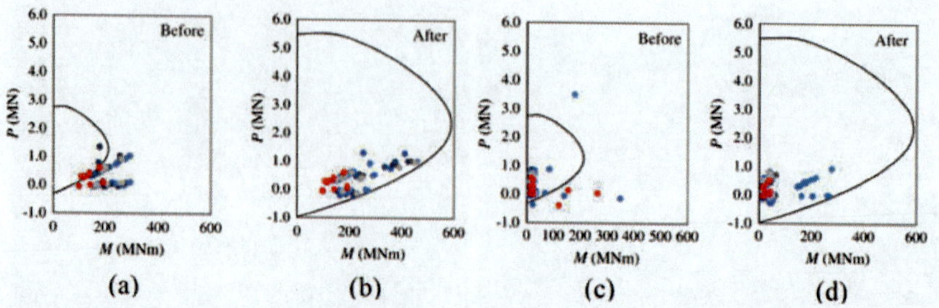


**Fig. 6** P–M interaction curves for all the columns of New Market building in comparison to C6-GR in the direction **a** X-axis (before strengthening), **b** X-axis (after strengthening), **c** Y-axis (before strengthening), and **d** Y-axis (after strengthening) [24]

A study shows that the city has encountered a construction boom in terms of multi-storied mid-rise building apartments with open ground story and no lateral load resisting systems [28]. Minor tremors of earthquake are felt almost every month in the region with the very recent being an earthquake of 4.4 magnitude at a 32 km depth in a region of Nongstoin, Meghalaya, which is at 150 km.

An evaluation was made to reduce the seismic vulnerability by introducing infill masonry walls to the reinforced concrete frame at certain locations of the ground story [28]. Two buildings were chosen for the analysis—(a) Sample building of five stories and (b) an irregular existing building. For each of the buildings, three different scenarios of modeling were considered—(i) Bare frame, (ii) Open ground story, and (iii) Masonry retrofitted building. For the retrofitting technique, 5 inches and 10 inches infill masonry walls were considered in the study. Plastic hinges were assigned as per the guidelines of FEMA 356 (inbuilt in SAP2000), and a pushover analysis was carried out. The pushover curve for the sample building as shown in Fig. 7a resulted that a considerable increase in horizontal base shear is observed when infill masonry walls of 5 inches and 10 inches are considered as a structural element. On the other hand, Fig. 7b shows that for the existing building, the horizontal shear for bare frame increased considerably as compared with the sample building but the horizontal shear for open ground story increased. The masonry infill wall at strategic

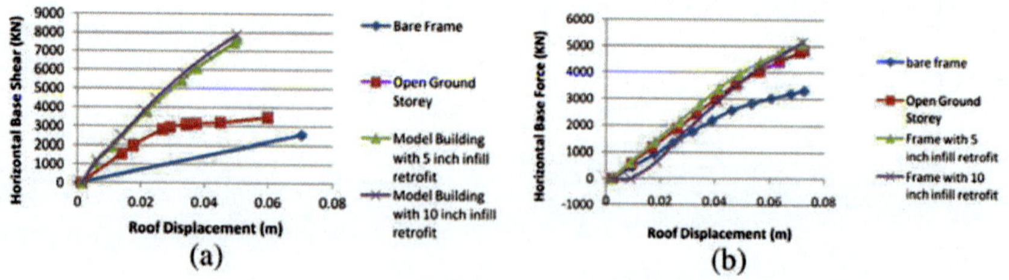


**Fig. 7** The static pushover curve for all the three considerations of—**a** sample building, **b** existing building [28]

locations further increased the horizontal base shear, but a significant difference was observed for 5 inches and 10 inches of thickness infill wall [28].

A study was conducted to compare the strength of the column size using the guidelines of IS 456:2000. The strength was evaluated using concrete jacketing and FRP wrapping using IS 15988:2013 guidelines [23]. The strength for concrete jacketing, one layer and two layers of FRP wrapping, was evaluated based on minimum and maximum conditions provided in the guidelines. It was found out that increase in strength of the column section with concrete jacketing was 1.78 times more than one layer of FRP wrapping and 1.40 times more than two layer of FRP wrapping. Another study was conducted to test the performance of a column in flexure after concrete jacketing using slant shear and compression test and lamellar analysis. The test showed that the moment capacities of the retrofitted specimen increased substantially. The retrofitted beam–column joint sub-assemblage showed an increase in lateral capacity, ductility, and energy dissipation. During the lamellar analysis, good behavior of stress versus strain and moment versus curvature was observed [30]. A similar study was conducted to study the performance of a beam in flexure using slant shear test and lamellar approach after strengthening the beam implementing concrete jacketing. A reference specimen and two retrofitted specimens were tested under monotonic load and half cyclic loads. The reference specimen showed failure in the region of soffit of the beam, whereas the retrofitted specimen under monotonic loading showed no buckling or spalling. The specimen under cyclic loading performed better than all the specimens in terms of strength and ductility. The lateral capacity of the retrofitted specimen under monotonic and cyclic loading than the reference specimen was 1.46 and 1.23 times, respectively. The strength and ductility of the retrofitted sub-assemblage increased significantly. Good performance was observed at the moment versus rotation behavior of the retrofitted specimen during the layered analysis (lamellar approach) [31].

A few of the existing retrofitting techniques have been found to be useful for the deformed RC components across India. For adopting a retrofitting scheme, a vital step is to assess the availability of materials that are required to implement during the entire process. As discussed in Sect. 1, the Himalayan mountain belt is the seismically active part of the entire Indian tectonic plate and majority of its land is on hilly terrain. In a developing country where reinforced concrete structures are in remote and hilly locations, and the locations are zoned seismically active by the Indian Standard guidelines, it is important to note the ease of implementation. The strengthening process of concrete jacketing can be executed economically, quickly, with the use of locally available materials and manual labors. Local retrofitting through concrete jacketing increases the shear strength, thereby increasing the confinement of concrete with less amount of required work.

## 4 Conclusions

This paper reviews some of the works completed across different earthquake zones of India based on the performance of the existing reinforced concrete buildings during the major earthquakes. It was observed that despite being on different seismic zones as defined by the IS 1893 (Part 1): 2016, the sixth edition, with different observed earthquake magnitudes and peak ground accelerations, the reinforced concrete buildings across the regions showed a similar form of deformations. The various governing failures of reinforced concrete buildings during the earthquakes have been discussed in the study. Due to the presence of open ground story, the reinforced buildings witnessed partial to complete collapse with common component deformations of flexural, flexural shear, lap splice, and sliding shear. According to the Indian Standard Code of Practice, about 60% of India's geographical land lies under zones III, IV, and V. The northeastern region of the country lies on the Himalayan Belt and, therefore, is one of the most active regions.

To prevent the existing RC structures from witnessing severe damages in life and property, strengthening the existing deformed reinforced concrete components is vital. Local retrofitting of the concrete members can be implemented using concrete jacketing to increase the strength and ductility of the member. The process of concrete jacketing is found out to be economically feasible and can be implemented locally available materials and operated with minimal manual labors. Local retrofitting of RC members through concrete jacketing increases the confinement of concrete, thereby increasing the shear strength of the member. Several reinforced concrete buildings deformed during the major earthquakes have been strengthened using concrete jacketing. Future works can be exploring the various retrofitting methods across the globe and comparing their feasibility in India. However, as pointed earlier, most of them might end up being futile because of the lack of proper resources.

## References

1. Bureau of Indian Standards (BIS) (2016) Criteria for earthquake resistant design of structures, 6th revision, vol 1893. New Delhi
2. Murty CVR, What are the Indian seismic codes? IIT Kanpur, pp 1–2. http://www.iitk.ac.in/nicee/EQTips
3. Kumar P (2007) The rapid drift of the Indian tectonic plate. Nature 449(7164):894897. https://doi.org/10.1038/nature06214
4. Reddy DV et al (2009) The great 1950 Assam earthquake revisited: field evidences of liquefaction and search for paleoseismic events. Tectonophysics 474(3–4):463472. https://doi.org/10.1016/j.tecto.2009.04.024
5. Tripathi JN (2006) Probabilistic assessment of earthquake recurrence in the January 26, 2001 earthquake region of Gujrat, India. J Seismol 10(1):119130. https://doi.org/10.1007/s10950-005-9004-9
6. Wikipedia, 2001 Gujarat earthquake—Wikipedia. https://en.wikipedia.org/wiki/2001_Gujarat_earthquake. Accessed 09 Dec 2020

7. Bokey PB, Pajgade PS (2004) Lessons from Jan, 26, 2001 Gujrat (India) Earthquake. In: 13th world conference on earthquake engineering, 2004, pp 1–6
8. Rajendran K (2011) The 18 September 2011, North Sikkim earthquake. JSTOR 101(11):1475–1479
9. Rai DC, Mondal G, Singhal V, Parool N, Pradhan T (2012) 2011 Sikkim earthquake: effects on building stocks and perspective on growing seismic risk
10. Kaushik HB (2007) Impact of great december 26, 2004 sumatra earthquake and Tsunami on structures inf port blair. J Perform Constr Facil 21(2):128142. https://doi.org/10.1061/(ASCE)0887-3828(2007)21:2(128)
11. Devi TK, Singh O (2017) A case study on Manipur earthquake Of 4th January 2016. In: Oinam Manoranjan Singh international journal of engineering technology science and research IJETSR, vol 4, no 3. www.ijetsr.com
12. Gahalaut VK (2016) The 4 January 2016 Manipur earthquake in the Indo-Burmese wedge, an intra-slab event. Geomatics. Nat Hazards Risk 7(5):15061512. https://doi.org/10.1080/19475705.2016.1179686
13. Rai DC (2017) M 6.7, 4 January 2016 Imphal earthquake: dismal performance of publicly-funded buildings. Curr Sci 113(12):23412350. https://doi.org/10.18520/cs/v113/i12/2341-2350
14. Mohapatra AK, Mohanty WK (2010) An overview of seismic zonation studies in India. In: Indian geotechnical conference, 2010, pp 175–178
15. Prajapati S, Kumar A, Chopra S, Bansal BK, Prajapati SK, Bansal BK (2013) Intensity map of Mw 6.9 2011 Sikkim–Nepal border earthquake and its relationships with PGA: distance and magnitude. Nat Hazards 69:1781–1801. https://doi.org/10.1007/s11069-013-0776-x.
16. Rai DC et al (2006) The effect of the December 2004 Great Sumatra earthquake and Indian Ocean tsunami on transportation systems in India's Andaman and Nicobar islands. Earthq Spectra 22(SUPPL 3):561579. https://doi.org/10.1193/1.2206809
17. Raisinghani BM (2016) Influence of parameters on performance evaluation of designed RC buildings: seismic hazard. Int J Struct Eng 7(4):378397. https://doi.org/10.1504/IJSTRUCTE.2016.079286
18. Humar JM (2001) Performance of buildings during the 2001 Bhuj earthquake. Can J Civ Eng 28(6):979991. https://doi.org/10.1139/cjce-28-6-979
19. Kaushik HB, Jain SK (2006) Performance of buildings in port blair (India) during the great sumatra earthquake of 26 December 2004. In: The 8th U.S. national conference on earthquake engineering, pp 18–22
20. Kam WY (2011) Seismic performance of reinforced concrete buildings in the 22 February Christchurch (Lyttelton) earthquake. Bull New Zeal Soc Earthq Eng 44(4):239278. https://doi.org/10.5459/bnzsee.44.4.239-278
21. Sreerama AK, Ramancharla K (2013) Earthquake behavior of reinforced concrete framed buildings on hill slopes
22. Pampanin S (2006) Controversial aspects in seismic assessment and retrofit of structures in modern times: understanding and implementing lessons from ancient heritage
23. Dingorkar P (2016) Retrofitting—Comparative study of RC jacketing and FRP wrapping. Int J Civ Eng Technol 7(5):304–310
24. Choudhury T (2020) Seismic vulnerability assessment and strengthening of new market building damaged after January 4, 2016, Imphal, India, Earthquake. J Perform Constr Facil 34(2):115. https://doi.org/10.1061/(ASCE)CF.1943-5509.0001392
25. Rani A (2014) Seismic retrofitting of a damaged school building. Int J Res Eng Technol 03(06):495500. https://doi.org/10.15623/ijret.2014.0306093
26. Sheth A, Chaudhari RD, Khan E, Gupta D, Saini M (2004) Seismic retrofitting of Mani Mandir complex at Morbi, Gujarat, India. In: 13th world conference on earthquake engineering Vancouver, B.C., Canada, vol 3228, pp 1–6
27. Paulay T, Priestly MJN (1992) Seismic design of reinforced concrete and Masonry buildings. Wiley, Hoboken

28. Bharali R, Deka B, Pathak J (2014) Retrofitting open ground storey building with masonry walls in Guwahati city. In: 15th symposium on earthquake engineering, 2014
29. Bureau of Indian Standards (BIS) (2013) IS 15988: 2013 Seismic evaluation and strengthening of existing reinforced concrete buildings—guidelines, Indian Stand, pp 1–25
30. Kaliyaperumal G (2009) Seismic retrofit of columns in buildings for flexure using concrete jacket. ISET J Earthq Technol 46(2):77–107
31. Badari Narayanan VT, Sengupta AK, Satish Kumar SR (2012) Seismic retrofit of beams in buildings for flexure using concrete jacket. ISET J Earthq Technol 49(1–2):1–22

# Review of Building Codes

# Scientific Perspectives to Earthquake Resistant Design of RC Buildings—A Global Approach

**B. M. Raisinghani, T. H. Bhoraniya, and E. Noroozinejad Farsangi**

## 1 Introduction

### *1.1 Background of Study*

Earthquakes are the natural hazards that overwhelm society in few (strong) minutes. It has become a global research arena where seismically active regions assemble the observations of prominent earthquakes and try to develop logical deductions to explain the occurrence of distress. The dynamic conditions generated in buildings due to the earthquake have significant implications on damage as visible in elements as well the complete response of structures. This has made the journey of earthquake engineering and safe design of structures to the next level. Each region has different seismicity and geological features that need a different approach to solve the same problem. Also, each building has a lot of many implications to consider apart from the location effects. The 'ring of fire' is titled due to the concentration of mega earthquakes along the plate lines in the pacific regions. Locations apart from these plate lines suffer moderate to big earthquakes. Thus, earthquake is a global phenomenon that has local effects and requires special considerations for different reasons to ensure safer built environment. However, there is a good reason to find the common grounds that make the buildings resilient and would present a global endeavor for mitigating earthquake disasters.

B. M. Raisinghani (✉)
Structural Engineer, Seismic Consultant, Ahmedabad, Gujarat, India

T. H. Bhoraniya
Government Engineering College, Kutch-Bhuj, Gujarat, India

E. N. Farsangi
Graduate University of Advanced Technology, Kerman, Iran

 
S. Kolathayar and S. C. Chian (eds.), *Recent Advances in Earthquake Engineering*,
Lecture Notes in Civil Engineering 175,
https://doi.org/10.1007/978-981-16-4617-1_32

## 1.2 Major Developments in Understanding of Earthquakes

**Geology.** The history that triggered the development of earthquake engineering in science and technology domains needs special mentions as below (see Fig. 1):

- Abraham Ortelius (1596)—published Thesaurus Geopgraphicus which W.J. Kious stated as "Americas were torn away from Europe and Africa by earthquake and floods, the vestiges of rupture shall reveal by the coast of three continents" [→] Deep seabed Challenger Expedition (1872–1876)—led to the knowledge that oceans are the permanent part of Earth's surface and did not change places with continents [→] Edward Suess (1885)—an Austrian geologist hypothesized two major former geographical features, the Gondwana and Tethys Ocean assuming a large-bridge between the present continents submerged in form of the geosyncline [→] John Perry (1895)—proposed that the earth's interior was fluid [→] Henry Fielding Reid (1906)—proposed elastic rebound theory to explain how energy is released in an earthquake [→] Alfred Wegner (1912)—on 6th January he proposed continental drift theory based on fossils on both sides of the Atlantic Ocean to the German Geological Society [→] Arthur Holmes (1928)—proposed that the Earth's mantle contains convective cells that dissipate radioactive head and moved the earth [→] Maurice Ewing (1947)—along with other scientists confirmed the existence of a rise in the Atlantic Ocean and observed that the seabed floor was different in composition to the continental crust [→] Bruce Heezen (1960)—mentioned the above discovery as the concept of Great Global Rift [→] Jack Oliver (1968)—developed theory of plate tectonics which superseded continental drift theory. It suggests that continental crust and oceanic crust are both floating above a much deeper plastic mantle. Also, the oceanic crust along with subduction drives the system of plates in a chaotic manner.

**Instrumentation**. Apart from the geological explorations, the technology has helped to capture the hostile behavior of earthquakes in form of displacement (D),

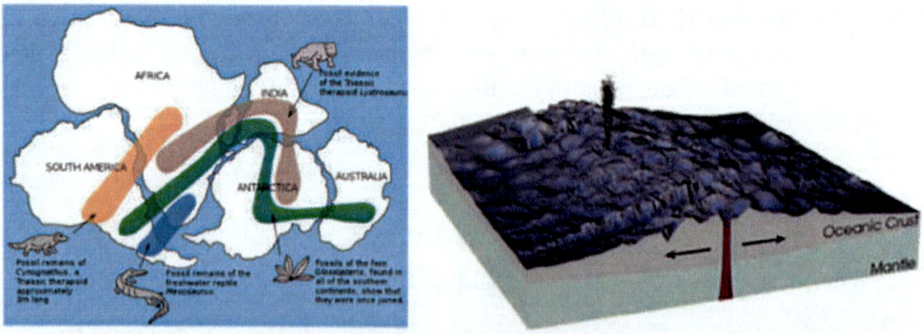


(a) Snider-Pellegrini Wegener fossil map (b) Seafloor spreading at mid-oceanic ridge

**Fig. 1** Knowledge of earth movements and genesis of earthquake science [1]

velocity (V) and acceleration (A). The discovery of recording instruments has significance in the engineering profession as they resemble the hazard (H), which forms the basis for estimating vulnerability (V) to determine seismic risk (R).

- The first seismoscope, Hou Feng Di Dong Yi (132AD), was of 8 chi (2 m diameter) and had Du Zhu (a pillar acting as a sensor), which had eight rods for eight directions to give location of earthquake by dropping a bronze ball in Toad's mouth for a sensing distance of 500 km (see Fig. 2a). This helped the ruler to send support in that direction for the recovery action. The old form, Wood-Anderson seismometer (see Fig. 2b), is still useful to get the damage potential of an earthquake as many buildings have resonant periods close to it (0.8 s). The electromagnetic seismograph introduced by Galitzin (1914) revolutionized the recording of earthquake motions. The first telemetered seismic networks were done in the early 1960s with 525 seismometers arranged in concentric patterns circulating an area of 200 km diameter (LASA) for detection of a nuclear explosion. The drum recorders (see Fig. 2c) had a sheet of paper of size 30 × 90 cm wrapped around a cylinder about 15 cm in diameter, which rotated (30–60 mm/min) and translated (2.5–5 mm/min). For monitoring of natural earthquakes, a telemetered seismic network was installed at Hawai (1962).

The USGS made the extensive installation of satellite telemetered networks with the Central California Microearthquake Network, which has an extensive role in understanding seismic wave propagation and the nature of source for earthquakes. The next development is seen in recording instruments for the continuous record to check arrival times of seismic waves and having records of PGA. The USGS records the daily earthquake data in digital form using Helicorders—each line read from left to right and top to bottom of the seismograph, represents 15 min. of the information, i.e. 4 lines/hr [4]. Currently, broadband seismometers are used for the collection of global earthquake data (three directions) by seismological societies (e.g. ISR Gujarat).

**Science.** Engineering is an applied science domain that has a dependency on pure science for genesis. The physicists and mathematicians have a crucial role toward the development of principles and theorems to support the complex calculations

a. Zhang Heng's seismoscope [1]  b. WA seismometer [2]  c. Drum recorder for daily monitoring of earthquake [3]

**Fig. 2** Historical instruments for sensing and recording of earthquake needed for engineering outcomes. Algebra made matrix methods, which have the most influence to modern-day programming packages. Differential equation, for dynamic analysis, is based on the equilibrium principles of the body in motion. Energy methods, based on equilibrium principles of work done, gave more base to structural calculations and behavior assessments [5]. Calculus gave way for the introduction of the calculation of forces and displacements in simpler problems. Thus, the mechanics that encompasses the principal domains of statics and dynamics has a basis in physics for motion in bodies and mathematics for numerical representation of structural problems [6]. With the advent of computers, the computational potential received its much-needed boost and complex equations could be solved with few clicks. The matrix method of analysis and knowledge of dynamic behavior of structural systems to complex excitations brought more capability to the domain [7]. The major earthquake events paved the way for the development of infrastructure facilities to understand the behavior of structures and develop principles for earthquake resistant design through an extensive state-of-art approach [8].

### *1.3 Outline of the Study*

At present, the global agendum toward mitigation of earthquakes is going through rapid transformation with the adoption of common protocol toward capacity building using performance-based design (PBD) methodologies. The founding principles for safe structural design and assessment of existing buildings for seismic actions are the constants. While each building type, local geology, technological base and knowledge base are the variables for regions around the globe.

In this paper, the emphasis is made toward the journey of development of earthquake-safe structures using scientific principles and technological advancements as the solid background. A 15-story regular RC building is designed for four seismic design codes with different methodologies for defining safety in buildings (see Fig. 3). The building is in seismic zone-III—250 km from the fault zone in Bhuj.

The performance of each case is evaluated using limits prescribed by respective codes and ASCE41. It is interesting to chalk out the changes in the matrix of common performance parameters that are brought about by each design code of practice using pushover analysis and time history for Indian earthquakes. Moreover, the possible inferences from seismic design codes will give a better review of implementation standards (US, EN, TEC, IS).

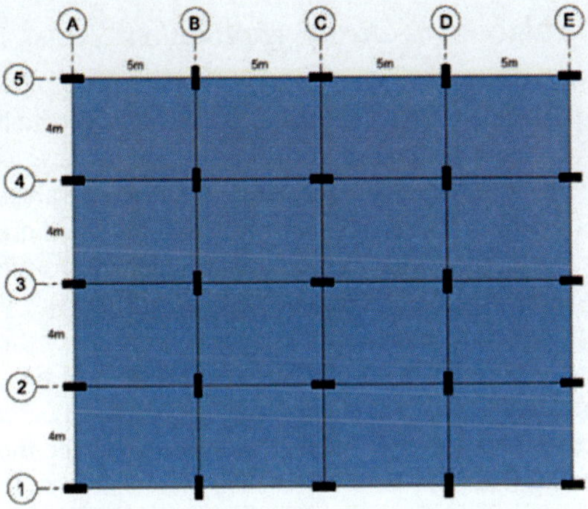


**Column (storey – D mm):**

- IS Code

1-3:700 ; 4-10:650 ; 11+: 600

- ACI Code

1-3:900 ; 4-10:850 ; 11+:800

- EC-8 & TSS-500

1-3:1000 ; 4-9:900 ; 10-14:800

- TSS-500

1-3:1000 ; 4-9:900 ; 10-14:800

**Beam (all cases):**

1-15: 300 x 500mm

**Slab (all cases):**

1-15: 120mm

**Fig. 3** Plan and section details of 15-story building

## 2 Review of Design Principles for Seismic Actions and Hazards

### *2.1 Basic Parameters for Safe Building Design*

Earthquakes by principle induce cyclic action to the building at the base making a building swing like an inverted pendulum. The concept of the weak link in a chain is the essence of distribution of forces and ductility in the system [9]. The equal displacement principle relates the elastic response of the structure with the actual response using a response reduction factor (R), which covers ductility (μ), overstrength (Ω), redundancy (r) and damping (ξ) [10]. Ductility is the muscle of any building, which brings toughness to a structure. The area under force–displacement curve widens with the introduction of ductility leading to higher energy dissipation capability of a building. Ductility is a function of the amount of transverse reinforcement and the anchorage length [11]. Another important aspect is the determination of failure mode of building, i.e. strong column-weak beam principle by capacity-based estimates [12]. The most significant parameters for the behavior of building depend on the inherent properties—stiffness (K), damping (ξ) and period (T). The parameters that make an excitation notable are the frequency content (f), period of action (T) and PGA (a). Hence, safety is a direct function of modeling assumptions, hazard estimation, dynamic characteristics, design philosophy and methodology, detailing and implementation mechanism.

## 2.2 Force-Based Versus Displacement-Based Approaches

Force is the quantity measurable through Newton's—second law of motion, which relates acceleration with the net force as directly proportional and with the mass as inversely proportional. The force can be calculated if acceleration and mass values are known. Thus, it leads to a basic principle that a higher mass will require more acceleration to cause motion and give higher force as the outcome. This forms the basis of current code-based procedures that are prescribed in nature. However, it is evident that the displacement, which is the measure of change in position of the object, is more relevant to describe the condition of the building under base excitation. The displacement leads to the generation of strain in elements which in turn develop forces in the structure, i.e. reverse process of the code-specified method for the design of structures. However, the displacement-based method is more elaborate to implement in terms of accurate behavior representation by model parameters [13]. The displacement coefficient method (DCM) and the capacity spectrum method (CSM) are used to evaluate the efficiency of design based on IS codes. The potential of displacement-based design methodology has been documented since 1982 [13–16]. It is possible to relate displacement with acceleration to find the design force. The following codes have included displacement-based design criteria: EN, ASCE41, NZSEE, TEC.

## 2.3 Evaluation of Seismic Hazard for Ahmedabad City

The city of mills or the Manchester of India are the titles given to Ahmedabad city, which has heritage and modern structures. Moreover, it is the international business center of Gujarat. In 2001, the Bhuj earthquake caused havoc in the city as the 7 M earthquake rocked all multi-story buildings and damaged 120 buildings [17]. The collapse of new buildings indicated the poor construction quality and lack of engineering in the built structures. The region of Ahmedabad and Gandhinagar falls within the Cambay sedimentary basin and the soft sediments of the Sabarmati river, which leads to amplification effects to the order of 1.5 [18]. The seismic microzonation study of Ahmedabad city has now marked areas with higher amplification levels and the overall hazard estimate (0.18 g) shows an underestimation of IS1893 (0.16 g) [19]. The PSHA study done under NDMA, New Delhi shows a hazard of 0.22 g for the city [20]. The mapping of hazards is done using anticipated seismic hazards and ground motion prediction equations (GMPE) for getting acceleration values (see Fig. 4) [21].

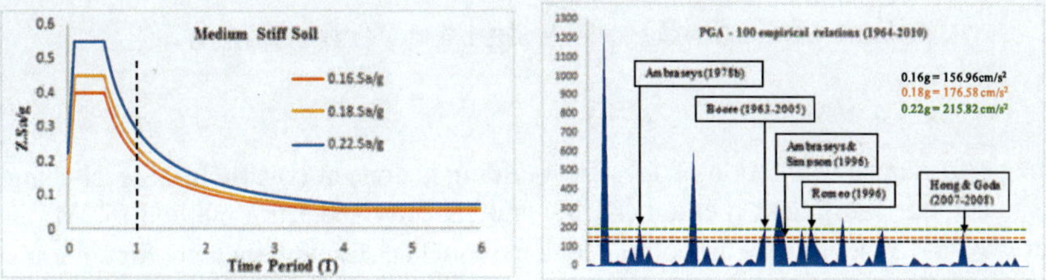


a. Response spectrum for anticipated EPGAs

b. Comparison of GMPEs for distance-acceleration-magnitude values

**Fig. 4** Mapping of seismic hazard for Ahmedabad city

# 3 Code-Based Design of Building

The building is designed considering the provisions of local governing bodies, which are developed based on experience or detailed structural evaluation through laboratory testing. Bureau of Indian Standards (BIS) is the governing body that regulates the professional minimum standards to be followed in India. The codes that are applicable for seismic design of buildings are IS456, IS1893-1 and IS13920. The building is designed using these provisions for the capacity part at the structure level enabling comparison with design standards of USA (ACI318), Turkey (TEC07) and Europe (EC8). The variation in section size and reinforcement requirement is important in terms of the distribution of stress and ductility in the structural elements (see Fig. 5).

The percentage of reinforcement in the lower segment is the highest in parabolic loading pattern as per IS1893 while the EC8 and ACI code-designed buildings have distributed reinforcement percentage along the building height. The section sizes for columns are highest in the case of EC8, followed by ACI and IS1893 design codes. Parabolic loading pattern for performance evaluation is good as it leads to higher reinforcement in the columns; however, it does not take care of mid-level elements as the percentage gets reduced due to higher distribution at lower elements. Thus, for the same level of earthquake force—1636kN, the three design codes have a dissimilar approach toward the safe design of buildings.

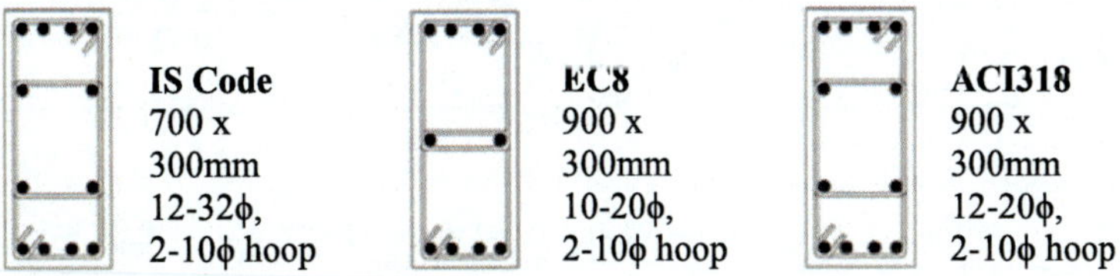


**Fig. 5** Design output for base column (S1) for three seismic design codes

# 4 Evaluation of Code-Based Design for Performance Review

The performance evaluation of 15-story building is done at (i) structural level using displacement coefficient method (DCM) and capacity spectrum method (CSM) as per ASCE41 guideline by monotonic load increments, (ii) element level for columns at ground floor to compare the performance as per the threshold limits matrix developed using seismic design codes and ASCE41 and (iii) non-structural level to estimate the effectiveness of code provisions for safety under seismic excitations using time history records for 7.0 M earthquakes available in India. Thus, a three-level performance evaluation scheme is implemented to understand the differences in design principles and code provisions towards safety for anticipated hazards.

## 4.1 First-Level Performance Evaluation—Global Inference

The monotonic responses of individual elements are defined in the ETABS model based on the backbone curves obtained by FEMA body in the USA. The ASCE41 mentions those threshold limits based on which overall assessment for a building is done by monitoring roof displacement under nonlinear static analysis (NSA) scheme [26]. The capacity curve of building is the intended outcome of POA (see Fig. 6) [27].

**DCM.** This displacement-controlled performance evaluation requires converting the capacity curve to bilinear curve following an equal area distribution profile. The effective time period ($T_e$) for the building is obtained based on the initial time period ($T_i$) and the ratio of initial stiffness ($K_i$) to effective stiffness ($K_e$). The software

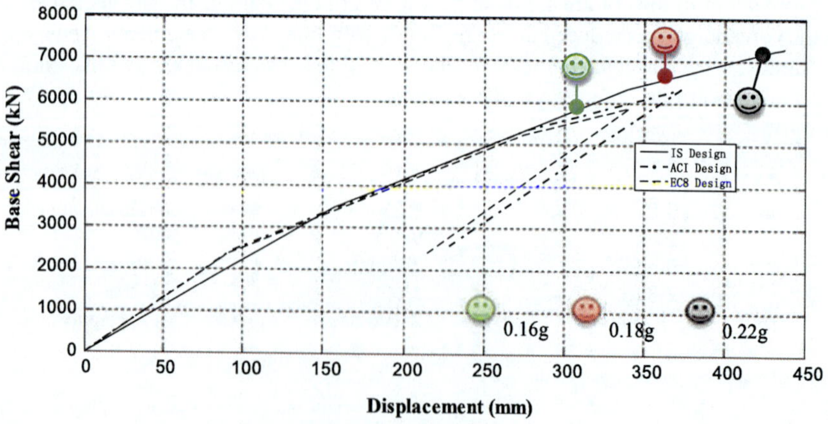


Fig. 6 The capacity curve for building with three design cases

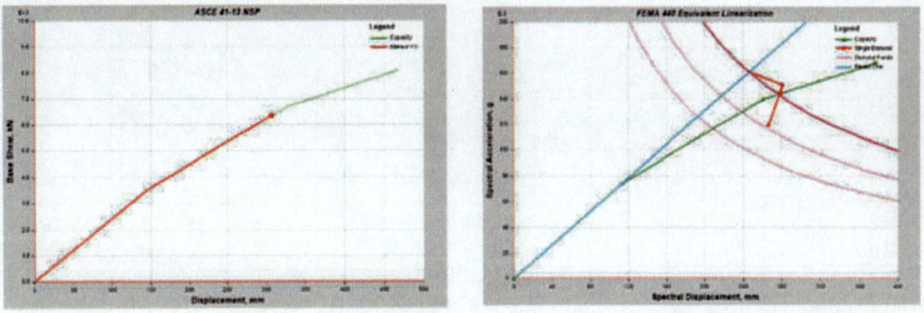

a. DCM results for IS design (307.2, 6342)    b. CSM results for IS design (348, 6900)

**Fig. 7** Structure-level performance assessment of building for three seismic hazards

calculates these properties to arrive at target displacement ($\delta_t$) for a given value of spectral acceleration ($S_a$). The performance level achieved for three hazard levels for three design cases indicates the extent to which building may meet the performance criteria.

**CSM.** This performance evaluation scheme requires the conversion of capacity curve to ADRS plot ($S_a - S_d$) [28]. The method is more suitable in case the hazard is accurately known, i.e. seismic micro-zonation in the current context. The Ahmedabad city is having accuracy toward hazard assessment as seismic micro-zonation is done by ISR. The second context for CSM is the equivalent viscous damping ($\beta_{eff}$), which has some inaccuracy to represent the nonlinear behavior of structures. The results of this method are presented below (see Fig. 7). The performance is for 0.16 g seismic hazard.

## 4.2 *Second-Level Performance Evaluation—Local Inference*

The elements of the building, i.e. the beams and columns, undergo inelastic deformations beyond the yield limit, which form the energy dissipation mechanism for them [29]. The performance of each element impacts global performance. Hence, it is of utmost importance to identify weak elements in a structure or design the elements for increased resilience. The threshold limits for failed columns (PMM hinges) are evaluated for all design cases using ASCE41, TEC07 and EC8 (see Fig. 8).

The software considers *Mander* model for the estimation of ultimate stress and strain in the element to compare with the threshold limits of ASCE41 [30]. However, there is some gap in section analysis and more flexibility is needed for which SEMAp tool suits the purpose (see Fig. 9) [31]. Furthermore, the map of threshold limits is made to do the element level performance estimation using the basics of moment–curvature-rotation ($M - \phi - \theta$) relations (see Table 1).

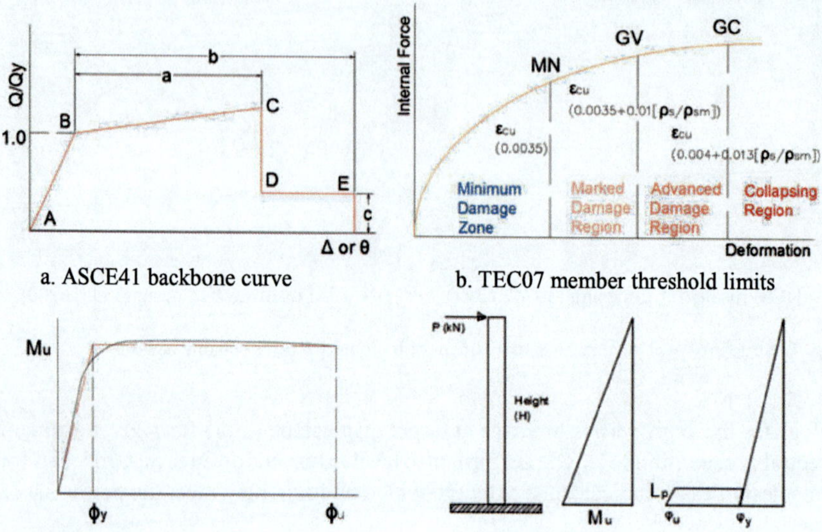


a. ASCE41 backbone curve

b. TEC07 member threshold limits

c. EC8 define yield moment

d. Deformed shape of beams []

**Fig. 8** Understanding of element behavior and threshold limits for model definition

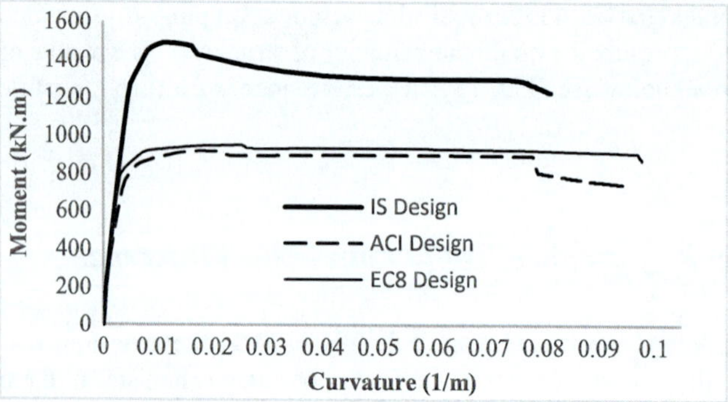


- **S1-IS Design**
  $M_u = 1493$kN.m
  $\varphi_u = 0.08$rad/m
  $\mu_\varphi = 11.57$
- **S1-ACI Design**
  $M_u = 918$kN.m
  $\varphi_u = 0.095$rad/m
  $\mu_\varphi = 13.6$
- **S1-EC8 Design**
  $M_u = 951$kN.m
  $\varphi_u = 0.098$rad/m
  $\mu_\varphi = 12.25$

**Fig. 9** Section analysis results of the base column using SEMAp tool for three design cases

## 4.3 Third-Level Performance Evaluation—Non-structural Elements

The non-structural elements considered in the current study are the masonry infills. The concentric single-strut model is the simplest way to include the infills in building design as they contribute to increased stiffness and higher ductility as compared with bare frame models [12, 22, 23]. Moderate-level earthquakes cause damage to infill and the fall of pieces on people standing on the ground may cause serious injury or

**Table 1** Threshold and performance map for section S1 as per *Mander* model

| Design case (1–3) | Tools or codes | $M_y$(kN.m) IO | $M_u$(kN.m) CP | $\phi_y$(rad/m) IO | $\phi_u$(rad/m) CP | $\theta_y$(rad) IO | $\theta_u$(rad) CP | $\varepsilon_{cu}$CP |
|---|---|---|---|---|---|---|---|---|
| IS 300 × 800 (@ 4%) | SEMAp | 1347 | 1493 | 0.007 | 0.081 | 0.005 | 0.035 | 0.032 |
| | ETABS | 1055 | 1386 | – | – | 0.002 | 0.04 | – |
| | TEC07 | 1485.5 | 1301 | 0.014 | 0.06 | 0.011 | 0.029 | 0.018 |
| | EC8 | 1347 | 1493 | 0.007 | 0.081 | 0.004 | 0.030 | 0.045 |
| ACI 300 × 900 (@ 1.4%) | SEMAp | 851.91 | 918 | 0.007 | 0.095 | 0.005 | 0.043 | 0.030 |
| | ETABS | 1093 | 1354 | – | – | 0.008 | 0.06 | – |
| | TEC07 | 918 | 893 | 0.018 | 0.080 | 0.013 | 0.041 | 0.018 |
| | EC8 | 851.91 | 918 | 0.007 | 0.095 | 0.004 | 0.029 | 0.048 |
| EC8 300 × 1000 (@1.1%) | SEMAp | 876 | 951 | 0.006 | 0.098 | 0.005 | 0.050 | 0.028 |
| | ETABS | 1112 | 1221 | – | – | 0.002 | 0.06 | – |
| | TEC07 | 951 | 922 | 0.022 | 0.080 | 0.016 | 0.045 | 0.018 |
| | EC8 | 876 | 951 | 0.006 | 0.098 | 0.003 | 0.030 | 0.040 |

fatal consequences. Also, the downtime of a building is related to damage to non-structural elements for which it shall have broad consideration than mentioned in the codes. The modes of failure of masonry infills [32] and Bhuj [36] damage survey have highlighted the damageability of OGS frames due to stiffness differences on the ground floor. The influence of masonry infills on performance evaluation is significant [35]. In the current building frame, infills are modeled to ensure that undue failure does not occur. The axial hinge is provided in the strut using stress–strain relation [33]. The non-linear time-history analysis (NLTHA) for the Chamoli earthquake and Uttarkashi earthquake is carried out using the matched-response history analysis feature in ETABS.

It is observed that the Chamoli earthquake ground motion history matches with the response spectrum curve for zone-III region, Ahmedabad city. The other records are reduced to match the required levels for design spectrum representation for location 250 km from the epicenter. Thus, the third-level evaluation is representing the response spectrum of 4–5 M earthquakes in the city, which will affect the overall behavior of the structure and affect the downtime. Also, it is necessary to get the force and acceleration at which such non-structural damages may occur (Fig. 10).

## 5 Observations

The observations are discussed below to summarize the important results obtained from the study in terms of a scientific basis for code development and implementation for earthquake safety:

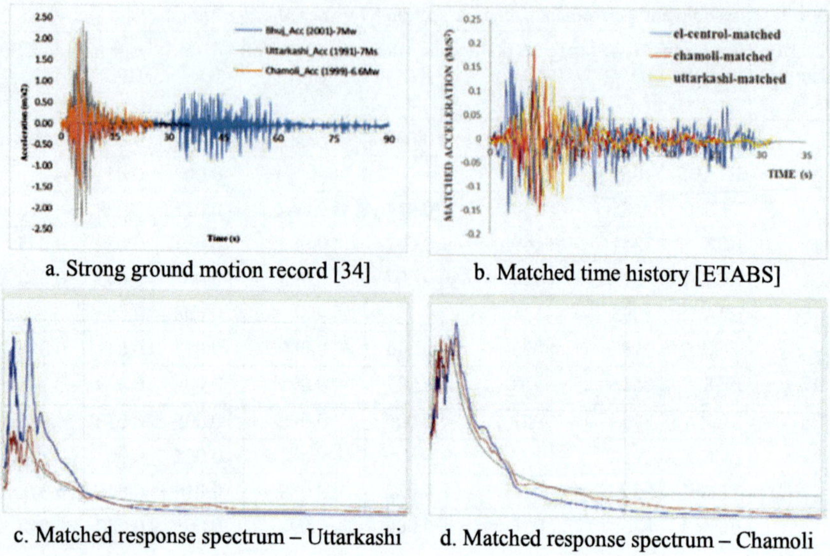


**Fig. 10** Matching of time history with response spectrum for seismic Zone-III

- The strong column—weak beam concept is the general basis for all design codes. However, the implementation norms are different with IS code demanding 40% higher capacity in columns compared with the combined capacity of connected beams.
- The capacity of a building designed with IS code has 5% higher strength and 6% higher ultimate displacement capacity as compared with ACI and EN codes. The ACI and EC8 are found to be in a similar range in terms of capacity generation in the RC building. The failure hinges formed in columns are at the base in ACI and EN design while in IS design, it is also found at mid-levels.

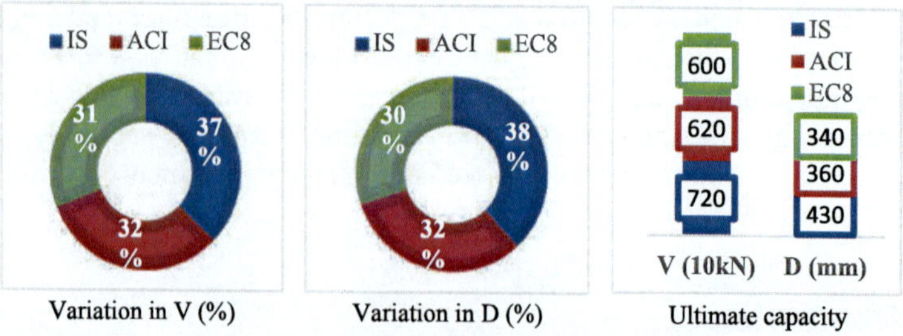


- The section analysis done in SEMAp for getting the moment–curvature rotation corresponding to the thresholds mentioned in ASCE41, EC8 and TEC07 shows the variation in regional deductions for safety. It is seen that TEC07 and EC8 have an efficient mechanism to dictate the performance.

- The performance of non-structural elements, i.e. masonry infill, is found to satisfy the magnitude range of 4–5 M in terms of strength and stability under the available records for strong ground motion. However, the concentric model is not the most accurate way to represent the effects of infill on structural elements, especially the columns.
- The three-level performance evaluation is very exhaustive and hence a comparative basis for putting the scientific principles in line with the performance standards.

## 6 Conclusion

The study focused on the design of the RC building, which is located at 250 km from fault location, has seismic micro-zonation done, has past observations repository for damages in Bhuj earthquake and is developing at a rapid pace. The next phase in the city toward mitigation is the efficient design of structures following the nuances of design principles and the latest implementation standards. Also, our focus was to give due regard to the contributions that have paved the way to the current level of science and technology to benefit the domain of earthquake-safe structures. The technology front in terms of ICT tools and artificial intelligence (AI) is bringing a better understanding toward safety in structures. Moreover, we would like to acknowledge more contributions, but the current limitations have bounded our outline.

The performance of buildings has improved from the past and newer methods are getting a place in design offices, yet the mechanism to control the performance of buildings shall be the way ahead in line with the disaster mitigation needs. Tools for displacement-based approaches need to be further developed. A more experimental basis would be required for modern structures and retrofitting of the existing structures.

## References


1. Wikipedia, Seismometer, https://en.wikipedia.org/wiki/Seismometer, Accessed 21 Dec 2020
2. Gutenber B, Richter CF (1956) Earthquake magnitude, intensity, energy and acceleration (second paper). Bull Seismol Soc Am 46(2):105–145
3. American Museum of Natural History (AMNH), Drum Recorder, https://www.amnh.org/exhibitions/permanent/planet-earth/why-are-there-ocean-basins-continents-and-mountains/earthquakes/monitoring-earthquakes-daily/drum-recorder, Accessed 21 Dec 2020
4. USGS website, About the Seismograms, https://earthquake.usgs.gov/monitoring/seismograms/about.php, Accessed 21 Dec 2020
5. Irodov IE (1981) Fundamental laws of mechanics, 1st edn. Mir Publications, Moscow. ISBN 13: 9780714716015
6. Heyman J (1996) Elements of the theory of structures, 1st edn. Cambridge University Press, Cambridge. https://doi.org/10.1017/CBO9780511526671
7. Bozorgnia Y, Bertero VV (2004) Earthquake engineering: from engineering seismology to performance-based engineering. CRC Press, New York. ISBN 978-0-8493-1439-1

8. Reitherman RK (2012) Five major themes in the history of earthquake engineering. In: 15th World Conference on Earthquake Engineering (17WCEE). IAEE Publications, Lisbon
9. Priestley MJN, Paulay T (1992) Seismic design of reinforced concrete and masonry buildings. Wiley, Hoboken. https://doi.org/10.1002/9780470172841
10. Reddy GR, Ghosh S, Mondal A (2013) Performance based evaluation of the response reduction factor for ductile RC frames. Eng Struct 56:1808–1819. Science Direct (2013)
11. Park R, Paulay T (1975) Reinforced concrete structures. Wiley, Hoboken, p 769. https://doi.org/10.1002/9780470172834.
12. Murthy CVR, Goswami R, Vijayanarayanan AR, Mehta VV (2012) Some concepts in earthquake behavior of buildings. GSDMA Project
13. Moehle JP (1992) Displacement based design of RC structures. In: 10th world conference on earthquake engineering (10WCEE), Rotterdam. ISBN 90-54100605
14. Moehle JP (1996) Displacement based seismic design criteria. In: 11th world conference on earthquake engineering (11WCEE), 2125. Elsevier Science Limited. ISBN 0-08 042822 3
15. Priestley MJN (1993) Myths and fallacies in earthquake engineering—conflicts between design and reality. Bull New Zealand Natl Soc Earthquake Eng (NZSEE) 26(3):329–341
16. Michael FN (2014) From performance and displacement-based assessment of existing buildings per EN1998-3 to design of new concrete structures in fib MC2010. In: Ansal A (eds) Perspectives on European earthquake engineering and seismology. geotechnical, geological and earthquake engineering, vol 34. Springer, Cham, pp 227–266. https://doi.org/10.1007/978-3-319-07118-3_7
17. Mishra PK (2004) The Kutch earthquake 2001: recollections, lessons & insights. Report published: National Institute of Disaster Management (NIDM), p 258. ISBN: 81-8347-000-9
18. Sairam B, Rastogi BK, Patel V, Pancholi V (2018) Site effects: case study of the 2001 Bhuj earthquake damages in the Ahmedabad city, Gujarat, India. Bull Seismol Soc Am 4:2170–2182. https://doi.org/10.1785/0120170266
19. Dwivedi V, Dubey RK, Pancholi V, Rout MM, Singh P, Sairam B, Chopra S, Rastogi BK (2019) Multi-criteria study for seismic hazard assessment of UNESCO world heritage Ahmedabad city, Gujarat, Western India. Bull Eng Geol Environ, Springer 79:1721–1733
20. The working committee experts (WCE) (2010) Development of seismic hazard map of India. NDMA, New Delhi
21. Douglas J (1964–2010) Ground motion prediction equations. Report by Pacific Earthquake Engineering Research Centre, PEER 2011/102, 2011
22. IS 1893-Part 1 (2016) Criteria for earthquake resistant design of structures: general provisions and buildings. Bureau Indian Standards (BIS), New Delhi
23. ASCE 41 (2017) Seismic rehabilitation of existing buildings. American Society of Civil Engineers, ASCE/SEI Standard 41, Reston, VA
24. CEN (2004) Design of structures for earthquake resistance—part 1: general rules, seismic actions and rules for buildings. EN 1998–1:2004, EC-8, Cmite Europeen de Normalisation, Brussels
25. Turkish Earthquake Code (TEC-2007) Specifications for buildings to be built in seismic areas. Ministry of Public Works and Settlement, Ankara, Turkey, 2007
26. Reinhorn AM (1997) Inelastic analysis techniques in seismic evaluations. In: Fajfar P, Krawinkler H (eds) Seismic design methodologies for the next generation of codes, Balkema, Rotterdam
27. Faella G (1998) Evaluation of the r/c structures seismic response by means of nonlinear static push-over analyses. In: Eleventh World Conference on Earthquake Engineering, Acapulco, Mexico. 1996. September. University of Ljubljana, Ljubljana, Slovenia
28. Fajfar P. Capacity spectrum method based on inelastic demand spectra. IKIPR Report EE3/98
29. Englekirk RE (2003) Seismic design of reinforced and precast concrete buildings. Wiley, Hoboken, p 848. ISBN: 978-0-471-08122-7
30. Mander JB, Priestley MJN, Park R (1988) Theoretical stress-strain model for confined concrete. J Struct Eng ASCE 114–8:1804–1825

31. Ozmen HB, Inel M, Bilgin H (2007) Modelling nonlinear behavior of reinforced concrete members. In: Proceedings of 6th national conference on earthquake engineering (Oct'07), 16–20 October 2007, Istanbul, Turkey, p 207–2015
32. Asteris PG, Antoniou ST, Sophianopoulos DS, Chrysostomou CZ (2011) Mathematical micro-modeling of infilled frames—state of the art. J Struct Eng ASCE 1508–1517
33. Kaushik HB, Rai DC, Jain SK (2008) A rational approach to analytical modeling of masonry infills in reinforced concrete frame buildings. In: 14th world conference on earthquake engineering (14WCEE), Beijing
34. Virtual data center (VDC) (2012) Consortium of organizations for strong motion observation systems (COSMOS). University of California Santa Barbara, https://strongmotioncenter.org/vdc
35. Raisinghani BM (2017) Evaluation of design parameters on PBD of RC buildings with masonry infills. Journal of Seismol & Earthq Eng (JSEE) 123–138. http://www.jsee.ir/article_240759.html
36. Jain SK (2004) Implications of 2001 Bhuj earthquake for seismic risk reduction in India, 13th World Conference on earthquake engineering (13WCEE), no. 3244. https://www.iitk.ac.in/nicee/wcee/article/13_3244.pdf

# Parametric Study on the Variation of Time Period of RC MRF Buildings

**Sasanka Sarma, Tridib Sundar Das, Ankita Bora, and Karabi Bharadwaj**

## 1 Introduction

The Indian subcontinent has faced some devastating earthquakes that have resulted in the loss of life and property. While no structure is entirely immune from earthquake damage, earthquake-resistant structures are intended to withstand the largest probable earthquake that is likely to occur at that location. Therefore, the design of structures prone to earthquakes needs to be safeguarded, which is governed by natural frequencies.

The design codes related to earthquake-resistant structures in India recommend empirical expressions for calculation of the fundamental natural period of the building, which is the reciprocal of natural frequency. The expressions are derived from various numerical and analytical computations, recordings of earthquake data taken, and laboratory experiments. These expressions are based on a function of building type, the overall dimension of the building especially the height. Several other parameters that affect the time period are not taken into consideration, and thus there exists room for improvement in the empirical expressions.

## 2 Scope and Objectives

The natural time period of a building is an intrinsic property, which is the time taken by a building to go completely through one back and forth motion, i.e., one complete oscillation cycle. It is broadly a function of mass and stiffness.

S. Sarma (✉) · T. Sundar Das · A. Bora · K. Bharadwaj
Department of Civil Engineering, Tezpur University, Napaam, Sonitpur District, Assam, India

K. Bharadwaj
e-mail: karabi@tezu.ernet.in



S. Kolathayar and S. C. Chian (eds.), *Recent Advances in Earthquake Engineering*, Lecture Notes in Civil Engineering 175,
https://doi.org/10.1007/978-981-16-4617-1_33

An increase in mass increases the natural time period and an increase in stiffness decreases the natural time period. Therefore, buildings having lesser stiffness and larger mass, i.e., flexible and heavier buildings, have time periods more than that for lighter and stiffer buildings. For bare MRF buildings, the fundamental natural time period can be evaluated using the following expression [1]

$$T_a = 0.075 \times h^{0.75} \tag{1}$$

where, h is the height of the building in metres; $T_a$ is the time period of the structure in seconds.

This study aims to find out how different parameters affect the fundamental natural time period of a structure. Performing eigenvalue analysis in various building models, we have tried to show the effect of varied parameters in the time period, which are not included in IS 1893 Part 1:2016.

To perform our analysis, we have used the academic licensed version of Seismostruct 2020 software. Seismostruct is a package of finite elements, used for structural analysis, which, under dynamic loadings, is successful in speculating the large displacement behaviour of space frames, considering both geometric non-linearities and material inelasticity.

The software consists of three most important modules: a pre-processor in which the input data of the structural model can be defined, a processor in which the evaluation and analysis are carried out, and subsequently a post-processor to get the output or the results; all dealt with through a completely visual interface [2].

## 3 Building Configuration

For the present study, two bare RC MRF building models are created as base models with all necessary specifications such as beam dimension, column dimension, etc. are as follows:

Model A—3 bays × 3 bays.
Model B—3 bays × 5 bays.
Total number of storey = 5 (G + 4).
Storey height = 3 m.
Therefore, the total height of the building = 15 m.

Time period using Eq. (1).

$T_a = 0.075 \times 15^{0.75}$
$= 0.572$ s.

Bay length = 4 m in both x and y-direction.
Beam dimension = 250 mm × 450 mm.
Column dimension = 300 mm × 350 mm.
Grade of concrete used = M 20.

Grade of steel used = Fe 415.

Several models are prepared in the Seismostruct 2020 software and the above specifications are entered to perform eigenvalue analysis to obtain various results for our study.

Two base models (3 × 3 and 3 × 5) are taken into account to compare the results after changing the parameters . Plan and elevation views of the base models are shown in the figures below (Figs. 1, 2, 3 and 4).

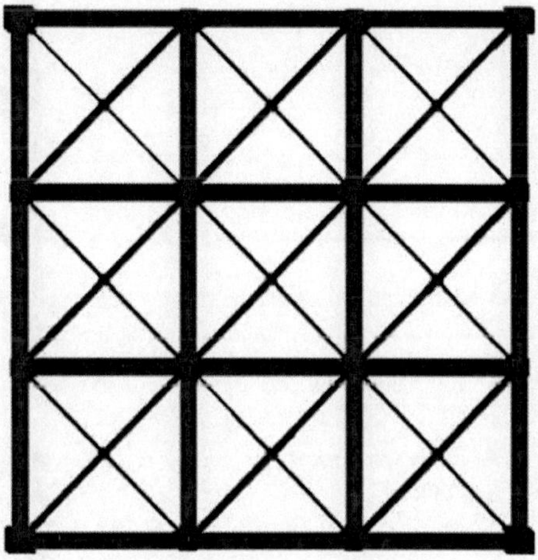

**Fig. 1** Plan view for 3 × 3 building

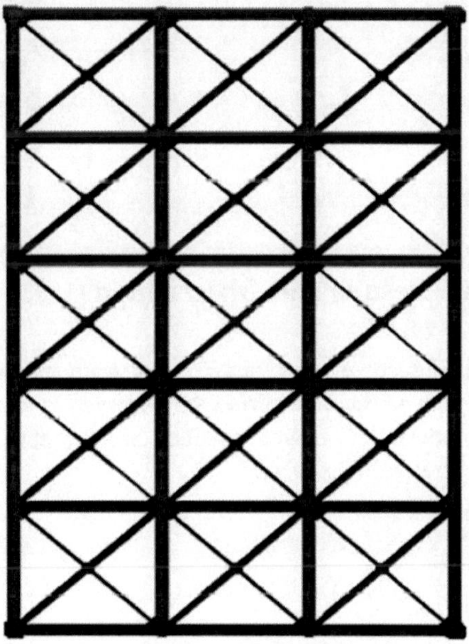

**Fig. 2** Plan view for 3 × 5 building

**Fig. 3** Elevation view for 3 × 3 building

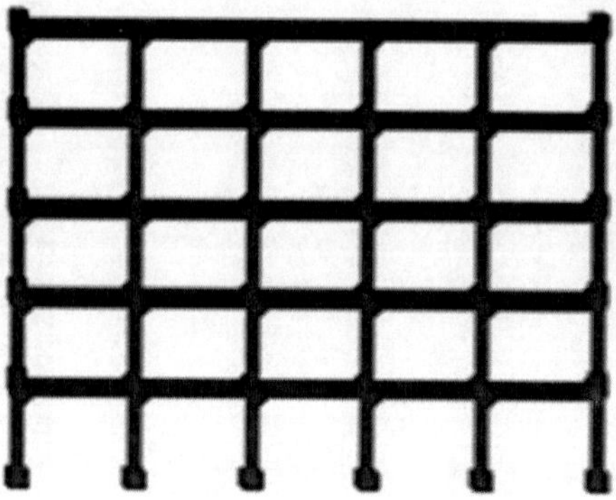

**Fig. 4** Elevation view for 3 × 5 building

## 4 Results and Observations

Using the Seismostruct 2020 software and performing eigenvalue analysis, the variations in the time period are observed by changing different parameters in the base models. The results obtained are presented in a tabular form followed by their corresponding charts.

## 4.1 Changing Column Size

To check the variation in time period due to change in column size, we have considered two models where different column sizes are 300 mm × 300 mm; 300 mm × 350 mm; 300 mm × 400 mm; 350 mm × 400 mm; 400 mm × 400 mm; 500 mm × 500 mm.

Here, buildings with column size 500 mm × 500 mm are relatively stiffer than the other column sizes (Table 1).

On changing the column dimensions, the time periods obtained from Seismostruct and as per IS 1893 Part1:2016 are not similar. For smaller column size, less difference is observed in the time period values (Fig. 5).

**Table 1** Time period values for various column dimensions

| Building | Column dimension (mm × mm) | Time period |
|---|---|---|
| A | 300 × 300 | 0.502 |
| | 300 × 350 | 0.488 |
| | 300 × 400 | 0.479 |
| | 350 × 400 | 0.427 |
| | 400 × 400 | 0.394 |
| | 500 × 500 | 0.36 |
| B | 300 × 300 | 0.501 |
| | 300 × 350 | 0.477 |
| | 300 × 400 | 0.468 |
| | 350 × 400 | 0.414 |
| | 400 × 400 | 0.395 |
| | 500 × 500 | 0.361 |

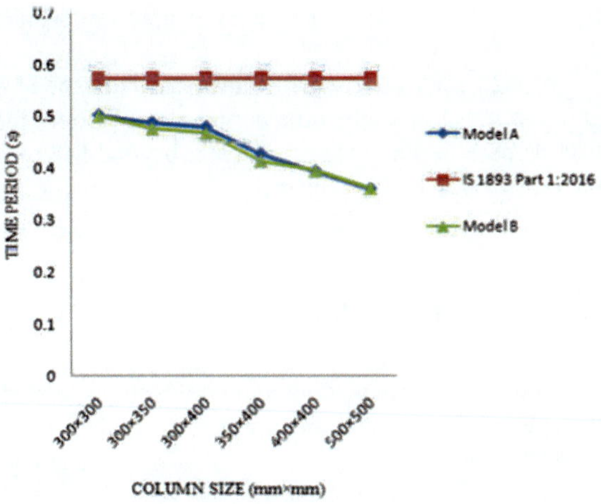


**Fig. 5** Time period values for different column sizes as per IS 1893 Part 1:2016 and Seismostruct

**Table 2** Time period values for various grades of concrete

| Building | Grade of concrete used | Time period (s) |
|---|---|---|
| A | M20 | 0.488 |
| | M25 | 0.467 |
| | M30 | 0.449 |
| | M35 | 0.435 |
| | M40 | 0.423 |
| B | M20 | 0.477 |
| | M25 | 0.456 |
| | M30 | 0.439 |
| | M35 | 0.425 |
| | M40 | 0.414 |

As the column size increases, the time period value decreases due to an increase in stiffness, and hence more difference in the time period is observed than values obtained from the code [5]. Therefore, we can say that when we increase the size of columns, the natural time period of the buildings reduces [4].

Also, when we consider square columns, the time period values for both the building types are almost equal.

### 4.2 Changing Grade of Concrete

The grade of concrete remains the same in all the concrete elements as they are cast monolithically. Columns are generally provided with a higher grade of concrete since they contribute to the strength of the structure.

To check the variation in time period due to change in grade of concrete, M20, M25, M30, M35 and M40 grade of concrete are considered for the two models (Table 2).

For M20 grade concrete, the time period obtained from the Seismostruct software is somewhat close to the time period obtained from the codal provisions. However, any increase in the grade of concrete beyond M20 leads to a decrease in the time period for both models (Fig. 6).

### 4.3 Changing Column Orientation

The columns are oriented along two directions to study its effect while finding the time period. In one case, the larger dimension of the column is along the x-axis and in another case, the larger dimension is oriented along the y-axis (Table 3).

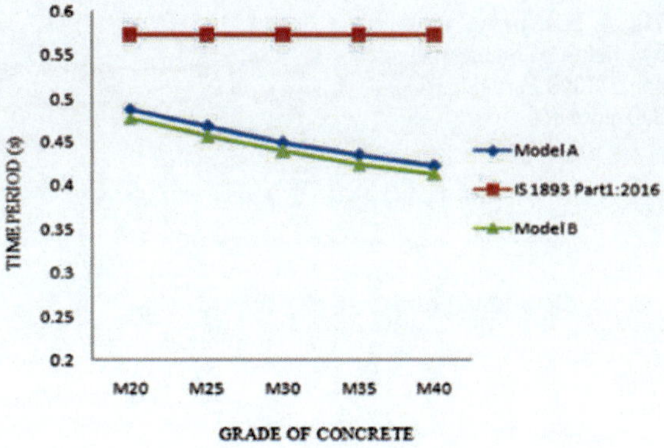


**Fig. 6** Time period values for different grades of concrete as per IS 1893 Part 1:2016 and Seismostruct

**Table 3** Time period values for change in column orientation

| Building | Column dimension (mm × mm) | Time period (s) |
|---|---|---|
| A | 300 × 350 | 0.488 |
| | 350 × 300 | 0.488 |
| | 300 × 400 | 0.479 |
| | 400 × 300 | 0.479 |
| | 350 × 400 | 0.427 |
| | 400 × 350 | 0.427 |
| B | 300 × 350 | 0.477 |
| | 350 × 300 | 0.49 |
| | 300 × 400 | 0.468 |
| | 400 × 300 | 0.48 |
| | 350 × 400 | 0.414 |
| | 400 × 350 | 0.428 |

Change in column orientation results in the same value of time period for a square plan. There is a minimal difference in the values for a rectangular plan but it won't have any considerable effect on the time period of the structure. Therefore, column orientation is not a deciding parameter for calculating the time period of a building (Fig. 7).

### 4.4 *Changing Bay Length*

To distinguish the effect of change in bay length on the time period, we have considered different bay lengths, keeping the other parameters constant. The bay lengths considered for the two base models are 3, 4, 5 and 6 m (Table 4).

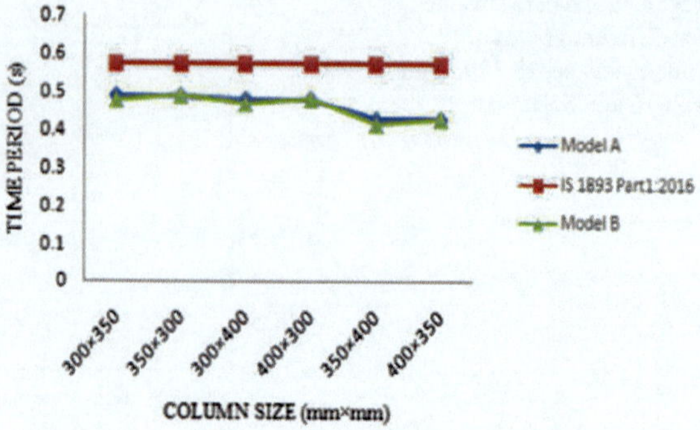


**Fig. 7** Time period values by orienting column size as per IS 1893 Part 1:2016 and Seismostruct

**Table 4** Time period values for change in bay length

| Building | Bay length (m) | Time period (s) |
|---|---|---|
| A | 3 | 0.431 |
| | 4 | 0.488 |
| | 5 | 0.546 |
| | 6 | 0.604 |
| B | 3 | 0.43 |
| | 4 | 0.477 |
| | 5 | 0.55 |
| | 6 | 0.609 |

For bay lengths of 3 and 4 m, the time period decreases and in case of 5 and 6 m, it is almost similar as obtained as per IS 1893 Part 1:2016. For both squares as well as rectangular plan, an increase in the bay length leads to an increase in time period. Similarly, a decrease in the bay length leads to a decrease in time period (Fig. 8).

We see that when we increase the length of the building keeping the number of bays constant, the mass of the building increases but the column stiffness remains

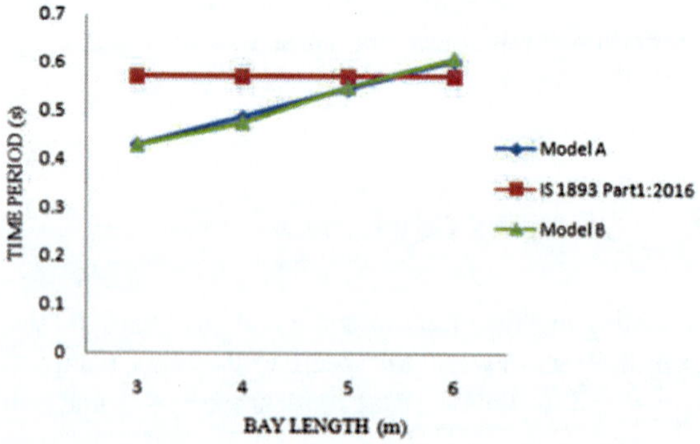


**Fig. 8** Time period values for bay length as per IS 1893 Part 1:2016 and Seismostruct

**Table 5** Time period values for change in building orientation

| Building | Number of bays | | Time period (s) |
|---|---|---|---|
| | X direction | Y direction | |
| A | 3 | 5 | 0.480 |
| B | 5 | 3 | 0.490 |

constant against any force from the lateral direction. Due to this, the flexibility in the bare RC MRF building increases, resulting in the increase of natural period.

The empirical equation (1) obtained from IS 1893 Part 1:2016 doesn't include the parameter of overall base dimension in it for an RC MRF building. Therefore, the time period predicted by it for a particular height is similar for any base dimension. It becomes important that the effect of the base dimension is included in the empirical equation for calculating the natural time period of a building.

### *4.5 Changing Building Orientation*

To note the effect of building orientation on the time period, the 3 × 5 building is oriented as 5 × 3 and its effect is analysed and observed (Table 5).

Although the values of the time period obtained from IS 1893 Part 1:2016 and that from Seismostruct have a slight difference, yet no notable change in the time period has been observed in changing the orientation of the building.

### *4.6 Changing Plan Area*

For finding the influence of the number of bays on the time period, we have considered plans of 3 × 3 bays; 5 × 3 bays; 5 × 5 bays and 7 × 5 bays (Table 6).

Although the time period obtained from IS 1893 Part 1:2016 and Seismostruct software differs from each other, no notable change in the time period has been observed by changing the plan area of the building. Therefore, we can deduce that changing the plan area of a building would not have any significant change in the time period of the building [3] (Fig. 9).

**Table 6** Time period values for change in plan area

| Plan dimension | | Time period (s) |
|---|---|---|
| X direction | Y direction | |
| 3 | 3 | 0.488 |
| 5 | 3 | 0.489 |
| 5 | 5 | 0.481 |
| 7 | 5 | 0.478 |

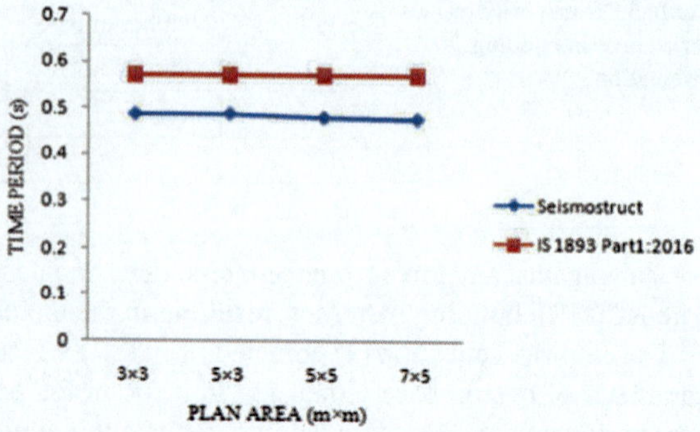


**Fig. 9** Time period values for change in plan area of building as per IS 1893 Part 1:2016 and Seismostruct

Similar studies are conducted for change in grade of steel and no notable variations in the time period are observed.

## 5 Comparison with STAAD Pro

Time period values of change in column size and change in bay length obtained from Seismostruct software are compared with that obtained from STAAD Pro. The results are tabulated below (Table 7).

On comparing the values, we observe that the percentage variations of the time period values are within 5%. The values obtained are almost similar (Fig. 10; Table 8).

**Table 7** Comparison of time period values for change in column size

| Building | Dimension (mm × mm) | Seismostruct | STAAD Pro |
|---|---|---|---|
| A | 300 × 350 | 0.488 | 0.497 |
| | 500 × 500 | 0.36 | 0.374 |
| B | 300 × 350 | 0.477 | 0.493 |
| | 500 × 500 | 0.361 | 0.38 |

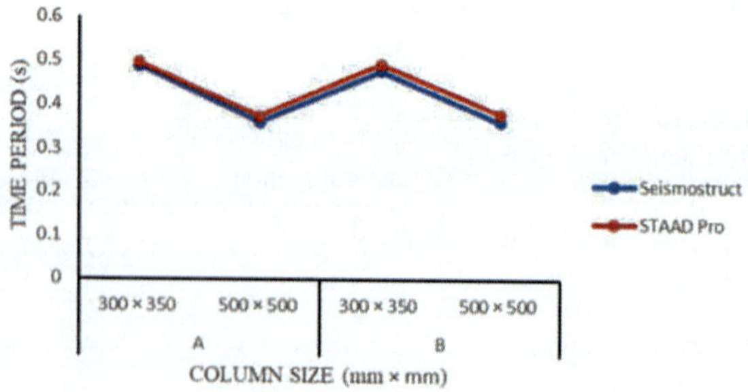


**Fig. 10** Time period comparison between Seismostruct and STAAD Pro for change in column size

**Table 8** Comparison of time period values for change in bay length

| Building | Dimension | Seismostruct | STAAD Pro |
|---|---|---|---|
| A | 4 m | 0.488 | 0.497 |
| | 5 m | 0.546 | 0.559 |
| B | 4 m | 0.477 | 0.493 |
| | 5 m | 0.55 | 0.556 |

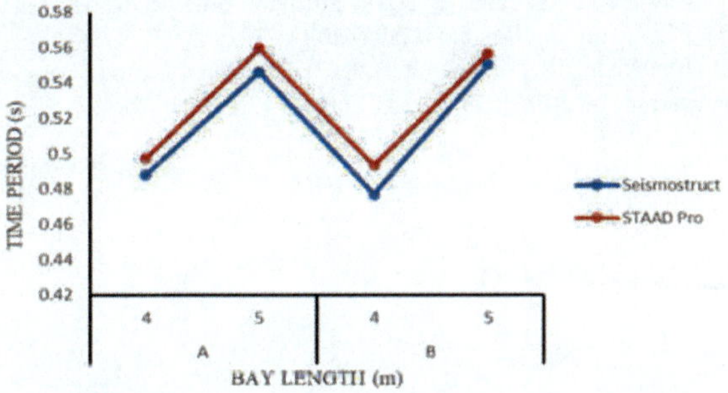


**Fig. 11** Time period comparison between Seismostruct and STAAD Pro for change in bay length

On comparing the values, we observe that the percentage variations of the time period values are within 3%. The values obtained are almost similar (Fig. 11).

## 6 Conclusions

It has been observed that apart from the height of the building, several other parameters may have a considerable effect on the time period of a building. Parameters such as the size of column, grade of concrete, bay length contribute to change in time period values. On the other hand, parameters such as plan area, building orientation, grade of steel do not affect the time period values. Also, on comparing with different software, we observed very little variations in the time period values.

From the results obtained from the analyses, we can conclude that time period of the building obtained as per empirical equations given in IS 1893 Part 1:2016 gives safe results in a maximum number of cases, but if the few parameters that were analysed are considered in addition to the height of the building, then the time period will be more precise, and more simplified relationships can be given for determining the time period of a building.

## References


1. IS 1893 (2016) Indian Standard Criteria for Earthquake Resistant Design of Structures-Part I: General Provisions and Buildings (Sixth Revision). Bureau of Indian Standards, New Delhi
2. Seismosoft (2020) SeismoStruct 2020—A computer program for static and dynamic nonlinear analysis of framed structures. https://seismosoft.com/
3. Sangamnerkar P, Dubey SK (2017) Effect of plan area of the building on period of vibration of RC framed structures. Electr J Struct Eng 17(1)
4. Cutinha LL, Karanth P (2018) Study on time period as per IS CODE using Etabs software. Int J Curr Eng Sci Res 5(5):2394–2697
5. Bhuskade SR, Sagane SC (2017) Effects of various parameters of building on natural time period. Int J Eng Res Technol 6(4). ISSN: 2278-0181

# Parametric Study of Performance-Based Seismic Design of Plan Irregular RC Frames—Indian Scenario

**Rajat Abhay Sirsikar, Ganesh D. Awchat, and J. S. Kalyana Rama**

## 1 Introduction

Earthquake-resistant design based on traditional approaches aims to attain the prescribed limits on strength and serviceability criteria as per code provisions. Even after practising those design practices, earthquakes incurred catastrophic damages to structures and led to huge loss of life and economy. Joshi et al. found that 52.86% of hospitals, 64.58% of schools and nearly 60% of other important surveyed buildings would collapse immediately after an earthquake in Uttarakhand [1]. Gautam et al. highlighted the urgent need to revise the Indian code by comparing the performance of structural forms during Bhuj, Chile, Kashmir, Haiti, L'Aquila and Turkey earthquakes from the past two decades worldwide [2]. Moreover, nonlinear analysis methods are not included in Indian building codes and no changes are expected in the near future [3]. Shukla and Dalal found that though the response reduction factor (R = 5) as per IS code is valid, but hinges are formed in columns and so the strong-column–weak-beam concept does not hold true [4]. The column/beam moment-capacity ratio (M.R.) in IS 13920:2016 is found uneconomical for regular frames [5].

In reality, many existing reinforced concrete (RC) buildings have irregular configurations especially in the plan [6]. Many researchers tried to study the effect of irregularity on the seismic performance of the structure. Chen et al. provided a framework for three-dimensional models to account for the torsional behaviour of asymmetric buildings [7]. It is found that the CSM-FEMA440 method better matched the

R. A. Sirsikar (✉) · G. D. Awchat
Shri Guru Gobind Singhji Institute of Engineering and Technology, Nanded 431606, India

J. S. Kalyana Rama
Vignana Bharathi Institute of Technology, Ghatkesar, Hyderabad, Telangana 501301, India

 
S. Kolathayar and S. C. Chian (eds.), *Recent Advances in Earthquake Engineering*, Lecture Notes in Civil Engineering 175,
https://doi.org/10.1007/978-981-16-4617-1_34

nonlinear dynamic analysis in plan irregular frame buildings [8]. The pushover analysis provides displacements within the range of ASCE but greater than the conventional code-based approach and must be considered when analysing irregular buildings [9]. An attempt has been made to know the difference in seismic response of two buildings having diaphragm discontinuity and without diaphragm discontinuity [10].

In engineering practice, standard and convenient member sizes are generally considered [11]. This results in a lower ductility capacity of the structure. The commitment of PBSD in earthquake engineering is to build structures of which the seismic performance is attained as desired. The potential loss of occupancy, repair costs and life safety impacts come under the desired seismic performance. The performance level is defined as the maximum acceptable damage state or condition caused by physical damage within a building, the threat to building occupant's life safety due to damage and serviceability of the structure post-earthquake. In this type of design, the decision-maker chooses the structural performance required as per his interest. The engineer uses his skills and provides a design that is capable of satisfying the decision-maker requirements. The combined effort of professionals and designers is required to make this commitment into reality. Mohd. Zameeruddin et al. reviewed recent developments in performance-based seismic design by defining the performance objectives (levels), evaluation techniques and assessment procedures [12]. Because of the advancements that took place recently in assessing seismic hazards, facilities for performing experiments and computer applications, many design engineers and developers in earthquake-prone regions attracted to performance-based seismic design. Hopefully, we can say that within a very short period of time PBSD becomes a conventional design method.

## 2 Description of Selected Frames

Six G + 4 RC frames which are having regular (R), torsional irregular (Ir-1), re-entrant corners (Ir-2), excessive cut-outs (Ir-3), out-of-plane offsets in vertical elements (Ir-4) and non-parallel lateral force system (Ir-5) confirming to clause 7.1 of IS 1893 (Part1):2016 [13] are modelled in Zone IV and Zone V. The software used for analyses is ETABS 2018. Dimensions of columns and beams are varied to find critical sections required in life safety and collapse prevention state. Figure 1 shows the geometry of all selected frames. The nomenclature of frames is done such that the Zone is followed by the regularity/irregularity conditions and then the type of irregularity. Hence, frame IV-Ir-1 represents an irregular frame of Type 1 (torsional irregularity) in Zone IV and V-R represents a regular frame in Zone V.

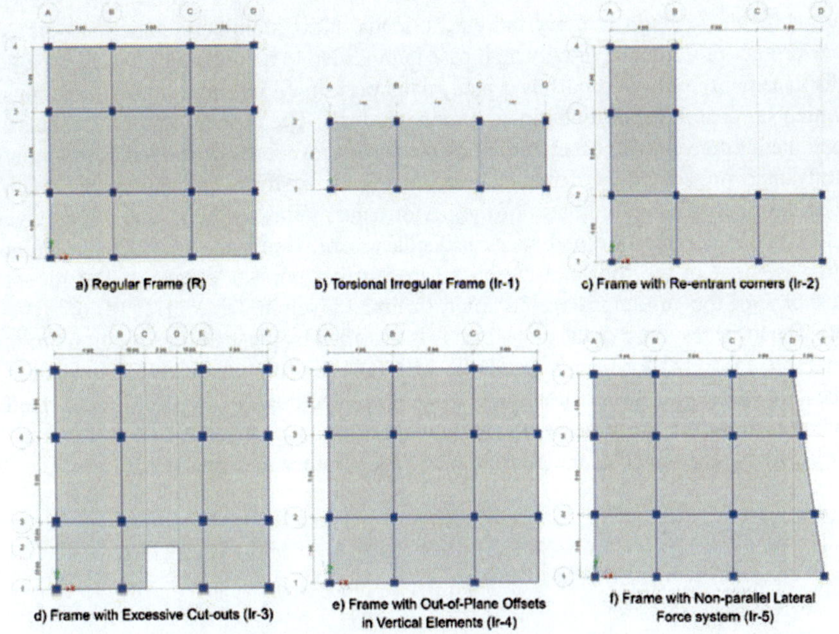


**Fig. 1** Various configurations of irregularities as per IS 1893(Part-I):2016

## 3 Modelling of RC Frames

Three-dimensional modelling of frames is done in ETABS 2018. A concrete hysteresis model is used. M25 concrete and Fe500 rebar is used [14]. A floor height of 3 m is considered in all selected frames. The loads are assigned to structural elements as per IS 875 [15, 16] and earthquake loads as per IS 1893:2016. Wall load is applied on beams as uniformly distributed load taking a 30% reduction for openings. Moment of inertia for beams and columns are taken in accordance with IS 13920:2016 [17]. Since the frames are in Zone IV and Zone V, the frames are considered as special moment-resisting frames. All frames have been given an importance factor of 1.5 for lifeline buildings.

## 4 Methodology

In this study, six G + 4 RC bare frames are modelled having a floor height of 3 m. The desired performance objectives of collapse prevention and life safety are selected. Preliminary design is done to the building for linear analysis where the dead load, imposed load and seismic loads are acting. The oscillation type in fundamental mode is verified by modal analysis. Then, it is verified whether all the structural elements are sufficient enough to carry the elastic loads. Next, nonlinear static analysis is performed. For this, the gravity load case is modified to a nonlinear load case. Now displacement-controlled pushover analysis is performed in both x and y directions which starts at the end of nonlinear gravity analysis. The capacity (pushover) curve obtained is converted to acceleration–displacement response spectra (ADRS) format and superimposed on the demand spectrum and the performance point is obtained. The corresponding coordinates of the performance point on the x- and y-axis give the spectral acceleration and spectral displacement. Using the performance point displacement of the roof and the base shear at that point is obtained. The global response of the building is verified by drift limits given in Table 11.2 [18] of ATC-40. The local response of the elements is also verified by plastic hinge rotation limits in Table 11.3 [18] for beams and Table 11.4 [18] for columns of ATC-40 whether they are within the acceptable limit or not. After verifying the limits, it is confirmed whether the structure possesses the required performance objective or not. If not, the structure is redesigned or the performance of the frames are enhanced.

## 5 Results and Discussions

### 5.1 *Cross-sections of Beams and Columns*

The primary aim of this study is to find the critical dimensions of beams and columns for life safety and collapse prevention state. Each type of plan irregular frame is compared with the regular frame. First, the preliminary design of members is done to check the minimum dimensions of beams and columns. The minimum dimensions of beams and columns are considered in Trial-1 as per IS 13920: 2016 since the frames are considered in Zone IV and Zone V. In Trial-1 Zone IV, Ir-1 and Ir-2 passed with lesser beams and columns dimension. Frame Ir-3 required the highest dimension among all considered frames to pass the preliminary design. This suggests that among all plan irregularities, buildings with excessive cut-outs would require more dimensions of beams and columns to pass the preliminary design as per codal provisions.

### 5.2 *Capacity Curve*

The capacity curves for all plan irregular frames are compared with that of the regular frame. It is found that the building with excessive cut-outs has maximum base shear among all plan irregular and regular frames.

**Table 1** Cross-sectional dimensions of beam and column for various cases

| | R | Ir-1 | Ir-2 | Ir-3 | Ir-4 | Ir-5 |
|---|---|---|---|---|---|---|
| Beam (Trial-1) | 350×450 | 300×350 | 300×350 | 500×500 | 350×400 | 420×450 |
| Beam (Trial-2) | 450×450 | 450×450 | 450×450 | 500×500 | 450×450 | 450×450 |
| Column (Trial-1) | 400×500 | 350×450 | 350×450 | 500×550 | 400×500 | 500×500 |
| Column (Trial-2) | 450×600 | 450×600 | 450×600 | 450×600 | 450×600 | 450×600 |

**Table 2** Trial-2 roof displacement and base shear for the frame using response spectrum method in Zone IV

| | R | Ir-1 | Ir-2 | Ir-3 | Ir-4 | Ir-5 |
|---|---|---|---|---|---|---|
| Roof displacement (mm) | 194.4 | 206.6 | 221.92 | 120.4 | 114.13 | 157 |
| Base shear (kN) | 5011.5 | 1556.92 | 2400.1 | 7501.94 | 4152.7 | 6258.3 |

## Zone IV

*Trial-1.*

An initial analysis is performed using the response spectrum method for all the irregularities as shown in Table 1. The maximum roof displacement and maximum base shear values for all the frames are shown in Table 2. The maximum roof displacement value for the Ir-2 frame is higher than Ir-1 for the same cross-sections and reinforcement of beams, columns and slabs as there is a higher force at the base generated due to seismic activity for a frame with a re-entrant corner (Table 3).

The maximum roof displacement for a regular and torsional irregular frame is nearly the same for different cross-sections of beams and columns. However, in the case of the response spectrum, the Ir-1 value was 6.3% higher than the frame R-value. It is interesting to know that in the case of a frame with out-of-plane offsets in vertical elements (Ir-4), the value of maximum roof displacement is 62.6% (highest) higher in the case of the time history method than the response spectrum method. While in all other frames, there is no significant deviation in values of maximum roof displacement for the time history and response spectrum method (Fig. 2).

*Trial-2*

From Tables 4 and 5, it is observed that the maximum roof displacement values of regular and frame with a non-parallel lateral force-resisting system (Ir-5) are the

**Table 3** Trial-1 roof displacement and base shear for the frame using time history analysis in Zone IV

| | R | Ir-1 | Ir-2 | Ir-3 | Ir-4 | Ir-5 |
|---|---|---|---|---|---|---|
| Roof displacement (mm) | 206.78 | 206.58 | 227.13 | 120.96 | 185.6 | 163.42 |
| Base shear (kN) | 5120.5 | 1556.92 | 2413.1 | 7519.1 | 4248.5 | 6636.5 |

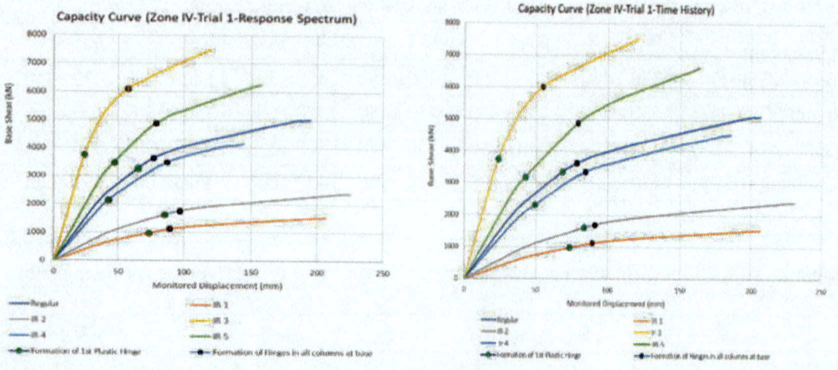


**Fig. 2** Trial-1 capacity curves for various irregularities in Zone IV

**Table 4** Trial-2 roof displacement and base shear for the frame using response spectrum analysis in Zone IV

| | R | Ir-1 | Ir-2 | Ir-3 | Ir-4 | Ir-5 |
|---|---|---|---|---|---|---|
| Roof displacement (mm) | 184.94 | 213.05 | 258.2 | 132.15 | 189.73 | 169.86 |
| Base shear (kN) | 7484.5 | 3298.9 | 5363.5 | 7232.8 | 6743.7 | 7616.1 |

**Table 5** Trial-2 roof displacement and base shear for the frame using time history analysis in Zone IV

| | R | Ir-1 | Ir-2 | Ir-3 | Ir-4 | Ir-5 |
|---|---|---|---|---|---|---|
| Roof displacement (mm) | 184.2 | 172.13 | 171.45 | 156.9 | 161.6 | 169.89 |
| Base shear (kN) | 7468.6 | 3413.9 | 5058.9 | 7574.53 | 6869.1 | 7616.1 |

same in response spectrum and time history method as the cross-sections of beams and columns of all frames are kept the same in Trial-2. However, the maximum roof displacement values are higher by 23.8, 50.6 and 17.4% for frames Ir-1, Ir-2 and Ir-4 in the case of response spectrum than the time history method. For a frame with excessive cut-outs (Ir-4), the maximum roof displacement was 18.73% lower in the case of response spectrum than the time history method (Figs. 3 and 4).

## Zone V

*Trial-1*

For Zone V similar trends are observed as that of Zone IV for trials 1 and 2, as shown in Tables 6, 7.

*Trial-2*

See Tables 8, 9 and Fig 5.

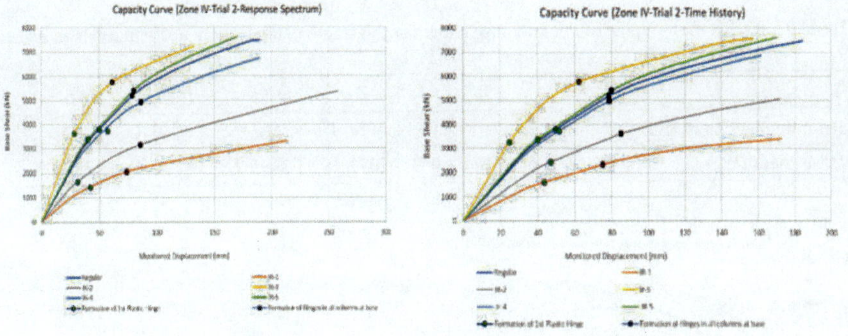


**Fig. 3** Trial-2 capacity curves for various irregularities in Zone IV

**Table 6** Trial-1 roof displacement and base shear for the frame using response spectrum analysis in Zone V

| | R | Ir-1 | Ir-2 | Ir-3 | Ir-4 | Ir-5 |
|---|---|---|---|---|---|---|
| Roof displacement (mm) | 211.3 | 230.2 | 249.4 | 120.3 | 140.6 | 161.5 |
| Base shear (kN) | 5153.5 | 1615.9 | 2482.2 | 7502.6 | 4095.6 | 6305.9 |

**Table 7** Trial-1 roof displacement and base shear for the frame using time history analysis in Zone V

| | R | Ir-1 | Ir-2 | Ir-3 | Ir-4 | Ir-5 |
|---|---|---|---|---|---|---|
| Roof displacement (mm) | 195 | 201.4 | 222 | 120.4 | 114.1 | 157 |
| Base shear (kN) | 5007.8 | 1548.5 | 2400.1 | 7502 | 4152.7 | 6258.3 |

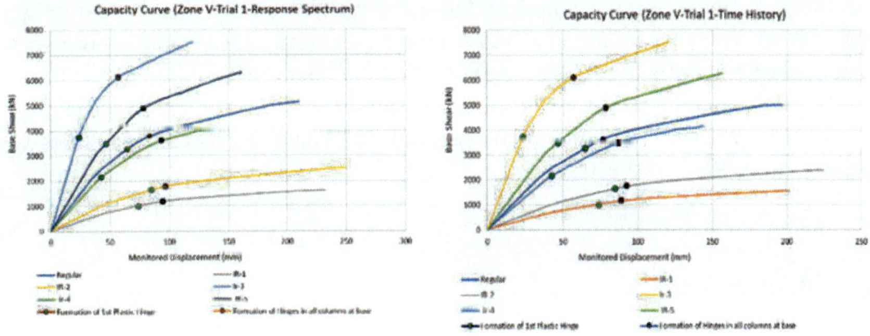


**Fig. 4** Trial-1 capacity curves for various irregularities in Zone V

**Table 8** Trial-2 roof displacement and base shear for the frame using response spectrum analysis in Zone V

| | R | Ir-1 | Ir-2 | Ir-3 | Ir-4 | Ir-5 |
|---|---|---|---|---|---|---|
| Roof displacement (mm) | 171.3 | 213.1 | 174.5 | 120.2 | 160 | 162.1 |
| Base shear (kN) | 7277.1 | 3298.9 | 5072.1 | 7388.4 | 6238.6 | 6930.8 |

**Table 9** Trial-2 roof displacement and base shear for the frame using time history analysis in Zone V

| | R | Ir-1 | Ir-2 | Ir-3 | Ir-4 | Ir-5 |
|---|---|---|---|---|---|---|
| Roof displacement (mm) | 219.3 | 172.1 | 174.5 | 123.3 | 156.3 | 167 |
| Base shear (kN) | 7829.8 | 3413.9 | 5072.1 | 7395.2 | 6773.4 | 7503.3 |

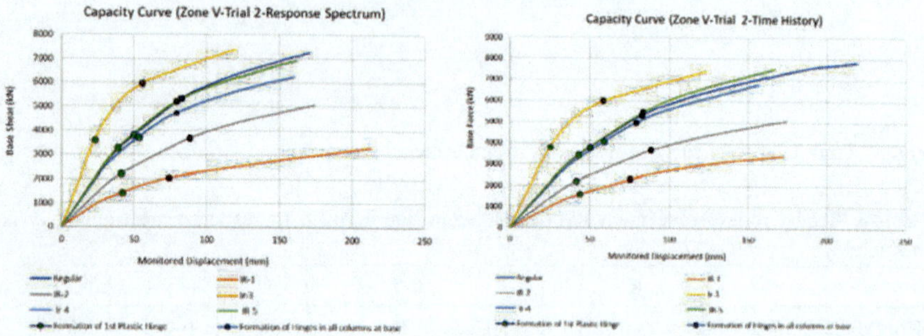


**Fig. 5** Trial-2 capacity curves for various irregularities in Zone V

## 5.3 *Formation of Plastic Hinges and Hinge Response*

From the initial analysis with constant cross-section dimensions of 300 mm × 450 mm, it is observed that all the categories of frames have reached a performance level of collapse prevention (CP) in Zone V. The corresponding performance points and spectral coordinates are shown in Table 10. In order to reduce the vulnerability of the chosen irregular frames, the cross-sectional dimensions have been modified as given in Table 1. From the analysis it is observed that the frame with torsional irregularity, non-parallel lateral force system and out of plane offsets have reached life safety level which proved to be safe for earthquake ground motion (Fig. 5; Tables 11 and 12).

# 6 Conclusions

Five-storied regular and plan irregular frames are analysed using nonlinear dynamic time history analyses and linear dynamic response spectrum methods in Zones IV and V for two trial cross-sections of primary structural members. The conclusions are as follows:

- It is found that out of all plan irregularities, Ir-2 and Ir-3 are safe in both Zones IV and V, and other configurations are vulnerable to seismic actions. Frame

**Table 10** Performance level and points for Zone V with preliminary c/s dimensions

| Type | Performance level | Performance point | Spectral coordinates |
|---|---|---|---|
| Regular | IO | (94.5 mm, 10,250.9 kN) | (74.8 mm, 1.08 g) |
| | LS | (102.9 mm, 8794.8 kN) | (83.7 mm, 0.92 g) |
| | CP | (120.8 mm, 4302.1 kN) | (99.60 mm, 0.52 g) |
| IR-1 | IO | (85.1 mm, 3724.0 kN) | (68.2 mm, 1.03 g) |
| | LS | (106.0 mm, 2822.8 kN) | (86.41 mm, 0.87 g) |
| | CP | (130.6 mm, 1340.8 kN) | (109.1 mm, 0.48 g) |
| IR-2 | IO | (102.4 mm, 5910.0 kN) | (80.31 mm, 1.00 g) |
| | LS | (118.8 mm, 4285.0 kN) | (91.41 mm, 0.81 g) |
| | CP | (142.2 mm, 2044.5 kN) | (116.23 mm, 0.44 g) |
| IR-3 | IO | (48.6 mm, 7899.3 kN) | (42.6 mm, 0.81 g) |
| | LS | (78.3 mm, 6452.9 kN) | (68.1 mm, 0.693 g) |
| | CP | (76.4 mm, 6575 kN) | (66.42 mm, 0.71 g) |
| IR-4 | IO | (101.3 mm, 8468.3 kN) | (86.7 mm, 0.96 g) |
| | LS | (108.9 mm, 5818.0 kN) | (92.9 mm, 0.72 g) |
| | CP | (117.2 mm, 3901.9 kN) | (102.0 mm, 0.52 g) |
| IR-5 | IO | (106.1 mm, 9054.7 kN) | (88.97 mm, 0.88 g) |
| | LS | (120.4 mm, 6561.2 kN) | (99.94 mm, 0.70 g) |
| | CP | (110.2 mm, 5991.1 kN) | (91.04 mm, 0.645 g) |

with excessive cut-outs is also vulnerable in both the zones using the preliminary and modified cross-sections based on maximum base shear, maximum roof displacement which is higher than the other regular and plan irregular frames.

**Table 11** Performance level and points Zone IV with modified c/s dimensions

| Type | Performance level | Performance point | Spectral coordinates |
|---|---|---|---|
| Regular | LS | (116.3 mm, 6264.4 kN) | (93.7 mm, 0.72 g) |
| | CP | (123.52 mm, 4290.4 kN) | (102.2 mm, 0.52 g) |
| IR-1 | LS | (132.8 mm, 1333.9 kN) | (111.7 mm, 0.47 g) |
| | CP | NA | NA |
| IR-2 | LS | (117.6 mm, 4265.9 kN) | (90.33 mm, 0.81 g) |
| | CP | (144.1 mm, 2047 kN) | (118 mm, 0.44 g) |
| IR-3 | LS | (83.08 mm, 6259.0 kN) | (70.7 mm, 0.705 g) |
| | CP | (76.82 mm, 6578 kN) | (66.81 mm, 0.71 g) |
| IR-4 | LS | (119 mm, 3837.7 kN) | (104.3 mm, 0.50 g) |
| | CP | NA | NA |
| IR-5 | LS | (118.2 mm, 5850.8 kN) | (100.4 mm, 0.63 g) |
| | CP | NA | NA |

A more detailed study on all vertical irregularities can provide a complete understanding of critical dimensions for each performance objective. Further, strengthening techniques for each type of irregularity can be suggested so as to improve the performance at each level.

**Table 12** Performance level and points Zone V with modified c/s dimensions

| Type | Performance level | Performance point | Spectral coordinates |
|---|---|---|---|
| Regular | LS | (115.7 mm, 6242.2 kN) | (93.2 mm, 0.72 g) |
| | CP | (120.8 mm, 4302.1 kN) | (99.60 mm, 0.52 g) |
| IR-1 | LS | (106.0 mm, 2822.8 kN) | (86.41 mm, 0.87 g) |
| | CP | NA | NA |
| IR-2 | LS | (118.8 mm, 4285.0 kN) | (91.41 mm, 0.81 g) |
| | CP | (142.2 mm, 2044.5 kN) | (116.23 mm, 0.44 g) |
| IR-3 | LS | (78.3 mm, 6452.9 kN) | (68.1 mm, 0.693 g) |
| | CP | (76.4 mm, 6575 kN) | (66.42 mm, 0.71 g) |
| IR-4 | LS | (108.9 mm, 5818.0 kN) | (92.9 mm, 0.72 g) |
| | CP | NA | NA |
| IR-5 | LS | (120.4 mm, 6561.2 kN) | (99.94 mm, 0.70 g) |
| | CP | (110.2 mm, 5991.1 kN) | (91.04 mm, 0.645 g) |

## References


1. Joshi GC, Ghildiyal S, Rautela P (2019) Seismic vulnerability of lifeline buildings in Himalayan province of Uttarakhand in India. Int J Disaster Risk Reduct 37: 101168. https://doi.org/10.1016/j.ijdrr.2019.101168
2. Gautam D, Chaulagain H (2016) Structural performance and associated lessons to be learned from world earthquakes in Nepal after 25 April 2015 (MW 7.8) Gorkha earthquake. Eng Fail Anal 68: 222–243. https://doi.org/10.1016/j.engfailanal.2016.06.002
3. Fajfar P (018) Analysis in seismic provisions for buildings: past, present and future In: The 5th professor Nicholas Ambraseys lecture, bull earthquake engineering. https://doi.org/10.1007/s10518-017-0290-8
4. Shukla KP, Dalal SP (2017) Evaluation of response reduction factor of a reinforced cement concrete building designed by performance based plastic design method and limit state design method. In: 11th international symposium on plasticity and impact mechanics, Implast 2016, Procedia Engineering, vol 173, pp 1854–1861
5. Manjula NK, Nagarajan P, Pillai TM (2018) Performance evaluation of RC buildings designed as per Indian seismic codes: a study on frames with vertical geometric irregularity. J Inst Eng India Ser A. https://doi.org/10.1007/s40030-018-0324-5
6. Hentri M, Hemsas M, Nedjar D (2018) Vulnerability of asymmetric multi-storey buildings in the context of performance-based seismic design. Eur J Environ Civ Eng. https://doi.org/10.1080/19648189.2018.1548380

7. Chen P, Collins KR (2001) Some observations on performance-based and reliability-based seismic design of asymmetric building structures. Eng Struct 23:1005–1010
8. Bhatt C Bento R (2012): Comparison of nonlinear static methods for the seismic assessment of plan irregular frame buildings with non-seismic details. J Earthq Eng 16(1):15–39. https://doi.org/10.1080/13632469.2011.586085
9. Al-Fadhli SKI (2020) Performance of multistory building under nonlinear static push-over. AIP Conf Proc 2213:020241.https://doi.org/10.1063/5.0000122
10. Ahirwal A, Gupta K, Singh V (2019) Effect of irregular plan on seismic vulnerability of reinforced concrete buildings. AIP Conf Proc 2158:020012. https://doi.org/10.1063/1.5127136
11. Xue Q, Chia-Wei Wu, Chen C-C, Chen K-C (2008) The draft code for performance-based seismic design of buildings in Taiwan. Eng Struct 30:1535–1547
12. Zameeruddin M, Sangle KK (2016) Review on recent developments in the performance-based seismic design of reinforced concrete structures. Structures 6:119–133.https://doi.org/10.1016/j.istruc.2016.03.001
13. IS 1893 (Part 1) (2016) Criteria for earthquake resisting design of structures—Indian Standard Code of Practice, Bureau of Indian Standards, New Delhi, India
14. IS 456 (2000) Plain and reinforced concrete—Indian standard code of practice, Bureau of Indian Standards, New Delhi, India
15. IS: 875 (Part 1) (1987) (Reaffirmed 2003) Dead loads—unit weights of building materials and stored materials—Indian standard code of practice for design loads (other than earthquake) for buildings and structures, Bureau of Indian Standards, New Delhi, India
16. IS: 875 (Part 2) (1987) (Reaffirmed 2008) Imposed loads—indian standard code of practice for design loads (other than earthquake) for buildings and structures, Bureau of Indian Standards, New Delhi, India
17. IS 13920 (2016) Indian standard ductile design and detailing of reinforced concrete structures subjected to seismic forces—code of practice (First Revision), Bureau of Indian Standards, New Delhi, India
18. ATC-40 (1996) Seismic evaluation and retrofit of existing concrete buildings, Redwood City (CA): Applied Technical Council

# Role of Openings in Seismic Resistance of Infill Walls: A Critical Review of Indian Codal Provision

**Karismita Pathak and Atanu Kumar Dutta**

## 1 Introduction

Reinforced concrete frames with masonry infill walls is a common construction practice. If infills are designed properly, these can increase the overall strength, lateral resistance and energy dissipation of the structure. Infill walls thus increase the lateral stiffness, and thereby reduce the lateral deflections and bending moments in the frame. This improved performance makes it more realistic to consider the infill wall as a structural element in the earthquake-resistant design of the structure.

However, infills are generally considered as non-structural elements and their stiffness contributions are generally ignored in practice. Of late, the Indian seismic code [1] has treated infill walls as a structural element, quantifying its effect as an equivalent bracing of the beam-column frame against lateral loadings. Advanced codes already have provisions for taking these infills into account in modelling the structure. According to FEMA-273 [2] and ATC-40 [3], the stiffness of the infill is modelled as that of an "Equivalent Diagonal Strut". In this procedure, the presence of openings is accounted for using a reduction factor on the width of the strut for infills walls calculated for without opening case. In contrast, the Indian seismic code [1] makes no distinction in modelling the provision of walls, whether these are with or without opening. It is worth enquiring whether the Indian standard provision is safe in one extreme or uneconomical in the other, in neglecting the reduction in strength of masonry infill walls with opening. The provision of FEMA-273 [2] and ATC-40 [3] in considering such opening is used as a benchmark here. The Indian provision is checked against this benchmark by conducting a comparative study of seismic performance of a model of a plan-irregular building in seismic Zone-V. The comparative seismic performance is done in SAP2000® using nonlinear static pushover analysis.

K. Pathak · A. K. Dutta (✉)
Jorhat Engineering College, Jorhat 785007, Assam, India



S. Kolathayar and S. C. Chian (eds.), *Recent Advances in Earthquake Engineering*, Lecture Notes in Civil Engineering 175,
https://doi.org/10.1007/978-981-16-4617-1_35

## 2 Literature Review

Surendran and Kaushik [4] observed that openings in infill walls significantly reduce the lateral strength and stiffness of RC frames, and alter their failure modes. They reviewed the relevant studies from the past and compared the seismic codes of different countries on in-plane lateral load behaviour and modelling approaches for masonry infill RC frames with openings.

Agrawal et al. [5] studied the performance of masonry-infilled reinforced concrete (RC) frames, including an open first story of with and without opening. A symmetrical frame of the college building (G + 5) located in the Indian seismic Zone-III was considered by modelling of the initial frame. It was observed that infill panels increased the stiffness of the structure, and the increase in the opening percentage led to a decrease in the lateral stiffness of the infilled frame.

Phadnis and Kulkarni [6] modelled a G + 10 reinforced concrete framed building, where widths of struts representing the infills were calculated using the equivalent strut method. Seismic performances of various configurations of infill in reinforced concrete frames were examined by performing nonlinear static analysis in SAP 2000®. The results of infilled frames were compared with the bare frame model.

It has been observed that while ample studies are conducted on comparing the structural behaviour of bare frame and frame with infill, no study has been observed which scrutinizes the provisions of the Indian seismic code regarding lack of voids in infill walls in strut action. This very aspect is the subject of the present work.

## 3 Description of the Frame Structure

The building under consideration is a plan-irregular double-storeyed block of Jorhat Engineering College building, as shown in Fig. 1.

Indian seismic code [1] defines plan irregularities as in Fig. 2.

In Fig. 2, along Y-direction $A/L_1$ is 9.2/29.96, i.e., 0.30 > 0.15. Thus, the classification as plan irregular is valid.

The structure is modelled as a 3-D frame using SAP2000®. The element sizes are as follows:

Beam1 (along Y-axis): 350 × 870 mm.

Beam2 (along X-axis): 380 × 970 mm.

Column1: 350 × 870 mm.

Column2: 640 × 640 mm.

Column3: 450 × 450 mm.

Column4: 350 × 350 mm.

Slab thickness = 150 mm.

Plan area (sqm) = 20.45 m × 29.96 m.

Dead loads are calculated considering the unit weight of concrete as 25 $kN/m^3$ [7].

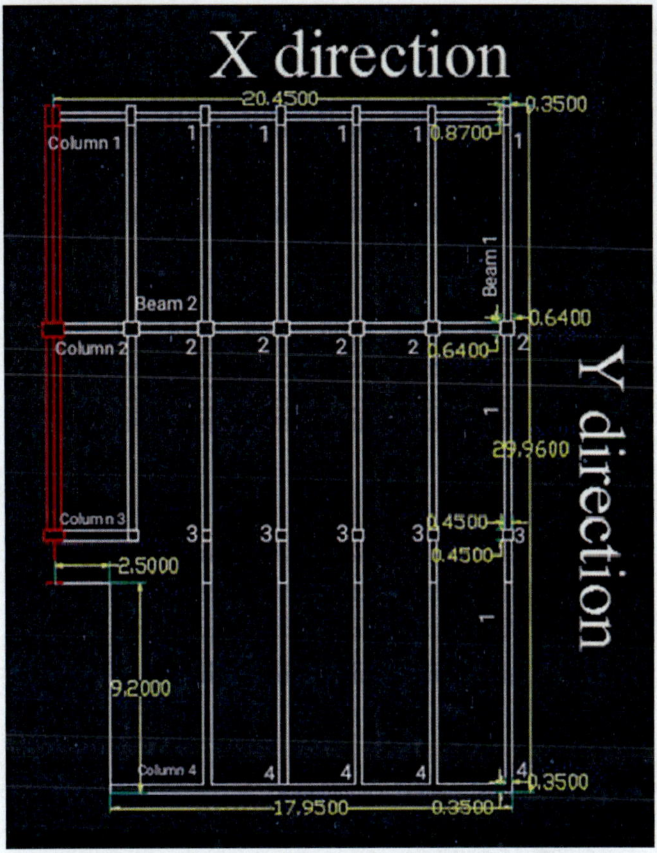


**Fig. 1** Plan view of the model nomenclature of beams and columns

Dead load of slab = $1 \times 0.15 \times 25 = 3.75\ \text{kN/m}^2$.

Imposed load for institutional buildings = 1.5 $\text{kN/m}^2$ (for roof) and 3 $\text{kN/m}^2$ (for floors) [8].

Floor to floor height = 4.48 m (ground storey) and 4.14 m (other floors).

Modulus of elasticity ($E_c$) = $5000\sqrt{f_{ck}} = 5000\sqrt{20} = 22{,}360.68\ \text{N/mm}^2$ [9]

The columns are assumed to be fixed at the base. Infill walls are modelled as equivalent concentric diagonal strut connected at the beam-column joints as a pinned connection. The 3-D model with schematic locations of the struts is shown in Fig. 3.

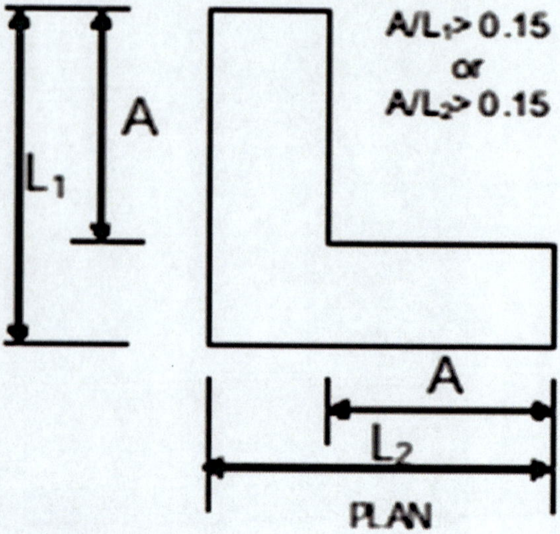


**Fig. 2** Definition of plan irregularity as per Indian seismic code [1]

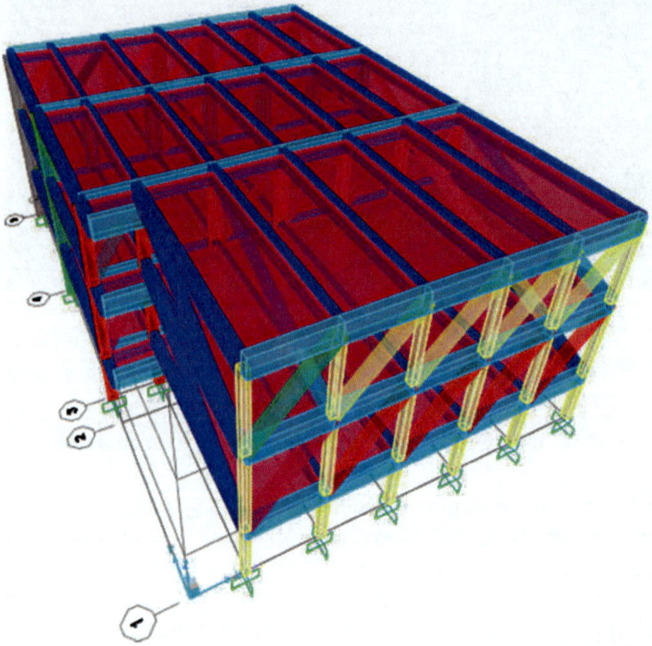

**Fig. 3** 3-D view of the structure

# 4 Modelling of Equivalent Strut

## *4.1 Modelling Using FEMA-273 rules (with and Without Openings)*

The infill properties of the equivalent strut are taken from Agrawal et al. [5] and presented in Table 1.

The infill panel is represented by an equivalent diagonal strut of width "a", and net thickness $t_w$ as shown in Fig. 4.

The width of equivalent struts without openings is calculated as per the following relations [3]:

$$\text{Width } (W_{eff}) = 0.175(\lambda_1 \times H)^{-0.4} \times w^{/\cdots} \quad (1)$$

$$\lambda_1 = \left[\frac{E_m t \sin(2\theta)}{4E_{fe} I_{col} H_{inf}}\right]^{1/4} \quad (2)$$

**Table 1** Properties of infill wall from Agrawal et al. [5]

| Infill material properties | Values |
|---|---|
| Modulus of elasticity | 5500 MPa |
| Shear modulus | 1018 MPa |
| Thermal coefficient | 0.0000081/C |
| Poisson's ratio | 0.15 |

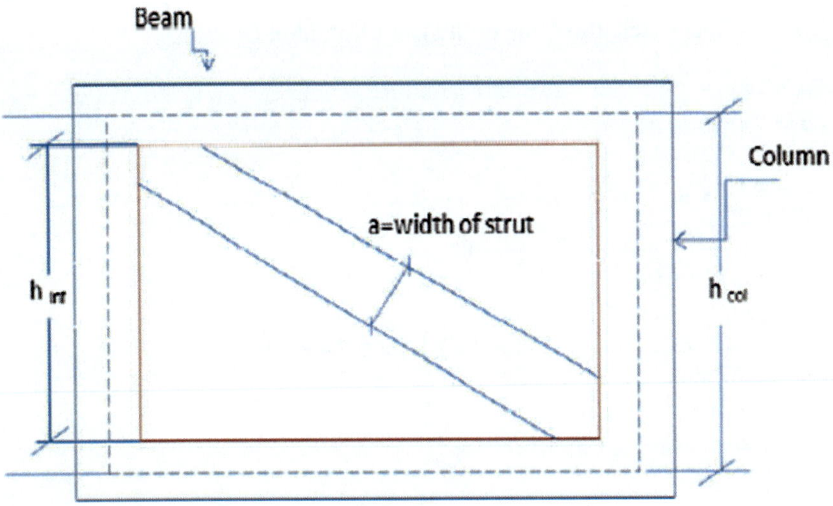


**Fig. 4** Equivalent strut idealization of infill walls [1]

where
H = column height between centre lines of beams.
$H_{inf}$ = height of infill panel.
$E_{fe}$ = expected modulus of elasticity of frame material.
$E_m$ = Expected modulus of elasticity of infill material.
$I_{col}$ = moment of inertia of column.
$w'$ = diagonal length of infill panel.
t = thickness of infill panel and equivalent strut.
$\theta$ = diagonal angle = $\tan^{-1}\frac{H}{L}$
Unit weight of masonry = 19 kN/m$^3$.
Unit weight of concrete = 25 kN/m$^3$.
L = bay length of the frame.
The thickness of the external wall = 230 mm.
Internal wall thickness = 230 mm.

In the presence of openings, a reduction factor is applied to reduce the width of the strut [3]:

$$\lambda = 1 - 2\alpha W^{0.54}\alpha W^{1.14} \tag{3}$$

where $\propto w$ = infill wall opening %

$$\text{Opening \%} = \frac{\text{Area of opening}}{\text{Area of infill wall}} \tag{4}$$

The location of equivalent struts is shown in Fig. 5.

### 4.1.1 Sample Calculations for Infill Wall 1 (Without Openings)

$w' = 5.74$ m.
$h = 4.48$ m.
$t_w = 0.230$ m.
$\theta = \tan^{-1}\frac{4.48}{3.59}$
$= 51.29°$
$\lambda_h = [\frac{5500 \times 230 \times \sin(2 \times 51.29)}{4 \times 22360.68 \times 0.875 \times 10^9 \times 4.48 \times 10^3}]^{1/4}$.
$= 1.369 \times 10^{-3}$/mm.
$W_{eff} = 0.175(\lambda_h \times H)^{-0.4} \times w'$
$= 0.175(1.369 \times 10^{-3} \times 4.48 \times 10^3)^{-0.4} \times 5.74 \times 10^3$.
$= 486.27$ mm.
$= 0.48627$ m.

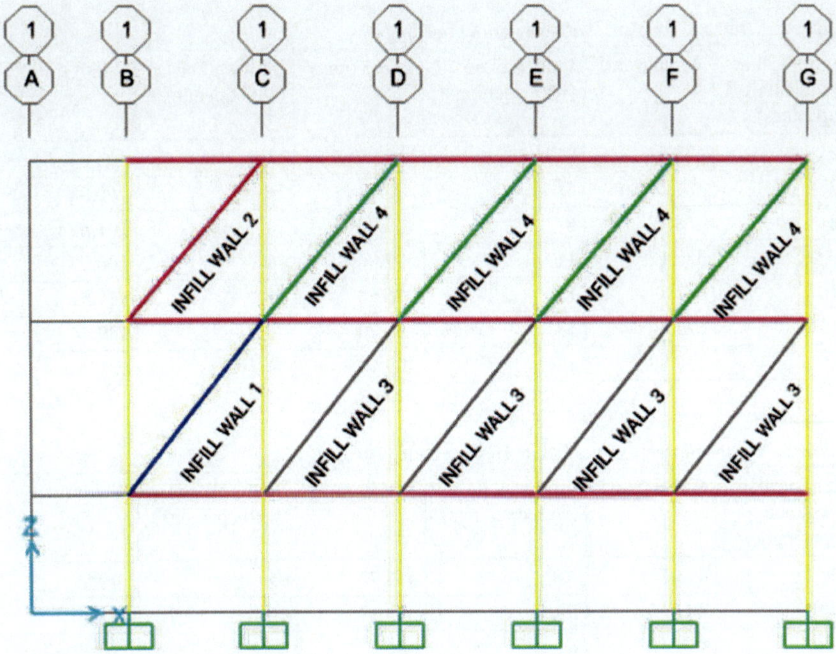


**Fig. 5** Location of equivalent struts

### 4.1.2 Sample Calculations Infill Wall 3 (with Openings)

$w' = 5.74$ m.

$h = 4.48$ m.

$t_w = 0.230$ m.

opening % = 4.32/16.083 = 0.268.

$\lambda = 1 - 2 \times (0.268)^{0.54} + (0.268)^{1.14}$.

= 0.241/mm.

$W_{eff} = 0.175(0.241 \times 4.48 \times 10^3)^{-0.4} \times 5.74 \times 10^3$.

= 61.465 mm.

The above procedure is repeated for 20 other infill walls and the summary is presented in Table 2.

## *4.2 Modelling as per Indian Seismic Code [1]*

As per Clause 7.9.9.2(b) [1], the procedure for calculation of the widths of equivalent strut without opening is the same as FEMA-273, i.e., as per Eqs. (1) and (2). However, no reduction of strut width is required as per Clause 7.9.2.2(c) [1]. The calculated

**Table 2** Widths of the strut (with or without openings)

| Infill wall no (without openings) | Width (mm) | Infill wall no (with openings) | Width (mm) | Infill wall no (with openings) | Width (mm) |
|---|---|---|---|---|---|
| 2 | 473.72 | 4 | 62.92 | 15 | 104.63 |
| 7 | 487.17 | 5 | 93.23 | 16 | 69.04 |
| 8 | 468.77 | 6 | 97.00 | 17 | 101.67 |
| 9 | 487.17 | 11 | 97.00 | 18 | 100.97 |
| 10 | 468.77 | 12 | 126.60 | 21 | 139.51 |
| 19 | 622.60 | 13 | 134.20 | 22 | 139.74 |
| 20 | 638.88 | 14 | 105.60 | | |

**Table 3** Widths of struts as per Indian seismic code [1]

| Infill wall no (without openings) | Width (mm) | Infill wall no (with openings) | Width (mm) | Infill wall no (with openings) | Width (mm) |
|---|---|---|---|---|---|
| 2 | 473.72 | 4 | 473.72 | 15 | 110.52 |
| 7 | 487.17 | 5 | 473.72 | 16 | 72.86 |
| 8 | 468.77 | 6 | 486.27 | 17 | 110.52 |
| 9 | 487.17 | 11 | 486.27 | 18 | 110.52 |
| 10 | 468.77 | 12 | 486.27 | 21 | 144.17 |
| 19 | 622.60 | 13 | 473.27 | 22 | 144.78 |
| 20 | 638.88 | 14 | 110.52 | | |

widths of struts as per the Indian seismic code [1] for the respective walls are given in Table 3.

## 5 Results and Discussion

### 5.1 Comparison of Dynamic Properties

Comparative dynamic properties between the two modelling approaches are presented in Table 4.

It has been observed that the Indian codal provisions result in a stiffer structure in comparison with the FEMA-273 model.

**Table 4** Comparative dynamic properties

| Model as per | Time period (s) | Frequency (Hz) |
|---|---|---|
| Indian code [1] | 0.44 (Mode-1) | 2.234 |
| | 0.42 (Mode-2) | 2.410 |
| | 0.39 (Mode-3) | 3.525 |
| FEMA-273 [2] | 0.49 (Mode-1) | 2.510 |
| | 0.45 (Mode-2) | 3.120 |
| | 0.40 (Mode-3) | s3.998 |

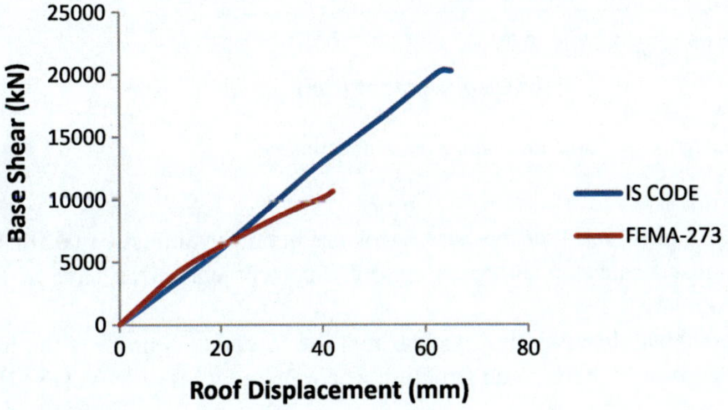


**Fig. 6** Comparative pushover curve along the X-direction

## 5.2 *Comparative Pushover Analysis*

Comparative nonlinear pushover analysis is conducted in SAP2000®. Results in terms of pushover curves are presented in this section.

### 5.2.1 Performance Along Shorter Direction

Figure 6 illustrates the comparison of base shear versus roof displacement along the shorter direction of the plan. It is clear from the comparative pushover curves that the IS code predicts bigger base shear capacity with more roof displacement as compared with the FEMA-273 model for the same structure.

### 5.2.2 Performance Along Longer Direction

In contrast with comparative shorter direction behaviour, the performance is almost similar in the longer direction as observed in Fig. 7.

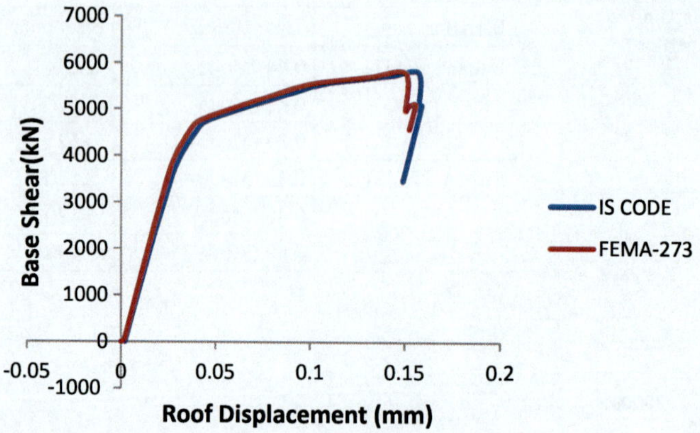


**Fig. 7** Comparative pushover curve along the Y-direction

It has been observed that the base shear prediction is almost equal for both the models against a marginal increase in roof displacement in the case of IS code-compliant model.

From the above discussion, it is clear that the IS code [1] predicts the increased base shear capacity for the same structure as compared with FEMA-273 [2].

## 6 Conclusion

The seismic behaviour of a structure for considering infills with and without openings is presented. It has been observed that the base shear of the structure considering IS codal provision is more. An increase in base shear in the latter case may be attributed to the bigger strut width as per IS provision.

## References

1. IS: 1893-Part I: Criteria for Earthquake Resistant Design of Structure, General Provisions and Buildings. Bureau of Indian Standard, New Delhi, (2016)
2. FEMA-273: NEHRP Guidelines for the seismic rehabilitation of buildings. Federal Emergency Management Agency, Washington, D.C (1997)
3. ATC 40: Seismic evaluation and retrofit of concrete buildings. Applied Technology Council, Seismic Safety Commission, State of California, volume 1 (1996)
4. Surendran S, Kaushik B (2012) Masonry infill RC frames with openings: review of in-plane lateral load behaviour and modeling approaches. The Open Construction and building Technology Journal
5. Agrawal N, Kulkarni P, Pooja R (2013) Analysis of masonary infilled R.C. frame with & without opening including soft storey by using equivalent diagonal strut method. Int J Sci Res Publ 3(9):3

6. Phadnis P, Kulkarni D (2016) Pushover analysis of high rise reinforced concrete building with and without infill walls. Int Res J Eng Technol IRJET 3(12)
7. IS:875-Part1: Code of practice for design loads for buildings and structures. Bureau of Indian standards, New Delhi (1897)
8. IS:875-Part 2: Code of practice for design loads for buildings and structures. Bureau of Indian standards, New Delhi (1897)
9. IS 456-Part I: Code of Design Reinforcement. Bureau of Indian Standards (2000)

# Simplified Macro Modeling Approach for Estimation of Nonlinear Response of Infilled RC Frames

**Panna Lal Kurmi and Putul Haldar**

## 1 Introduction

Reinforced concrete (RC) frame buildings with masonry infills are the most common building typology in modern cities around the world. The infills serve the purpose of interior partition walls and exterior claddings. Infills are treated as non-structural elements and often ignored in the design and analysis stage of RC frames. Past earthquakes [1–3] and laboratory experiments [4–7] have demonstrated that the complex interaction between infill and bounding frames can never be neglected under lateral load. The presence of infill can alter the load transfer mechanism, affecting the seismic behavior of the structures, leading to undesirable structural performance, and catastrophic failure of frames and infills [8–10]. The inherent uncertainty in infill–frame interaction leading to the complex failure of infilled frames compared to bare RC frames still makes modeling the infilled frame a challenging task for practicing engineers [8, 11]. The modeling approaches used to simulate the presence of infill are classified into two major categories, namely micro models and macro models. The micro model approach [12–14] is based on the complicated nonlinear finite element (FE) analysis. It requires appropriate constitutive models for describing the nonlinear behavior of the materials involved (concrete, steel, unit, mortar, and unit–mortar interface) and their interaction [15]. This approach can accurately predict the nonlinear behavior and the failure mechanisms of the infilled frames in a more detailed manner, such as crack propagation along with the masonry and RC frames at various load stages and internal redistribution of stresses. The material parameters of the micro model further require extensive experimental validation, which is not commonly available in the case of existing buildings [16]. Moreover, micro modeling

P. L. Kurmi (✉) · P. Haldar
Department of Civil Engineering, IIT Ropar, Rupnagar, India
e-mail: 2016CEZ0002@iitrpr.ac.in



S. Kolathayar and S. C. Chian (eds.), *Recent Advances in Earthquake Engineering*,
Lecture Notes in Civil Engineering 175,
https://doi.org/10.1007/978-981-16-4617-1_36

involves intensive computational effort, thus limiting their application to practical structures for practicing engineers.

The concept of the macro model approach [5, 17–20] arrived when experimental and conceptual observations have indicated that a diagonal strut with appropriate geometrical and mechanical characteristics could possibly represent the behavior of infill under lateral loading conditions. The macro model approach uses to represent the infill using an equivalent diagonal strut analogy that can capture the global response of infilled RC frames with reasonable accuracy. The mechanical properties of the strut are kept the same as infill and adopt empirical equations to estimate the width of the strut. International standards [21, 22] also recommend the use of a single equivalent diagonal strut to simulate the infill. Single equivalent strut analogy has become the most widely used approach for infill among researchers and practicing engineers for its simplicity to model with minimal computational efforts. However, some researchers have also identified that using a single diagonal strut connecting the two loaded diagonals cannot represent the distribution of bending moments and shear forces in the frame members and proposed multiple strut models [19, 23]. The primary advantage of the multiple strut models over a single strut is the ability to represent the actions in the frame more accurately, despite an increase in modeling complexity. Another drawback of the concentric diagonal strut is that it fails to capture the infill–frame interaction, which may lead to shear failure of seismically deficient columns. To overcome this drawback, Haldar et al. [24] demonstrated the efficacy of 1-strut, 2-strut, and 3-strut macro modeling approaches for simulation of infill. The authors have identified that using a single eccentric equivalent diagonal strut, the failure of RC columns observed due to exceedance of shear capacity under the lateral action of infill observed during the experiential studies can be simulated efficiently. However, in most cases, the strut models are based on case-specific data, and their accuracy in predicting the lateral behavior of infilled frames tends to vary significantly [16, 25]. Moreover, the widely used equivalent strut analogy of ASCE 41-06 [22] to simulate the infill in infilled RC frame underestimates the stiffness and strength of the infilled frame considerably [25]. Martin and Stavridis [26] proposed a simplified analytical framework for the seismic assessment of infilled RC frames which have been included in ASCE 41-17 [27]. The authors have also discussed that assessment guidelines in ASCE 41-06 [22], and 41-13 [28] are not fully validated with the experimental results.

The present paper proposes a simplified macro modeling approach to estimate the in-plane nonlinear response of single-bay, single-story physically tested infilled RC frames. The procedure uses the simulation of infill using a single equivalent eccentric diagonal strut. The width of the diagonal strut is estimated as per the guidelines of ASCE 41-06 [22]. The current methodology modifies the elastic modulus of strut and RC columns based on the requirement of the experimentally tested infilled RC frame's initial stiffness. The proposed procedure's accuracy is checked with load-deformation envelopes of 12 single-bay, single-story physically tested infilled RC frames. A comparative seismic performance assessment of four-story uniformly infilled RC frame building is also carried out with the proposed infill model and ASCE 41-06 [22] infill model through nonlinear static pushover analysis.

## 2 Methodology Used for the Proposed Procedure

The first step requires obtaining the backbone envelope of the infilled bay of the RC frame of interest. These curves can be obtained through an analytical procedure or more accurately by actual testing, and detailed FE analysis [26].

The second step requires the modeling of infilled RC frames using a structural analysis program. The RC frames shall be modeled using line elements, and the cross-section of the frame members shall be designed to represent the test specimen. The infill shall be modeled using a pin-jointed eccentric diagonal compressive strut. The material properties and thickness of the strut remain the same as infill, and the ASCE 41-06 [22] procedure can be used to estimate the width of the strut. Among the various failure modes and strength of infills available in the literature, Haldar et al. [24] identified that the strength of infills is usually minimum in shear, and thereby the shear failure governs the inelastic modeling of infills. The sliding shear strength is estimated using ASCE 41-17 [27], and the same is used to assign the nonlinear backbone curve is lumped at the center of the strut. For RC members, lumped plasticity models have been used to model the inelastic behavior of beams and columns. Flexural hinges (M3) and interacting hinges (P-M-M) are assigned at both the ends of beams and columns, respectively, as per ASCE 41-17 [27]. The cracked section properties are also assigned to RC members as per ASCE 41-17 [27]. The shear strength of the leeward column is governed by the minimum shear or the flexural capacity of the column. It shall be assumed equal to the strength of the column to shear failure, ($V_n$) for non-ductile frames and equal to the shear strength caused by plastic hinge formation, ($V_p$) for ductile frames. The plastic shear capacity ($V_p$) of the RC column is estimated from its plastic moment-carrying capacity [27] and the shear strength ($V_n$) is estimated from ASCE 41-17 [27].

The third step requires obtaining the capacity curve of the infilled RC frame model under gravity load of actual test condition through nonlinear static pushover analysis. The initial stiffness ($K_{strut-model}$) of the capacity curve shall be estimated through a bilinear approach.

The fourth step requires estimating the factors $\alpha$ and $\beta$, where $\alpha$ is the ratio of initial stiffness of the tested infilled RC frame ($K_{infilled-test}$) and initial stiffness ($K_{strut-model}$) obtained from numerical modeling of the same test structure using a single eccentric equivalent diagonal strut. $\beta$ is the ratio of peak strength obtained from the test result ($V_{infilled-test}$) to the sliding shear strength of the infill panel ($V_{\text{Inf}}$). After obtaining $\alpha$ and $\beta$, $\alpha$ shall be used to multiply the elastic modulus of RC column and strut material, and $\beta$ shall be used to multiply the strut strength.

The fifth and final step requires performing pushover analysis again with the modified elastic modulus properties and the strength of the strut to obtain the modified capacity curve of the infilled RC frame model.

## 3 Evaluation of the Proposed Methodology

The proposed methodology uses 12 single-bay single-story test specimens to evaluate their accuracy, out of which 10 specimens from Mehrabi et al. [29] named as M#, single AAC infilled RC frame specimen from Bose and Rai [30] and single S1A1 specimen from Cavaleri and Trapani [7] have been considered. The geometry and material properties of all test structures are summarized in Tables 1 and 2. The nomenclatures reported in Tables 1 and 2 are according to Martin and Stavridis [26].

**Table 1** Geometry of test specimens

| Specimen | $L_w$ | $L_{inf}$ | $h_w$ | $h_{inf}$ | $t_{inf}$ | $A_s$ | $A_v$ | s | $h_c$ | $b_c$ | $b_b$ | $h_b$ |
|---|---|---|---|---|---|---|---|---|---|---|---|---|
| | mm | mm | mm | mm | mm | mm$^2$ | mm$^2$ | mm | mm | mm | mm | mm |
| M3 | 2311 | 2134 | 1537 | 1422 | 92 | 1013 | 63 | 64 | 178 | 178 | 152 | 229 |
| M4 | 2311 | 2134 | 1537 | 1422 | 33 | 1013 | 63 | 64 | 178 | 178 | 152 | 229 |
| M5 | 2311 | 2134 | 1537 | 1422 | 92 | 1013 | 63 | 64 | 178 | 178 | 152 | 229 |
| M6 | 2337 | 2134 | 1537 | 1422 | 33 | 1583 | 63 | 38 | 203 | 203 | 152 | 229 |
| M7 | 2337 | 2134 | 1537 | 1422 | 92 | 1583 | 63 | 38 | 203 | 203 | 152 | 229 |
| M8 | 2311 | 2134 | 1537 | 1422 | 33 | 1013 | 63 | 64 | 178 | 178 | 152 | 229 |
| M9 | 2311 | 2134 | 1537 | 1422 | 92 | 1013 | 63 | 64 | 178 | 178 | 152 | 229 |
| M10 | 3124 | 2946 | 1537 | 1422 | 33 | 1013 | 63 | 64 | 178 | 178 | 152 | 229 |
| M11 | 3124 | 2946 | 1537 | 1422 | 92 | 1013 | 63 | 64 | 178 | 178 | 152 | 229 |
| M12 | 3124 | 2946 | 1537 | 1422 | 92 | 1013 | 63 | 64 | 178 | 178 | 152 | 229 |
| AAC | 2400 | 2200 | 1430 | 1330 | 125 | 628 | 57 | 50 | 200 | 200 | 200 | 200 |
| S1A1 | 1800 | 1600 | 1800 | 1600 | 210 | 314 | 57 | 100 | 200 | 200 | 200 | 400 |

**Table 2** Material properties and vertical load

| Specimen | $P_T$ | $f_m$ | $f_c$ | $f_y$ | $f_{yv}$ | $E_m$ | $E_c$ | $E_s$ | C |
|---|---|---|---|---|---|---|---|---|---|
| | kN | N/mm$^2$ | N/mm$^2$ | N/mm$^2$ | N/mm$^2$ | N/mm$^2$ | N/mm$^2$ | N/mm$^2$ | |
| M3 | 293.6 | 15.1 | 30.9 | 420.6 | 367.5 | 9522 | 21,925 | 199,948 | 0.34 |
| M4 | 293.6 | 10.6 | 26.8 | 420.6 | 367.5 | 4599 | 17,237 | 199,948 | 0.34 |
| M5 | 293.6 | 13.6 | 20.9 | 420.6 | 367.5 | 8949 | 18,064 | 199,948 | 0.34 |
| M6 | 293.6 | 10.1 | 25.9 | 420.6 | 367.5 | 4199 | 19,857 | 199,948 | 0.34 |
| M7 | 293.6 | 13.6 | 33.4 | 420.6 | 367.5 | 9074 | 18,616 | 199,948 | 0.34 |
| M8 | 293.6 | 9.5 | 26.8 | 420.6 | 367.5 | 5102 | 17,237 | 199,948 | 0.34 |
| M9 | 293.6 | 14.2 | 26.8 | 420.6 | 367.5 | 8239 | 17,237 | 199,948 | 0.34 |
| M10 | 293.6 | 10.6 | 26.9 | 420.6 | 367.5 | 3944 | 20,133 | 199,948 | 0.34 |
| M11 | 293.6 | 11.4 | 25.7 | 420.6 | 367.5 | 9604 | 18,133 | 199,948 | 0.34 |
| M12 | 440.4 | 13.9 | 26.9 | 420.6 | 367.5 | 7336 | 20,133 | 199,948 | 0.34 |
| AAC | 110 | 2.38 | 37.6 | 417.6 | 417.6 | 2400 | 27,600 | 222,000 | 0.31 |
| S1A1 | 400 | 2.67 | 25 | 450 | 450 | 3933 | 25,500 | 200,000 | 0.73 |

## 4 Results and Discussion

Figure 1 compares the experimental load-deformation envelope, capacity curves obtained from infilled RC frame models, and backbone curves developed from Martin and Stavridis [26] method. The Martin and Stavridis [26] method adopts the shear-beam model [31] to predict the initial stiffness of the infilled RC frame, and ASCE 41-06 [22] guideline simulates the effect of infill using equivalent diagonal strut. The comparison of predicted stiffness and the test result shows ASCE 41-06 [22] model underestimates the initial stiffness significantly in all the cases, and Martin and Stavridis [26] method gives the best approximation of initial stiffness. However, the proposed model, which requires modification of elastic modulus of strut and

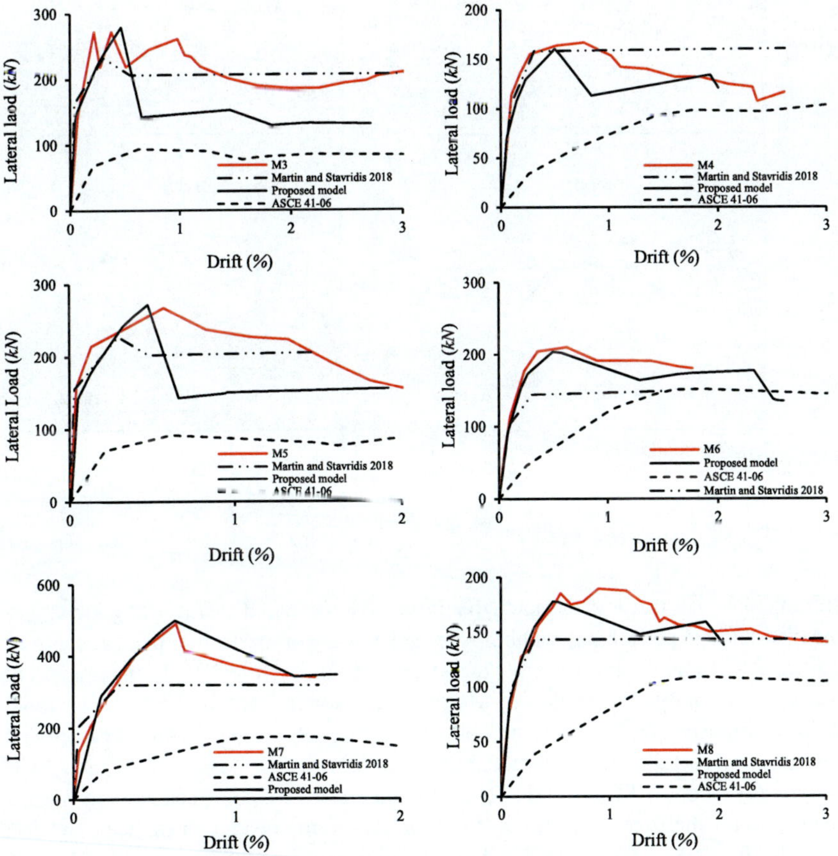


**Fig. 1** Comparison of capacity curves of infilled RC frame with the proposed model, ASCE 41-06 model [22], experimental load-deformation envelope, and backbone curve obtained from Martin and Stavridis [26] method

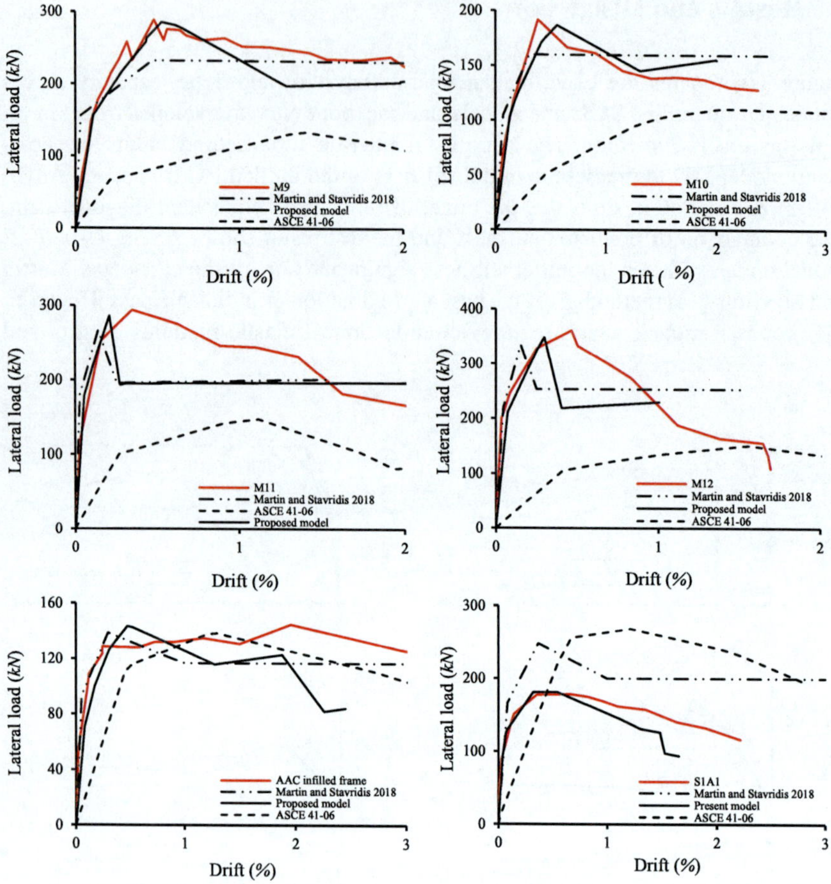


**Fig. 1** (continued)

column material, is found to match the initial and post-yield stiffness in good agreement with the experimental backbone envelope. The comparison of peak strength of infilled RC frame observed from experimental results, infilled RC frame numerical models, and analytical method of Martin and Stavridis [26] are reported in Table 3. The proposed model gives the best match of peak strength with experimental results as compared to the analytical method of Martin and Stavridis [26] and ASCE 41-06 [22] model. ASCE 41-06 [22] model underestimates the peak strength significantly in all the cases except the AAC infilled–frame specimen of Bose and Rai [30]. In the case of specimen S1A1 of Cavaleri and Trapani [7], the ASCE 41-06 [22] model overestimates the peak strength. The reason for overestimating the peak strength may be attributed to the high value of masonry shear strength, leading to the high value of sliding shear strength of the infill panel. The specimens M3, M5, M9, M11, and M12 are classified as the strong-infill weak-frame (SIWF) category.

**Table 3** Comparison of the proposed model with experimental data along with Martin and Stavridis model [26] and ASCE 41-06 model [22]

| Specimen | Max peak test load (kN) | Martin and Stavridis (2018) | Infill–frame classification | Peak load (kN) | | Error (%) | | Failure mechanism* | |
|---|---|---|---|---|---|---|---|---|---|
| | | | | Proposed model | ASCE 41-06 | Proposed model | Martin and Stavridis (2018) | Experiment | Proposed model |
| M3 | 277.7 | 236.7 | SIWF | 279.5 | 93.6 | + 0.64 | – 14.76 | B | B |
| M4 | 162.4 | 159.8 | WIDF | 160.5 | 105.3 | – 1.16 | – 1.6 | A | A |
| M5 | 266.9 | 230.4 | SIWF | 271.5 | 91.7 | + 1.69 | – 13.67 | B | B |
| M6 | 207.3 | 152.6 | WIDF | 203.7 | 151.7 | – 1.73 | – 26.38 | A | A |
| M7 | 488.9 | 313.8 | WIDF | 498.4 | 174 | + 1.9 | – 35.8 | A | A |
| M8 | 189.9 | 143.2 | WIDF | 177.8 | 109.4 | – 6.37 | – 24.59 | A | A |
| M9 | 292.5 | 234.3 | SIWF | 285.7 | 133.1 | – 2.32 | – 19.89 | B | B |
| M10 | 191.2 | 159.8 | WIDF | 186.7 | 110.6 | – 2.35 | – 16.43 | B | B |
| M11 | 293.9 | 277.0 | SIWF | 286.1 | 148.2 | – 2.65 | – 5.75 | B | B |
| M12 | 360.8 | 335.3 | SIWF | 348.1 | 151.3 | – 3.51 | – 7.06 | B | B |
| AAC | 145 | 139 | WIDF | 143.4 | 138.5 | – 1.1 | – 4.14 | A | A |
| S1A1 | 178.7 | 248 | SIWF | 180.9 | 198.9 | + 1.21 | + 27.9 | B | B |

***Failure mechanism A**: This failure mechanism indicates sliding along the bed joints distributed along with the height of the infill, crushing of the infill near the frame corners, and columns failing in flexure. No shear failures of columns were observed in the test. The failure mechanism obtained from the proposed model shows the failure of the diagonal strut and the formation of plastic hinges at column ends. **Failure mechanism B**: This failure mechanism indicates failure mode dominated by major shear cracks in the infill and RC columns observed during the test. The failure mechanism obtained from the proposed model shows the failure of the diagonal strut, formation of plastic hinges at column ends, and exceedance of plastic shear capacity at column ends.

A sudden post-peak strength drop can be seen in the capacity curves of these specimens obtained using the proposed model. This sudden drop in strength is due to the exceedance of the shear capacity of the column due to the lateral force component of the eccentric diagonal strut. All the SIWF category specimens failed due to shear crack in infill and shear crack/failure of columns observed during the experiment. The proposed model's failure modes are also in good agreement with the failure mechanism observed during the experimental studies. The specimens M4, M6, M7, M8, and M10 are classified as weak-infill ductile-frame (WIDF) and hence relatively ductile failure pattern is anticipated indicating sliding along the bed joints distributed along with the height of the infill, crushing of the infill near the frame corners, and columns failing in flexure, and no shear failure of the column is observed in the test as reported by Mehrabi et al. [29]. The failure mode observed in the proposed model is the failure of the diagonal strut due to exceedance and sliding shear capacity of the infill and the formation of plastic hinges at the column ends. Compared to the experimental backbone envelope, the post-peak nonlinear response of the proposed model is observed on the conservative side. However, the drift at which the proposed model estimates the peak strength is in fair accuracy compared to the experimental backbone envelopes, despite its simplicity.

## 5 Analysis and Design of Uniformly Infilled RC Frame Building

The buildings considered in the present study have a generic plan, as shown in Fig. 2. The plan is symmetric with significantly different redundancy in the two directions. The corridor is free from the transverse beams with different beam span along both the directions, representing the characteristics of a wide range of real buildings in India. The story height is considered as 3.3 m. The buildings have been assumed to be situated on medium soil in seismic zone V in India.

For the design, M25 grade concrete and Fe500 grade steel have been used. The revised ductile design and detailing standard with its amendment (2017) [32] recommends that the minimum dimension of the column shall not be less than 20 times the diameter of the largest beam longitudinal bar or 300 mm, whichever is greater. Hence, the selection of the column dimension is directly related to the selection of the largest beam longitudinal bar diameter. The beam longitudinal bar diameter is selected as per IS 456:2000 [33], to keep sufficient space between the adjacent bars so that the needle vibrator can be immersed during concrete casting. The present study assumes to have at least 50 mm clear space between the adjacent bars to select the appropriate diameter of the longitudinal beam bar. The beam sections have been proportioned to have a maximum of 1.2% demand steel on each face. The dead load (DL) and live load (LL) are calculated using the Indian standard IS 875, Part 1 [34] and Part 2 [35], respectively. The slab thickness is assumed to be 150 mm and found safe against the limit state design criteria [33]. External brick masonry

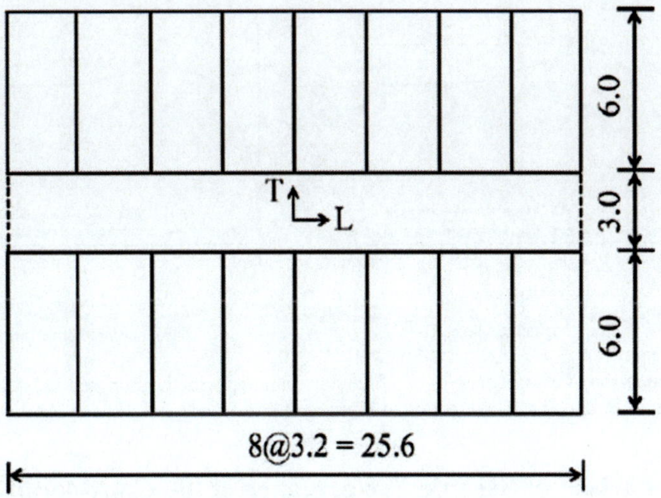


**Fig. 2** Plan of the considered building

wall thickness is considered to be 230 mm and the internal wall as 110 mm as per the prevailing practices in India. Also, a 230 mm thick parapet wall of 1 m height is considered along the roof periphery. A three-dimensional four-story space frame with the slab as a rigid diaphragm has been designed as per the revised Indian seismic standards [32, 36]. The infills have been modeled using eccentric single equivalent diagonal strut as per ASCE 41-06 [22] guidelines and the proposed model of the present study. In the proposed model, the modified elastic modulus for infill and column material is assigned as per the nonlinear response of Bose and Rai [26] AAC infilled frame, which is a typical representative single-story single-bay frame of typical Indian infilled RC frame building. The compressive strength of infill panels is considered 4.1 MPa considering the fair quality of masonry, consistent with the typical compressive strength of masonry in northern India [37]. For nonlinear analysis, ASCE 41-17 [27] default flexural (M) concentrated hinges are assigned at both ends of the beams, and axial force–moment interaction (P-M-M) concentrated hinges are assigned to both ends of columns considering conforming transverse reinforcement. The effective stiffness values, as suggested in IS 1893:2016 [36] have been used for concrete frames. P-delta analysis is included in both linear and nonlinear analysis. The analysis and design have been performed in the structural analysis program SAP2000 V 22.2.0 [38] (Fig. 3).

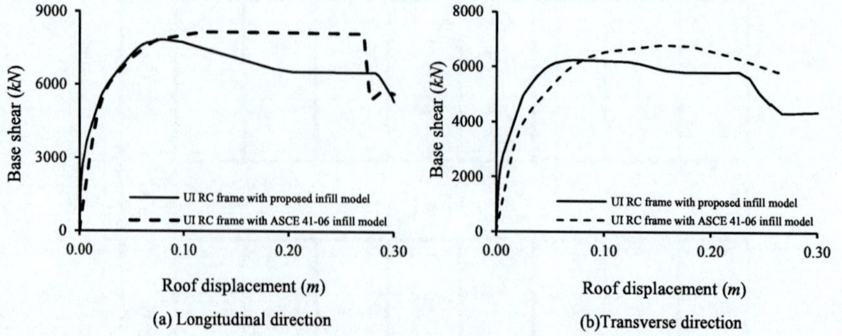


**Fig. 3** Comparison of capacity curves of a four-story building with the proposed and ASCE 41-06 [22] infill mode. **a** Longitudinal direction, **b** Transverse direction

## 6 Comparison of Seismic Performance of the Considered Buildings

The RC frame building with the proposed model shows an increase in effective stiffness along both the directions as shown in Table 5. This is mainly because, as shown in Fig. 1, the ASCE 41-06 [22] model underestimates the initial stiffness compared with the experimental backbone envelope and may lead to inaccurate estimation of seismic response of the practical structures. Thus, the proposed model is an alternative to the traditional strut approach of ASCE 41-06 [22], which requires modification of the elastic modulus of strut and column material to establish a match with the experimental backbone curve of single-story single-bay infilled RC frame. The effective stiffness is found to be increased by 56 and 54%, along longitudinal and transverse directions, respectively. The proposed model increases the overall stiffness of the infilled frame, resulting in a decrease in the fundamental translational modal time period, yielding of frame and infill at a lower displacement level as shown in Tables 4 and 5, respectively. The comparison of strength at yield and ultimate levels shows a marginal increase of 3–6% along both the considered directions. This is because a typical Indian construction practice leaves a gap between frame and masonry infill, and under lateral loading, the gap remains open. The gap between the frame and masonry infill prevents the transfer of any gravity load from beam to infill. Thus, in both the modeling cases, infill fails when its sliding shear capacity exceeds, and lateral load shifts to the RC frame. The complete failure of infill is observed, followed by exceedance of collapse prevention plastic hinges at first and second story beams and ground story level columns in both the modeling cases.

**Table 4** Dynamic properties of the considered buildings

| Building model | Fundamental time period obtained from modal analysis (sec) | | Design time period obtained from IS 1893 (sec) | | Modal participation factor | | Modal mass participation factor (%) | |
|---|---|---|---|---|---|---|---|---|
| | Long | Trans | Long | Trans | Long | Trans | Long | Trans |
| RC frame with proposed infill model | 0.159 | 0.195 | 0.234 | 0.306 | 1.19 | 1.22 | 0.85 | 0.84 |
| RC frame with ASCE 41-06 infill model | 0.336 | 0.41 | 0.234 | 0.306 | 1.29 | 1.31 | 0.85 | 0.84 |

## 7 Conclusions

The present study concludes that simulation of infill using the ASCE 41-06 strut concept underestimates the initial stiffness, peak strength, and overall load-deformation response when compared with the backbone envelopes of 12 single-story single-bay infilled RC frame. However, simulation of infill with equivalent strut for analysis of infilled RC frame is the most widely used technique among researchers and practicing engineers because of its simplicity, and requires minimal computing effort. The proposed model extends the same practice by modifying the elastic modulus of strut and column material. The nonlinear response of the infilled RC frame test specimen can be simulated using the proposed macro modeling approach with reasonable accuracy. The comparison of numerically developed capacity curves with the experimental envelopes show that the peak strength predicted by the proposed model is in good agreement with experimental data. Further, the proposed model is used to simulate the infills in a four-story uniformly infilled RC frame building, and the result shows a significant difference in terms of effective stiffness, yield displacement, and fundamental translational modal time period.

**Table 5** Performance parameters of the considered buildings

| Model | Yield displacement (m) | | Ultimate displacement (m) | | Yield strength (kN) | | Ultimate strength (kN) | | Effective stiffness (kN/m) | | Increase in stiffness (%) | |
|---|---|---|---|---|---|---|---|---|---|---|---|---|
| | Long | Trans | Long | Trans | Long | Trans | Long | Trans | Long | Trans | Long | Trans |
| RC frame with proposed infill model | 0.014 | 0.019 | 0.237 | 0.218 | 6266 | 5832 | 6266 | 6044 | 447,571 | 306,947 | 56 | 54 |
| RC frame with ASCE 41-06 infill model | 0.03 | 0.04 | 0.297 | 0.242 | 5912 | 5605 | 5912 | 5821 | 197,067 | 140,125 | | |

## References


1. Jain SK et al (2002) Bhuj, India Earthquake of January 26, 2001, Reconnaissance Report. Earthquake Spectra
2. Ozcebe G et al (2003) Bingöl earthquake engineering report May 2003. Tubitak, Turkey, pp 75–100
3. Mayorca JJP, Leon TTE (2007) A reconnaisance report on the Pisco, Peru Earthquake of August 15, 2007
4. Mehrabi AB et al (1996) Experimental evaluation of masonry-infilled RC frames. J Struct Eng 122(3):228–237
5. Smith BS (1967) Methods for predicting the lateral stiffness and strength of multi-storey infilled frames. Build Sci 2(3):247–257
6. Basha SH, Kaushik HB (2016) Behavior and failure mechanisms of masonry-infilled RC frames (in low-rise buildings) subject to lateral loading. Eng Struct 111:233–245
7. Cavaleri L, Di Trapani F (2014) Cyclic response of masonry infilled RC frames: experimental results and simplified modeling. Soil Dyn Earthq Eng 65:224–242
8. Haldar P (2013) Seismic behavior and vulnerability of Indian RC frame buildings with URM infills. In: Department of Earthquake Engineering, Indian Institute of Technology, Roorkee, India
9. Haldar P, Singh Y (2012) Modelling of URM infills and their effect on seismic behaviour of RC frame buildings. The open construction and building technology journal, Bentham Science Publishers, vol 6, (Suppl 1-M1) (Special Issue on Advances in Infilled Framed Structures: Experimental & Modelling Aspects), pp 35–41
10. Haldar P, Singh Y, Paul DK (2012) Effect of URM infills on seismic vulnerability of Indian code designed RC frame buildings. Earthq Eng Eng Vib 11(2):233–241
11. Haldar P, Singh Y, Paul D (2015) Design guidelines for URM infills and effect of construction sequence on seismic performance of code compliant RC frame buildings. Advances in Structural Engineering. Springer, pp 1055–1069
12. Penava D, Sigmund V, Kožar I (2016) Validation of a simplified micromodel for analysis of infilled RC frames exposed to cyclic lateral loads. Bull Earthq Eng 14(10):2779–2804
13. Deng H, Sun B (2016) Finite element modeling and mechanical behavior of masonry-infilled RC frame. Open Civil Eng J 10(1):76–92
14. Mehrabi AB, Shing PB (1997) Finite element modeling of masonry-infilled RC frames. J Struct Eng 123(5):604–613
15. Asteris PG et al (2013) Mathematical micromodeling of infilled frames: state of the art. Eng Struct 56:1905–1921
16. Bose AS (2018) Modeling of the seismic performance of buildings with infilled RC frames. in Proceedings of 11th national conference on earthquake engineering
17. Asteris PG et al (2011) Mathematical macromodeling of infilled frames: state of the art. J Struct Eng 137(12):1508–1517
18. Di Trapani F et al (2015) Masonry infills and RC frames interaction: literature overview and state of the art of macromodeling approach. Eur J Environ Civ Eng 19(9):1059–1095
19. Crisafulli FJ, Carr AJ (2007) Proposed macro-model for the analysis of infilled frame structures. Bull N Z Soc Earthq Eng 40(2):69–77
20. Mainstone RJ (1971) Summary of paper 7360. On the stiffness and strengths of infilled frames. Proc Inst Civil Eng 49(2):230
21. FEMA-356 (2000) Prestandard and commentary for the seismic rehabilitaion of buildings. Fedeal Emergency Management Agency
22. ASCE/SEI 41-06 (2007) Seismic rehabilitation of existing buildings. American Society of Civil Engineers
23. Chrysostomou C, Gergely P, Abel J (2002) A six-strut model for nonlinear dynamic analysis of steel infilled frames. Int J Struct Stab Dyn 2(03):335–353
24. Haldar P, Singh Y, Paul DK (2013) Identification of seismic failure modes of URM infilled RC frame buildings. Eng Fail Anal 33:97–118

25. Stavridis A, Martin J, Bose S (2017) Updating the ASCE 41 provisions for infilled RC frames. In: Proceedings of the 2017 SEAOC Convention, San Diego, CA
26. Martin J, Stavridis A (2018) Evaluation of a simplified method for the estimation of the lateral resistance of infilled RC frames. In: Proceedings 16th European conference on earthquake engineering
27. ASCE/SEI 41-17 (2017) Seismic evaluation and retrofit of existing buildings. American Society of Civil Engineers
28. ASCE/SEI 41-13 (2013) Seismic rehabilitation of existing buildings. American Society of Civil Engineers
29. Mehrabi AB, Shing PB, Schuller MP, Noland JL (1994) Performance of masonry-infilled R/C frames under in-plane lateral loads, Report No. CU/SR-94/6, University of Colorado at Boulder
30. Bose S, Rai DC (2016) Lateral load behavior of an open-ground-story RC building with AAC infills in upper stories. Earthq Spectra 32(3):1653–1674
31. Fiorato AE, Sozen MA, Gamble WL (1970) An investigation of the interaction of reinforced concrete frames with masonry filler walls. In University of Illinois Engineering Experiment Station. College of Engineering. University of Illinois at Urbana-Champaign
32. IS 13920 (2016) Ductile design and detailing of reinforced concrete structures subjected seismic forces; Bureau of Indian standards
33. IS 456 (2000) Plain and reinforced concrete—code of practice; Burea of Indian standard
34. IS 875-1 (1987, Reaffirmed 2008) Code of practice for design loads (other than earthquake) for buildings and structures. Part 1: Dead loads-unit weights of building materials and stored materials
35. IS 875-2 (1987) Code of practice for design loads (other than earthquake) for buildings and structures, Part 2: Imposed loads [CED 37: Structural Safety]
36. IS 1893 (Part1) (2016) Criteria for earthquake resistant design of structures, Bureau of Indian standards
37. Pisode M et al (2017) Comparative assessment of seismic fragility of RC frame buildings designed for older and revised Indian standards. ISET J Earthq Technol 54(1):17–29
38. Computers and Structures Inc. (CSI) (2020) Integrated software for structural analysis and design, SAP2000, Berkeley, USA

# Field Surveys, Geospatial Tools and Others

# Seismic Vulnerability of Buildings in Palghar Area: Observations and Remedial Measures

Rohan M. Shinde and Ravi Sinha

## 1 Seismicity in Maharashtra

Different parts of Maharashtra are located in seismic zones II, III and IV as per the Indian standards [1]. Seismic zone IV corresponds to those areas where design earthquakes with the intensity of VIII on the MSK intensity scale may occur, while seismic zone III corresponds to areas where earthquakes with intensity VII may occur. The region of Maharashtra adjoining Koyna has been assigned with seismic zone IV, while the other parts of the state are located in seismic zones II and III. There have been a number of damaging earthquakes in Maharashtra in the past. The most devastating earthquake after independence occurred in 1993 in the Killari region of Marathwada, near the town of Latur. The earthquake magnitude was estimated as 6.2 and resulted in the deaths of nearly 9,000 people, while another 30,000 people suffered injuries. Prior to the occurrence of that earthquake, the Marathwada region was considered to be free from seismic risk and was assigned with seismic zone I by the then applicable Indian Standard (IS 1893: 1984). The existence of the seismic fault where the Killari earthquake occurred was also not known prior to the earthquake.

The Killari earthquake was preceded by a large number of smaller magnitude earthquakes in the area for several months before the main event. Some of these pre-shocks events were strong enough to damage buildings in that area. However, since the region was considered to be free from seismic sources, the necessary scientific instrumentation to monitor the seismicity and thus identify the cause of the pre-shocks was not installed. An earlier large earthquake (magnitude 6.6) in the Koyna region of Maharashtra in 1967 resulted in the death of nearly 300 people and caused

R. M. Shinde (✉) · R. Sinha
Indian Institute of Technology Bombay, Mumbai, India
e-mail: rohan_s@iitb.ac.in

R. Sinha
e-mail: rsinha@civil.iitb.ac.in



S. Kolathayar and S. C. Chian (eds.), *Recent Advances in Earthquake Engineering*, Lecture Notes in Civil Engineering 175,
https://doi.org/10.1007/978-981-16-4617-1_37

damage to a large number of buildings in the area. This earthquake was also preceded by pre-shocks for several years before the main earthquake shock occurred. The Koyna earthquake is scientifically remarkable since it was the first earthquake where reservoir-induced seismicity (RIS) was conclusively established. The Killari and Koyna earthquakes and the large number of earthquake shocks that are recorded or felt in the state every year have their origins in the complex system of faults and lineaments that exist in the region. The location of many of these faults and lineaments are known and are mapped in the Seismicity Atlas of India [2]. However, a number of these faults have been identified only after the occurrence of large earthquakes. The current scientific knowledge of the potential seismic sources in Maharashtra, including Palghar district, is thus, at best, incomplete.

## 2 Recent Seismic Activities in Palghar District

The Palghar district, which has been recently carved out of erstwhile Thane district, has been experiencing frequent ground shaking. These shocks are generally in the magnitude range between 3 and 4, while a few shocks are between magnitude 4.0 and 4.5. Frequent tremors were first reported after monsoons in 2017, which eventually subsided after a few months. But again since November 2018, regular ground shaking is observed in the Palghar region which continues to date. The district administration informed that some buildings have been reportedly damaged during the ground shaking.

Palghar district has numerous seismic sources through a complex network of faults and lineaments. The seismic sources in the vicinity of the reported recent earthquakes are shown in Fig. 1 [2]. The figure also shows the locations of some earthquakes that occurred in the region. It can be seen that a number of potential seismic sources exist in the region. It can also be seen that the region has a past history of seismic events, and several earthquakes have been felt in the region and mentioned in the earthquake catalog of the Seismotectonic Atlas of India.

After the frequent ground shaking was reported to the Government of Maharashtra in late 2018, the India Meteorological Department was requested by the state government to monitor the ground shaking and advise the government on the cause of the ground shaking. The National Centre for Seismology (NCS) under India Meteorological Department sets up three mobile seismic recording stations to monitor the ground shaking. The NCS has recorded a large number of shocks in which the ground shaking to have a magnitude of less than 4.0, except for a couple of shocks that exceed magnitude 4.0. The NCS also advised the government that the ground shaking appears to be very shallow, with depth not exceeding 5 km, and appear to be of non-seismic origin. They also reported that the shocks are likely to be swarms and may not increase their strength significantly in the future.

Subsequently, the Government of Maharashtra also requested the National Geophysical Research Institute (NGRI), Hyderabad, to carry out monitoring of the

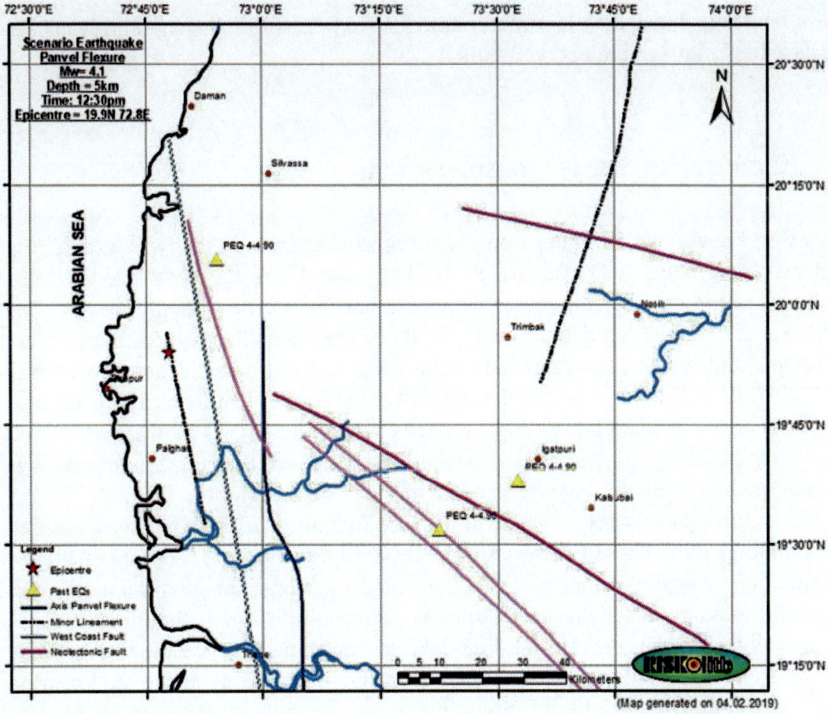


**Fig. 1** Seismic sources in Palghar region based on the Seismotectonic Atlas of India [2]

ground shaking and provide their advice regarding the causes and possible consequences. The NGRI has set up five mobile seismic stations in the area. The NGRI also recorded a number of the shocks that have occurred in the region after their installation. The magnitude of shaking, as per NGRI, is consistent with the report of NCS that almost all shocks are of magnitude less than 4.0. The NGRI has also determined the depth of the shocks and other seismicity parameters based on their recording. However, as per NGRI, the shocks are occurring at a depth between 5 and 15 km. Due to the relatively deeper locations of the shocks, the NGRI has stated that the origin may be tectonic, and represent energy release from the seismic faults in that region.

The district lies in seismic zone III according to IS 1893 (Part 1): 2016. Therefore, it was felt that these low-magnitude earthquakes should not be taken lightly as these may be a precursor to larger earthquakes, like in the case of the Koyna and Killari earthquakes. The government also recognized that, in seismic zone III, earthquakes may result in damage with an intensity of VII. Intensity VII represents the condition wherein poorly built buildings may be destroyed in large numbers, ordinary load-bearing buildings may be moderately damaged, while well-built buildings may suffer minor damage. Intensity VII may therefore lead to widespread damage and casualties in case the buildings are not well-built.

## 3 Earthquake Safety Considerations

It is well known that the impact of an earthquake depends on the types of building constructions in the affected area. If the buildings are strong, the loss due to the earthquake is much lower than in cases where the buildings are inherently weaker. The strength of the buildings against an earthquake depends on the material of constructions, particularly the materials used for the construction of vertical members such as columns and walls. To an extent, the strength of buildings against an earthquake also depends on the local soil conditions, type of foundation, type of roof, etc. However, in typical building constructions in Maharashtra, the safety against earthquakes is predominantly influenced by the strength of the vertical members.

From earthquake safety considerations, the Indian Standards [1] have classified the buildings into Type A (weakest), Type B (stronger) and Type C (strongest). Type A buildings typically include those constructed using stone, unburnt bricks, clay, etc. Type B buildings include ordinary brick buildings while Type C buildings include well-built reinforced concrete buildings. Within each building type, the strength may vary based on variables such as the building layout, locations of the load-bearing members, inherent strength of the building materials, level of maintenance, etc.

An important factor that influences the effect of the consequences of damage is the use of the building. While the safety of all buildings in the affected region is important, the safety of buildings that are used for important purposes such as schools, hospitals, important government functions, etc. is even more important. In order to ensure the safety of the people, it is therefore important to assess the safety of the buildings in the affected area.

## 4 Field Survey of Buildings in Palghar

The District Administration of Palghar conducted multiple field surveys in the areas affected by earthquake tremors to assess the potential seismic vulnerability of government buildings. The surveys were carried out as per the methodology provided by IIT Bombay. The government officers were also provided training to carry out the surveys by IIT Bombay. The training program included understanding building types, their vulnerability to earthquakes and how to conduct a rapid visual survey of buildings using the methodology provided by IIT Bombay.

Rapid visual survey (RVS) of buildings is intended for rapid seismic vulnerability assessment of buildings. The RVS of buildings is a visual examination of damage and condition assessment of the building. It requires examination of the interiors and the exteriors of the buildings for existing damages and vulnerable features. It establishes the overall seismic risk to the building. The main objectives of RVS are to understand the extent of the seismic vulnerability of a building and also to help prioritize the buildings for detailed assessment based on their seismic vulnerability.

The district administration along with the government engineers from various departments carried out this rigorous exercise of the field survey and conducted the survey of around 850 government as well as few private buildings.

The survey data shows that the building stock in the urban epicentral region exhibits a rich mix of several different building typologies and construction materials. All three types of building classes were found to be present. There are a large number of Type A buildings that are typically non-engineered. The area has a number of Type B buildings consisting of load-bearing masonry buildings and also a much smaller number of reinforced concrete buildings, which are under Type C.

During the survey, around 850 buildings were covered. Based on the survey, 11 reinforced concrete buildings and 35 load-bearing masonry buildings were found to be in an extensive damage state and potentially unsafe under the effect of future tremors. For the extensively damaged RC buildings, the age varies from 12 to 29 years, while for extensively damaged masonry buildings, the age varies from 11 to 56 years [3].

The key observations were made regarding the buildings in extensive damage state.

## 5 Survey Details of Extensively Damaged Buildings

In order to assess the likely strength of the buildings in the Palghar district against earthquakes and to understand the prevalent building typologies, the team of engineers from IIT Bombay and PWD carried out a detailed survey of buildings near the epicentral area. Moment-resisting frame RC buildings possess sufficient strength against earthquake shaking. It was observed that the area had a very small percentage of RC buildings. Most of these buildings are located in urban areas. The predominant building typologies in the rural area are load-bearing buildings with unreinforced brick masonry walls with cement mortar and asbestos sheet roofs. These buildings have moderate strength against earthquake shaking. In this category of buildings, the damage was observed for few buildings. There were cracks in the 90° joint between the walls. There were also deep cracks observed in the walls. There were no lintel bands or roof bands in the buildings. Poor connectivity at wall corners may have resulted in the cracks. There are few buildings with wooden walls (Kavri) with mud mortar with an asbestos sheet. The buildings in this category were not damaged.

Overall, there were 11 reinforced concrete buildings that were found to be extensively damaged during the survey. For these buildings, it was observed that beams were hanging, slabs were heavily damaged, and foundation settlement issues were there. The survey data also reveals that all these RC buildings are quite small in size, with the largest building area of less than 100 $m^2$.

Damage to RC buildings, which are expected to possess high seismic resistance, under normal service loads (without having been subjected to code-level earthquake shaking) can be due to two possible reasons as discussed below.

**Load capacity of critical members being exceeded**: The load capacity of critical elements can be exceeded resulting in their damage if forces applied on the building are larger than design forces or if the members are weaker than expected. In either of these cases, the load capacity of the damaged members needs to be enhanced. Sometimes, the load capacity of thereto undamaged members may also need to be enhanced depending on the distribution of loads when code-level forces are applied. Retrofitting of structural members as required may be taken up under such situations to repair damage and enhance the load capacity of the building.

**Damage induced due to ground movement**: RC buildings may also be damaged if their foundations experience excessive deformations due to ground movement. There may be a number of reasons such as insufficient depth of foundation, changes in the groundwater table, presence of certain types of soil such as black cotton soil at foundation level, etc. The damage pattern for this condition is different from that due to the exceeding load capacity of the members. The locations, nature and sizes of damage can be used to distinguish this type of damage from other causes. When buildings are damaged due to ground movement-induced damage, repair to the RC elements does not remove the underlying causes. As a result, the same damage may recur again and again.

For the 11 buildings that are found to be in extensive damage condition, detailed engineering information is required to distinguish these two categories of damage.

There were 35 load-bearing masonry buildings that were found to be extensively damaged during the survey. The survey data reveals that for these load-bearing buildings walls were collapsed or had major deep cracks, slabs were heavily damaged and foundation settlement issues were there. Also, the sizes of the walls were variable and could range from small buildings to large ones and are exceeding 230 $m^2$.

## 6 Seismic Strengthening of Buildings

There are many deficient buildings that fail to meet the seismic requirements which may undergo major or extensive damage in case of an earthquake. It may cause casualties for the occupants as well as people residing in the vicinity of the building. So, it becomes important to identify the seismic vulnerability of the buildings, and then seismic strengthening is required to enhance the capacity of weak structures to resist earthquakes. It has been extensively studied that for buildings damaged during the earthquake, seismic strengthening techniques may be used to enhance the capacity of the building so that there is no hazard to property or life safety. This is also a better alternative catering to the economic considerations rather than demolition and reconstruction even in cases of severe structural damage.

## 6.1 *Reinforced Concrete Buildings*

For reinforced concrete buildings, it is important to prevent possible side sway collapse of moment resisting frame for given ground shaking. It is also to be noted that even though walls are normally not supported to carry the load, due to deflection of beams, the loads may partially be transferred to walls and it may cause a partial or full collapse of infill wall panels during a seismic event. Thus, considering these issues, strengthening measures for RC buildings are proposed.

During the field survey in the Palghar district, it was observed that for damaged buildings, structural members such as beams and columns and also non-load bearing walls were damaged. Thus, remedial measures for all these members are suggested below.

**Remedial Measures for Non-load Bearing Walls**

1. Large cracks (≥5 mm): Require breaking and reconstruction of the wall panel (slab to be temporarily supported).
2. Medium cracks (2–5 mm): Require stitching with steel reinforcement bars and re-plaster.
3. Small cracks (≤2 mm): Require re-plaster after filling cracks with plaster mortar.

**Remedial Measure for Structural Members (Beams/Columns)**

Any work should be taken up only after temporary support to slabs in the vicinity of the damaged member taken up for strengthening. Strengthening of structural members is a specialized activity and should be taken up after engaging a suitable structural engineer. In the case of cracks, they cannot be simply filled. The deficient strength needs to be restored. The strength of a column can be enhanced by jacketing. Typical strengthening techniques include the following:

(i) Jacketing at beam/column joint
(ii) Filling cracks
(iii) Injection of cracks
(iv) Steel plates for column strengthening
(v) External post-tensioning of beams
(vi) Steel jacketing of columns
(vii) Putting additional stirrups and ties for beams and columns
(viii) Supporting beams and slabs using steel plate

## 6.2 *Seismic Strengthening of Load-Bearing Masonry Buildings*

It was observed during the field survey that most of the damage for the load-bearing masonry buildings was at the wall. In the walls where the crack width is less than 5 mm, a repair can be considered. If cracks larger than 5 mm have already occurred under normal loads, in the event of a moderate earthquake, these walls may perform very poorly even after repair. Therefore, repair of wall panels with crack width larger than 5 mm should be decided in consultation with engineers on a case-by-case basis. Remedial measures for strengthening of load-bearing walls based on the size of the cracks are mentioned below:

i. Large cracks (≥5 mm): Require breaking and reconstruction of the wall panel (slab to be temporarily supported)
ii. Medium cracks (2–5 mm): Require stitching with steel reinforcement bars and re-plaster.
iii. Small cracks (≤2 mm): Require re-plaster after filling cracks with plaster mortar.

The following repair methods can also be used in consultation with engineers on a case-by-case basis.

(i) Grout injections
(ii) Braced masonry walls
(iii) Reinforcing of unreinforced reinforced masonry foundations
(iv) Repairing cracks by injection of cement/epoxy grout depending on the thickness
(v) Addition of new bracing walls
(vi) Improvement of the connections between the resisting elements (e.g. tying of intersecting wall etc.)
(vii) Application of vertical and horizontal confining elements to the walls
(viii) Strengthening of walls by means of reinforced concrete jackets.

## 7 Conclusions

Based on the survey carried out, it was observed that the housing stock in the rural epicentral region is mostly non-engineered which is quite vulnerable even to low-intensity (less than VI MSK intensity) earthquakes. The building stock in the urban epicentral region exhibits a rich mix of several different building technologies and construction materials such as reinforced concrete, brick/stone load-bearing masonry and non-engineered buildings. These buildings in urban areas are vulnerable to moderate intensity (greater than VII MSK intensity) earthquakes.

An earlier survey of important government buildings is completed by the district administration. A similar rapid visual survey can be conducted for all the remaining buildings. This survey can be done by engineers and students from engineering and science backgrounds. Also, the help of the private sector can be taken for this. Engineers and architects from the government as well as the private sector can have sensitization training for earthquake-resistant construction practices.

Similarly, the industries in the Palghar region can be asked to do a preliminary seismic vulnerability assessment.

**Acknowledgments** This study was carried out with support from the Government of Maharashtra. The authors are also grateful to the District Collector's Office, Palghar, DDMA Palghar and Government Engineers from various departments for their valuable onfield input and support.

## References

1. BIS. IS 1893 (Part 1)-2016 (2016) Indian Standard Criteria for Earthquake Resistant Design of Structures, Part 1—General Provisions and Buildings (Sixth Revision), Bureau of Indian Standards, New Delhi
2. GSI (2000) Seismotectonic Atlas of India and its Environs, Geological Survey of India
3. Sinha R, Goyal A, Shinde R (2020) Recent earthquake tremors experienced in Palghar region of Maharashtra: note on building surveys and follow-up action plan. Report submitted to the Government of Maharashtra India

# Evaluating the Seismic Performance of Domestic and Historical Masonry Structures in Himachal Pradesh Region of India

**Ashwani Kumar Sharma, Ashutosh Kumar, and Vasilis Sarhosis**

## 1 Introduction

The state of Himachal Pradesh in India is located along the foothills of the Himalayas comprising steep terrain, mountainous region and valleys. The state is divided into 12 districts in which their altitude ranges from 350 m to 7,000 m as per Jaswal et al. [15]. As per the 2011 census, around 90% of the population of the state lives in rural areas are located mainly at higher altitudes and rugged terrain. Many of the villages lack road connectivity leading to a limited supply of modern materials for construction purposes. Buildings are mainly constructed by local masons using traditional methods and locally available construction materials such as mud, stone and timber. Thus, many of the domestic and historic buildings were made of unreinforced rubble masonry and rammed earth.

Singh et al. [21] reported the use of earthen material and stone rubble masonry for the construction of one or two stories residential buildings. Such buildings were constructed using flat and sloping roofs supported by timber beams. Sood et al. [22] reported the use of rammed earthen masonry construction in the Lahaul and Spiti district where walls thicknesses are ranging from 300 to 500 mm. Thick walls were used not only to sustain the load from above but also to maintain the temperature in the building during the cold winters. Such construction type was in practice for 800 years; see the monastery in the village Kii in Fig. 1a. Sood et al. [23] reported

A. K. Sharma (✉) · A. Kumar
School of Engineering, IIT Mandi, Kamand 17500, Himachal Pradesh, India
e-mail: t19013@students.iitmandi.ac.in

A. Kumar
e-mail: ashutosh@iitmandi.ac.in

V. Sarhosis
School of Civil Engineering, University of Leeds, Leeds LS2 9JT, UK
e-mail: V.Sarhosis@leeds.ac.uk

 
S. Kolathayar and S. C. Chian (eds.), *Recent Advances in Earthquake Engineering*,
Lecture Notes in Civil Engineering 175,
https://doi.org/10.1007/978-981-16-4617-1_38

**Fig. 1** **a** Kii monastery in village Kii in Lahaul and Spiti district of Himachal Pradesh. **b** Typical stone masonry house with sloping roof in Himachal Pradesh

the construction of dry stone masonry walls with a sloping roof. The stone masonry was confined using timber beams. This is one of the common construction typologies observed in the region, where dry stone masonry is often accompanied by timber members. Such style of construction was named Kath-Kuni architecture and is prevalent in Shimla, Kinnaur, Kullu and Mandi districts. The literal meaning of "Kath-kuni" in Sanskrit is a "wooden corner", representing the indigenous building style of walls in which wooden beams are interlocked at corners (see Sect. 2). As stated by Dave et al. [8], Kath-kuni building style is commonly found in the Himalayan region mainly in Kashmir, Himachal Pradesh, Sikkim and Uttarakhand in India while its variations could also be found in other places, e.g., in Pakistan, Nuristan to Baltistan province in Afghanistan and East Himalayan region of Bhutan. This may be mainly because of the availability of stone and timber in these regions. In addition, such an architectural style is very quick to construct. Das [7] stated that the Kath-kuni building style extensively uses timber connection as a bonding element. There are several structures with the Kath-kuni architectural style. These range from traditional palaces of Maharaja of Kullu to towering temples present in district Mandi, to some simple houses in the old Manali region. However, only recently the rural population is discarding the traditional methods of construction and instead build domestic houses using the new construction materials. In particular, the mud or lime mortar is now replaced by cement and such combinations of newer and traditional building materials can be disastrous. In most cases, sun-dried earthen bricks and mud mortar were used for the construction of walls. So, during a seismic event, mortar can lose its strength and allow dissipation of the stress on the building. However, in the case where cement mortar with higher strength than bricks was used, during a seismic event, low-strength mud bricks will fail while high-strength mortar does not, as described by Langenbach [19]. Figure 1b shows the typical stones masonry structures constructed using the dressed stone masonry interlocked using cement mortar in the district Mandi.

## 2 Seismicity of the Himachal Pradesh Region and Its Impact on Lives and Livelihoods

Another important aspect to consider is the seismicity of the Himachal region. In fact, the region has experienced many devastating earthquakes (1905 in Kangra, $M_w = 7.8$; 1975 in Kinnaur, $M_w = 6.8$; 1986 Dharamshala Earthquake, $M_w = 5.7$; 1991 in Uttarkashi, $M_w = 6.8$; 1997 in Sundernagar Earthquake, $M_w = 4.7$ and 1999 in Chamoli Earthquake, $M_w = 6.6$), and it is believed that these areas could experience further devastating earthquakes ($M_w \geq 6$) in future. In addition to huge harm to life and livelihoods, the 1905 Kangra earthquake had caused extensive damage to centuries-old medieval temples made of unreinforced masonry [16]. Singh et al. [21] reported evidence of the collapse of mud and stone masonry structures during the 1975 Kinnaur Earthquake. Along with these major events, moderate earthquakes like Uttarkashi (1991), Chamoli (1999), Kashmir (2005) and Sikkim (2011) have also caused significant damages to various masonry structures in the Himalayan region. The generalized tectonic features of Himachal Pradesh are shown in Fig. 2. From Fig. 2, it is evident that seismically active faults surround Himachal Pradesh. These faults contribute to tectonic activity in the region, primarily because of the continent–continent collision of the Indian and Eurasian plates as stated by Finch [11]. At present, there is no clear-cut information about the last great earthquake in this region. The recent research on the topic suggested that the currently available strain in the central seismic gap could drive one or more great earthquakes [2, 3]. Bilham [4] has reported that the Central Seismic Gap (CSG) region has accumulated strain levels enough to cause an earthquake of the maximum magnitude of 8.7 in the near future. Hence, the state is facing severe seismic risk as it is located in the CSG region.

One of the most recent earthquakes that have occurred in the Himalayan region is the 2015 Nepal (Gorkha) earthquake. The earthquake has claimed the lives of around 8,800 people, displaced about 2.8 million people and destroyed 5,00,000 houses [5, 12, 26]. The earthquake also caused the collapse of 190 heritage structures and damaged 663 monuments [25]. This underlines the importance of regular maintenance and structural health monitoring of the heritage structures present in the region. Many heritage monuments, structures and temples are also situated in the regions of Himachal Pradesh, one of which is the 470-year-old Pagoda style "Hadimba Devi temple" at Manali. Kumar et al. [17] reported the damaged pattern of heritage structures within the World heritage sites of Kathmandu during the 2015 Gorkha Earthquake and stated the age-related deterioration and out-of-plane collapse of masonry wall were the main reason. Kumar et al. [18] studied the collapse of Jaisidewal temple (historical Pagoda style architecture) in Nepal during the 2015 Gorkha earthquake. The study showed that the seismic vulnerability in a 325-year-old temple arose due to deteriorating masonry wall and the presence of discontinuous columns in the temple, while there was no failure in substructure, i.e., no remarkable differential settlement in the foundation. Investigations undertaken demonstrated that the exposed foundation or plinth of temples significantly affects the structural

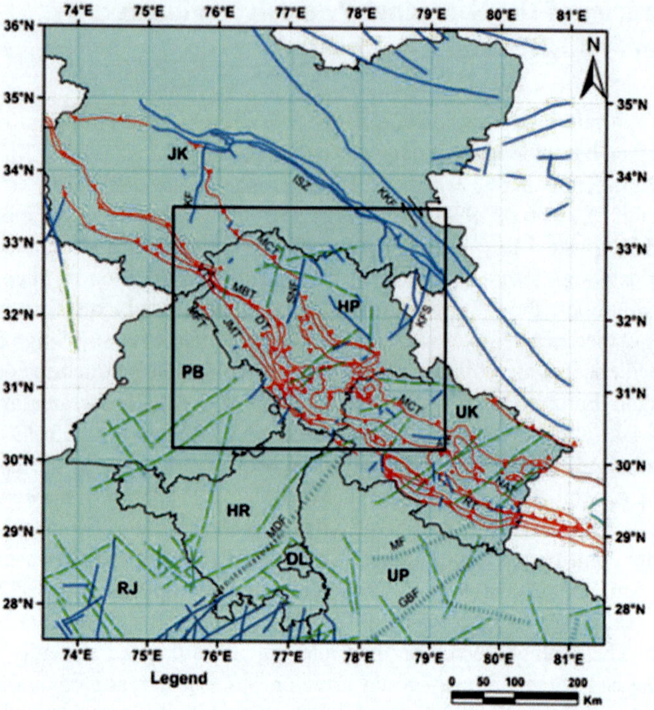


**Fig. 2** Tectonic features present in and around Himachal Pradesh, MCT: Main Central Thrust, MBT: Main Boundary Thrust, SNF: SunderNagar Fault, KKF: Karakoram Fault, KF: Kishtwar-Fault, KFS: Kaurik Fault System, SS: Shyok Suture, JF: Jhelum Fault, JMT: JwalaMukhi Thrust, MDF: Mahendragarh-Dehradun Fault, MF: Moradabad Fault, DT: Drang Thrust, NAT: North Almora Thrust, RT: Ramgarh Thrust, GBF: Great Boundary Fault, AF: Alaknanda Fault (*Source* Seismotectonic Atlas of India and its environs by GSI [14])

behavior of the temples during earthquakes. In a study reported by Arya [1], it was demonstrated that the damage potential of magnitude 8.0 earthquake in the same area of Kangra as that of the 1905 earthquake is still quite high and, in turn, increased many folds after the 1905 event. It is reported that there will be high structural damage and potential deaths in case of repeatability of such events. As per estimates, it can cause the collapse of 145,000 houses and partial collapse of 268,000 houses in an area of 7,900 $km^2$ and loss of lives could range from 88,000 to 344,000. Gopal et al. [13] reported that the unreinforced masonry structures are the most vulnerable in case of seismic events experiencing full or partial collapse. The Vulnerability Atlas of India [24] shows that approximately 70% of the houses in Himachal Pradesh are unreinforced and made of rammed earth, low-strength brick and stone constructions. This percentage includes the Kath-kuni building style from districts Chamba, Kullu, Kinnaur, Mandi, Shimla and Sirmaur. So, the coexistence of high seismicity and unreinforced vulnerable constructions in the area can increase the potential damage to life and livelihoods during a strong earthquake event.

## 3 Aim of the Study

The disastrous consequence of several earthquakes in the Himalayan region has proven that masonry structures are vulnerable to seismic events. This may be due to a lack of knowledge of local masons to install seismic strengthening measures in such structures, age-related deterioration of the structural members and lack of periodic maintenance of the structure. Figure 3 shows the damage evidence in an inhabited typical unreinforced stone masonry structure where cracks in the buildings developed due to the differential settlement in the foundation components of the building.

Figure 3a shows the out-of-plane failure of the wall and a tilt of approximately 9° was observed, and in Fig. 3b, a weakened bonding in the wall and resulting racking crack which is widest at the top can be noticed. However, the exact reason for such damage is unknown. This study aims at understanding the construction methodology and the performance of the Kath-kuni style structure during past earthquake events by performing a reconnaissance survey. First, the study investigates the common construction typologies for the domestic and heritage structures of the Himachal Pradesh state. Then, it focuses on identifying the performance of the structure from a seismic standpoint by collecting associated damage evidence present in the Kullu region. Finally, precautionary measures to be adopted to safeguard the structures from future earthquakes are suggested.

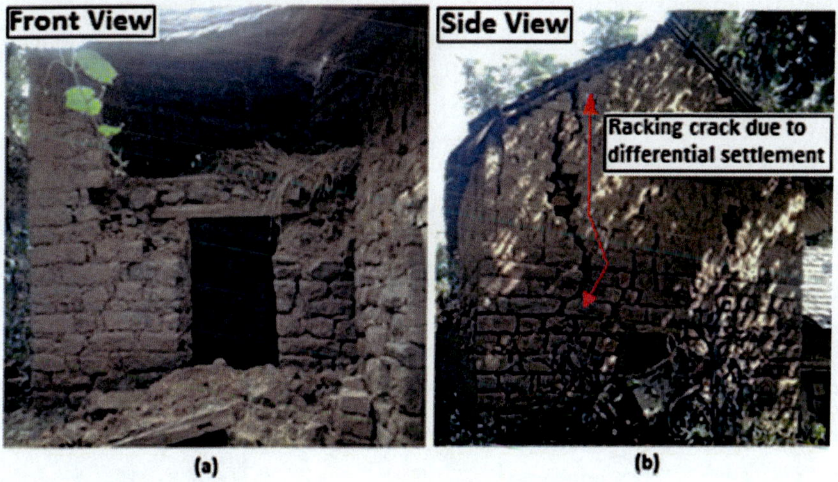


**Fig. 3** **a** Out-of-plane failure of the wall. **b** Development of cracks from openings

## 4 Field Investigation

A field reconnaissance survey was carried out in the villages of the Kullu district of Himachal Pradesh during November 2020. These villages consist of a mix of historic, domestic and new buildings and temples which follow the Kath-kuni architectural style. A total of 35 structures were surveyed in villages, namely Jaa, Meeteura Jari, Jari and Chenni Kothi comprising domestic (25.7%) and ancient heritage structures (74.3%) (Fig. 4). Many local masons and village dwellers were interviewed to understand the engineering strategy to construct, preserve and maintain these structures. Residential structures were used for dwelling, whereas heritage structures named in this study were mainly used for religious purposes.

### 4.1 *Typical Structural Configuration and Seismic Adaptation*

Most of the structures mainly low-rise buildings (e.g., 1–3 stories) were constructed using rubble stone masonry. Figure 5a shows a 300-year-old two-storied building based on the Kath-kuni architectural style and a typical arrangement of stones and timber members along the wall. Dry stone–timber composite masonry walls were the main load-bearing system in the Kath-kuni architecture. Wall openings were supported by wooden lintel beams. Buildings had alternate layers of complete dry stone masonry and composite layers of timber as shown in Fig. 5b. These two different layers of dry stones and timber beams also run alternatively in the transverse wall in a way that the complete stone masonry layer meets the composite layers of another wall at the corners. The composite section of load-bearing walls had two timber

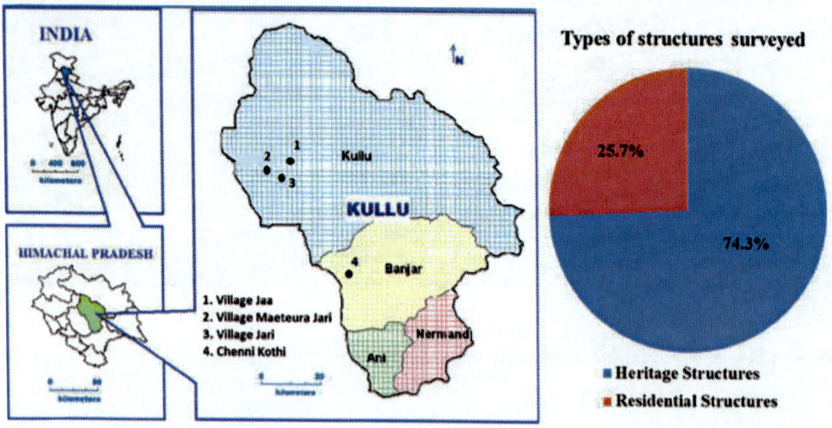


**Fig. 4** Places visited for site investigation in the Himachal Pradesh region and a pie diagram showing the types of structures surveyed

**Fig. 5** **a** Kath-kuni tower-like structure. **b** Enlarged view of Kath-kuni sidewall. **c** Corner details showing interlocking of transverse timber beams

beams (usually of deodar) with dry stone masonry in between and a masonry layer comprising well-dressed stones. The thickness of walls varied from 460 to 610 mm, depending on the height of the building. The thickness of one timber beam in the wall section was around 125 mm. Such even distribution of masonry–timber composite would help distribute the stresses generated during the seismic event. Symmetrical geometrical configuration was the main feature of the structures; thus, minimizing loading eccentricities and reducing extra stress resultants that may arise if geometry was asymmetric. The bonding of stone within the timber members was ensured by providing the connection at the corner and mid of the two timber beams of the walls. Figure 5c shows the interlocking arrangement at one of the corners of the wall where the alternate composite sections of transverse walls meet. It is to be noted that the bonding between the structural members might provide restrain against the failure during the earthquake loading by distributing the shear forces uniformly within the timber members. However, the authors are currently investigating the performance of this building under earthquake loading by using a numerical modeling program to ascertain this analogy.

Figure 6 shows the samples collected during the site visit and a model of a typical connection of the timber elements. Figure 6a shows the stones collected from a deteriorating masonry structure. It is to be noted that the walls of the building use two horizontal beams in composite action which are connected by a double dovetail timber connection and their thickness is approximately 100 mm. The connections between the beams were made in the middle length of the wall and at every 1.8 m after grooving to restrict horizontal movement. This wooden connection in between the composite section of the wall provides additional stiffness and thereby resists the transverse movement of the beams. Figure 6b shows a model describing double dovetail timber connection in the wooden framed structure which is made by a local mason who has constructed more than 70 Kath-kuni style structures. The timber beams of cross-section 40 mm × 40 mm were used in this model and the double dovetail connection has a length of 100 mm and a thickness of 10 mm. The model

**Fig. 6** **a** Rubble stone collected from the site. **b** Model of wooden double dovetail connection made by a local mason. **c** Wooden dowel in a timber beam obtained from the site. **d** Modeling of connections in wooden frame present in Kath-kuni walls

uses nails to tie the beams instead of wooden dowels. In the field, two perpendicular beams meeting at corners were connected using a timber dowel by drilling a hole through two beams, as shown in Fig. 6c (sectional view). The diameter of these dowels was generally kept as 50 mm. In the case of a seismic event, dowels provided lateral stiffness to resist lateral drift between two adjacent layers of the wall. Figure 6d shows the overall arrangement of timber beams and connection in the structure.

These structures were built on isolated footing made up of dry rubble masonry and the depth of foundation mainly relied on the level at which hard rocky strata were encountered. Considering the rocky terrain, the depth of the footing usually varied from 0.5 to 2 m. The plinth level of the building was approximately 150 mm above the ground and a timber beam with dimension 300 mm × 300 mm was placed on it in the longitudinal direction while another beam of the same dimensions was placed in transverse direction before starting the construction of a wall. These wooden beams placed at the plinth level helped in the even distribution of building load to the foundation level. Besides, most of these buildings were located on the sloping ground which increases the risk of failure in case of slope subsidence during the seismic events. Although there was no proper bonding between stone masonry and timber, such a construction sequence suggests a greater seismic adaptation for these buildings in terms of providing greater stiffness by the timber members and the connection conditions at the center and corners. However, the top floors of these buildings were cantilevered and made of timber (Fig. 7a). The space available on the top floor was for lightening during the winter season. Herein, overhang beams were projected from walls to support the loads of balcony where columns made of timber were used to distribute loads of a thick sloped roof as shown in Fig. 7b. It has to be pointed out that a column discontinuity existed in the structure. This reduces the ability of the upper portion of the structure to withstand the lateral forces during a seismic event. Also, the roofing material consisted of slates that are abundantly available in the region while proper anchorage from the roof to the walls was not observed. The non-uniform mass distribution and heavyweight of slates covering the roof will attract greater inertia forces that need to be resisted in case of an earthquake event. The poor connectivity between the walls and the top portion of the building will reduce the ability of the upper portion of the structure to withstand the lateral forces during the earthquake. Furthermore, age-related deterioration of the structural members and lack of proper maintenance could further reduce the seismic resilience. The failure of the top 2 stories of a 500-year-old building is the evidence describing the seismic vulnerability of the structure (discussed in Sect. 5).

It has to be pointed out that many of these structures survived during the 1905 Kangra and 1975 Kinnaur earthquakes. Rautela and Joshi [20] also reported similar seismic adaptation by performing an equivalent seismic analysis of a five-storied building constructed using a similar architectural style in the Uttarakhand region of India where Kath-kuni architectural style is known as Koti Banal architectural style. However, age-related deterioration of structural members, lack of periodic maintenance and loss of connection rigidity is evident in these structures and may

**Fig. 7** **a** Cantilevered balcony in old Kath-kuni buildings. **b** Cantilevered balcony having columns to support the slate roof

be detrimental for future earthquake hazards. It was also noted during the post-survey 2015 Gorkha earthquake that some of the masonry structures which survived the 1934 Great Nepal Bihar earthquake and the 2011 Sikkim-Nepal earthquake, collapsed during the 2015 Gorkha earthquake having similar structural features, lack of proper maintenance [17, 18].

# 5 Types of Kath-Kuni Structures

Kath-kuni structures were used for domestic purposes besides the existing heritage structures that were primarily used as religious places. Construction of new structures was undertaken using the locally available construction materials and the ancient structures are maintained using modern construction materials. We have divided Kath-kuni style structures into two major types depending on their usage.

## 5.1 *Domestic Structures*

Domestic Kath-kuni style structures are used by local people for several generations as shown in Fig. 8. They were usually 1–3 stories where the ground story was meant for cattle and served as storage purpose. The first story was for living and dining and the topmost story was often used as a place for worship and kitchen. The sloped roof was made of timber truss on which slate tiles were resting while the other structural configurations remain the same as discussed in the earlier section. The lateral dimensions of the buildings were usually dependent on the number of family members living and the financial resources of the residents. Figure 8a shows the repair of the traditional structures with modern construction materials, i.e., brick

**Fig. 8** Residential Kath-kuni structure

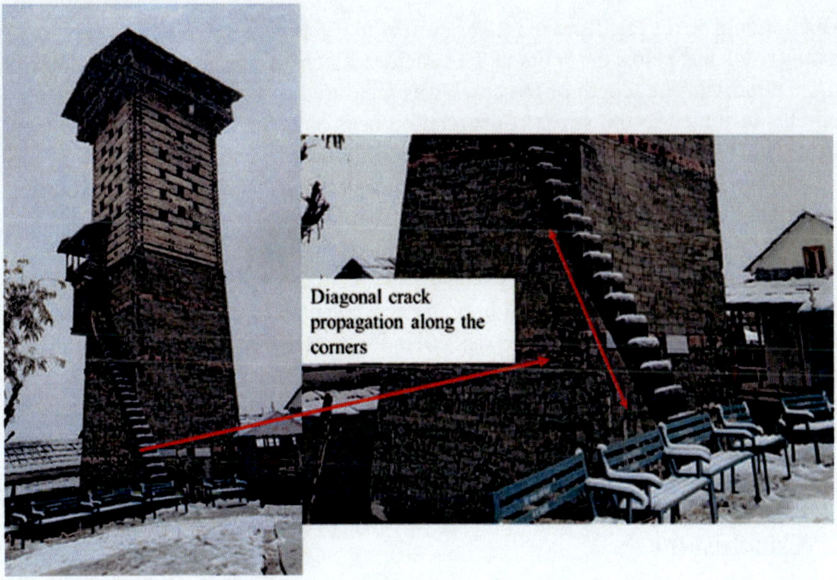


**Fig. 9** Chenni Kothi heritage tower

and cement mortar. Such repair practice may have a negative impact on the seismic resilience of the building as one portion of the structure becomes stiffer compared to the rest altering the global stiffness compatibility of the structure.

## 5.2 *Heritage Structures*

Figure 9 shows a 30 m high classical tower type Kath-kuni architecture located at Chenni Kothi village of Kullu district. The structure is believed to be more than 500 years old and survived most of the disastrous earthquakes. This ancient tower was once the palace of the Queen but now transformed into a religious place. It has a symmetrical plan and walls with a narrow opening for windows and doors. The average width of the tower was 5 m. The structure was four-storied having a cantilevered balcony in all directions. The length of the cantilever varied from 1.2 m to 1.5 m. The tower had initially nine stories and during the seismic event of the 1905 Kangra earthquake, the top two stories collapsed. The heavy mass lumped at such height and with lack of maintenance caused the historical structure to fail under the seismic event. Irrespective of the cultural and heritage value of the structure, there was no maintenance work done and there is also no documentation of the damage that happened to the structure. During the 2015 Gorkha earthquake, there was the same direct economic and historical loss in case of emergency measures taken by local people after the seismic event by clearing the parts of the collapsed structure which should be recycled and reused or preserved [6]. Figure 9 also highlights a wide visible crack just below the stairs of the structure which propagates toward the corner of the building. The length of the crack was measured as 2 m. This damage pattern may be attributed to the stress concentration near of corners of the wall and their inability to bear the shear stresses. Also, the possibility of rotation of the building was not observed because of the intact rocky bearing strata. The maintenance, retrofitting and rebuilding work of such ancient marvels should be done by skilled engineers.

Some of the ancient structures are maintained using modern construction materials without assessing the primary cause of the damage. Such practise may lead to irreversible destruction of cultural heritage and potential loss of seismic adaptation being developed over the years and risk reduction strategies without highlighting the research that modern construction materials would offer enhanced resilience. This issue of discarding the usage of locally available material and disrupting the cultural heritage with modern construction material was highlighted by Davis et al. [9].

## 6 Conclusions

This study presents an investigation of structural configuration and deterioration of the structural members of Kath-kuni style domestic and ancient structures of the Himachal Pradesh region of India. A total of 35 buildings were surveyed in the Kullu district of Himachal Pradesh that comprises both domestic as well as ancient structures. Following conclusions are drawn from the study:

(a) The Kath-kuni structures have a great traditional value, seismic adaptation and heritage importance attached to them. The construction of the domestic and ancient structures from a seismic standpoint, i.e., symmetrical geometric configuration (boxed wall) to avoid loading eccentricity, location of the plinth on the rocky stratum, alternate layers of stone-timer arrangements and connections are the corner and center of the walls were beneficial to withstand seismic forces.

(b) The lack of column continuity at the top portion of the building, greater concentration of mass at the top (driving huge inertia forces), degradation in the connection at the corners of the wall and between the composites and reducing joint strength may be a weak zone to induce collapse. The evidence of which was recorded during the 1905 Kangra earthquake that collapsed the top 2 stories of the Chenni Kothi heritage tower.

(c) It has been observed that the seismic events have affected these buildings. While interviewing the local dwellers, it was found that some of the stones from the buildings got dislodged due to out-of-plane collapse and they hammered back these stones in the frame.

(d) Diagonal and vertical line cracks near corners of the walls and deterioration of the timber members were the main damage pattern observed during the survey.

Inspection within the buildings reveals the cracking of many walls due to the loss of bonding between the stones and timber members.

(e) The out-of-plane collapse noted in some of the structures was due to a lack of the ability to resist the lateral forces induced by the previous earthquakes rather than the failure of founding material. Age-related deterioration of timber material, loosening of connection rigidity and lack of proper maintenance may increase the seismic vulnerability of all the buildings.

A risk reduction strategy is required to be undertaken to safeguard these structures from future earthquake hazards and to save the loss of life and property. A periodic inspection and maintenance are suggested together with detailed structural assessment at least for heritage structures to increase their seismic resilience. Seismic strengthening measures are required to be adopted as well. The establishment of data acquisition techniques and remote sensing using (structure from motion and terrestrial laser scanners) could help to record the damage on a large scale as suggested by Dhonju et al. [10]. In such a condition, community participation, consisting of heritage documentation, could potentially contribute to the preservation of this deteriorating heritage.

## References

1. Arya AS (1990) Damage scenario of a hypothetical 8.0 magnitude earthquake in Kangra region of Himachal Pradesh. Bull Ind Soc Earthq Tech. Paper No 297, 27:121–132
2. Bilham R, Gaur VK (2000) Geodetic contributions to the study of seismotectonics in India. Curr Sci-Bangalore 79(9):1259–1269
3. Bilham R, Gaur VK, Molnar P (2001) Himalayan seismic hazard. Science 293(5534):1442–1444
4. Bilham R (2019) Himalayan earthquakes: a review of historical seismicity and early 21st century slip potential. Geol Soc London Spec Pub 483(1):423–482
5. Coningham RAE, Acharya KP, Davis CE, Kunwar RB, Simpson IA, Schmidt A, Tremblay JC (2016) Preliminary results of post-disaster archaeological investigations at the Kasthamandap and within Hanuman Dhoka, Kathmandu Valley UNESCO World Heritage Property (Nepal). Ancient Nepal 191–192:28–51
6. Coningham RAE, Acharya KP, Barclay C, Barclay R, Davis CE, Graham C, Hughes PN, Joshi A, Kelly L, Khana IS, Kilic A, Kinnaird T, Kunwar RB, Kumar A, Maskey PN, Lafortune-Bernard A, Lewer N, Mccaughie D, Mirnig N, Roberts A, Sarhosis V, Schmidt A, Simpson IA, Sparrow T, Toll DG, Tull B, Weise K, Wilkinson S, Wilson A (2019) Reducing disaster risk to life and livelihoods by evaluating the seismic safety of Kathmandu's historic urban infrastructure: enabling an interdisciplinary pilot. J Br Acad 7(s2):45–82
7. Das R (2002) Standing firm: traditional aseismic architecture of western-central Himalayas. Peoples' Science Institute, Dehradoon
8. Dave B, Thakkar J, Shah M (2013) Prathaa: Kath-kuni architecture of Himachal Pradesh. SID Research Cell, School of Interior Design, CEPT University, Ahmedabad
9. Davis C, Coningham R, Acharya KP, Kunwar RB, Forlin P, Weise K, Maskey PN, Joshi A, Simpson I, Toll D, Wilkinson S, Hughes P, Sarhosis V, Kumar A, Schmidt A (2020) Identifying archaeological evidence of past earthquakes in a contemporary disaster scenario: case studies of damage, resilience and risk reduction from the 2015 Gorkha Earthquake and past seismic

events within the Kathmandu Valley UNESCO World Heritage Property (Nepal). J Seismol 2020(24):729–751
10. Dhonju H, Xiao W, Mills J, Sarhosis V (2018) Share Our Cultural Heritage (SOCH): worldwide 3D heritage reconstruction and visualization via web and mobile GIS. ISPRS Int J Geo Inf 7(9):360
11. Fitch TJ (1970) Earthquake mechanism in the Himalayan, Burmese, and Andaman regions and continental tectonics in Central Asia. J Geophys Res 75:2699–2709
12. Gautam D, Chaulagain H (2016) Structural performance and associated lessons to be learned from world earthquakes in Nepal after 25 April 2015 (Mw 7.8) Gorkha earthquake. Eng Fail Anal 68:222–243
13. Gopal N, Sankaranarayanan KM, Sivan PP (2017) Behaviour of masonry buildings under seismic action—a review. Int Res J Eng Technol 6(4):2881–2884
14. GSI (2000) Seismotectonic Atlas of India and its environs; Geological Survey of India.
15. Jaswal AK, Bhan SC, Karandikar AS, Gujar MK (2015) Seasonal and annual rainfall trends in Himachal Pradesh during 1951–2005. Mausam 66(2):247–264
16. Joshi M, Thakur VC (2016) Signatures of 1905 Kangra and 1555 Kashmir earthquakes in medieval period temples of Chamba region, northwest Himalaya. Seismol Res Lett 87(5):1–11
17. Kumar A, Hughes PN, Toll DG, Coningham RAE, Davis C (2019) Damage assessment within the Kathmandu valley's World Heritage Monument Zones after 2015 Gorkha earthquake. In: XVII European conference ECSMGE-2019 on soil mechanics and geotechnical engineering
18. Kumar A, Hughes PN, Sarhosis V, Toll D, Wilkinson S, Coningham R, Acharya KP, Weise K, Josh A, Davis C, Kunwar RB, Maskey PN (2020) Experimental, numerical and field study investigating a heritage structure collapse after the 2015 Gorkha earthquake. Nat Hazards 2020(101):231–253
19. Langenbach R (1986) The uses of masonry in earthquake zones: a perspective on traditional and contemporary construction practices utilizing unreinforced masonry. University of California, Berkeley
20. Rautela P, Joshi GC (2008) Earthquake-safe Koti Banal architecture of Uttarakhand, India. Curr Sci 95(5):475–481
21. Singh S, Jain AK, Singh VN, Shrivastava LS (1976) The Kinnaur earthquake of January 19, 1975: a field report. Bull Seismol Soc Am 66(3):887–901
22. Sood A, Rahul A, Singh Y, Lang DH (2012) Housing Report: Mud wall construction in Spiti Valley (Himachal Pradesh). World Housing Encyclopedia, Report 171
23. Sood A, Rahul A, Singh Y, Lang DH (2013) Housing report: dry stone construction in Himachal Pradesh. World Housing Encyclopedia, Report 172
24. Vulnerability Atlas of India (2019) http://bmtpc.org/topics.aspx?mid=56&Mid1=180. Last accessed 30 Dec 2020
25. Weise K, Gautam D, Rodrigues H (2017) Response and rehabilitation of historic monuments after Gorkha earthquake. In: Gautam D, Rodrigues HFP (eds) Impacts and insights of the Gorkh earthquake. Elsevier, Amsterdam, pp 65–94
26. Zhao B (2016) April 2015 Nepal earthquake: observations and refection. Nat Hazards 80(2):1405–1410

# Mapping of Earthquake-Induced Land Deformation on Urban Area Using Interferometric Synthetic Aperture Radar Data of Sentinel-1

**Fathoni Usman, Agusril Syamsir, and Jihan Melasari**

## 1 Introduction

Earthquake is a natural hazard that disastrously impacted the community. The destruction of infrastructure and loss of life is the major issue. The early warning system's emerging technology monitors closely in near real-time earthquake events. Moreover, remote sensing data becomes an important material in civil engineering applications to determine earth surface deformation. A combination of ground sensors and remote sensing data from the active sensor with higher temporal resolution provides more applications in earth observation activities [1].

The capability to measuring deformation to millimetre-scale on static features of the earth's surface is the advantage of interferometry of the synthetic aperture radar (SAR) [2–4]. Two common techniques in utilizing the SAR data are interferometric SAR (InSAR) and differential interferometric SAR (DInSAR) [4, 5]. Interferometry application on surface changing ranges from land subsidence induced by natural hazards, human-made natural resources on a regional scale and the small scale of deformation of slope and tunnel [2–5]. The workflow in processing the SAR data has enormously enhanced the quality of images for further applications [6]. The processes have become more reliable in time and quality due to the high-performance computing of personal desktop and mobile workstations affordability.

This paper presents earthquake-induced land deformation and the variation velocity in an urban area using single-look complex (SLC) data of synthetic aperture

F. Usman (✉) · A. Syamsir
Institute of Energy Infrastructure, College of Engineering, Universiti Tenaga Nasional, Kajang, Malaysia
e-mail: fathoni@uniten.edu.my

J. Melasari
Civil Engineering Department, Universitas Putra Indonesia "YPTK", Lubuk Begalung, Padang 25221, Indonesia



S. Kolathayar and S. C. Chian (eds.), *Recent Advances in Earthquake Engineering*, Lecture Notes in Civil Engineering 175,
https://doi.org/10.1007/978-981-16-4617-1_39

radar (SAR) data. Using the active remote sensing sensor from the Sentinel-1 mission mapping area subjected to surface changing could be conducted. The result can be used in imposing mitigation plans for future earthquake events.

## *1.1 Sentinel-1 Mission*

The Copernicus Programme initiative of the European Space Agency (ESA) has exposed SAR imagery. The first of the Copernicus Programme satellites is called Sentinel-1. The Sentinel-1 satellite performs synthetic aperture radar imaging in the C-band. The first satellite in the constellation was launched in 2014, and the second in 2016. The satellite orbits and tracks the earth from 632 km altitude at 20 m ground resolution covering 250 by 170 km. The tandem orbital mission gives more frequent site visits of 6 days for the Sentinel-1A and Sentinel-1B satellites. It collects data using an interferometric wide (IW) mode instrument in three slices across the instrument swath. Each sub-slice of the collected images is a series of processed SLC images, and the resolution of the single-look data is 5 and 20 m in square range and azimuth. They have level 1 SLC images with single-polarization of HH, HV, VV and VH; dual-polarization of HH + HV and VV + VH. Figure 1 shows debursted and terrain corrected of a GRD image converted from the SLC image, and the red rectangular is a single burst.

The ESA provides open-source software to process satellite image data from the Copernicus Program's different missions from Sentinel-1 to Sentinel-5. The software is the Sentinel Application Platform (SNAP), jointly developed by Brockmann Consult, SkyWatch, and C-S, and currently in version 8.0. The SAR data from ESA is available at https://scihub.copernicus.eu/. The early collection of images are available

**Fig. 1** The debursted and terrain corrected images of a ground range detected (GRD) converted from the single-look complex (SLC) wide swath (IW) synthetic aperture radar (SAR) data

offline and require a request to initiate download when the file is already available. Another alternative to download SAR data is from the Google Earth Engine (GEE). In the GEE, the SLC SAR data already converted to ground range detected (GRD) in log scaling.

## 1.2 Interferometry and Differential Interferometry

The SAR processing technologies such as interferometric processing (InSAR) operate with signal coherence calculated from SAR acquisition pairs. The basic concept of determining changing on a measured surface in a region is subtracting pairs of interferograms with the same geometry and high coherence [3]. A SAR image is produced by an active sensor emitting microwave radiation and received the reflected wave, the backscatterer [2]. The backscattered signal contains amplitude ($A$) and phase ($\varphi$) information in complex number (c) and is expressed in Eq. (1):

$$c = Ae^{i\varphi} \tag{1}$$

Euler's number ($e$) is an exponential function and $i$ is an imaginary number in the expression. The interferogram ($I$) is generated by pairing two SAR images, master ($m$) and slave ($s$), obtained from different ingestion dates. The interferogram is computed by exploiting the amplitude and calculating the difference between phases as Eqs. (2) and (3).

$$I = c_m c_s = A_m A_s e^{i(\varphi_m - \varphi_s)} \tag{2}$$

$$\varphi_m - \varphi_s = \varphi_{ip} = \varphi_{\text{elev}} + \varphi_{\text{def0}} + \varphi_{\text{flat}} + \varphi_{\text{atm}} + \varphi_{\text{noise}} \tag{3}$$

The interferometric phase ($\varphi_{ip}$) summarizes the topographic phase ($\varphi_{elev}$), the surface deformation phase ($\varphi_{def0}$), the flat earth phase ($\varphi_{flat}$), the atmospheric phase ($\varphi_{atm}$) and the phase noise ($\varphi_{noise}$). The value of coherence plays a big role in providing a high-quality InSAR application.

The DInSAR system is used to exploit the phase in InSAR to determine the variation velocity on the earth surface. In the SNAP software, the topography-induced phase is removed using the Shuttle Radar Topography Mission (SRTM) Digital Elevation Model (DEM) 30 m data. The multilook process continues by applying Lee's filter on the debursted interferogram. The multilook process is intended to reduce speckle, provide improved image recognition and improve image quality [1].

### 1.3 Earthquake and Seismicity on Sumatra Island

The seismic activities trigger major natural disasters on Sumatra Island that sit on the Sunda Plate adjacent to the Indo-Australian plate. The length of fault length dictates the possible size of an earthquake on geological formation [7]. Two maps in Fig. 2 show the estimated surface strain (a) and the earthquake events for the magnitude, *M* above 4.5. The plate convergence velocity is 14 mm/year. Moreover, the Sumatra Island subjects to convergence velocity obliquely about 50–60 mm/year from the Indo-Australian plate and 23 mm/year of geodetic slip across the Great Sumatra Fault (GSF) [8]. Moreover, the Mentawai Fault zone's forearc and the GSF move as separate microplate or sliver plate along Sumatra Island [9].

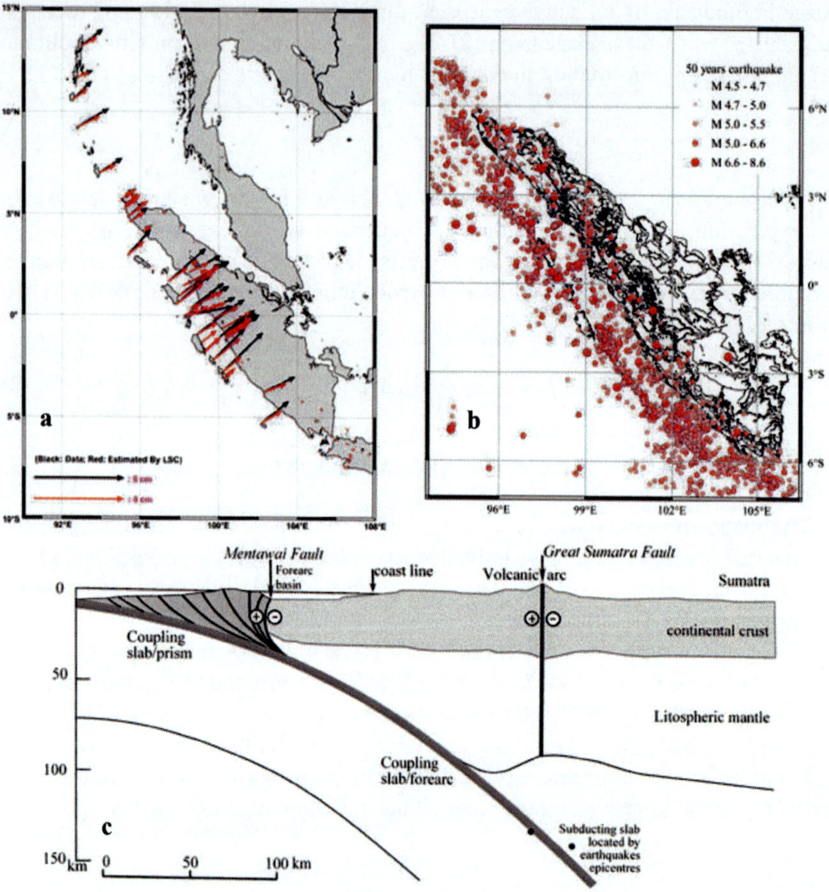


**Fig. 2** **a** The direction and magnitude of surface strain as in [7], **b** the 50 years earthquake events from the USGS earthquake catalogue with the magnitude of $M = 4.5$ and above, and **c** the Mentawai Fault zone act as sliver plate along the Sumatra Island [9]

## 2 Study Area

The study area is the urban area of Padang City, West Sumatra Province, Indonesia. In 2019, Padang City's population reached 950,871, with a population growth rate ranging from 1.25 to 1.47% per year since 2014. Padang is situated between 0°44′00″ and 1°08′35″ South Latitude and 100°05′05″ and 100°34′09″ East Longitude on the western coast of the Sumatra Island. Padang Pariaman Regency borders it in the north, Solok Regency in the east, Pesisir Selatan Regency and the Indonesian Ocean in the south. The Indonesian Ocean borders the western portion. The area of Padang City is 694.96 km$^2$ or 1.65% of the province of West Sumatra. Padang City is made up of 11 districts and has 19 islands and 104 villages (Fig. 3).

Padang City is located on the coast of a flatland across a river from mountains. The region is composed of tertiary sedimentary rocks in its interior regions, while metamorphic rocks are exposed at the surface. The alluvial plain extends around the base of the mountains and is about 10 km wide east–west and about 20 km wide north–south as in Fig. 4. Regarding the topography of Padang, alluvium is loose soil or sediments eroded and carried in suspension by flood or river water before being deposited. For alluvium, some material is unconsolidated, i.e., not formed together into solid rock. The alluvial can be picked up and carried away by flowing water before depositing elsewhere when the flow stops.

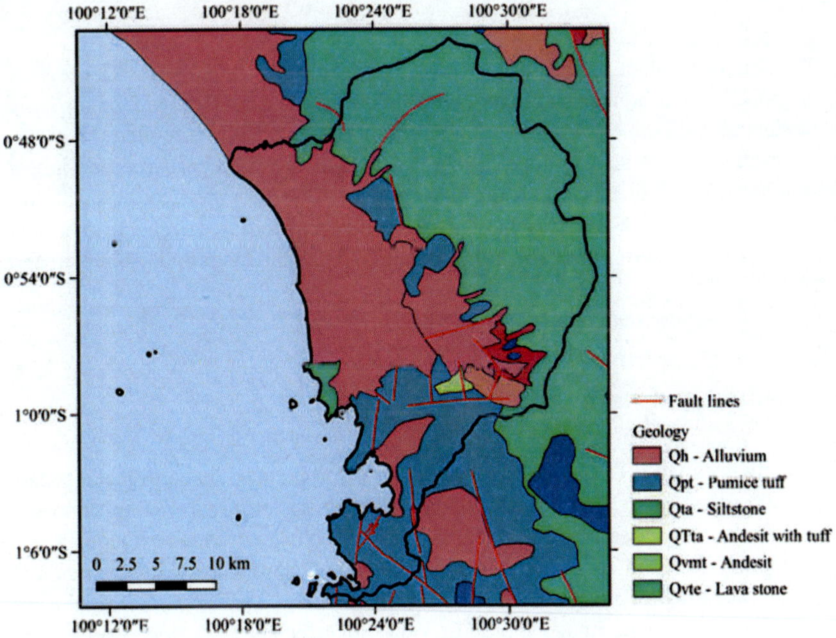


**Fig. 3** The geological formation of Padang City and its structural fault lines

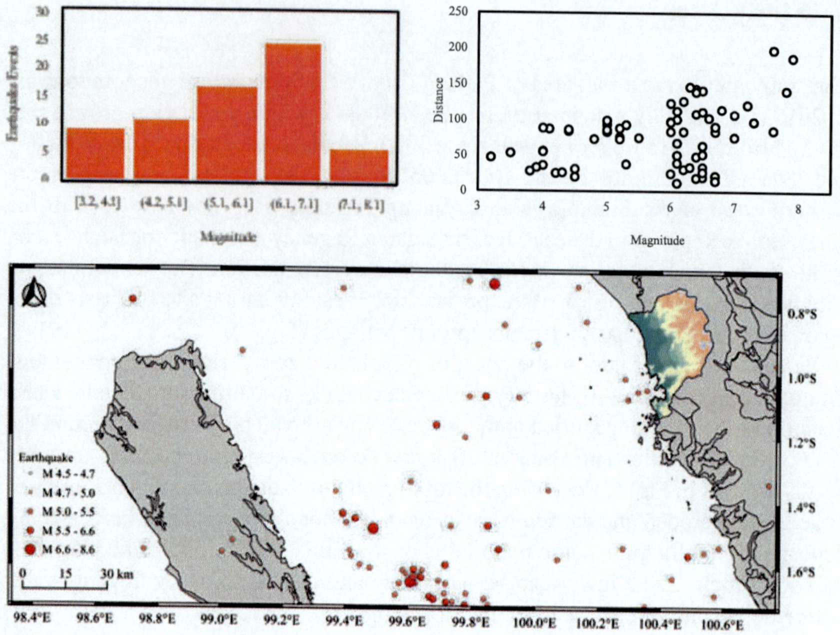


**Fig. 4** Earthquake data from USGS catalogue at a distance within 200 km from Padang City

The seismicity data in Fig. 4 were collected from the United States Geological Survey (USGS) earthquake catalogue. Earthquake events from 1900 to January 2021 was selected from the catalogue with a magnitude, $M \geq 2.0$. There were 68 earthquake events in the distance about 200 km from Padang City recorded. The highest magnitude was 7.9 that occurred on 12 September 2007 and was located 188 km from Padang City at Mentawai Island. There were 25 earthquake events triggered at the Mentawai Island Region Fault zone since 1970 with the magnitude, $M = 5.0$ and above.

## 3 Method and Data

This study is aimed to determine surface deformation induced by earthquake events that occurred in Padang City. Remote sensing data of SLC from a 1-year mission and 1-month pre and post-earthquake events in August 2020 are used. The image's basic information is listed in Table 1 for a pair of master and slave images. There are three steps in processing the image pairs using SNAP version 8.0 software. Before the co-registration process, the image should be split into selected bursts and swath where Padang City is located and increase its accuracy by applying the orbit file. The InSAR technique generates the interferogram by multiplying the amplitudes and calculates

**Table 1** Pairs of SLC IW data for interferometric process

| | | |
|---|---|---|
| Spatial resolution | 5 m (ground range) × 20 m (azimuth) | |
| Pixel spacing | 2.3 m (slant range) × 14.1 m (azimuth) | |
| Incidence angle | 29°–46° | |
| Polarization | VH, VV | |
| Total swath width | 250 km | |
| Product class | SAR Standard L1 Product | |
| Product composition | Slice | |
| Product-level | L1 | |
| Product type | SLC | |
| Slice number | 3 | |
| Instrument | Synthetic Aperture Radar (C-band) | |
| Instrument mode | Interferometric Wide | |
| Instrument swath | IW1 IW2 IW3 | |
| Data | Master | Slave |
| Data size | 6.98 GB | 6.98 GB |
| Relative orbit | 91 | 91 |
| Ingestion date | 2019-09-15T03:45:02.529Z | 2020-09-21T04:56:21.167Z |
| Mission data take ID | 147,593 | 192,071 |
| Orbit number | 20,988 | 26,238 |

the difference of phases as in Eqs. (2) and (3) with image improvement by applying the back geocoding and enhanced spectral diversity (ESD). The output of the first step is a debursted interferogram where the gap between the bursts is removed.

The second step is conducted by reading the debursted interferogram, removing the topographic phase, setting the multilook parameters, and enhancing the image by applying Lee's filter in Goldstein phase filtering. The product of the second step is a wrapped interferogram. The created file has a long suffix that represents the completed processes (i.e., *_Split_Orb_Coreg_ESD_Ifg_Deb_DInSAR_ML_Flt*). From the second step, a file was also created for the phase unwrapping process in SNAPHU software, statistical cost, and a network flow algorithm for phase unwrapping was developed by Stanford University researchers [10]. The complete flow is shown in Fig. 5.

## 4 Results and Discussion

Surface displacement is observed in the urban area of Padang City. The surface variation velocity for a period of one year from September 2019 to September 2020 is 10–30 cm/y. There are seven earthquake events recorded within the period in an

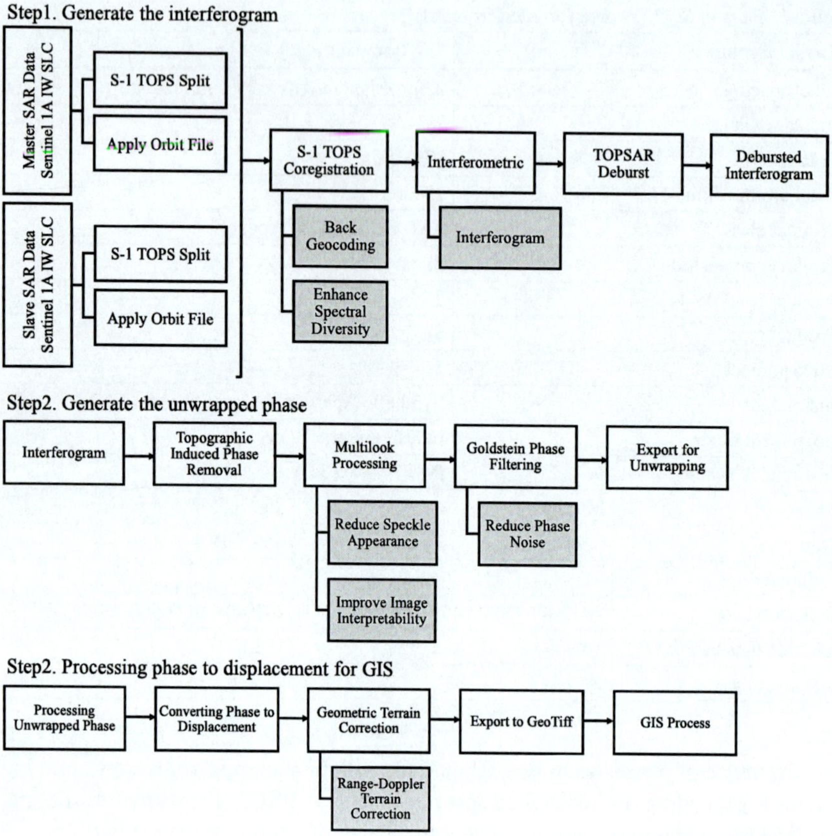


**Fig. 5** The flow of processing SAR data using InSAR and DInSAR with SNAP and SNAPHU

average distance of 61.36 km from Padang City, with the magnitude ranged from 3.2 to 5.3 and depth from 10 to 102 km. Figure 6a plots the coherence with the value ranging from 0 to 1. The area with less static features will have a lower coherence value. To reduce computational time, removing the low coherence value less than 0.2 and replace with NaN value was applied in this study. The area with a low coherence value is mostly due to non-stationary objects of vegetation cover and water body [1, 6, 11]. Consecutively, Fig. 6b, c shows the wrapped phase image and the displacement.

Figure 7 plots the result of interferometry on pre-and post-seismic activity recorded in August 2020. The earthquake is recorded on 4 August 2020, 92 km south of Padang City with a magnitude of 5.1, 52 km depth. The SAR image pairs are taken on 23 July 2020 and 16 August 2020, 24 days different of ingestion days. It shows an obvious variation velocity between a long and short period of image pairs. The 24-day interval of image pairs gave uplifted variation that is relative to long interval surface variation.

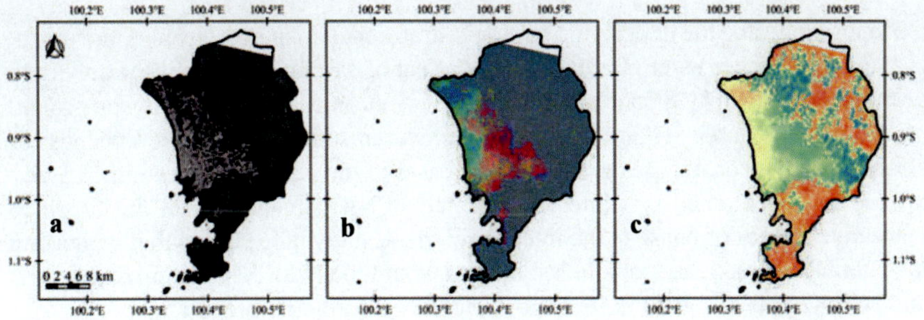

**Fig. 6** The plot of **a** coherence, **b** phase, and **c** displacement for one year SAR image pairs

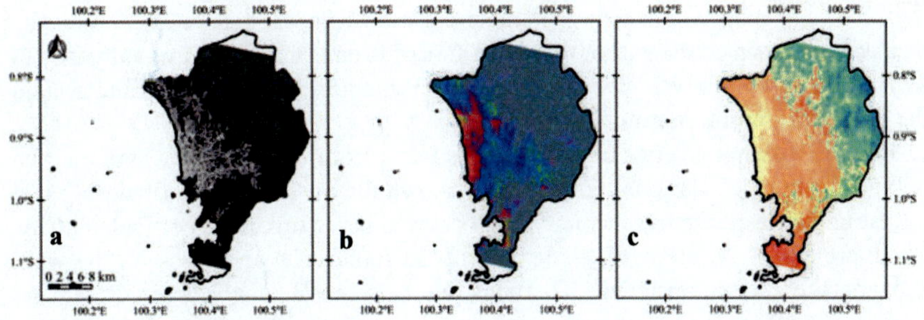

**Fig. 7** The plot of **a** coherence, **b** phase, and **c** displacement from 1 month SAR image pairs pre- and post-seismic in August 2020

The different surface variation is shown in Fig. 8. The uplifted surface variation of 0.1–0.3 m is measured based on the interferometric analysis which shows that the urban area of Padang City is quite responsive to seismic activity. The location of infrastructures on the alluvial layer is of varying thickness. The variations between

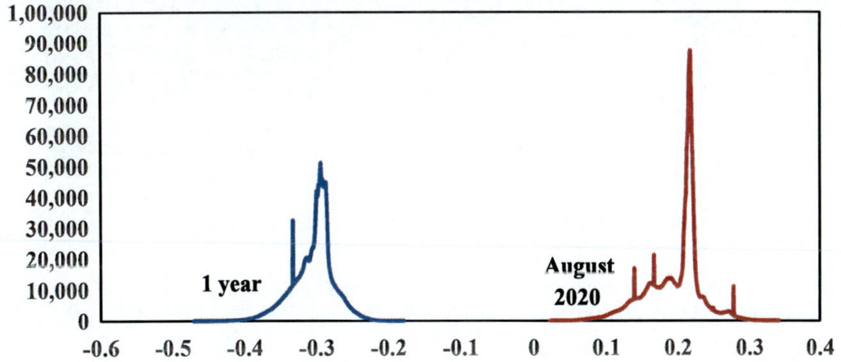


**Fig. 8** Surface variation velocity for one-year observation and after the earthquake in August 2020

the topography and the underlying geologic strata result in uneven ground movement [12]. On the thicker layer of alluvium, the effect of earthquake amplification would be more. A study in [13] concluded that there is a strong correlation between ground motion, shaking intensity and thickness of a layer of soft soil where the thickness of alluvium on the coastal area gradually decreases to the landward direction. A study in [14] also mentioned that interval of a pair of SAR images is not the factor to inaccurate fringes of phase in the interferogram. A study in [6] found that increasing the interval between the first and last images would likely further minimize data loss and optimize the signal-to-noise ratio during interferometric processing.

Figure 9a shows the overlying of displacement data on the coherence and vector data of urban roads in Padang City. In Fig. 9b, c, the vector data of buildings from Open Street Map (OSM) is overly on the surface displacement image exposing buildings that were subjected to change in their elevation. The last item in Fig. 9 plots the displacement map on the part of the urban area of Padang City. The area subjected to high surface variation velocity is located near Padang Beach's coastline and around the Batang Kuranji River mouth.

Figure 10 shows a series of photographs taken from the location where the two polygons coincide. The photographs have shown the possible sign of building and infrastructure response to seismicity that occurred in the urban area of Padang City. There are cracks on the road, settlement of the foundation and crack on the wall, showing the surface variation. The coastline between Batang Kuranji River and Batang Arau River was the liquefaction area during the earthquake in 2009 [15].

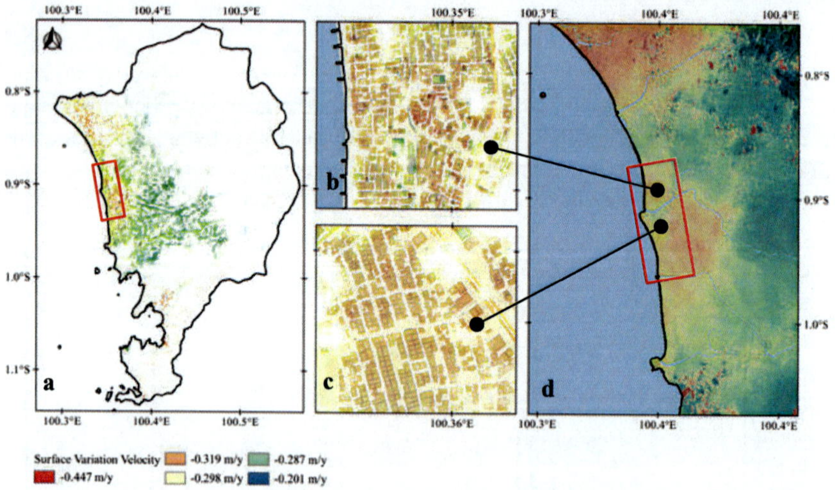


**Fig. 9** **a** Surface displacement on high coherence area, **b** and **c** buildings on the area that subjected to displacement. **d** One-year surface variation velocity (m/y) in a period of 2019–2020

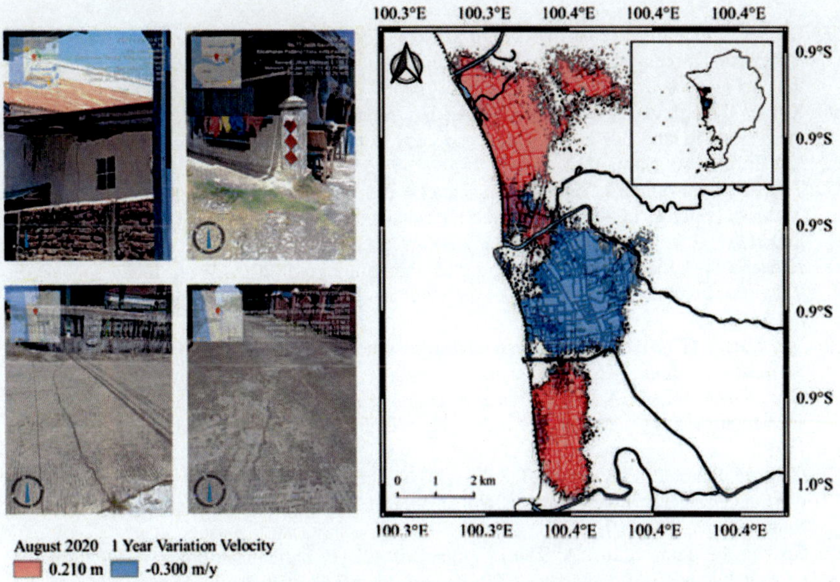


**Fig. 10** Photographs were taken from the site where the two polygons are overlapped

## 5 Conclusion

The interferometric analysis of DInSAR in this study has found that the urban area of Padang City is subjected to a surface variation velocity of 30 cm/year. After the August 2020 earthquake, surface deformation has occurred in the urban area of 10–30 cm. The location and geological formation of alluvium characteristically lose rock formation, predominantly affecting the earthquake's surface variation of the urban area of Padang City. Further study is recommended to include pairs of SAR images that its interval is covering more earthquake events.

## References

1. Amariena Hamim S, Usman F, Kurnia Shalihat A (2019) Determination of Land Subsidence Caused by Land-use Changing in Palembang City using Remote Sensing Data. Adv Eng Res 187(IcoSITE):101–106
2. Ramirez R, Lee SR, Kwon TH (2020) Long-term Remote Monitoring of Ground Deformation using Sentinel-1 Interferometric Synthetic Aperture Radar (INSAR): Applications and Insights into Geotechnical Engineering Practices. Appl Sci 10(21):1–20
3. Miky YH (2019) Multitemporal sentinel-1SAR Interferometry for surface deformation monitoring near high dam in Aswan, Egypt. Am J Geogr Inf Syst 8(2):96–102
4. Téllez-Quiñones A, Salazar-Garibay A, Valdiviezo-Navarro JC, Hernandez-Lopez FJ, Silván-Cárdenas JL (2020) DInSAR method applied to dual-pair interferograms with Sentinel-1 data: a study case on inconsistent unwrapping outputs. Int J Remote Sens 41(12):4662–4681

5. Usman F, Ibrahim E (2020) Detecting Seasonal Extent of Inundated Area of River Body in Banyuasin Regency using Radar Data of Sentinel-1A. Lecture Notes in Civil Engineering, vol 53, pp 771–784
6. Kovács IP et al (2019) How to avoid false interpretations of Sentinel-1A TOPSAR interferometric data in landslide mapping? A case study: recent landslides in Transdanubia, Hungary. Nat Hazards 96(2):693–712
7. Triyoso W, Suwondo A, Yudistira T, Sahara DP (2020) Seismic Hazard Function (SHF) Study of Coastal Sources of Sumatra Island: SHF evaluation of Padang and Bengkulu Cities. Geosci Lett 7(1)
8. Acocella V, Bellier O, Sandri L, Sébrier M, Pramumijoyo S (2018) Weak Tectono-magmatic Relationships along an Obliquely Convergent Plate Boundary: Sumatra, Indonesia. Front Earth Sci 6
9. Van Gorsel JT (2018) Sumatra – Sundaland. Bibliography of the Geology of Indonesia and Surrounding Areas
10. Chen CW, Zebker HA (2002) Phase unwrapping for large SAR interferograms: statistical segmentation and generalized network models. IEEE Trans Geosci Remote Sens 40(8):1709–1719
11. Chini M, Pelich R, Pulvirenti L, Pierdicca N, Hostache R, Matgen P (2019) Sentinel-1 InSAR coherence to detect floodwater in urban areas: Houston and Hurricane Harvey as a test case. Remote Sens 11(2):107
12. Rosyidi AP, Jamaluddin TA, Sian LC, Taha MR (2011) Earthquake Impacts of the Mw 7.6, Padang, Indonesia, 30 September 2009 - Kesan Gempa 7.6 Mw Padang Indonesia, 30 September 2009. Sains Malaysiana 40(12):1393–1405
13. Putra RR, Kiyono J (2012) Shaking Characteristic of Padang City, Indonesia. In: 15th World Conference on Earthquake Engineering, vol 1779
14. Pepe A, Calò F (2017) A review of interferometric synthetic aperture RADAR (InSAR) Multi-track approaches for the retrieval of earth's surface displacements. Appl Sci 7(12):1–39
15. Gratchev I, Irsyam M, Towhata I, Muin B, Nawir H (2011) Geotechnical aspects of the Sumatra earthquake of September 30, 2009, Indonesia. Soils Found 51(2):333–341

# Seismic Data Mapping of South India from 1820 to 2020 Using Geographic Information System

**Mrudula Madhukumar and M. Helen Santhi**

## 1 Introduction

Earthquake may be a natural disaster because of passing of unstable wave through the ground that causes vibration on earth surface. A geographic system is a technological tool for comprehending earth science and creating intelligent selections. GIS organizes geographic knowledge so an individual reading a map will choose knowledge necessary for a particular project or task. A decent GIS program is in a position to process geographic knowledge from a spread of sources and integrates it into a map project. Accurate seismic data visualization is of greatest importance for seismic interpretations. The software used is QGIS that is open source software. For many basic GIS tasks QGIS is enough. During this paper, an attempt was created into consideration of the spatial distribution knowledge of earthquake magnitude, intensity, and depth of South India from the year 1820 to 2020 by detailed mapping using open software known as QGIS.

Miles and Carlton [1] studied the applications and problems with GIS as tool for technology modeling and finished by the importance of GIS for Civil engineers. The common edges of using GIS as a supplement to engineering modeling helped heaps in interpretations and hazard management. Vipin et al. [2] studied the seismic hazard of South India using probabilistic approach. Seismic hazard was evaluated using the earthquake data obtained till December 2006, and the details of seismic sources were obtained form SEISAT-2000. The variation of PHA as well as SA for 1 s were estimated for 2 and 10% probability of exceedance in 50 years. It is observed that the difference in PHA values and the SA values for the probability of exceedance of 2 and 10% is very less for major portion of the study area. Further, local site effects have been considered and a simple methodology is presented to estimate ground

M. Madhukumar (✉) · M. H. Santhi
School of Civil Engineering, Vellore Institute of Technology, Chennai, India
e-mail: mrudula.madhukumar2020@vitstudent.ac.in



S. Kolathayar and S. C. Chian (eds.), *Recent Advances in Earthquake Engineering*,
Lecture Notes in Civil Engineering 175,
https://doi.org/10.1007/978-981-16-4617-1_40

level response spectrum for any local site based on PSHA and site class. Yanger and William [3] did a summary on the unstable zonation and small zonation studies in India. The map has been changed and updated frequently with the prevalence of major harmful earthquakes additionally of recent knowledge. Map is not the ultimate output map, because it will still be updated at later stage with more computer file. Naveen et al. [4] prepared the pre and post-earthquake landslide condition zonation maps for Uttarakhand, India. The result of unstable displacements on landslide condition zonation in Garhwali Himalayan region of Bharat victimisation GIS and remote sensing techniques was studied and finished the importance and application of QGIS in technology and easiness in assessment of natural disaster. Chen et al. [5] have done a brand new unstable knowledge image methodology and located that the present approaches are restricted to supply sturdy ability to examine macroearth science structure. They conjointly found the importance for unstable interpretation. Mandal [6] studied the variations of unstable velocities within the Kachchh rift zone, Gujarat, India, throughout 2001–2013 and found that variations of seismic velocities within the main rupture zone and increase in unstable velocities is because of the healing method. Kumaran and Bansal [7] studied a GIS-based methodology for safe website selection of a building in an exceedingly unsmooth region. This study explored the applying of geographic data systems (GIS) in modeling the locational and geographics aspects to spot areas of suitableness and GIS-based methodology for locating a secure website that satisfies varied spatial safety aspects was developed. Ningthoujam and Radhikesh [8] evaluated the advantages of open code QGIS compared to business code and earthquake vulnerability methodology integrated with an open supply GIS package, provides an answer for each prediction and vulnerability estimation. Basu and Pal [9] mapping in Gish River Basin, province, India and concluded the Importance of mapping to know the assorted metamorphic parameters and benefits of study and conclusion by visualization and application of GIS in hazard management. Mazumder and Salman [10] studied the seismic harm assessment exploitation RADIUS and GIS: A case study of Sylhet town, Bangladesh. The study reveals that the two situation earthquakes (Mw 8.0 from Dauki Fault and Mw 7.6 from Plate Boundary Fault 2) may end up in intensive harm. The integrated GIS and RADIUS provides data for deciding within the unstable risks. Ozbay et al. [11] studied the image of unstable vulnerability of buildings with the employment of a mobile knowledge transmission and an automatic GIS based mostly tool and located the speedy visual screening analysis of concrete buildings and image in GIS are extremely counseled to be a regular part of the unstable analysis procedure.

The objective of the current study is to map spatial distribution data of earthquake magnitude, intensity, and depth from 1820 to 2020 of South India.

## 2 Study Area

South India includes Indian states of Andhra Pradesh, Kerala, Karnataka, Tamil Nadu, and Telangana occupying India's 19.31% of area (635,780 km$^2$), covering

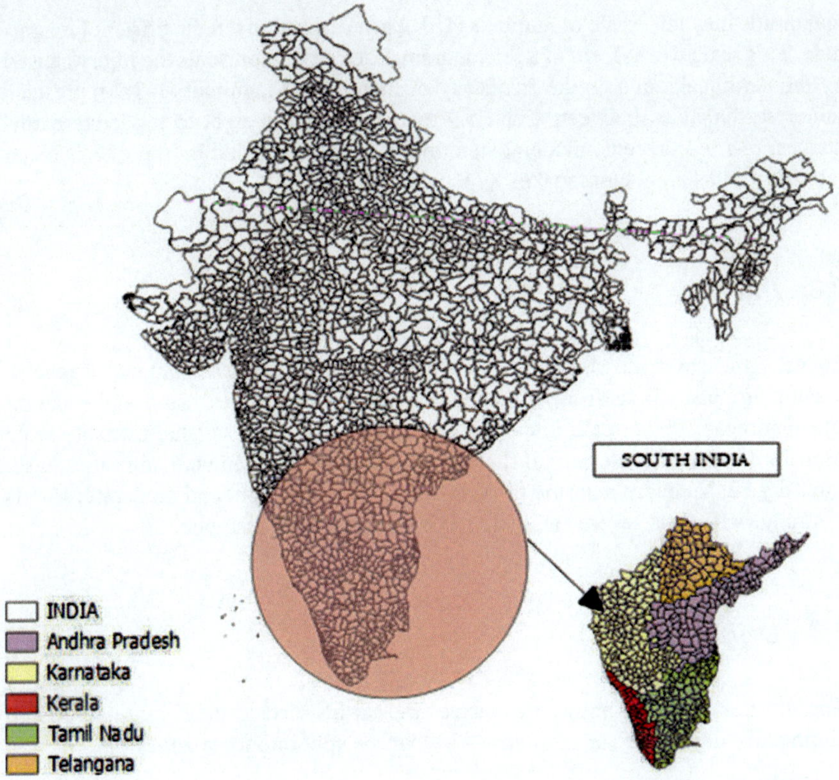


**Fig. 1** Study area

the southern part of the Peninsular Deccan Plateau. The region is bounded by the Bay of Bengal, Arabian Sea, and the Indian Ocean. In this study seismic data such as magnitude, intensity, and depth from 1820 to 2020 was collected from different sources. Population and soil properties of the five states are also collected for risk and hazard assessment. Figure 1 shows the study area where the seismic data are collected.

## 3 Parameters Considered

### *3.1 Magnitude*

The magnitude characterizes the relative size of earthquake. Magnitude relies on activity of the foremost motion recorded by a measuring instrument. The Richter's magnitude may be a scale of numbers (1–10) accustomed to tell the power of magnitude. This scale works form of a seismogram. Each scale represents the injury caused by the earthquake and so the frequency of incidence. Magnitude 3–3.9 represents minor earthquakes; between 4 and 5.9 represents light-weight to moderate earthquakes; 6–7.9 represents moderate and major earthquakes and on top of 8 is taken into account as great earthquakes.

## 3.2 Intensity

Intensity measures the strength of shaking created by the earthquake at a precise location. Intensity is set from effects on people, human structures, and so the natural surroundings. The Mercalli intensity scale (I–XII) may be a seismal intensity scale used for measure the intensity of shaking created by Associate in nursing earthquake. Intensity but V square measure thought-about as light-weight and moderate; V–VII is strong; VIII–X is severe and extreme; XI and XII is catastrophe.

## 3.3 Depth

Earthquakes can occur anywhere between the earth's surface and regarding 700 km. Earthquake depth that varies of zero—700 kms is split into three zones:

- Shallow earthquakes—between 0 and 90 km.
- Intermediate earthquakes—between 90 and 350 km.
- Deep earthquakes—between 350 and 700 km.

## 3.4 Population

Population is a major concern in seismic risk and hazard management. Population density and urbanization increases vulnerability to disasters. The total population in South India is around 24 crores.

## 3.5 Soil Type

Soils are class classified based upon the texture, color, and moisture. In 1956, soil survey of India was established and was classified as Alluvial soil, Red soil, Black soil, Laterite soil, Forest soil, Desert soil, Marshy, and Saline soil. Among these alluvial, red sand black soils are commonly seen in the several parts of India.

**Table 1** Description of data

| Data | Description |
|---|---|
| Date | Date when the event occurred |
| Lat | Latitude of the location |
| Long | Longitude of the location |
| Location | Place where the event occurred |
| D | Depth of the earthquake |
| M | Magnitude of the earthquake |
| I | Intensity of the earthquake |
| S | Soil type |
| DS | Desert soil |
| BS | Black soil |
| RS | Red soil |
| AS | Alluvial soil |
| FS | Forest soil |

## 4 Data Collected

A reliable catalog of earthquake is not readily available for South India due to various reasons such as lack of detailed historical records in the past. In this paper, seismic data from the year 1820 to 2020 was collected from various sources such as Incorporated Research Institutions for Seismology (IRIS), India Meteorological Department (IMD) Earthquake Report, National Earthquake Information Center (NEIC), United Stated Geological Survey and the map was plotted using the open software known as QGIS. Tables 1 and 2 show the description of data and the data collected, respectively.

## 5 Mapping of Seismic Data

In this project, seismic visualization of South India from the year 1820 to 2020 was mapped. 75 points were evaluated based on earthquake happened in past 200 years. Data from different sources were collected by considering magnitude, intensity, and depth of earthquake. These sources have generated earthquakes in the magnitude range of 2.8–6.2 and the depth range of 10–70 km. From the map, we can understand the earthquake prone area and carry out safe and economic design of structures. Intensity values were found from the magnitude values therefore, intensity and magnitude maps will be same. Mapping of magnitude, number of earthquakes in each magnitude bin and mapping intensity is shown in Figs. 2, 3, and 4, respectively.

The depth of 75 points was between 0 and 90 km which is shallow depth. Therefore, shallow depth is again classified in to small ranges with an interval of 10 km for easy understanding. Figure 5 shows the mapping of depth.

**Table 2** Data collected

| Sl no | Lat | Long | Location | D (m) | M | I | S | Population |
|---|---|---|---|---|---|---|---|---|
| 1 | 14.5 | 80 | Nellore | 10 | 5.2 | V | AS | 405,000 |
| 2 | 12.5 | 79.7 | Vandavasi | 15 | 5 | VI | AS | 74,320 |
| 3 | 13 | 80 | Sriperumbudur | 20 | 5.2 | VI | RS | 486,063 |
| 4 | 9.5 | 76.6 | Karukachal | 22 | 5 | V | AS | 20,133 |
| 5 | 17.7 | 83.4 | Vishakapatanam | 10 | 4.9 | V | LS | 173,000 |
| 6 | 13 | 75 | Bantwal | 25 | 6 | VII | AS | 40,155 |
| 7 | 13 | 75.5 | Manglore | 20 | 5.7 | VII | RS | 485,000 |
| 8 | 13 | 77.6 | Banglore | 10 | 5.8 | V | LS | 844,000 |
| 9 | 9.5 | 76.6 | Karukachal | 10 | 4 | V | LS | 20,133 |
| 10 | 15.2 | 76.9 | Bellary | 25 | 6 | VIII | LS | 1,400,000 |
| 11 | 9.5 | 76.6 | Bellary | 10 | 4.9 | V | BS | 1,400,000 |
| 12 | 8.5241 | 76.93 | Thiruvananthapuram | 15 | 5.2 | V | AS | 958,000 |
| 13 | 8.731 | 76.94 | Thiruvananthapuram | 12 | 5 | V | RS | 958,000 |
| 14 | 9.5 | 76.6 | Karukachal | 10 | 5.2 | V | RS | 20,133 |
| 15 | 11.4 | 76 | Kevadur | 10 | 4.7 | V | LS | 19,703 |
| 16 | 11.4 | 76 | Kevadur | 15 | 5.9 | VI | FS | 19,703 |
| 17 | 13 | 77.6 | Banglore | 20 | 5.8 | V | LS | 8,440,000 |
| 18 | 18.3 | 84 | Srikakulam | 15 | 4.4 | V | LS | 228,000 |
| 19 | 12.5 | 79 | Kadaladi | 10 | 5.8 | VI | RS | 13,135 |
| 20 | 16.299 | 80.5 | Guntur | 10 | 5.9 | VI | BS | 4,890,000 |
| 21 | 16.299 | 80.5 | Guntur | 25 | 5.4 | V | BS | 4,890,000 |
| 22 | 16.299 | 80.5 | Guntur | 22 | 5.2 | V | AS | 4,890,000 |
| 23 | 13.7 | 79.4 | Tirupati | 15 | 4.9 | V | AS | 281,000 |
| 24 | 16.4 | 77.3 | Raichur | 12 | 5.2 | V | RS | 232,000 |
| 25 | 16.4 | 77.3 | Raichur | 10 | 5 | V | AS | 232,000 |
| 26 | 16 | 80.3 | Paruchuru | 20 | 5.2 | V | LS | 13,375 |
| 27 | 12 | 79.6 | Villuppuram | 20 | 6 | VII | AS | 96,253 |
| 28 | 17.747 | 80.77 | Hyderabad | 20 | 5.7 | V | RS | 3,945,323 |
| 29 | 14.5 | 80 | Nellore | 10 | 5.9 | V | LS | 405,000 |
| 30 | 17 | 82.3 | Kakinada | 15 | 5.4 | V | FS | 443,028 |
| 31 | 17.7 | 83.4 | Vishakapattanam | 15 | 5.2 | V | LS | 173,000 |
| 32 | 18.86 | 80.1 | Mahadeopur | 10 | 5.9 | VI | BS | 38,451 |
| 33 | 17.45 | 78.45 | Secunderabad | 15 | 5.8 | VI | LS | 214,000 |
| 34 | 11.46 | 76.7 | Udagamandalam | 20 | 6 | VII | LS | 88,430 |
| 35 | 11.4 | 76 | Kevadur | 25 | 5.9 | VI | LS | 19,703 |
| 36 | 13.1 | 80.3 | Chennai | 15 | 5.8 | VI | AS | 7,090,000 |

(continued)

**Table 2** (continued)

| Sl no | Lat | Long | Location | D (m) | M | I | S | Population |
|---|---|---|---|---|---|---|---|---|
| 37 | 10.8 | 76.8 | Coimbatore | 22 | 6 | VII | BS | 1,600,000 |
| 38 | 12 | 75 | Off coast kerala | 18 | 5 | VI | RS | 64,473 |
| 39 | 15.505 | 80.04 | Ongole | 25 | 6.2 | VII | LS | 203,000 |
| 40 | 13 | 77.3 | Banglore | 15 | 5 | V | FS | 8,440,000 |
| 41 | 9.9 | 76.3 | Tekkumuri | 10 | 5 | VI | LS | 16,945 |
| 42 | 15.68 | 80.07 | Vallempalle | 20 | 5 | VI | AS | 66,789 |
| 43 | 16 | 80.3 | Paruchuru | 15 | 4.3 | V | AS | 13,375 |
| 44 | 11.3 | 75.8 | Kozhikode | 12 | 4.3 | V | RS | 609,000 |
| 45 | 14.7 | 80 | Alluru | 10 | 4.5 | V | RS | 11,656 |
| 46 | 15.62 | 80.16 | Ongole | 15 | 5.2 | V | RS | 203,000 |
| 47 | 17.81 | 80.67 | Kichanapalle | 25 | 5.7 | VII | LS | 1029 |
| 48 | 13 | 76.1 | Hassan | 12 | 5 | V | LS | 227,000 |
| 49 | 15.6 | 80.1 | Inamanamelluru | 20 | 4.9 | V | LS | 6871 |
| 50 | 12.4 | 77 | Malavalli | 10 | 4.6 | V | BS | 37,601 |
| 51 | 12.4 | 77 | Malavalli | 10 | 4.5 | V | BS | 37,601 |
| 52 | 11 | 77 | Coimbatore | 20 | 5 | VI | AS | 1,600,000 |
| 53 | 17 | 76.3 | Almel | 10 | 4 | IV | FS | 18,667 |
| 54 | 13.8 | 75.3 | Shimoga | 15 | 5 | V | LS | 323,000 |
| 55 | 17.5 | 81.4 | Chodavaram | 54 | 4.4 | V | BS | 12,412 |
| 56 | 17.9 | 78.5 | Masaipet | 33 | 4.9 | V | BS | 6146 |
| 57 | 12.55 | 77.77 | Denkanikota | 20 | 4.6 | V | AS | 31,868 |
| 58 | 12.87 | 78 | Masti | 15 | 4.5 | V | AS | 6409 |
| 59 | 9.8 | 77.2 | Kalar | 10 | 4.5 | V | AS | 19,414 |
| 60 | 13 | 75.5 | Dharmasthala | 10 | 4.6 | V | BS | 9818 |
| 61 | 18.09 | 76.47 | Killari | 10 | 6.2 | VII | AS | 15,958 |
| 62 | 12.2 | 77.05 | Tallakad | 10 | 4.5 | V | AS | 8539 |
| 63 | 17.707 | 78.29 | Hyderabad | 15 | 2.8 | III | AS | 6,810,000 |
| 64 | 9.824 | 76.76 | Idukki | 10 | 4.6 | V | RS | 1,090,000 |
| 65 | 9.801 | 76.54 | Idukki | 25.4 | 4.7 | V | AS | 1,090,000 |
| 66 | 12.595 | 77.22 | Bangalore | 15 | 4.3 | IV | LS | 8,440,000 |
| 67 | 11.984 | 80.22 | Off coast of Pondycherry | 10 | 5.5 | VI | AS | 2365 |
| 68 | 15.34 | 76.16 | Koppal | 11 | 3.5 | IV | RS | 70,698 |
| 69 | 10.7 | 76.14 | Kunnakulam | 10.5 | 2.8 | III | LS | 54,071 |
| 70 | 10.698 | 76.14 | Kunnakulam | 10 | 3 | III | LS | 54,071 |
| 71 | 12.8 | 78.8 | Palar valley | 33 | 3.8 | IV | LS | 865 |
| 72 | 11.1 | 79.1 | Aariyalur | 33 | 3.5 | IV | BS | 28,902 |

(continued)

**Table 2** (continued)

| Sl no | Lat | Long | Location | D (m) | M | I | S | Population |
|---|---|---|---|---|---|---|---|---|
| 73 | 18.41 | 83.28 | Salur | 10 | 4.7 | V | LS | 49,500 |
| 74 | 19.44 | 79.5 | Sirpur | 10 | 4.8 | V | LS | 1467 |
| 75 | 15.54 | 78.14 | Kurnool | 10 | 4 | IV | LS | 425,000 |

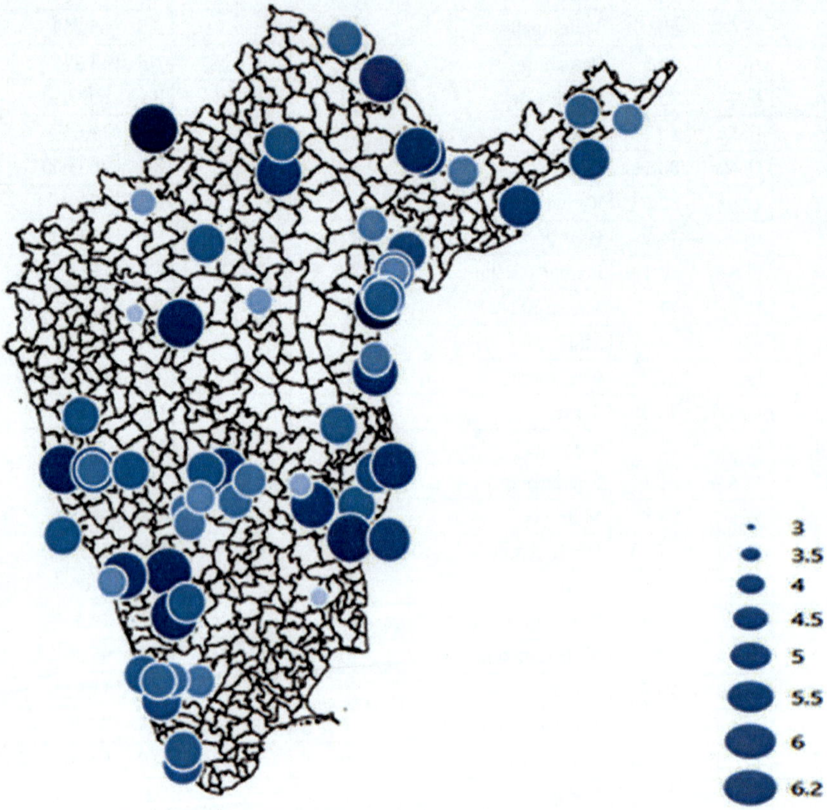


**Fig. 2** Mapping of magnitude

The soil type and population of 75 points was plotted to understand the risk of earthquake in those regions in the future. Figure 6 and 7 show the different soil types and population in South India.

Overlay of map helps us to understand the result from a single map showing all the values which gives an overall idea of the study area. Figure 8 shows the final result of the study by showing the overlay map of South India.

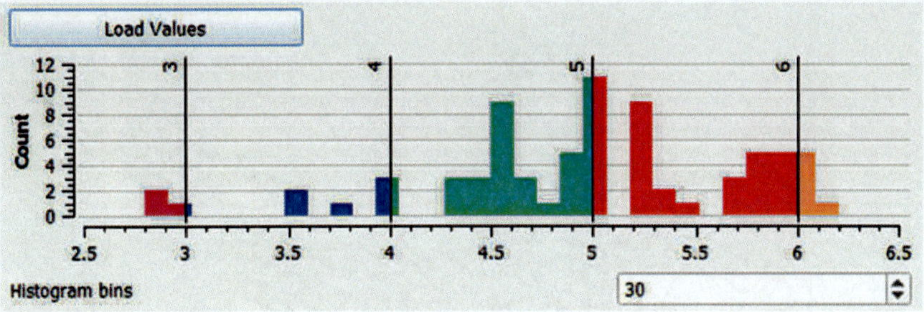


**Fig. 3** Number of earthquakes in each magnitude bin

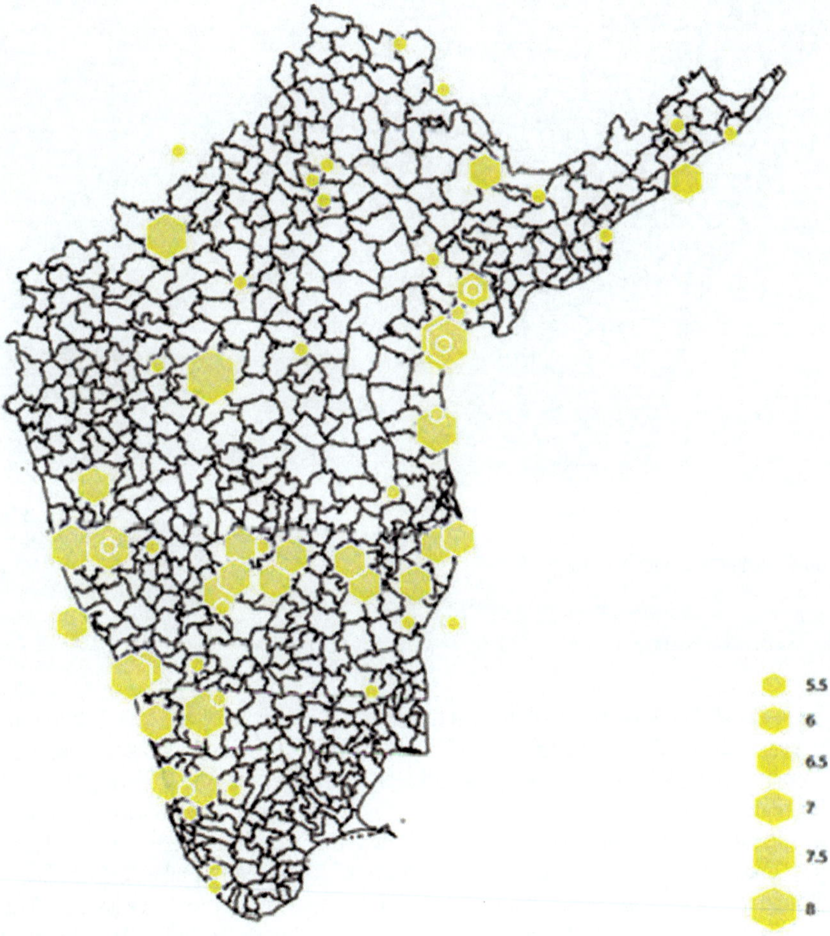


**Fig. 4** Mapping of intensity

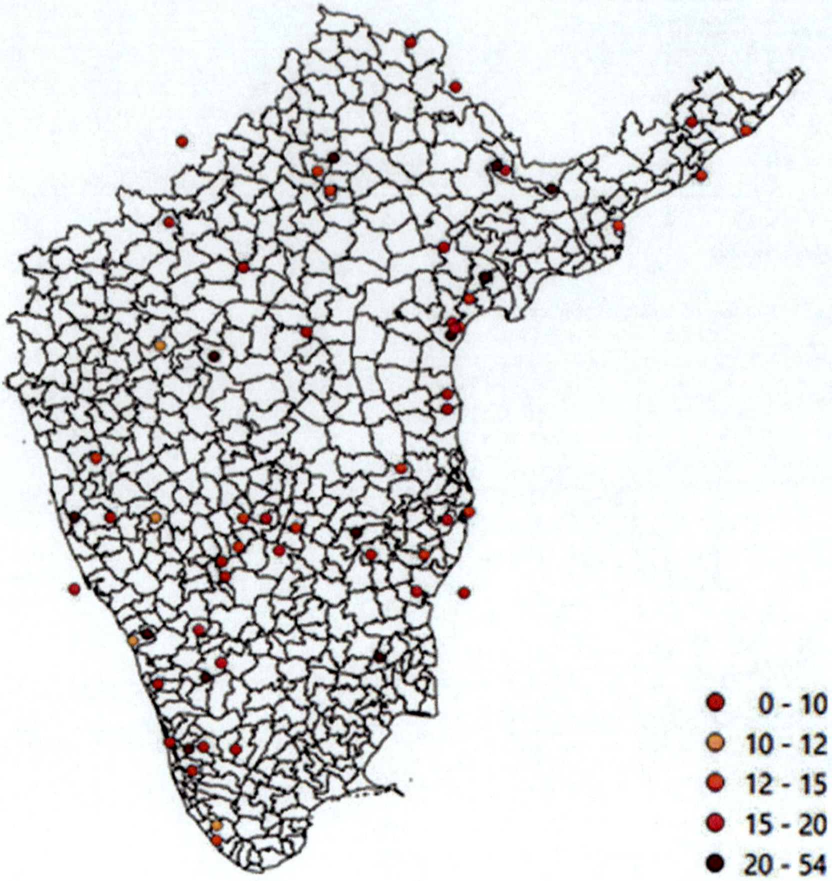


**Fig. 5** Mapping of depth

## 6 Conclusion

The scope of this study was to carry out the seismic visualization of South India from the year 1820 to 2020 using QGIS, and the study indicates that the different parts of states are characterized by low to strong level of seismic activities. During the past 200 years, 35 earthquakes of magnitude ranging from 5 to 5.9 and 6 earthquakes of magnitude ranging from 6 to 6.5 occurred in South India. Earthquakes with magnitude more than 6 can cause lots of damage in an area. These earthquakes release huge amount of energy causing vibration in ground which results in large losses. The southern part of the Tamil Nadu and northern part of Andra Pradesh was frequently affected ny earthquake in the past 200 years. Based on the present study, the following conclusions have been reached.

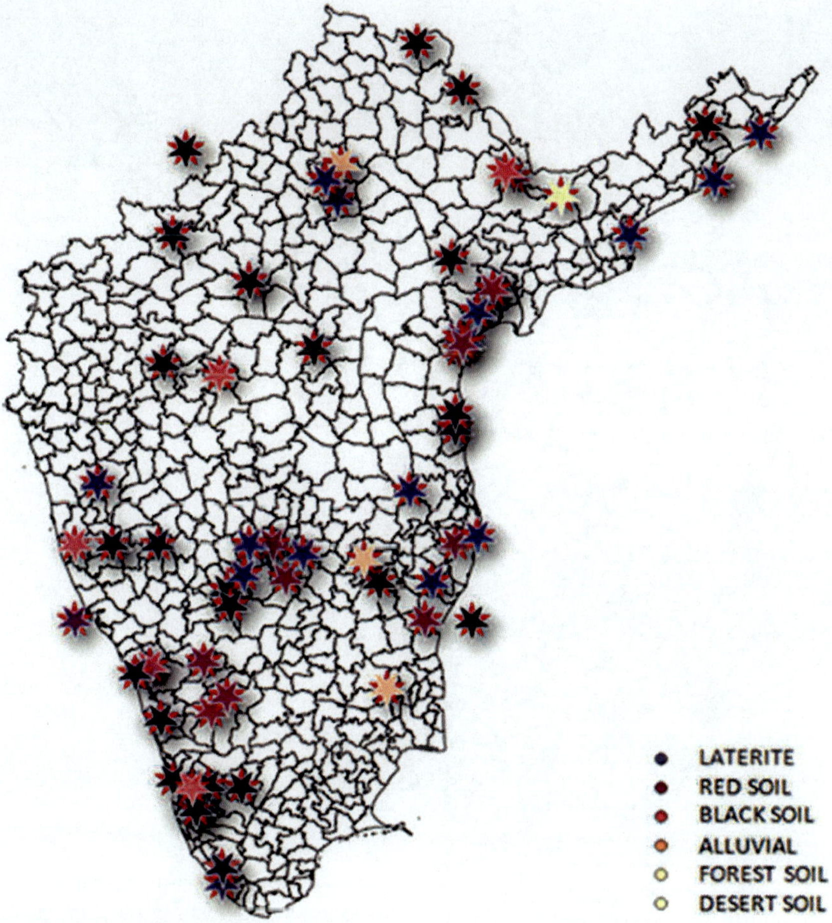


**Fig. 6** Mapping of soil type

- Bantwanl (Karnataka), Villupuram (Tamil Nadu), Ongole (Andhra Pradesh), Bellary (Karnataka), Udayamandam (Tamil Nadu), and Coimbatore (Tamil Nadu) are the six places where earthquakes with magnitude ranging from 6 to 6.5 occurred. All these earthquakes caused great damages and around thousands of people died. Among these the population density is high in Bellary which is located in Karnataka.
- In Kerala, Thiruvananthapuram, Karukachal, Kavadur, and Idukki have experienced more than one earthquake of magnitude ranging from 5 to 5.9 in different years. Therefore, there is a chance for earthquake to happen in this area in future. Poorly constructed structures maybe can be affected easily.
- Tamil Nadu is one of the states in South India where natural hazards are occurring every year. Moderate to severe earth quakes also happen where great amount of

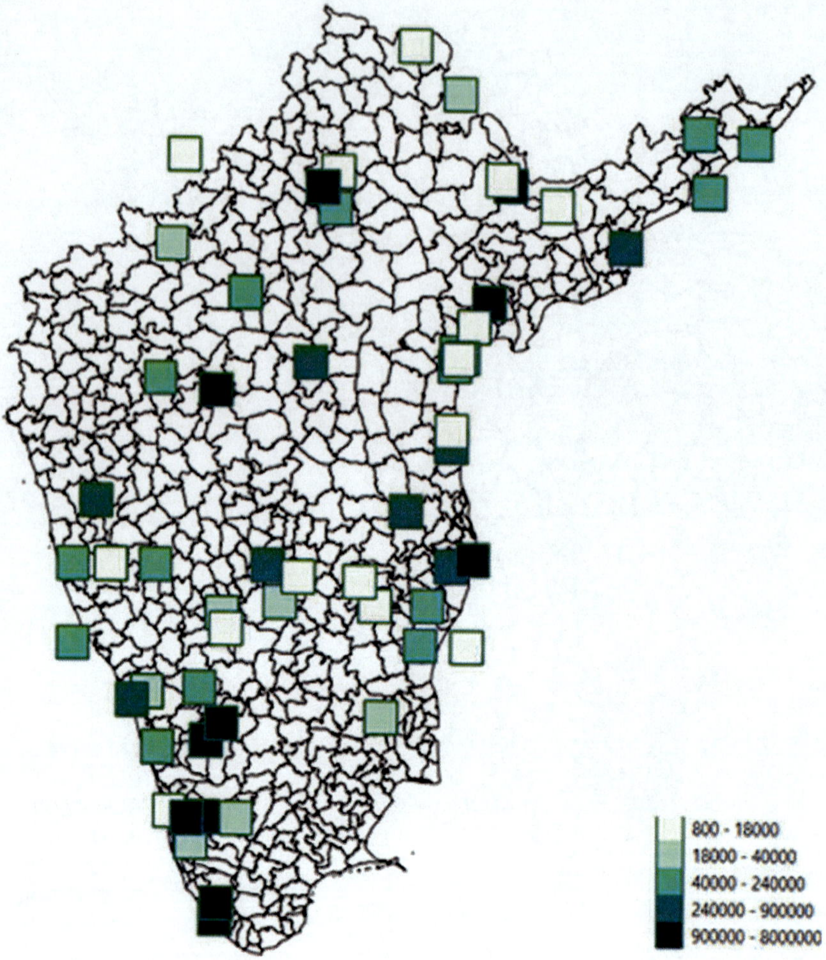


**Fig. 7** Mapping of population

losses and damages occurs. All the earthquakes happened in Tamil Nadu have a magnitude greater than 5. Among these Villupuram, Udayamandam, Coimbatore had affected the worst earthquake in the history where magnitude is greater than 6. The population in Coimbatore is 16 lakhs; therefore, proper precautions can save many lives in the future.

- In the past 200 years more than 20 earth quakes happened in Andhra Pradesh where magnitude ranges from 2.8 to 6.2. The population density in Andhra Pradesh is very high compared to other states in South India.
- Karnataka is also having high population with weak soil type. Therefore, while constructing huge building proper study of soil should be done to avoid huge

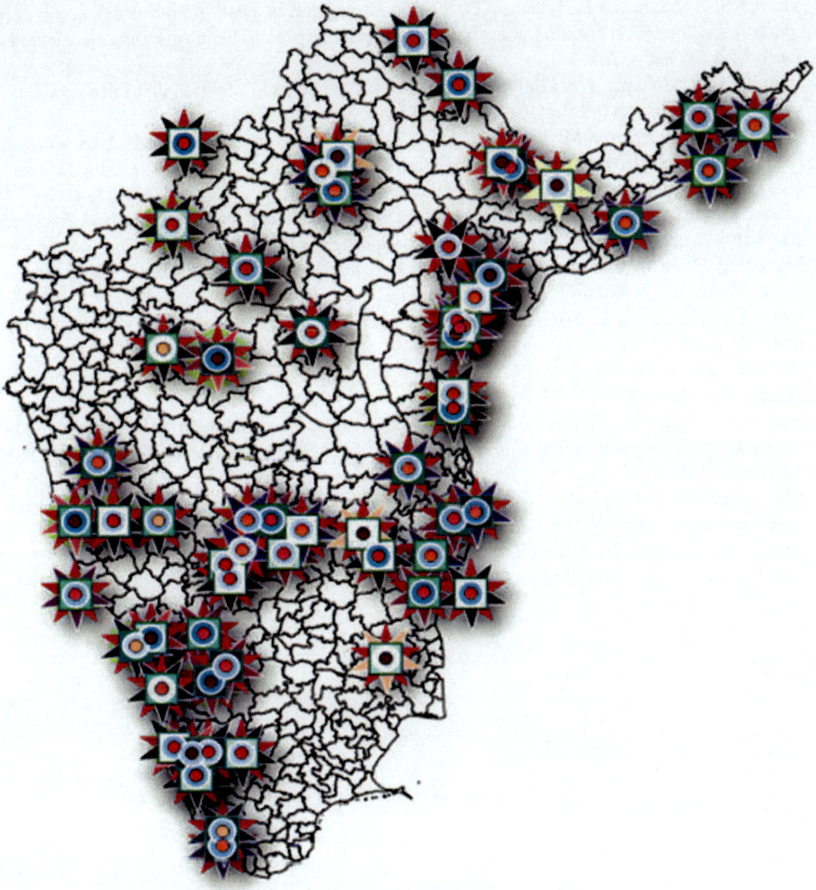

**Fig. 8** Overlay of map

losses. The magnitude ranging from 4 to 6 had happened in Karnataka from the year of 1820 to 2020.

- Telangana has only affected by earthquake 3 times in past 200 years. Among these five states Telangana has less risk on earthquakes.

## References


1. Miles SB, Carlton LH (2000) Applications and issues of GIS as tool for civil engineering modeling. J Comput Civil Eng 13(3):98–120

2. Vipin KS, Anbazhagan P, Sitharam TG (2019) Estimation of peak ground acceleration and spectral acceleration for South India with local site effects: probabilistic approach. Nat Hazards Earth Syst Sci 865–878
3. Walling YM, Mohanty WK (2009) An overview on the seismic zonation and micro zonation studies in India. Earth Sci Rev 96(1–2):197–220
4. Pareek N, Pal S, Sharma ML, Arora MK (2013) Study of effect of seismic displacements on landslide susceptibility zonation (LSZ) in Garhwal Himalayan region of India using GIS and remote sensing techniques. Comput Geosci 61:50–63
5. Chen L, Yu J, Ditta A, Xiao C, Li X (2016) A new seismic data visualization method. In: 22nd International conference on automation and computing, Colchester, pp 467–472
6. Mandal P (2016) Variations of seismic velocities in the Kachchh rift zone, Gujarat, India, during 2001–2013. Techno Phys 672–673:68–86
7. Kumaran S, Bansal VK (2016) A GIS-based methodology for safe site selection of a building in a hilly region. Front Architect Res 5:39–51
8. Ningthoujam MC, Radhikesh P (2018) Nanda: a GIS system integrated with earthquake vulnerability assessment of RC building. Structures 15:329–340
9. Basu T, Pal S (2018) RS-GIS based morphometric and Geological multi-criteria approach to the landslide susceptibility mapping in Gish River Basin, West Bengal, India. Adv Space Res 63:1253–1269
10. Mazumder RK, Salman AM (2019) Seismic damage assessment using RADIUS and GIS; a case study of Sylhet City, Bangladesh. Int J Disaster Risk Reduct 34:243–254
11. Ozbay AE, Karpinar IS, Unen HC (2020) Visualization of seismic vulnerability of buildings with the use of a mobile data transmission and an automated GIS-based tool. Structures 24:50–58

# Earthquake Loss Estimation Using State-of-the-Art Tools

**Leanda J. Payyappilly and Surendra Nadh Somala**

## 1 Introduction

Uncertainty quantification in seismic risk assessment has received limited attention in the literature. Particularly, works would focus on one type of building when multitude of earthquakes are considered [1] while those studies which used several building models are limited in terms of earthquakes variability accounted for. In this study, we explore variability in ground motion using stochastic earthquake simulations [2] for time histories of intensities that are of engineering interest, considering variability in the shallow shear wave velocity as uniform distribution.

Building models covering all possible building materials used in civil engineering have been explored. Wood, steel, concrete, unreinforced masonry, and even mobile homes have been considered. High response reduction factor moment resisting frames are also considered. Increasingly more precast concrete structures are also explored. Earthquake resistant technologies like shear walls have also been dealt with in this study.

Furthermore, many studies employ codal provisions that are based on simple coefficients to estimate the seismic damage and losses. In this study, rigorous nonlinear time history analysis for all possible building classes, listed in the HAZUS [3], is done using a large number of stochastically simulated earthquakes and the variability in economic losses and injuries is quantified in a normalized sense.

L. J. Payyappilly (✉) · S. N. Somala
Indian Institute of Technology Hyderabad, Sangareddy, India
e-mail: ce20mtech12003@iith.ac.in



S. Kolathayar and S. C. Chian (eds.), *Recent Advances in Earthquake Engineering*,
Lecture Notes in Civil Engineering 175,
https://doi.org/10.1007/978-981-16-4617-1_41

## 2 NHERI SimCenter Workflow

The Natural Hazards Engineering Research Infrastructure (NHERI) workflow for Performance-Based Engineering [4] integrates the sophisticated time history analysis capabilities of OpenSees (Open System for Earthquake Engineering Simulation) together with uncertainty quantification engine Dakota, developed by the Sandia National Laboratories, USA. This seamless coupling allows us to understand how the variability in structural parameters (storey height, stiffness, hardening ratio, etc.) influence the response (inter-storey drift) and risk (losses and injuries). Both HAZUS-based and FEMA P-58 [5] based damage and loss computations are available in NHERI tools. However, due to limitation of availability of input parameters for FEMA P-58 [5], we use HAZUS-based approach in this study.

### 2.1 *Building Details*

All possible building types, i.e., wood, steel, reinforced concrete, precast as well as masonry are considered in this study. In addition to these standard building materials, earthquake resistant structures like braced frames with steel, shear walls with concrete are also included. Even mobile homes have also been considered. The symbolic notation of building types used in this study is listed in Table 1. All buildings have been limited to symmetric low-rise, designed for high code, throughout this work. The properties of all the building models used in the study are as listed in Table 2.

### 2.2 *Finite Element Method*

An explicit Newmark's time integration scheme is used for transient analysis solved with unsymmetric multifrontal solver interfacing through scipy, a python library for scientific computation. Rayleigh damping is assumed. OpenSees, the engine behind structural response computation, gives inter-storey drift as well as peak floor accelerations. As the focus of this study is restricted to structural components, we only use the peak inter-storey drift due to base excitation.

### 2.3 *Uncertainty Propagation*

A Latin Hypercube Sampling approach is adopted with a random seed simulating a large number of samples for each random variable used in this work. A multi-variate log-normal distribution is fit to the engineering demand parameter to 2000 realizations of structural response. In particular, we explore the uncertainty on the

**Table 1** Building structure (model building) types (adopted from HAZUS)

| No | Building label | Building description | Range | |
|---|---|---|---|---|
| | | | Name | Stories |
| 1 | W1 | Wood, light frame (≤5,000 ft$^2$) | | 1–2 |
| 2 | W2 | Wood, commercial, and industrial (>5,000 ft$^2$) | | All |
| 3 | S1L | | Low-rise | 1–3 |
| 4 | S1M | Steel moment frame | Mid-rise | 4–7 |
| 5 | S1H | | High-rise | 8+ |
| 6 | S2L | | Low-rise | 1–3 |
| 7 | S2M | Steel braced frame | Mid-rise | 4–7 |
| 8 | S2H | | High-rise | 8+ |
| 9 | S3 | Steel light frame | | All |
| 10 | S4L | | Low-rise | 1–3 |
| 11 | S4M | Steel frame with cast-in-place concrete shear walls | Mid-rise | 4–7 |
| 12 | S4H | | High-rise | 8+ |
| 13 | S5L | | Low-rise | 1–3 |
| 14 | S5M | Steel frame with unreinforced masonry infill walls | Mid-rise | 4–7 |
| 15 | S5H | | High-rise | 8+ |
| 16 | C1L | | Low-rise | 1–3 |
| 17 | C1M | Concrete moment frame | Mid-rise | 4–7 |
| 18 | C1H | | High-rise | 8+ |
| 19 | C2L | | Low-rise | 1–3 |
| 20 | C2M | Concrete shear walls | Mid-rise | 4–7 |
| 21 | C2H | | High-rise | 8+ |
| 22 | C3L | | Low-rise | 1–3 |
| 23 | C3M | Concrete frame with unreinforced masonry infill walls | Mid-rise | 4–7 |
| 24 | C3H | | High-rise | 8+ |
| 25 | PC1 | Precast concrete tilt-up walls | | All |
| 26 | PC2L | | Low-rise | 1–3 |
| 27 | PC2M | Precast concrete frames with concrete shear walls | Mid-rise | 4–7 |
| 28 | PC2H | | High-rise | 8+ |
| 29 | RM1L | Reinforced masonry bearing walls with wood or metal deck diaphragms | Low-rise | 1–3 |
| 30 | RM1M | | Mid-rise | 4+ |
| 31 | RM2L | | Low-rise | 1–3 |
| 32 | RM2M | Reinforced masonry bearing walls with precast concrete diaphragms | Mid-rise | 4–7 |
| 33 | RM2H | | High-rise | 8+ |
| 34 | URML | Unreinforced masonry bearing walls | Low-rise | 1–2 |
| 35 | URMM | | Mid-rise | 3+ |

(continued)

**Table 1** (continued)

| No | Building label | Building description | Range | |
|---|---|---|---|---|
| | | | Name | Stories |
| 36 | MH | Mobile homes | | All |

**Table 2** Building details

| | |
|---|---|
| Storey height | 14 ft |
| Number of stories | 1 |
| Plan area | 900 $ft^2$ |
| Floor weight | 144 kips |
| Storey stiffness | 100 kips/in |
| Yield strength | $1 \times 10^6$ ksi |
| Occupancy class (adopted from HAZUS) | RES4 (temporary lodging) |

most unconstrained parameter $V_{s30}$, the top 30 m shear wave velocity [6] and how it influences the damage and loss estimates.

## 3 Stochastic Ground Motions

Stochastic ground motion models are utilized to obtain simulated earthquake motions for analyzing the seismic demand of the model building types. Simulation models are formulated to realistically incorporate the features of real earthquake ground motions such as frequency content and time-varying intensity. These characteristics can greatly influence the structural response [7]. Thus, using stochastic models synthetic ground motions can be created for a target seismic event. Two stochastic motion models available in the PBE workflow are Vlachos et al. [8] and Dabaghi and Kiureghian [9]. The Vlachos et al. model which is used in this study makes use of regression relations to generate seismic ground acceleration time histories for the site of interest when an earthquake scenario description is provided. The user can enter values for a number of earthquake scenarios describing parameters that are used to generate a seismic event. The various parameters include

(1) moment magnitude of the earthquake $M_w$,
(2) closest-to-site rupture distance $R_{rup}$, and
(3) average shear-wave velocity in top 30 m $V_{s30}$ at the site of interest.

The details of the ground motion used in this study are given in Table 3.

**Table 3** Earthquake scenario describing parameters

| Parameter | Symbol | Value |
|---|---|---|
| Shear-wave velocity in top 30 m | $V_{s30}$ | Uniform distribution<br>Min: 500 m/s<br>Max: 800 m/s |
| Moment magnitude | $M_w$ | 40 km |
| Closest-to-site rupture distance | $R_{rup}$ | 7 |

# 4 Results and Discussion

Earthquake loss assessment studies have been conducted using PBE software considering the Hazus Earthquake Model. All structures are subjected to stochastic ground motions. Vlachos et al. [8] model is used and uncertainty is introduced by varying $V_{s30}$. All models are considered as single storey structures having high code level of design (adopted from Hazus) and occupancy class RES4 (Temporary Lodging).

## 4.1 *Peak Inter-storey Drift*

The Peak Inter-storey Drift (PID) parameter is used as the Engineering Demand Parameter (EDP). The EDP ranges from as low as 0.0015–0.0085. Steel frame with cast-in-place shear walls (S4L) has the highest median PID. The variability is highest for wood buildings (W1) and Steel Light Frames (S3). Interesting, the uncertainty in PID for mobile homes (MH) is quite low for the range of uncertainty considered in earthquake loading and structural parameters. Figure 1 shows the variation of Peak Inter-storey Drift (PID) for the different building model types.

## 4.2 *Economic Losses*

The economic losses considered in this study are normalized with respect to the repair costs. The lowest median economic losses are observed for Steel Light Frames (S3). Incidentally, this is also the building type whose PID is considerably high compared to that of other model building classes. Figure 2 shows the variation of economic loss for the different building model types.

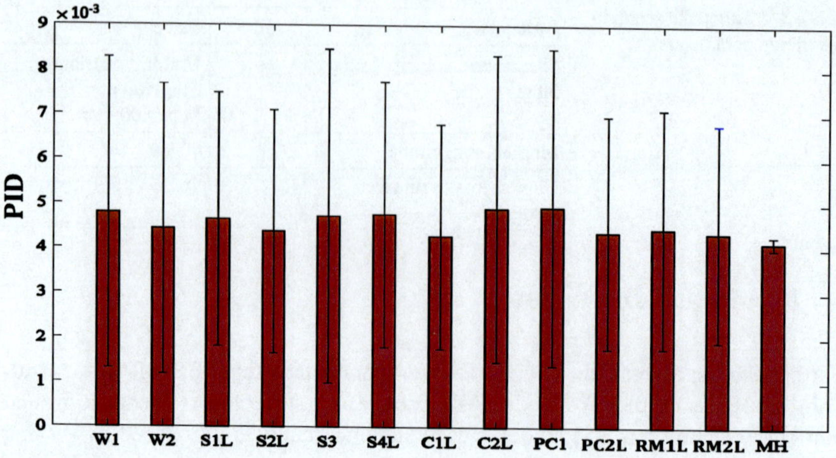


**Fig. 1** Variation of peak inter story drift for the various building models along with the associated uncertainty

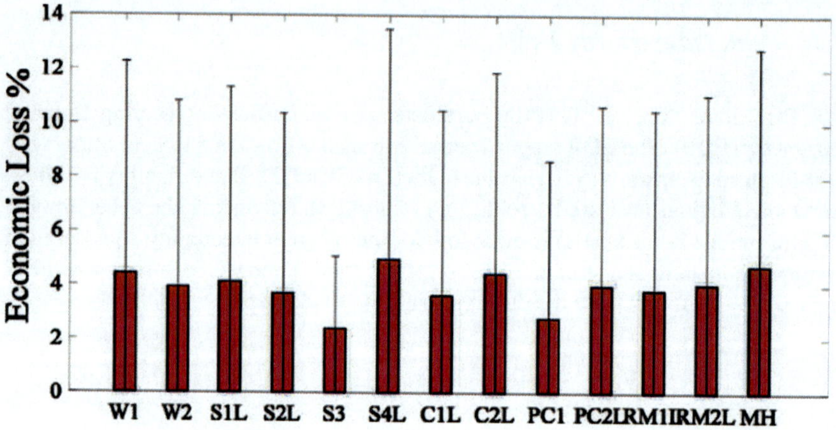


**Fig. 2** Variation of economic loss (percentage of replacement cost) for the various building models along with the associated uncertainty

## 4.3 *Injuries*

The percentage of injuries are found to be quite high for Steel frame with cast-in-place shear walls (S4L) and Precast concrete frames with concrete shear walls (PC2L) compared to other low-rise building types. Interestingly, Precast concrete when used with tilt-up walls (PC1) has the lowest percentage of injuries. Figure 3 shows the variation of percentage of injuries for the different building model types.

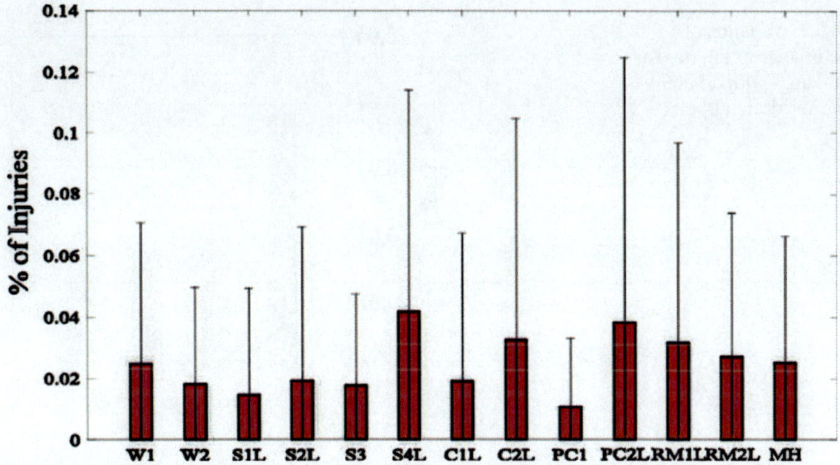


**Fig. 3** Variation of percentage of injuries for the various building models along with the associated uncertainty

## 4.4 Design Code Level

The design code level to which buildings in earlier sections were designed is for high code. We further examine the influence of designing one particular building type of pre code, low code, and moderate code also in terms of economic losses and injuries. A two-storey building of model type W2 (Wood, Commercial and Industrial) and occupancy class RES4 (Temporary Lodging) is used. Figures 4 and

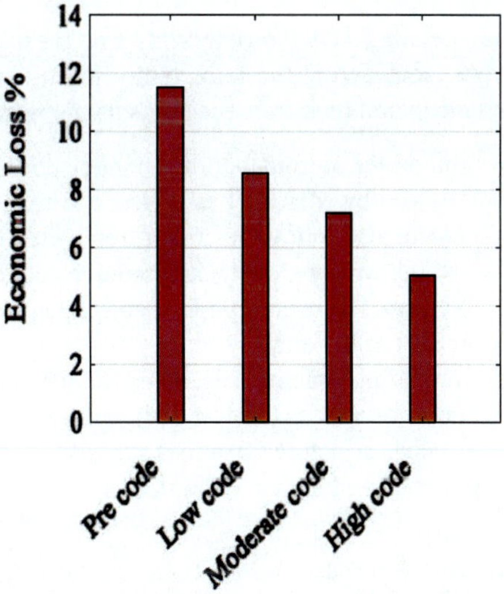


**Fig. 4** Variation of economic loss for buildings with different seismic design levels

**Fig. 5** Variation of percentage of injuries for buildings with different seismic design levels

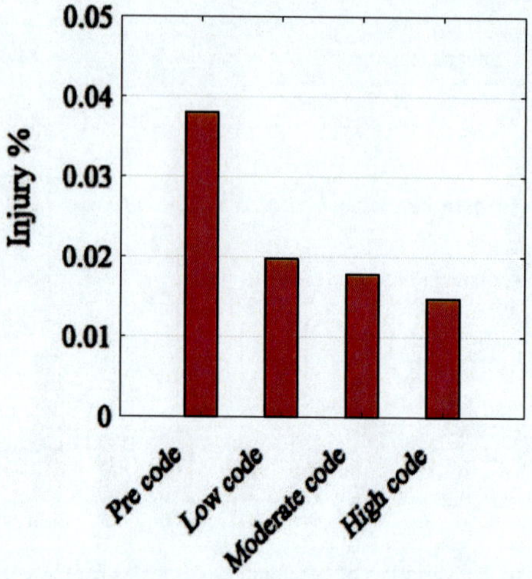


5 show a comparison of economic loss and number of injuries for buildings having different seismic design levels. Figure 4 shows that there is roughly a linear decrease in economic losses by ~2% as we move from one code level to higher code level. The injury percentage decreases drastically from pre-code to low code (Fig. 5). With further improvements in code levels of designs, injuries percentages have only marginally decreased (Fig. 5) this comparison.

## 5 Conclusion

Our detailed study on rigorous nonlinear dynamic analysis of various building types, considering the uncertainty in building parameters and earthquake loading to understand their influence on seismic risk, leads us to the following inferences:

- Among the various building models, S4L has the highest economic loss percent followed by MH. S3 has lowest economic loss percent followed by PC1. Injury percent also follows a similar trend with S4L having the highest percent and PC1 having lowest percent. Compared to economic loss the injury percent is very less. Compared to economic loss and injury percent peak inter-storey drift has lesser amount of uncertainty.
- Comparing structures designed for different seismic design levels, most difference in economic loss (almost 3%) and injury percent is found between pre-code and low code structures. There is a total difference of 6.5% for economic loss between pre-code and high code structures.

**Acknowledgements** Funding from the Ministry of Earth Sciences, India, through grant MoES/P.O.(Seismo)/1(304)/2016 is greatly acknowledged.

## References

1. Ellingwood BR, Kinali K (2009) Quantifying and communicating uncertainty in seismic risk assessment. Earthq Saf 31:179–187
2. Pavel F (2021) Seismic risk assessment of on-ground circular reinforced concrete and prestressed concrete wa7ter tanks using stochastic ground motion simulations. Bull Earthq Eng 19:161–178
3. HAZUS (2020) Earthquake model technical manual
4. Zsarnoczay A, McKenna F, Wang C, Elhaddad W, Gardner M (2019) NHERI-SimCenter/PBE: Release v2.0.0 (Version v2.0.0). Zenodo
5. FEMA P-58 (2018) Seismic performance assessment of buildings—methodology. Volume 1. Federal Emergency Management Agency, 1st ed
6. Passeri F, Foti S, Cox BR, Rodriguez-Marek A (2018) Influence of epistemic uncertainty in shear wave velocity on seismic ground response analyses. Earthq Spectra 35(2):929–954
7. Li Y, Conte J, Barbato M (2016) Influence of time-varying frequency content in earthquake ground motions on seismic response of linear elastic systems. Earthq Eng Struct Dyn 45(8)
8. Vlachos C, Papakonstantinou K, Deodatis G (2017) Predictive model for site specific simulation of ground motions based on earthquake scenarios: predictive model for site specific simulation of ground motions. Earthq Eng Struct Dyn 47(1):195–218
9. Dabaghi M, Kiureghian A (2018) Simulation of orthogonal horizontal components of near-fault ground motion for specified earthquake source and site characteristics. Earthq Eng Struct Dyn 47(6):1369–1393

# Post-Earthquake Safe Shelter for Shimla City

**Hemant Kumar Vinayak, Abhinandan, and Shashidhara**

## 1 Introduction

As urbanization is taking place at a faster pace in India, it has become difficult to regulate the past haphazard planning which has been done in many cities in India. With having a vast hilly area lying under seismic zones IV and V which has plain as well as high altitude regions of the north and north-east India. These regions are vulnerable to the danger of earthquakes. Small earthquakes of Intensity between Richter scales III to IV have started occurring on a regular basis, which may be an indication of upcoming major earthquakes.

Among all the natural disasters like tsunami, volcanoes, landslides, floods, and wildfires, earthquakes are the most devastating that encompass a large area around their occurrence. Some of the locations where the devastating earthquakes have occurred in India over the two past centuries are Kutch, Assam, Kangra, Bihar-Nepal Quetta, Anjar, Koyna, Uttarkashi, Killar, Jabalpur, Chamoli, and Bhuj Earthquake which have caused deaths of around 0.1 million [1]. With thousands of people have losing their hard-earned permanent shelter in these earthquakes, developing earthquake safe shelter is one step forward in the direction of disaster mitigation. Many vulnerable regions can be made safer in terms of less damage to human lives by adopting appropriate models for disaster mitigation.

The work conducted under this research was not about constructing new structures as earthquake safe but to develop the strategy of a safe shelter as the per requirement in post-earthquake situations in a hilly region. Although, the strategies of having

H. K. Vinayak (✉)
National Institute of Technical Teachers Training and Research Chandigarh, Chandigarh, India

Abhinandan
Tech-Pecific, Mohali, Punjab, India

Shashidhara
Municipal Council, Himachal Pradesh Nalagarh, India

 
S. Kolathayar and S. C. Chian (eds.), *Recent Advances in Earthquake Engineering*,
Lecture Notes in Civil Engineering 175,
https://doi.org/10.1007/978-981-16-4617-1_42

emergency shelter are being worked upon for more than five decades to have important assembly buildings as emergency shelters [2]. But the concept of shelter is yet to get absorbed in the administrative setup in the context of post-earthquake emergency scenario. Over the years, learning tools have developed as per the requirement across the globe to support improved responses to crises [3]. However, uncertainty in the occurrence of earthquakes does create scenarios beyond control [4] which further lead to the development of ready reckoners for disaster relief shelter [5]. The individual concentrated efforts are localized. Hence, an overview of the design of relief shelters has been compared based on various parameters environmental, economic, technical, and socio-cultural. Thus, replication of standard plans for safe shelter cannot be feasible [6]. The study done on cities will have to be more planned than ever and to aid this planning, shelters can be based on the location and population count of a particular area [7]. Further, strategies can be developed to have time-varying shelter demands met [8].

There are guidelines issued by the Ministry of Home Affairs (MoHA), Government of India (GoI) on the design and construction of cyclone/tsunami shelters under the GoI-UNDP Disaster Risk Management Programme [9]. As the occurrence of cyclones is a much frequent phenomenon, many cyclone shelters which have a multipurpose role have been constructed at the required designated location [10]. Thus, there is a concern for states located in the hilly regions which lack such guidelines on safe shelter for the disasters they are subjected to.

## 2 Conceptualization of Safe Shelter Site

Earthquake safe shelter model consists of three phases, planning, development, and procurement, respectively. The planning phase includes the study of the type of disaster, anticipated intensity, existing evacuation and mitigation action plan, Rapid Visual Screening of existing critical infrastructure like Institutions (school, college, and university), key Govt. offices (district headquarter, food and supply office, local MC office), hospitals, telecommunication exchange and centres, police stations, fire stations, community centres, water pumping stations, electrical sub-stations, and public transport terminals. Data collected through RVS is analyzed and based on results, suitable sites are selected to develop the safe shelter. Also, geotechnical investigation of required sites is done. Especially, sites that are situated on steep slopes.

The second phase includes preparation of model layouts for selected sites, construction of inventory storeroom to store the procured items, connecting safe shelter with district headquarter through wireless communication, conducting awareness campaigns among nearby people, conducting mock drills, and replacing inventory items on required period. Thus, the development phase continues till the lifespan of the safe shelter and is repeated periodically as per requirement. Developing a safe shelter is not just about constructing a building or putting tents in the open ground. It is much more beyond this.

The third phase is the procurement of inventory items that are required while accommodating people post-earthquake. This includes food items (ration items like pulses, rice, sugar, editable oil, and dry milk), medical supplies (first aid kits), electricity equipment (solar LED lights, solar-powered generators), water purifications items (water purifiers and chlorine tablets), water storage items (PVC water storage tanks), food cooking equipment (stoves and fuel), and fire safety devices (fire blankets) [7].

This study was about developing a safe shelter for Shimla city, the capital of Himachal Pradesh.

## 3 Positioning and Planning a Safe Shelter

The first step led to the identification, selection, and earmarking of buildings and open spaces which could be used as a post-earthquake safe shelter. As Shimla is an old city and it has become very congested due to the unplanned development of the city, leading to have any evacuation or mitigation plan difficult to implement. It was necessary to prepare a list of buildings structures and available open spaces which if needed would be used as safe shelter. To gather basic data about the safe shelter site, the most effective and time-saving method was to conduct Rapid Visual Survey (RVS) which would be helpful to extract maximum information about the site in less time. There were basically three categories of sites. The first category involved existing buildings, the second category involved open grounds, and the third category contained multilevel parking sites. In total, 41 sites were surveyed through RVS. The parameters that were considered for non-structural components of buildings and open spaces are the year of construction, location, history of landslide, no. of storey, area, ground condition, presence of non-structural components, distance from the nearest natural water source. The following parameters were considered for structural components of buildings and multilevel parkings—presence of vertical irregularity; fire Safety devices; infill walls and brick on edge, staircase, and elevators in the building plan; checks of building plan aspect ratio is more than 3.0; beam-column junctions centrally positioned; distance from adjacent buildings; re-entrant corner; out of plane offsets in the building plan; diaphragm opening; dampness; and presence of cracks [11].

### 3.1 Sites Selection for the Safe Shelter

The space available belonged to govt. agencies, schools, and private ownership. Sites that have a probability of accommodating people of nearby areas were considered to be the locations of safe shelter project by MC Shimla. Soil samples were collected for detailed geotechnical investigation from each multilevel car parking site as these sites are constructed on steep slopes, and stability of slopes is ensured before selecting and developing a particular site for safe shelter. The slopes were found to be stable but needed some arrangement and treatment for more stability. In case of an earthquake, there is a huge probability that the municipal water supply gets interrupted, and alternative naturally available water bodies can be the source. In Shimla city, such streams do exist people are using this water for their daily needs, although some water bodies were earmarked as 'Non-Drinkable' by MC Shimla. RVS of safe shelter sites were further extended to the basic survey of water bodies with special attention on the following details: availability of water, turbidity of water, discharge, and GPS location of the water body.

### 3.2 Sites Selected for Safe Shelter in Shimla City

The followings sites were surveyed as per the list given by Municipal Corporation Shimla.

*Sites under open ground category*:

Colony Ground at Nabha,
D.A.V School at New Shimla,
Forest Ground at Khalini,
GHS at Malyana, Krishna Nagar,
GMS at RulduBhatta,
Engine Ghar Ground at Sanjauli,
GSSS at Chota Shimla, at Phagli, at Totu, at Bhatakufer,
GolPahari Ground at Kaithu,
Hotel Gable at Mashobra,
HPU Ground at Summerhill,
Hotel Woods Ville at Benmore,
HPCA at Krishna Nagar,
Jangli Ground at Tutikandi,
Loreto School at Tarahall,
Judicial Comp. Ground-Kachhighati,
Milk Plant at New Totu,
Police Line at Bharari,
Rani Park at Parimahal,
Rothney Castle at Jakhu,
SD School at AnajMandi,
SVM School at Vikas Nagar,
Tibetans School at Chhota Shimla,
Water Reservoir Ground at Sanjauli,
Waqf Board Ground at RulduBhatta,
Waqf Board Ground at Boileauganj, Satsang Ground at Benmore.

*Sites under car parking on ground category*: MC Car Parking at Khalini, Kasumpti, and New Shimla.

*Sites under multilevel car parking category*: Multilevel car parking at Sanjauli, Chhota Shimla, and Lift.

Sites Required Special Permission before Conducting RVS: There were few sites belonging to the Indian air force, CPWD, private owners, HPPWD, and Indian railway which required special permission to conduct RVS. These sites were Annandale Ground at Annandale, CPWD Ground at Badhai, Ridge Maidan at Lower Bazar, Raj Bhawan Ground at Benmore, Subzi Mandi Ground at Dhalli, Marketing Board Ground at Khalini, and Railway Station Ground at Majath [7]. Sites where the RVS was carried out have been shown in Figs. 1, 2, 3, 4, 5, and 6. The GPS location of all the designated sites for safe shelter is shown in Fig. 7.

**Fig. 1** Police line, Bharari

**Fig. 2** HPU ground, Summerhill

**Fig. 3** Parking, SDA, Kasumpti

**Fig. 4** Multilevel parking, lift

**Fig. 5** Sarswati School, Vikasnagar

**Fig. 6** Bawari, Forest Colony Ground

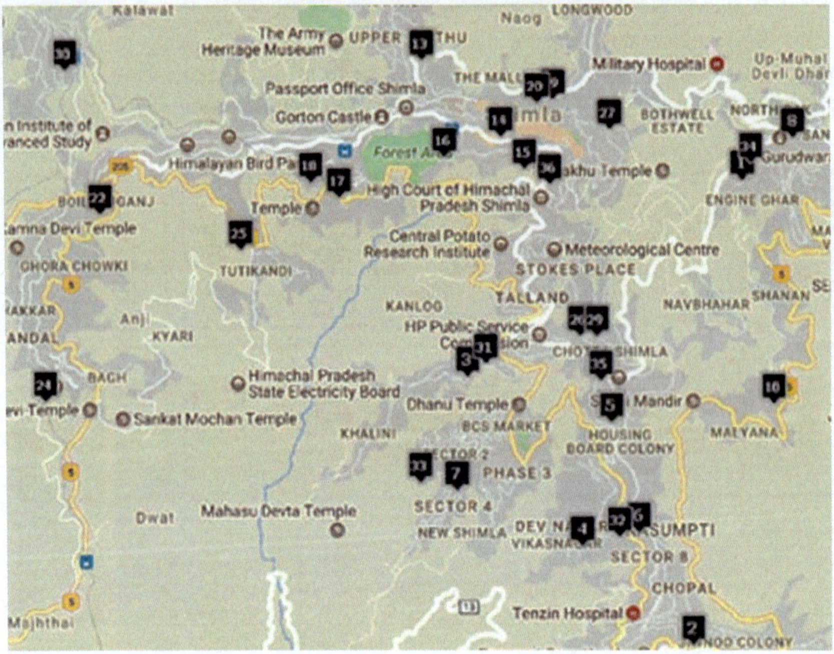


**Fig. 7** GPS location of safe shelter sites

### 3.3 Layout Design of Safe Shelter in Open Ground

The component of the safe shelter is defined based on the minimum requirements which are gathering area, administrative office, space for first aid, storeroom for inventory, power generation point, area for cooking food, serving food, water tank/supply point, boarding place, and sanitation area. The ideal layout of open ground safe shelter for smooth operation is given in Fig. 8. The layout plan can be modified as per the land availability, geographical condition of the area, number of people to be accommodated, and funds availability. There are predefined points where activities or accommodation have been assigned to make a centralized system so that operations can be managed and controlled from one place. The administrative office is connected with the district HQ through wireless communication and in case the regular telecommunication system is damaged in disaster which is most probable, mobile vehicles having telecommunication facilities shall connect various safe shelter sites. Particular areas marked in Fig. 10 are for representation and are not fixed. The state and district administration shall plan the layout as per the actual requirements of the area where the safe shelter is going to be developed. In this depicted layout, the inventory store has been placed in the front row to make the loading and unloading of the inventory items runs in a hassle-free and smooth manner.

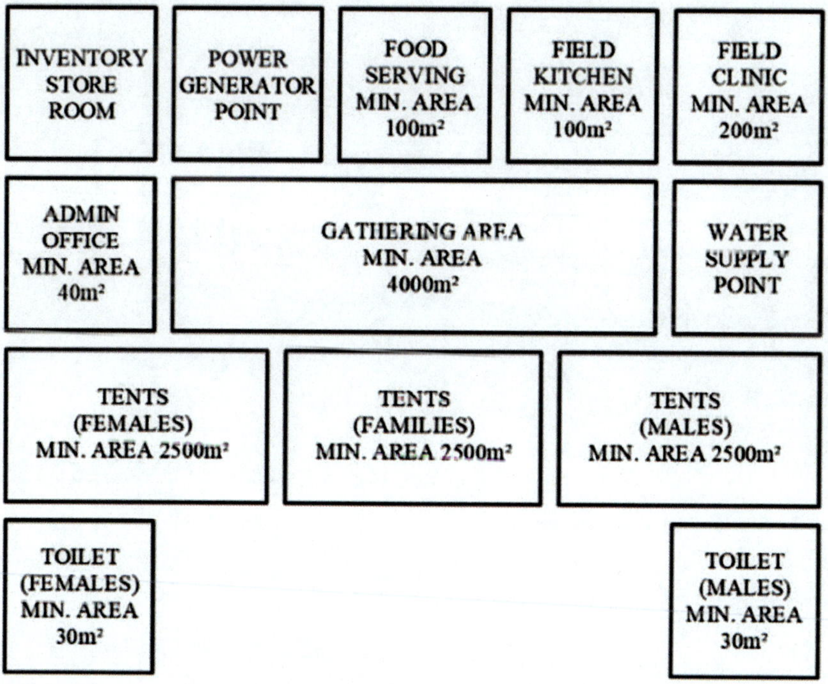


**Fig. 8** Layout for open ground safe shelter

Areas like Kitchen, Clinic, and Living areas are placed adjoining the inventory storeroom. Admin office has been kept in the second row to be connected with people and staff working in the inventory storeroom, field clinic, food serving area, field kitchen, and power generators room. The gathering area has been kept in the centre to be equally and easily connected with the entire safe shelter site. The water supply point is placed near the gathering area to ensure uninterrupted supply and availability of the water. Accommodation tents are placed in the third row so they can stay away from the operations activities going on. Toilets for women and men have been placed in the last row. Inventory vehicle and ambulance arriving point have been planned in accordance to the ease of working and operations. This is an ideal layout for reference and while planning for the open ground which has a different shape, the flow of working and operations must be followed and practiced to increase the efficiency of the safe shelter [12].

### *3.4 Layout Design of Safe Shelter in Multilevel Parking*

The components of the safe shelter for multilevel parking structures as shown in Fig. 9 are the same as in the layout planning for open spaces. However, the layout is totally different but follows the same principles which have been adopted in the Open Ground case as shown in Fig. 8. In the case of multilevel structures, it will be difficult to approach every area so circulation flow should be planned to minimize the efforts. Different levels are marked for separate activities and accommodation.

**Level 1**—Administrative office is on the ground for keeping control of the whole safe shelter as the only entry point is on the same floor. Inventory storeroom is also placed on the same floor so as to have less obstruction in loading and unloading. Working on this level would be much easier for the food and supply department during the regular replacement of inventory items. The power supply of all buildings is given on the ground floor so that the power generator or alternative power supply arrangement can be made at this level. Notice board point is also placed at this level to make sure that authorities and local administration are aware of the current status of people staying at the safe shelter.

**Level 2**—Field clinic is placed on this level which will not create any obstruction in the administrative operation and also only required people will be given access to this floor to avoid any kind of congestion during the operations. The ambulance access point is also given on this floor to shift the patients to be treated in the city hospital. To keep the cooking activities away from the accommodation of people and administrative activities, cooking space is also provided on this level. As the inventory store is on the ground floor (Level 1), it is easier to bring food and ration items from Level 1 for sorting or cooking at level 2 and then as per requirement, food can be distributed on the desired floor levels. To make the cooking process easier, the water supply or storage location is given near to the cooking station or field cooking area.

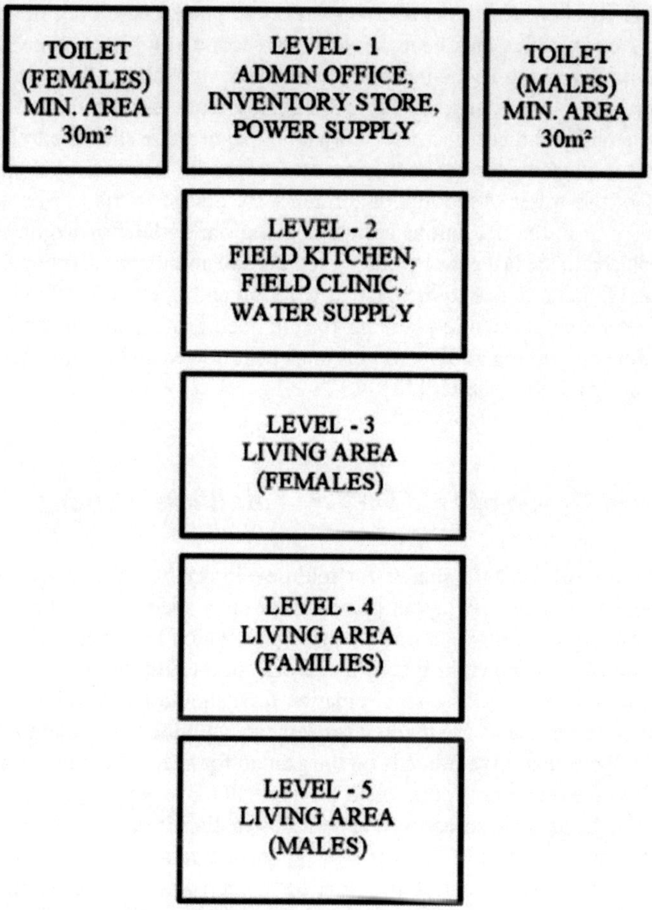


**Fig. 9** Layout for multilevel car parking safe shelter

**Level 3**—The upper level floors shall be utilized for the accommodation of the patients or the people. However, privacy shall be one of the main concerns in such shared congested accommodation. At Level 3, females can be accommodated, At Level 4, families or particular groups of people can be accommodated, and at Level 5, males be accommodated. The overall planning of multilevel or built-up structures has been done with reference to a 5-storey building, the planning can be modified as per requirement, the number of persons to be accommodated [12].

## 3.5 Design of Inventory Storeroom

The inventory storeroom has to be earthquake-resistant design, could be disassembled, lightweight, economic in design, construction with moisture-resistant material, and insects and rat-free area. The size of the storeroom can be adopted based on inventory to be stored but must not be less than 2.4 m × 2.4 m (8′-0″ × 8′-0″) shown in Fig. 10.

Steel Racks must be provided for customized space and tied up with the frame of the storeroom structures to avoid any damage to the stored items during earthquakes and other such natural calamities. Only one or two entry/exit points shall be provided to eliminate the security threats and to have better control on inventory distribution. Enough circulation area must be given so that two persons at a time can pass each other while operating inventory items. Autoclave Aerated Concrete blocks can be used in the place of conventional brick to make the structural weight lighter. In the case of steel frame prefabricated structure, moisture-resistant particle board can be used for covering the structures. The storeroom must be above the ground level with a minimum height of 300 mm (1′-0″). A platform above the ground can be made with the framing of hollow steel pipes to provide the desired height of the storeroom

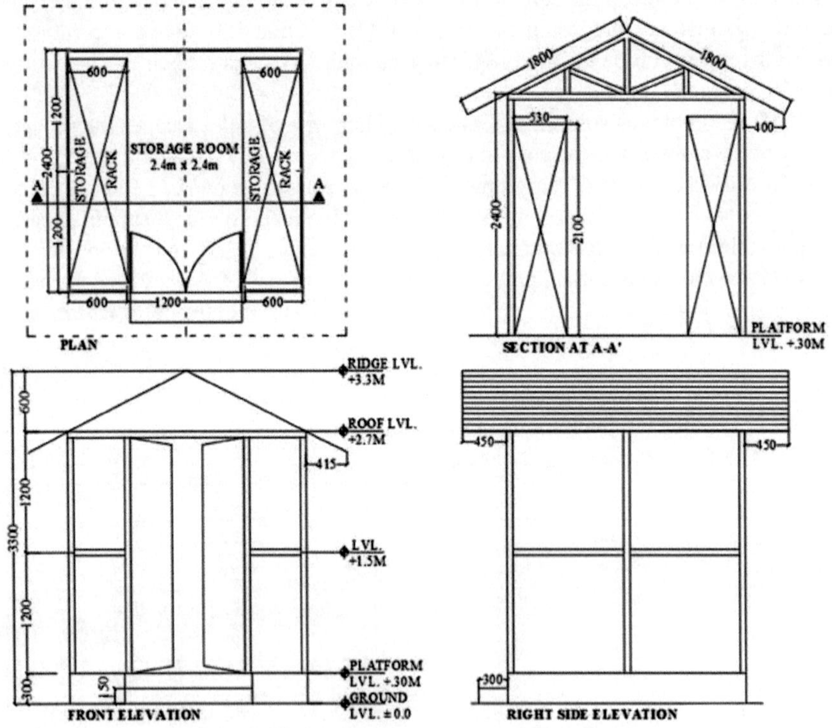


**Fig. 10** Inventory store plan, elevation, and section

floor. Provision for ventilation and natural lighting shall be done like grain storage godowns. Solar panels can be installed on the roof of the storeroom to generate electricity for its own usage [12].

## 4 Essential Items and Inventory for the Safe Shelter

The higher risk constructions can leave the society in a huge psychological trauma after an earthquake which will force the administration to provide some essential items as relief material at the designated safe shelters.

### 4.1 Arrangement for Ration

The guidelines on 'Food and Nutrition needs in emergencies' issued by World Health Organization indicate the minimum requirement of rations for one person during the emergency situations as shown in Table 1 [13]. In case of Shimla safe shelter, example 3 was adopted. To keep the safe shelter more feasible and effective, an initial count of 2000 persons was suggested by Municipal Corporation Shimla and accordingly, the arrangement can be ensured regarding the quantity of rations in the inventory store.

WHO recommends sourcing the most familiar type of pulses as per the acceptability of the people. The purpose of providing the food with such a defined quantity is to fulfil the daily energy requirement of one person which he/she needs in a day. The quantity of ration items shall be provided by the local food and supply department and since each ration/food item has an expiry date or best usage period. Hence, it will be the most appropriate practice to replace the items before they become non-consumable. It will be appropriate to replace the item two months before the expiry

**Table 1** Nutrition values for one person during an emergency situation as per WHO

| Items | Rations (quantity in grams) | | | | |
|---|---|---|---|---|---|
| | Example 1 | Example 2 | Example 3 | Example 4 | Example 5 |
| Cereal | 400 | 450 | 350 | 400 | 400 |
| Pulses | 60 | 60 | 100 | 60 | 50 |
| Oil | 25 | 25 | 25 | 30 | 30 |
| Fish/Meat | – | 10 | – | 30 | – |
| Fortified blended food | 50 | 40 | 50 | 40 | 45 |
| Sugar | 15 | – | 20 | – | 25 |
| Iodized salt | 5 | 5 | 5 | 5 | 5 |
| Energy: kcal | 2,113 | 2,075 | 2,113 | 2,146 | 2,100 |

**Table 2** Ration quantity for 2000 persons

| Food item | Quantity (kg) | Shelf life (months) |
|---|---|---|
| Cereal | 700 | 6–8 |
| Pulses | 200 | 24 |
| Oil | 50 | 12 |
| Fortified Food | 100 | 24 |
| Sugar | 20 | 24 |
| Iodized Salt | 10 | 60 |

and be put it into the food distribution system. The total quantity required for 2000 persons along with shelf life is given in Table 2 [12].

### 4.2 *Procurement of Inventory Items Other Than Ration*

Apart from the ration, arrangement of medical facility, water, electricity, safety, and sanitization requirements at safe shelter site are fulfiled. These items can be procured in desired quantity and stored in the safe shelter inventory storeroom. In this case of Shimla safe shelter, several samples of inventory items were procured, tested, and later, the most suitable ones in each with cost range and specification were shown to the authorities so that the state govt. can procure the inventory items. The ollowing items were procured as the sample for safe Shimla inventory.

First aid kit, water purifier bottle, chlorine tablet, compressed tissue, foldable PVCvc pillow type water tank, PVC carry bag, biological inoculum, odour control agent, solar-powered LED light lantern, solar power generator, fibreglass fire blanket, camouflage toilet tent, woven fabric storage bag, portable stove, and hexamine fuel cubes. The quantity of such items shall be based on the desired number of persons to be accommodated in a safe shelter [12].

## 5 Conclusion

The implementation of technological advancement of disaster-resistant construction is questionable leading to the possibility of having safe shelters.

- Adopting effective models and customized development of safe shelters in each state would improve the disaster mitigation system.
- Important buildings can be designated as safe shelters with appropriate coordination between the owner department and state disaster management authority due to land constraints and economic viability.
- Integrated network of safe shelter models would improve the efficiency of the mitigation process.

- Safe shelter concept lead to the city having alternative arrangements.
- Living experience in a safe shelter can be improved by observing the working of the safe shelter model.
- A list of volunteers can be trained in advance by Municipal Corporations who will be responsible for initial operations of safe shelter, if not injured.

**Acknowledgements** The author would like to thank the Municipal Corporation Shimla for sponsoring the project 'Identification, Plan and Development of Post-Earthquake Safe Shelter for Shimla City' under which the presented work was carried out.

## References

1. Murty CVR (2005) IITK-BMTPC earthquake tips—learning earthquake design and construction. NICEE, IIT Kanpur, India
2. Davis I (1975) Emergency shelter and natural disasters. Lecture, sponsored by INTERTECT, given at Carnegie-Mellon University on Friday, October 17, 1975
3. Ashmore J, Fowler J, James Kennedy J, Argenal E, Funes M, Hodgkin D, Rhyner K (2008) Shelter Projects UN Habitat. http://shelterprojects.org/shelterprojects2008/ref/UN_Habitat_shelter_projects_2008.pdf
4. Fachrurrazi A (2009) Rapid shelter assessment. CARE International Indonesia
5. Asfaw W, Headley D, Liza N, Nederhoed D (2013) Disaster relief shelter. Calvin College Engineering, USA
6. Bashawria A, Garritya S, Moodleya K (2014) An overview of the design of disaster relief shelters. In: 4th international conference on building resilience, building resilience 8–10 September, Salford Quays, United Kingdom
7. Jinghai X, Xiaozhe Y, Dingchao C, Jiwen A, Gaozong N (2016) Multi-criteria location model of earthquake evacuation shelters to aid in urban planning International. J Disast Risk Reduct 20:51–62
8. Zhao L, Li H, Sun Y, Huang R, Hu Q, Wang J, Gao F (2017) Planning emergency shelters for urban disaster resilience: an integrated location-allocation modeling approach. Sustainability 9:1–20
9. Arya AS, Agarwal A (2006) Guideline for design and construction of cyclone/tsunami shelters. NDMA, MoHA New Delhi
10. Dash B, Walia A (2020) Role of multi-purpose cyclone shelters in India: last mile or neighbourhood evacuation. Trop Cycl Res Rev 9:206–217
11. Ningthoujam MC, Nanda RP (2018) Rapid visual screening procedure of existing building based on statistical analysis. Int J Disast Risk Reduct 28:720–730
12. Vinayak HK (2018) Identification, plan and development of post-earthquake safe shelter for Shimla City. Himachal Pradesh State Disaster Management Authority
13. WHO (2004) Food and nutrition needs in emergencies. World Health Organization, Geneva, Switzerland